Classics in Mathematics

Michel Ledoux · Michel Talagrand

T0280574

Michel Ledoux

Michel Ledoux held first a research position with CNRS, and since 1991 is Professor at the University of Toulouse. He is moreover, since 2010, a senior member of the Institut Universitaire de France, having been also a junior member from 1997 to 2002. He has held associate editor appointments for various journals, including the Annals of Probability and Probability Theory and Related Fields (current). His research interests centre on probability, random matrices, logarithmic Sobolev inequalities, probability in Banach spaces.

Michel Talagrand

Michel Talagrand has held a research position with the CNRS since 1974. His thesis was directed by Gustave Choquet and his interests revolve around the theory of stochastic processes and probability in Banach spaces, as well as the mathematical theory of spin glasses. He was invited to deliver a lecture at the International Congress of Mathematicians in 1990, and to deliver a plenary lecture at the same congress in 1998. He received the Loeve Prize (1995) and the Fermat Prize (1997) for his work in probability theory.

Michel Ledoux · Michel Talagrand

Probability in Banach Spaces

Isoperimetry and Processes

Reprint of the 1991 Edition

 Springer

Michel Ledoux
Institut de Mathématiques de Toulouse
Université de Toulouse
31062 Toulouse
France

Michel Talagrand
Équipe d'Analyse de l'Institut Mathématique
Université Paris VI
4 Place Jussieu
75230 Paris Cedex 05
France

Originally published as Vol. 23 of the series *Ergebnisse der Mathematik und ihrer Grenzgebiete 3. Folge*

ISSN 1431-0821
ISBN 978-3-642-20211-7   e-ISBN 978-3-642-20212-4
DOI 10.1007/978-3-642-20212-4

Library of Congress Control Number: 2011930191

Cover design: deblik, Berlin

Printed on acid-free paper

9 8 7 6 5 4 3 2 1

springer.com

Ergebnisse der Mathematik und ihrer Grenzgebiete

3. Folge · Band 23

A Series of Modern Surveys in Mathematics

Michel Ledoux   Michel Talagrand

Probability in Banach Spaces

Isoperimetry and Processes

Springer-Verlag

Berlin Heidelberg New York
London Paris Tokyo
Hong Kong Barcelona
Budapest

Michel Ledoux
Institut de Recherche Mathématique Avancée
Département de Mathématique, Université Louis Pasteur
F-67084 Strasbourg, France

Michel Talagrand
Equipe d'Analyse, Université de Paris VI
F-75252 Paris, France
and
Department of Mathematics
The Ohio State University
Columbus, OH 43210, USA

Mathematics Subject Classification (1980):
60-02, 60 Bxx, 60 E15, 60 Fxx, 60 G15, 60 G17, 60 G50, 46 Bxx,
28 C20, 49 G99, 52 A40, 62 G30

First Reprint 2002

ISBN 3-540-52013-9 Springer-Verlag Berlin Heidelberg New York
ISBN 0-387-52013-9 Springer-Verlag New York Berlin Heidelberg

Springer-Verlag Berlin Heidelberg New York
a member of BertelsmannSpringer Science+Business Media GmbH

© Springer-Verlag Berlin Heidelberg 1991
Printed in Germany

41/3111-54321   Printed on acid-free paper

To Marie-Françoise and WanSoo

Preface

This book tries to present some of the main aspects of the theory of Probability in Banach spaces, from the foundations of the topic to the latest developments and current research questions. The past twenty years saw intense activity in the study of classical Probability Theory on infinite dimensional spaces, vector valued random variables, boundedness and continuity of random processes, with a fruitful interaction with classical Banach spaces and their geometry. A large community of mathematicians, from classical probabilists to pure analysts and functional analysts, participated to this common achievement.

The recent use of isoperimetric tools and concentration of measure phenomena, and of abstract random process techniques has led today to rather a complete picture of the field. These developments prompted the authors to undertake the writing of this exposition based on this modern point of view.

This book does not pretend to cover all the aspects of the subject and of its connections with other fields. In spite of its ommissions, imperfections and errors, for which we would like to apologize, we hope that this work gives an attractive picture of the subject and will serve it appropriately.

In the process of this work, we benefited from the help of several people. We are grateful to A. de Acosta, K. Alexander, C. Borell, R. Dudley, X. Fernique, E. Giné, Y. Gordon, J. Kuelbs, W. Linde, M. B. Marcus, A. Pajor, V. Paulauskas, H. Queffélec, G. Schechtman, W. A. Woycziński, M. Yor, J. Zinn for fruitful discussions (some of them over the years), suggestions, complements in references and their help in correcting mistakes and kind permission to include some results and ideas of their own. M. A. Arcones, E. Giné, J. Kuelbs and J. Zinn went in particular through the entire manuscript and we are mostly indebted to them for all their comments, remarks and corrections. Finally, special thanks are due to G. Pisier for his interest in this work and all his remarks. His vision has guided the authors over the years.

We thank the Centre National de la Recherche Scientifique, that gave to the authors the freedom and opportunity to undertake this work, the University of Strasbourg, the University of Paris VI and the Ohio State University. The main part of the manuscript has been written while the first author was visiting the Ohio State University in autumn 1988. He is grateful to this institution for this invitation that moreover undertook the heavy typing job.

We thank T. H. England and the typing pool at OSU for the typing of the manuscript. Carmel and Marc Yor kindly corrected our poor English and we warmly thank them for their help in this matter. We thank the editors for accepting this work in their Ergebnisse Series and Springer-Verlag for their kind and efficient help in publishing.

Columbus, Paris, Strasbourg   Michel Ledoux
January 1991   Michel Talagrand

Table of Contents

Introduction

Probability in Banach spaces is a branch of modern mathematics that emphasizes the geometric and functional analytic aspects of Probability Theory. Its probabilistic sources may be found in the study of regularity of random processes (especially Gaussian processes) and Banach space valued random variables and their limiting properties, whose functional developments revealed and tied up strong and fruitful connections with classical Banach spaces and their geometry.

Probability in Banach spaces started perhaps in the early fifties with the study, by R. Fortet and E. Mourier, of the law of large numbers and the central limit theorem for sums of independent identically distributed Banach space valued random variables. Important contributions to the foundations of probability distributions on vector spaces, towards which A. N. Kolmogorov already pointed out in 1935, were at the time those of L. Le Cam and Y. V. Prokhorov and the Russian school. A decisive step to the modern developments of Probability in Banach spaces was the introduction by A. Beck (1962) of a convexity condition on normed linear spaces equivalent to the validity of the extension of a classical law of large numbers of Kolmogorov. This geometric line of investigation was pursued and amplified by the Schwartz school in the early seventies. The concept of radonifying and summing operators and the landmark work of B. Maurey and G. Pisier on type and cotype of Banach spaces considerably influenced the developments of Probability in Banach spaces. Other noteworthy achievements of the period are the early book (1968) by J.-P. Kahane, who systematically developed the crucial idea of symmetrization, and the study by J. Hoffmann-Jørgensen of sums of independent vector valued random variables. Simultaneously, the study of regularity of random processes, in particular Gaussian processes, saw great progress in the late sixties and early seventies with the introduction of entropy methods. Processes are understood here as random functions on some abstract index set T, in other words as families $X = (X_t)_{t \in T}$ of random variables. In this setting of what might appear as Probability with minimal structure, the major discovery of V. Strassen, V. N. Sudakov and R. Dudley (1967) was the idea of analyzing regularity properties of a Gaussian process X through the geometry of the index set T for the L_2-metric $\|X_s - X_t\|_2$ induced by X itself. These foundations of Probability in Banach spaces led to a rather intense activity for the last fifteen years. In particular, the Dudley-Fernique theorems on boundedness and continuity of Gaussian and stationary Gaussian processes allowed the definitive treatment by M. B. Marcus and G. Pisier of regularity of random Fourier series, initiated in this line by J.-P. Kahane. With the concepts of type and cotype, limit theorems for sums of independent Banach space valued random variables were appropriately described. Under the impulse, in particular, of the local theory of Banach spaces, isoperimetric methods and concentration of measure phenomena, put forward most vigorously by V. D. Milman, made a strong entry in the subject during the late seventies and eighties. Starting from Dvoretzky's theorem on almost Euclidean sections of convex bodies, the isoperimetric inequalities on spheres and in Gauss space proved most powerful in the study of Gaussian measures and processes, in particular through the work by C. Borell. They were useful too in the study of limit theorems through the technique of randomization. An important recent development was the discovery, motivated by these results, of a new isoperimetric inequality for subsets of a product of probability spaces that is closely connected to the tail behavior of sums of independent Banach space valued random variables. It gives in particular today an almost complete description of various strong limit theorems like the classical laws of large numbers and the law of the iterated logarithm. In the mean time, almost sure boundedness and continuity of general Gaussian processes have been completely understood with the tool of majorizing measures.

One of the fascinations of the theory of Probability in Banach spaces today is its use of a wide range of rather powerful methods. Since the field is one of the most active contact points between Probability and Analysis, it should be no surprise that many of the techniques are not probabilistic but rather come from Analysis. The book focuses on two connected topics – the use of isoperimetric methods and regularity of random processes – where many of these techniques come into play and which encompass many (although not all) of the main aspects of Probability in Banach spaces. The purpose of this book is to give a modern and, at many places, seemingly definitive account on these topics, from the foundations of the theory to the latest research questions. The book is written so as to require only basic prior knowledge of either Probability or Banach space theory, in order to make it accessible from readers of both fields as well as to non-specialists. It is moreover presented in perspective with the historical developments and strong modern interactions between Measure and Probability theory, Functional Analysis and Geometry of Banach spaces. It is essentially self-contained (with the exception that the proof of a few deep isoperimetric results have not been reproduced), so as to be accessible to anyone starting the subject, including graduate students. Emphasis has been put in bringing forward the ideas we judge important but not on encyclopedic detail. We hope that these ideas will fruitfully serve the further developments of the field and hope that their propagation will influence other or new areas.

This book emphasizes the recent use of isoperimetric inequalities and related concentration of measure phenomena, and of modern random process techniques in Probability in Banach spaces. The two parts are introduced by chapters on isoperimetric background and generalities on vector valued random variables. To explain and motivate the organization of our work, let us briefly analyze one fundamental example. Let (T, d) be a compact metric space and let $X = (X_t)_{t \in T}$ be a Gaussian process indexed by T. If X has almost all its sample paths continuous on (T, d), it defines a Gaussian Radon probability measure on the Banach space $C(T)$ of all continuous functions on T. Such a Gaussian measure or variable may then be studied for its own sake and shares indeed some remarkable integrability and tail behavior properties of isoperimetric nature. On the other hand, one might wonder (before) when a given Gaussian process is almost surely continuous. As we have seen, an analysis of the geometry of the index set T for the L_2-metric $\|X_s - X_t\|_2$ induced by the process yields a complete understanding of this property. These related but somewhat different aspects of the study of Gaussian variables, which were historically the two main streams of developments, led us thus to divide the book into two parts. (The logical order would have been perhaps to ask first when a given process is bounded or continuous and then investigate it for its properties as a well defined infinite dimensional random vector; we have however chosen the other way for various pedagogical reasons.) In the first part, we study vector valued random variables, their integrability and tail behavior properties and strong limit theorems for sums of independent random variables. Successively, vector valued Gaussian variables, Rademacher series, stable variables and sums of independent random variables are investigated in this scope using recent isoperimetric tools. The strong law of large numbers and the law of the iterated logarithm, for which the almost sure statement is shown to reduce to the statement in probability, complete this first part with extensions to infinite dimensional Banach space valued random variables of some classical real limit theorems. In the second part, tightness of sums of independent random variables and regularity properties of random processes are presented. The link with the Geometry of Banach spaces through type and cotype is developed with applications in particular to the central limit theorem. General random processes are then investigated and regularity of Gaussian processes characterized via majorizing measures with applications to random Fourier series. The book is completed with an account on empirical process methods and with several applications, especially to local theory of Banach spaces, of the probabilistic ideas presented in this work. A diagram describes some of the interactions between the two main parts of the book and the natural connections between the various chapters.

We would like to mention that the topics of Probability in Banach spaces selected in this book are not exhaustive and actually only reflect the tastes and interests of the authors. Among the topics not covered, let us mention especially martingales with values in Banach spaces and their relations to Geometry. We refer to [D-U] (on the Radon-Nikodym Property), [Schw3], [Pi16] (on convexity and smoothness) and [Bu] (on Unconditional Martingale Differences and ζ-convexity) for an account on this deep and fruitful subject as well as for detailed references for further reading. Empirical processes are only briefly treated in this book and the interested reader will find in [Du5], [Ga],

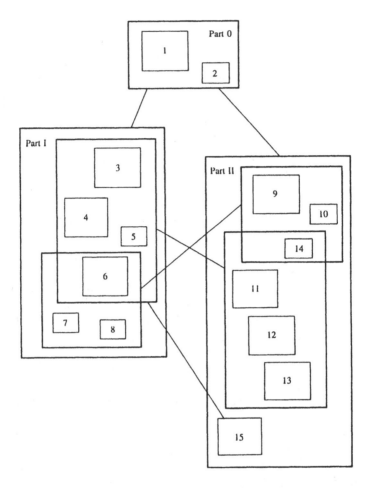

[G-Z3], [Pol]... various expositions on this subject. Infinitely divisible distributions in Banach spaces and the general central limit theorem are treated in [Ar-G2] and [Li] and rates of convergence in the vector valued central limit theorem are investigated in [P-R2]. Large deviations in the context of Probability in Banach spaces are introduced for example in [Az], [Ja3], [De-St], etc. Recent investigations on stable measures are conducted by the Polish school. Finally, we refer to the list of proceedings and seminars on Probability in Banach spaces and related topics for a complete picture of the whole field.

On our exposition itself, we took the point of view that completness is sometimes prejudiciable to clarity. With a few exceptions, this work is however self-contained. Actually, many of our choices, reductions or simplifications were motivated only by our lack of resistance. We did not try to avoid repetitions and we use from time to time results that are proved or stated only further in the exposition. In general, we do not give references in the text but in the Notes and References at the end of each chapter where credit is to be found. We apologize for possible errors or for references that have been omitted for lack of accurate information.

Proceedings and Seminars
on Probability in Banach Spaces and Related Topics

Séminaire L. Schwartz 1969/70. Applications Radonifiantes. Ecole Polytechnique, Paris

Séminaire Maurey-Schwartz 1972/73. Espaces L^p et Applications Radonifiantes. Ecole Polytechnique, Paris

Séminaire Maurey-Schwartz 1973/74, 1974/75, 1975/76. Espaces L^p, Applications Radonifiantes et Géométrie des Espaces de Banach. Ecole Polytechnique, Paris

Séminaire sur la Géométrie des Espaces de Banach 1977/78. Ecole Polytechnique, Paris

Séminaire d'Analyse Fonctionnelle 1978/79, 1979/80, 1980/81. Université de Paris VI

Séminaire de Géométrie des Espaces de Banach 1982/83. Université de Paris VI

Séminaire d'Analyse Fonctionnelle 1983/84, 1984/85, 1985/86/87. Université de Paris VI

Les Probabilités sur les Structures Algébriques, Clermont-Ferrand 1969. Colloque C.N.R.S. (1970)

Aspects Statistiques et Aspects Physiques des Processus Gaussiens, St-Flour 1980. Colloque C.N.R.S. (1981)

Colloque International sur les Processus Gaussiens et les Distributions Aléatoires, Strasbourg 1973. Ann. Inst. Fourier 24 (1974)

Probability in Banach Spaces, Oberwolfach 1975. (Lecture Notes in Mathematics, vol. 526). Springer, Berlin Heidelberg 1976

Probability in Banach Spaces II, Oberwolfach 1978. (Lecture Notes in Mathematics, vol. 709). Springer, Berlin Heidelberg 1979

Probability in Banach Spaces III, Medford (U.S.A.) 1980. (Lecture Notes in Mathematics, vol. 860). Springer, Berlin Heidelberg 1981

Probability in Banach Spaces IV, Oberwolfach 1982. (Lecture Notes in Mathematics, vol. 990). Springer, Berlin Heidelberg 1983

Probability in Banach Spaces V, Medford (U.S.A.) 1984. (Lecture Notes in Mathematics, vol. 1153). Springer, Berlin Heidelberg 1985

Probability in Banach Spaces 6, Sandbjerg (Denmark) 1986. (Progress in Probability, vol. 20). Birkhäuser, Basel 1990

Probability in Banach Spaces 7, Oberwolfach 1988. (Progress in Probability, vol. 21). Birkhäuser, Basel 1990

Probability Theory on Vector Spaces, Trzebieszowice (Poland) 1977. (Lecture Notes in Mathematics, vol. 656). Springer, Berlin Heidelberg 1978

Probability Theory on Vector Spaces II, Błażejewko (Poland) 1979. (Lecture Notes in Mathematics, vol. 828). Springer, Berlin, Heidelberg 1980

Probability Theory on Vector Spaces III, Lublin (Poland) 1983. (Lecture Notes in Mathematics, vol. 1080). Springer, Berlin Heidelberg 1984

Probability Theory on Vector Spaces IV, Lancut (Poland) 1987. (Lecture Notes in Mathematics, vol. 1391). Springer, Berlin Heidelberg 1989

Vector Spaces Measure and Applications I, Dublin 1977. (Lecture Notes in Mathematics, vol. 644). Springer, Berlin Heidelberg 1978

Vector Spaces Measure and Applications II, Dublin 1977. (Lecture Notes in Mathematics, vol. 645). Springer, Berlin Heidelberg 1978

Probability on Banach Spaces. (Advances in Probability, vol. 4). Edited by J. Kuelbs. Dekker, 1978

Martingale Theory in Harmonic Analysis and Banach Spaces, Cleveland 1981. (Lecture Notes in Mathematics, vol. 939). Springer, Berlin Heidelberg 1982

Banach Spaces, Harmonic Analysis and Probability Theory, Univ. of Connecticut 1980-81. (Lecture Notes in Mathematics, vol. 995). Springer, Berlin Heidelberg 1983

Banach Spaces, Missouri 1984. (Lecture Notes in Mathematics, vol. 1166). Springer, Berlin Heidelberg 1984

Probability Theory and Harmonic Analysis. Edited by J. A. Chao and W. A. Woyczyński. Dekker, 1986

Geometrical and Statistical Aspects of Probability in Banach Spaces, Strasbourg 1985. (Lecture Notes in Mathematics, vol. 1193). Springer, Berlin Heidelberg 1986

Probability and Analysis, Varenna (Italy) 1985. (Lecture Notes in Mathematics, vol. 1206). Springer, Berlin Heidelberg 1986

Probability and Banach Spaces, Zaragoza (Spain) 1985. (Lecture Notes in Mathematics, vol. 1221). Springer, Berlin Heidelberg 1986

Geometric Aspects of Functional Analysis. Israel Seminar, 1985-86. (Lecture Notes in Mathematics, vol. 1267). Springer, Berlin Heidelberg 1987

Geometric Aspects of Functional Analysis. Israel Seminar, 1986-87. (Lecture Notes in Mathematics, vol. 1317). Springer, Berlin Heidelberg 1988

Geometric Aspects of Functional Analysis. Israel Seminar, 1987-88. (Lecture Notes in Mathematics, vol. 1376). Springer, Berlin Heidelberg 1989

Notation

Here, we present some general and basic notation which we have *tried* to keep coherently throughout the book.

Usually, $c, C, K...$ denote positive constants which very often may vary at each occurence (in general, c represents a small constant, C, K large constants). The letter K almost always denotes a numerical constant. We never attempt to find sharp numerical constants. $\mathbb{N}$ is the set of all positive integers (usually starting at 1). $\mathbb{Z}$ is the set of all integers. We write $A \underset{C}{\sim} B$ to signify that $C^{-1}A \leq B \leq CA$ for some constant C. We also use the symbol $\sim$ to signify that two sequences (of real numbers) are equivalent, or only of the same order of growth. Usually, less (resp. larger) than means less (resp. larger) than or equal to. When $1 \leq p < \infty$, we usually denote by $q = p/p - 1$ the conjugate of p.

$(\cdot)^+, (\cdot)^-, [\cdot]$ denote respectively the positive, negative and integer part functions. Card A is the cardinality of a (finite) set A. A^c is the complement of A. I_A is the indicator function of the set A. For subsets A, B of a vector space and λ a scalar, we let $A + B = \{x + y; x \in A, y \in B\}$, $\lambda A = \{\lambda x; x \in A\}$. Similarly, $x + A = \{x + y; y \in A\}$. Conv A denotes the convex hull of A. If A is a subset of a metric space (T, d), $d(t, A)$ is the distance from the point t to A, i.e. $d(t, A) = \inf\{d(t, a); a \in A\}$. For every $r > 0$, A_r is the neighborhood of order r of A: $A_r = \{t; d(t, A) < r\}$. The Lipschitz norm $\|f\|_{\mathrm{Lip}}$ of a real Lipschitz function f on a metric space (T, d) is given by

$$\|f\|_{\mathrm{Lip}} = \sup\left\{\frac{|f(s) - f(t)|}{d(s, t)}; s \neq t, s, t \in T\right\}.$$

A Banach space B is a vector space over the field of the real or complex numbers equipped with a norm $\|\cdot\|$ for which it is complete. The dual space of all continuous linear functionals is denoted by B' and duality is denoted by $f(x) = \langle f, x \rangle$, $f \in B'$, $x \in B$. The norm on B' is also denoted by $\|\cdot\|$ ($\|f\| = \sup_{\|x\| \leq 1} f(x)$) and by the Hahn-Banach theorem, for all x in B, $\|x\| = \sup_{\|f\| \leq 1} f(x)$. For simplicity, we also use the notation $\|\cdot\|$ for quotient norms. If F is a closed subspace of B, we let B/F denote the quotient of B by F and let $T_F : B \to B/F$ denote the quotient map. Accordingly, for any x in B, $\|T(x)\| = d(x, F) = \inf\{\|x - y\|; y \in F\}$. The Banach space B is separable if its topology is countably generated or equivalently if there is a countable dense subset in B. For simplicity (and with the exception of Chapter 13), we deal with Banach spaces over the field of real numbers although most of the results presented in this book hold precisely similarly in the complex case.

$\mathbb{R}^N$ is the Euclidean space of dimension N with canonical basis $(e_i)_{i \leq N}$ ($\mathbb{R}^N_+$ is its positive cone). A point x in $\mathbb{R}^N$ has coordinates $x = (x_1, \ldots, x_N)$. These notation extend to $\mathbb{R}^{\mathbb{N}}$. We let $\langle \cdot, \cdot \rangle$ be the Euclidean scalar product and $| \cdot |$ be the Euclidean norm. $B_2 = B_2^N$ and $S_2 = S_2^N$ denote respectively the Euclidean unit ball and unit sphere of $\mathbb{R}^N$. These notation also extend to the infinite dimensional Hilbert space ℓ_2 of all sequences $x = (x_i)$ of $\mathbb{R}^{\mathbb{N}}$ such that $|x| = (\sum_i |x_i|^2)^{1/2} < \infty$.

More generally, for $0 < p \leq \infty$, ℓ_p will denote the space of all real sequences $x = (x_i)$ for which $\|x\|_p = (\sum_i |x_i|^p)^{1/p} < \infty$ ($\sup_i |x_i| < \infty$ if $p = \infty$). (Occasionally, we use the same notation for sequences of complex numbers.) For $1 \leq p \leq \infty$, ℓ_p is a Banach space. c_0 is the subspace of ℓ_∞ consisting of those elements x for which $\lim_{i \to \infty} x_i = 0$. ℓ_p^N denotes $\mathbb{R}^N$ equipped with the ℓ_p-norm. If B is a Banach space, $\ell_p(B)$ is the space of all sequences (x_i) in B such that $\sum_i \|x_i\|^p < \infty$ ($\sup_i \|x_i\| < \infty$ if $p = \infty$).

If $0 < p < \infty$, and $(x_i)_{i \geq 1}$ is a sequence of real numbers, set

$$\|(x_i)\|_{p,\infty} = \left(\sup_{t > 0} t^p \text{Card}\{ i \geq 1; |x_i| > t \} \right)^{1/p} .$$

It is easily seen that if $(x_i^*)_{i \geq 1}$ is the non-increasing rearrangement of the sequence $(|x_i|)_{i \geq 1}$, then $\|(x_i)\|_{p,\infty} = \sup_{i \geq 1} i^{1/p} x_i^*$. It is known that the quasi-norm $\|(\cdot)\|_{p,\infty}$ is equivalent to a norm when $p > 1$ (cf. [S-W]) and that, as is easily seen, for $r < p$,

$$\|(x_i)\|_{p,\infty} \leq \left(\sum_i |x_i|^p \right)^{1/p} \leq \left(\frac{p}{p-r} \right)^{1/p} \|(x_i)\|_{r,\infty} .$$

$\ell_{p,\infty}$ denotes the space of all sequences (x_i) with $\|(x_i)\|_{p,\infty} < \infty$. If B is a Banach space, one defines similarly $\ell_{p,\infty}(B)$ and set $\|(x_i)\|_{p,\infty} = \|(\|x_i\|)\|_{p,\infty}$. By the preceding inequalities, $\ell_{r,\infty} \subset \ell_p \subset \ell_{p,\infty}$ for all $r < p$.

If (T, d) is a compact metric space, we denote by $C(T)$ the Banach space of all continous functions on T equipped with the sup-norm that we denote by $\| \cdot \|_\infty$ or simply $\| \cdot \|$ when no confusion is possible.

If m is a measure, $|m|$ denotes its total mass. δ_x is point mass at x.

The canonical Gaussian measure on $\mathbb{R}^N$ is denoted by γ_N; it is the probability measure on $\mathbb{R}^N$ with density

$$(2\pi)^{-N/2} \exp(-|x|^2/2) .$$

$\gamma = \gamma_\infty$ is the probability measure on $\mathbb{R}^{\mathbb{N}}$ which is the infinite product of the canonical one-dimensional Gaussian measure on each coordinate. We let Φ denote the distribution function of the canonical Gaussian measure on $\mathbb{R}$, i.e.

$$\Phi(t) = (2\pi)^{-1/2} \int_\infty^t \exp(-x^2/2) dx , \quad t \in [-\infty, \infty] .$$

Φ^{-1} is its inverse function and $\Psi = 1 - \Phi$. Note the classical estimate: if $t \geq 0$,

$$\Psi(t) \leq \tfrac{1}{2} \exp\left(-t^2/2\right).$$

In Chapter 1, Theorem 1.3, and Chapter 4, we use the letters μ_N and μ for the canonical probability measures (Haar measures) $\left(\tfrac{1}{2}\delta_{-1} + \tfrac{1}{2}\delta_{+1}\right)^{\otimes N}$ and $\left(\tfrac{1}{2}\delta_{-1} + \tfrac{1}{2}\delta_{+1}\right)^{\otimes \mathbb{N}}$ respectively defined on $\{-1, +1\}^N$ and $\{-1, +1\}^{\mathbb{N}}$.

Throughout this book, $(\Omega, \mathcal{A}, \mathbb{P})$ is a probability space which is always assumed to be large enough to support all the random variables we are dealing with. This is justified by Kolmogorov's extension theorem. For simplicity, we may assume $(\Omega, \mathcal{A}, \mathbb{P})$ to be complete, i.e. all the negligible sets for $\mathbb{P}$ are measurable. $\mathbb{P}_*$ and $\mathbb{P}^*$ are the inner and outer probabilities. $\mathbb{E}$ denotes the expectation with respect to $\mathbb{P}$. If $\mathcal{B}$ is a sub-σ-algebra of $\mathcal{A}$, $\mathbb{E}^{\mathcal{B}}$ is defined as the conditional operator with respect to $\mathcal{B}$. Random variables (on $(\Omega, \mathcal{A}, \mathbb{P})$), with real or vector values, are denoted by $X, Y, Z, \ldots, \eta, \xi, \zeta \ldots$.

If X and X' are independent random variables, we can always assume that they are built on different probability spaces as above, say $(\Omega, \mathcal{A}, \mathbb{P})$ for X and $(\Omega', \mathcal{A}', \mathbb{P}')$ (with expectation $\mathbb{E}'$) for X'. This notation will be used together with Fubini's theorem. Furthermore, we shall make the following abuse of notation: if (X_n), (ξ_n) are independent sequences of random variables, $\mathbb{P}_X$, $\mathbb{E}_X$ (respectively $\mathbb{P}_\xi$, $\mathbb{E}_\xi$) denote conditional integration with respect to (ξ_n) (respectively (X_n)).

The following remarkable sequences of (real valued) random variables will be used extensively: a sequence (ε_i) of independent random variables taking the values ± 1 with equal probability (for example constructed on $\{-1, +1\}^{\otimes \mathbb{N}}$ and (ε_i) is distributed according to μ); a sequence (g_i) of independent standard normal variables (with law γ on $\mathbb{R}^{\mathbb{N}}$); a sequence (θ_i) of independent standard p-stable random variables, $0 < p \leq 2$; p is usually understood and when $p = 2$, (θ_i) is just (g_i). Accordingly, we shall use $\mathbb{P}_\varepsilon$, $\mathbb{E}_\varepsilon$, $\mathbb{P}_g$, $\mathbb{E}_g$, $\mathbb{P}_\theta$, $\mathbb{E}_\theta$, etc. for partial integration with respect to (ε_i), (g_i), (θ_i).

For $0 < p < \infty$, $L_p = L_p(\Omega, \mathcal{A}, P) = L_p(\Omega, \mathcal{A}, \mathbb{P}; \mathbb{R})$ is the space of all real valued random variables X on $(\Omega, \mathcal{A}, \mathbb{P})$ such that $\|X\|_p = \left(\int |X|^p \, d\mathbb{P}\right)^{1/p} = (\mathbb{E}|X|^p)^{1/p} < \infty$. (If we deal with L_p-spaces on some measure space (S, Σ, μ), we use the notation $L_p(\mu)$ with norm $\|\cdot\|_{L_p(\mu)}$, or just, as above, when no confusion arises, L_p and $\|\cdot\|_p$.) L_0 is the space of all random variables equipped with the topology of convergence in probability, L_∞ is the space of all bounded random variables X with norm $\|X\|_\infty = \inf\{c > 0; \mathbb{P}\{|X| \leq c\} = 1\}$. For $1 \leq p \leq \infty$, L_p is a Banach space, and a metric space when $0 \leq p < 1$, with, when $p = 0$, the metric e.g. $\mathbb{E}\min(1, |X - Y|)$. We let $L_{p,\infty}$, $0 < p < \infty$, be the space of all real valued random variables X (on $(\Omega, \mathcal{A}, \mathbb{P})$) such that

$$\|X\|_{p,\infty} = \left(\sup_{t>0} t^p \mathbb{P}\{|X| > t\}\right)^{1/p} < \infty.$$

(Note that the span for this functional $\|\cdot\|_{p,\infty}$ of the step random variables are the random variables X such that $\lim_{t\to\infty} t^p \mathbb{P}\{|X| > t\} = 0$.) We have, when $r > p$,

$$\|X\|_{p,\infty} \le \|X\|_p \le \left(\frac{r}{r-p}\right)^{1/p} \|X\|_{r,\infty} ,$$

hence $L_{r,\infty} \subset L_p \subset L_{p,\infty}$ (actually, if X is in L_p, then $\lim_{t\to\infty} t^p \mathbb{P}\{|X| > t\}$ $= 0$). $\|\cdot\|_{p,\infty}$ is a quasi-norm; it is equivalent to a norm when $p > 1$; take for example

$$N_p(X) = \sup\left\{\mathbb{P}(A)^{-1/q} \int_A |X|\, d\mathbb{P}; A \in \mathcal{A}; \mathrm{P}(A) > 0\right\}$$

where $q = p/p - 1$ is the conjugate of p and we have $\|X\|_{p,\infty} \le N_p(X) \le q\|X\|_{p,\infty}$ as is easily seen by integration by parts (see below). For $0 < p, q < \infty$, the interpolation space $L_{p,q}$ is a further intermediate space which is defined as the space of all random variables X such that

$$\|X\|_{p,q} = \left(q \int_0^\infty \left(t^p \mathbb{P}\{|X| > t\}\right)^{p/q} \frac{dt}{t}\right)^{1/q} .$$

$L_{p,p}$ is just L_p and $L_{p,q_1} \subset L_{p,q_2}$ if $q_1 \le q_2$ (cf. [S-W]).

A function $\psi : \mathbb{R}_+ \to \mathbb{R}_+$ is a Young function if it is convex, increasing and satisfies $\psi(0) = 0$, $\lim_{x\to\infty} \psi(x) = \infty$. We define the Orlicz space $L_\psi = L_\psi(\Omega, \mathcal{A}, \mathbb{P}) = L_\psi(\Omega, \mathcal{A}, \mathbb{P}; \mathbb{R})$ as the vector space of all random variables X such that $\mathbb{E}\,\psi(|X|/c) < \infty$ for some $c > 0$. It is a Banach space for the norm

$$\|X\|_\psi = \inf\{c > 0; \mathbb{E}\,\psi(|X|/c) \le 1\}$$

(cf. [K-R]). Note that since ψ is convex and satisfies $\lim_{x\to\infty} \psi(x) = \infty$, $L_\psi \subset L_1$. When $\psi(x) = x^p$, $1 \le p < \infty$, L_ψ is just L_p. Besides the power functions, we shall mostly be interested in the exponential functions $\psi_q(x) = \exp(x^q) - 1$, $1 \le q < \infty$, and $\psi_\infty(x) = \exp(\exp x) - e$. We can consider similarly L_{ψ_q} when $0 < q < 1$ with however some (trivial) modifications to handle the convexity problem at the origin. (Since $\mathbb{P}$ is a probability measure, the space L_ψ depends essentially on the behavior of ψ at infinity.)

If B is a Banach space, we may define in the same way the spaces $L_p(B)$, $L_{p,\infty}(B)$, $L_\psi(B)$, etc. (cf. Section 2.2). Similarly, we agree to denote their norms by $\|\cdot\|_p$, $\|\cdot\|_{p,\infty}$, $\|\cdot\|_\psi$. If (X_n) is a sequence of (real or) vector valued random variables, it is said to be bounded in $L_0(B)$ or to be bounded in probability, or also to be stochastically bounded, if for each $\varepsilon > 0$ one can find $M > 0$ with

$$\sup_n \mathbb{P}\{\|X_n\| > M\} < \varepsilon.$$

The sequence (X_n) is bounded in $L_p(B)$ if $\sup_n \mathbb{E}\|X_n\|^p < \infty$.

Now we recall the integration by parts formula. If $X \ge 0$ and F is increasing on $\mathbb{R}_+$,

$$\mathbb{E}F(X) = F(0) + \int_0^\infty \mathbb{P}\{X > t\}\, dF(t) .$$

In particular, when $0 < p < \infty$,

$$\mathbb{E}X^p = \int_0^\infty \mathbb{P}\{X > t\}\, dt^p = p \int_0^\infty \mathbb{P}\{X > t\} t^{p-1}\, dt .$$

Further, a simple comparison between series and integrals indicates, as an example among many, that $\mathbb{E}X^p < \infty$ if and only if $\sum_n \mathbb{P}\{X > \varepsilon n^{1/p}\} < \infty$ for some, or all, $\varepsilon > 0$, or also $\sum_n 2^n \mathbb{P}\{X > \varepsilon 2^{n/p}\} < \infty$. By the Borel-Cantelli lemma, if (X_n) is a sequence of independent copies of X (i. e. with the same law, or distribution), and if (and only if) $\sup_n X_n/n^{1/p} < \infty$ almost surely, then $\sum_n \mathbb{P}\{X_n > \varepsilon n^{1/p}\} < \infty$ for some $\varepsilon > 0$, and thus, using the identical distribution of the X_n's, we have $\mathbb{E}X^p < \infty$. Of course, this type of classical argument also applies to (smooth) normalizations other than $n^{1/p}$. Recall also Jensen's inequality: if F is convex, then $\mathbb{E}F(X) \geq F(\mathbb{E}X)$.

If (X_i) is a sequence of random variables, we usually let $S_n = X_1 + \cdots + X_n$, $n \geq 1$. We sometimes abbreviate independent and identically distributed in iid. SLLN, CLT and LIL denote in short respectively strong law of large numbers, central limit theorem and law of the iterated logarithm. For the LIL, we shall use the (iterated logarithm) function $LLt = L(Lt)$, $Lt = \max(1, \log t)$, $t \geq 0$. Introduced in Chapter 5 and used throughout the book, the sequence $(\Gamma_j)_{j \geq 1}$ is such that, for all j, $\Gamma_j = \lambda_1 + \cdots + \lambda_j$ where (λ_j) is a sequence of independent random variables with the standard exponential distribution $\mathbb{P}\{\lambda_j > t\} = \exp(-t)$, $t \geq 0$.

A pseudo-metric d on a set T is a metric that does not necessarily separate points (i. e. $d(s,t) = 0$ does not always imply that $s = t$). Given a pseudo-metric space (T, d), $D = D(T)$ will usually denote its diameter (finite or not): $D = \sup\{d(s,t); s, t \in T\}$. If S is a subset of T, it is always assumed to be equipped with the induced pseudo-metric so that (S, d) defines a new pseudo-metric space. Given a metric or pseudo-metric space (T, d), $B(t, \varepsilon)$ denotes the open ball with center t and radius $\varepsilon > 0$. When several metric spaces are under consideration, we agree that $B(t, \varepsilon)$ denotes the ball in the space which contains its center t. We will usually deal with open balls but actually closed balls would not make any difference in our study and we usually do not specify this point with care. For example, this remark applies to the entropy numbers. For each $\varepsilon > 0$, $N(T, d; \varepsilon)$ is the minimal number of (open) balls of radius $\varepsilon > 0$ which are necessary to cover T. In $\mathbb{R}^N$, or more generally in a (Abelian) group G, we may consider the covering numbers $N(A, B)$ where A, B are subsets of G as the minimal number of translates of B (by elements of A) which are necessary to cover B, i.e.

$$N(A, B) = \inf\left\{N \geq 1; \exists t_1, \ldots, t_N \in A, A \subset \bigcup_{i=1}^{N}(t_i + B)\right\}$$

where $t_i + B = \{t_i + s; s \in B\}$. (The fact here that the translates are translates by elements of A is not really basic and for our purposes one can consider almost equivalently translates by elements of G.) For example, if d_2 is the Euclidean metric on $\mathbb{R}^N$, $T \subset \mathbb{R}^N$ and $B_2 = B_2^N$ is the (open! for consistency) Euclidean unit ball, then $N(T, d_2; \varepsilon) = N(T, \varepsilon B_2)$.

A contraction φ between two metric (or pseudo-metric) spaces (T, d), (S, δ) is a map $\varphi : T \to S$ such that $\delta(\varphi(t), \varphi(t')) \leq d(t, t')$ for all t, t' in T.

A metric space (U, δ) is called ultrametric if its distance satisfies the improved triangle inequality:

$$\delta(u,v) \leq \max\big(\delta(u,w),\delta(w,v)\big)$$

for all u,v,w in U. The nice feature of ultrametric spaces is that two balls of the same radius are either disjoint or equal. The structure of ultrametric spaces is that of a tree.

If T is any set, $\ell_\infty(T)$ is the space of all bounded functions on T with sup-norm denoted by $\|\cdot\|_\infty = \|\cdot\|_T$. With some abuse of notation, if h is some (bounded) function on T, we sometimes let $\|h(t)\|_T = \|h\|_T = \sup_{t\in T}|h(t)|$. A stochastic or random process indexed by T is a collection $X = (X_t)_{t\in T}$ of real (or vector) valued random variables indexed by T (cf. Section 2.2). It is also denoted by $X = (X_t)_{t\in T} = (X(t))_{t\in T}$.

Part 0

Isoperimetric Background
and Generalities

1. Isoperimetric Inequalities and the Concentration of Measure Phenomenon

In this first chapter, we present the isoperimetric inequalities which now appear as the crucial concept in the understanding of various concentration inequalities, tail behaviors and integrability theorems in Probability in Banach spaces. These inequalities often arise as the final and most elaborate forms of previous, weaker (but already efficient) inequalities which will be mentioned in their framework throughout the book. In these final forms however, the isoperimetric inequalities and associated concentration of measure phenomena provide the appropriate ideas for an in depth comprehension of some of the most important theorems of the theory.

The concentration of measure phenomenon, which roughly describes how a well-behaved function is almost a constant on almost all the space, can moreover be seen as the explanation for the two main parts of this work: the first one deals with "nice" functions applying isoperimetric inequalities and concentration properties, the second tries to determine conditions for a function to be "nice".

The concentration of measure phenomenon was mainly put forward by the local theory of Banach spaces in the study of Dvoretzky's theorem on almost Euclidean sections of convex bodies. Following [G-M], [Mi-S], the basic idea may be described in the following way. Let (X, ρ, μ) be a (compact) metric space (X, ρ) with a Borel probability measure μ. The concentration function $\alpha(X, r), r > 0$, is defined as

$$\alpha(X, r) = \sup \left\{ 1 - \mu(A_r) ; \ \mu(A) \geq \tfrac{1}{2}, \ A \subset X, A \text{ Borel} \right\}$$

where A_r denotes the ρ-neighborhood of order r of A i.e.

$$A_r = \left\{ x \in X ; \ \rho(x, A) < r \right\}.$$

For many families (X, ρ, μ), the concentration function $\alpha(X, r)$ turns out to be extremely small when r increases to infinity. A typical and basic example is given by the Euclidean unit sphere S^{N-1} in $\mathbb{R}^N$ equipped with its geodesic distance ρ and normalized Haar measure σ_{N-1} for which it can be shown (see below) that

$$\alpha(S^{N-1}, r) \leq \left(\frac{\pi}{8} \right)^{1/2} \exp\left(-(N-2)r^2/2\right) \ (N \geq 3).$$

Hence, the complement of the neighborhood of order r of a set of probability bigger than $1/2$ decreases extremely rapidly when r becomes large. This is what is now usually called the concentration of measure phenomenon. In the presence of such a property, any nice function is very close to being a constant (median or expectation) on all but a very small set, the smallness of which depends on $\alpha(X, r)$. For example, if f is a function on S^{N-1}, denote by $\omega_f(\varepsilon)$ its modulus of continuity, $\omega_f(\varepsilon) = \sup\{|f(x) - f(y)|\,;\ \rho(x, y) < \varepsilon\}$, and let M_f be a median of f. Then, for every $\varepsilon > 0$,

$$\sigma_{N-1}\big(|f - M_f| > \omega_f(\varepsilon)\big) \le \left(\frac{\pi}{2}\right)^{1/2} \exp\big(-(N-2)\varepsilon^2/2\big) \,.$$

While the concentration of measure phenomenon deals with the behavior of $\mu(A_r)$ for *large* values of r, the quintessence of the isoperimetric inequalities rather concerns *small* values of r, leading, via Minkowski contents and surface measures (cf. [B-Z], [Os]) to the more familiar formulations of isoperimetry. For our purposes in this work, we only deal with concentration and the consequences of isoperimetric inequalities to concentration. (Let us mention here the recent work [Ta19], see also [Mau4], where some new ideas on the distinct aspects of isoperimetry and concentration are developed.)

The concentration of measure phenomenon is thus usually derived from an isoperimetric inequality (see however [G-M], [Mi3], [Led7], [Led8]). This chapter describes various isoperimetric inequalities, which we develop here in their abstract and measure theoretical setting. Their application to Probability in Banach spaces will be one of the purposes of Part I of this book. The first section presents the isoperimetric inequalities and related concentration properties in the classical cases of the sphere, Gauss space and the cube. The main object of the second section is an isoperimetric theorem for product measures (independent random variables) while the last section is devoted to some well-known and useful martingale inequalities. Proofs of some of the deepest inequalities like the isoperimetric inequality on spheres and that for product measures are omitted and replaced by (hopefully) accurate references. However, various comments and remarks try to describe some of the ideas involved in these results as well as some of their consequences which will be useful in the sequel.

1.1 Some Isoperimetric Inequalities on the Sphere, in Gauss Space and on the Cube

Concentration properties for families (X, ρ, μ) are thus usually established via isoperimetric inequalities. We present some of these inequalities in this section and we start with the isoperimetric inequality on the sphere already alluded to above (cf. [B-Z], [F-L-M]...).

Theorem 1.1. *If A is a Borel set in S^{N-1} and if H is a cap (i.e. a ball for the geodesic distance ρ) with the same measure $\sigma_{N-1}(H) = \sigma_{N-1}(A)$, then, for any $r > 0$,*

$$\sigma_{N-1}(A_r) \geq \sigma_{N-1}(H_r)$$

where we recall that $A_r = \{x \in S^{N-1} ; \ \rho(x, A) < r\}$ *is the neighborhood of order* r *of* A *for the geodesic distance. In particular, if* $\sigma_{N-1}(A) \geq 1/2$ *(and* $N \geq 3$*), then*

$$\sigma_{N-1}(A_r) \geq 1 - \left(\frac{\pi}{8}\right)^{1/2} \exp\left(-(N-2)r^2/2\right) .$$

The main interest of such an isoperimetric theorem is of course the possiblility of an estimate (or even an explicit computation) of the measure of a cap (the neighborhood of a cap is again a cap), a particularly important estimate being given in the second assertion of the theorem.

Our interest in Theorem 1.1 lies in its connections with and consequences for a similar isoperimetric result for Gaussian measures. The close relationships and analogies between uniform measures on spheres and Gaussian measures have been noticed in many contexts. In this isoperimetric setting, it turns out that the isoperimetric inequality on S^{N-1} leads, in the Poincaré limit when N increases to infinity, to an isoperimetric inequality for Gauss measure. Poincaré's limit expresses the canonical Gaussian measure in finite dimension as the limiting distribution of projected uniform distributions on spheres of radius $\sqrt{N}$ when N tends to infinity. This is one example of the deep relations mentioned previously; it is also a way of illustrating the common belief that the Wiener measure can be thought of as a uniform distribution on a sphere of infinite dimension and of radius square root of infinity (cf. [MK]).

To be more precise about this observation of Poincaré (although it does not seem to be due to H. Poincaré – see the Notes and References), denote, for every N, by $\sigma_{N-1}^{\sqrt{N}}$ the uniform normalized measure on the sphere $\sqrt{N}S^{N-1}$ with center the origin and radius $\sqrt{N}$ in $\mathbb{R}^N$. Denote further by $\Pi_{N,d}(N \geq d)$ the projection from $\sqrt{N}S^{N-1}$ onto $\mathbb{R}^d$. Then, the sequence $\{\Pi_{N,d}(\sigma_{N-1}^{\sqrt{N}}); N \geq d\}$ of measures on $\mathbb{R}^d$ converges weakly when N goes to infinity to the canonical Gaussian measure on $\mathbb{R}^d$. For a sketch of the proof, simply note that by the law of large numbers $\rho_N^2/N \to 1$ almost surely (or only in probability) where $\rho_N^2 = g_1^2 + \cdots + g_N^2$ and (g_i) is a sequence of independent standard normal random variables. Now, clearly, $(N^{1/2}/\rho_N) \cdot (g_1, \ldots, g_N)$ is equal in distribution to $\sigma_{N-1}^{\sqrt{N}}$; hence $(N^{1/2}/\rho_N) \cdot (g_1, \ldots, g_d) = \Pi_{N,d}(\sigma_{N-1}^{\sqrt{N}})$ and the conclusion follows. Note that we will actually need a little more than weak convergence in this Poincaré limit, namely convergence for *all* Borel sets. This can be obtained similarly with some more efforts (see [Fer9], [D-F]).

With this tool, it is simple to see how it is possible to derive an isoperimetric inequality for Gaussian measures from Theorem 1.1. The basic result which is obtained in this way concerns the canonical Gaussian distribution in finite dimension; as is classical however, the fundamental feature of Gaussian distributions then allows to extend these results to general finite or infinite dimensional Gaussian measures. Let us denote therefore by γ_N the canonical Gaussian probability measure on $\mathbb{R}^N$ with density

$$\gamma_N(dx) = (2\pi)^{-N/2} \exp\left(-|x|^2/2\right) dx \ .$$

Observe the simple, but essential, fact that γ_N is the product measure on $\mathbb{R}^N$ when each factor is endowed with γ_1. Denote further by Φ the distribution function of γ_1, i.e.

$$\Phi(t) = (2\pi)^{-1/2} \int_{-\infty}^{t} \exp(-x^2/2) dx, \quad t \in [\infty, +\infty] \ .$$

Φ^{-1} is the inverse function and $\Psi = 1 - \Phi$ for which we recall the classical estimate:

$$\Psi(t) \leq \tfrac{1}{2} \exp(-t^2/2), \quad t \geq 0 \ .$$

The next theorem is the isoperimetric inequality for $(\mathbb{R}^N, d, \gamma_N)$ where d denotes the Euclidean distance.

Theorem 1.2. *If A is a Borel set in $\mathbb{R}^N$ and if H is a half-space $\{x \in \mathbb{R}^N;$ $\langle x, u \rangle < \lambda\}$, $u \in \mathbb{R}^N$, $\lambda \in [-\infty, +\infty]$, with the same Gaussian measure $\gamma_N(H) = \gamma_N(A)$, then, for any $r > 0$, $\gamma_N(A_r) \geq \gamma_N(H_r)$ where, accordingly, A_r is the Euclidean neighborhood of order r of A. Equivalently,*

$$\Phi^{-1}\left(\gamma_N(A_r)\right) \geq \Phi^{-1}\left(\gamma_n(A)\right) + r$$

and in particular, if $\gamma_N(A) \geq 1/2$,

$$1 - \gamma_N(A_r) \leq \Psi(r) \leq \tfrac{1}{2} \exp(-r^2/2) \ .$$

Proof. The equivalence between the two formulations is easy since the Gaussian measure of a half-space is being computed in dimension one (take H to be orthogonal to a coordinate axis and remenber that γ_N is a product measure) and thus

$$\gamma_N(H_r) = \Phi\left(\Phi^{-1}\left(\gamma_N(H)\right) + r\right) = \Phi\left(\Phi^{-1}\left(\gamma_N(A)\right) + r\right) \ .$$

The case $\gamma_N(A) \geq 1/2$ simply follows from the fact that $\Phi^{-1}(1/2) = 0$. Turning to the proof itself, since $\Phi^{-1}(0) = -\infty$, we may assume that $a = \Phi^{-1}(\gamma_N(A)) > -\infty$. Let then $b \in]-\infty, a[$. By Poincaré's observation, for every large enough $k\ (\geq N)$,

$$\sigma_{k-1}^{\sqrt{k}}\left(\Pi_{k,N}^{-1}(A)\right) > \sigma_{k-1}^{\sqrt{k}}\left(\Pi_{k,1}^{-1}(]-\infty, b])\right) \ .$$

It is easy to see that $\Pi_{k,N}^{-1}(A_r) \supset (\Pi_{k,N}^{-1}(A))_r$ where the neighborhood of order r on the right is understood with respect to the geodesic distance on $\sqrt{k} S^{k-1}$. Since $\Pi_{k,1}^{-1}(]-\infty, b])$ is a cap on $\sqrt{k} S^{k-1}$, by the isoperimetric inequality on spheres (Theorem 1.1),

$$\sigma_{k-1}^{\sqrt{k}}\big(\Pi_{k,N}^{-1}(A_r)\big) \geq \sigma_{k-1}^{\sqrt{k}}\Big(\big(\Pi_{k,N}^{-1}(A)\big)_r\Big)$$

$$\geq \sigma_{k-1}^{\sqrt{k}}\Big(\big(\Pi_{k,1}^{-1}(]-\infty,b])\big)_r\Big)\,.$$

Now $(\Pi_{k,1}^{-1}(]-\infty,b]))_r = \Pi_{k,1}^{-1}(]-\infty, b+r(k)])$ for some $r(k) \geq 0$ satisfying $\lim_{k\to\infty} r(k) = r$. Therefore, in the Poincaré limit, we get $\gamma_N(A_r) \geq \Phi(b+r)$, and hence the result since $b < \Phi^{-1}(\gamma_N(A))$ is arbitrary.

Note that half-spaces, such as caps on spheres (that are intersections of the sphere and of a half-space), are extremal sets for the Gaussian isoperimetric inequality since they achieve equality in the conclusion.

Thus, the isoperimetric inequality for Gauss measure follows rather easily from the corresponding inequality on spheres. However, this latter inequality requires quite an involved proof which is based on extensive use of powerful symmetrization techniques (in the sense of Steiner). One of the remarkable observations of A. Ehrhard was to show how one can introduce a similar symmetrization procedure which is adapted to Gauss measure (with half-spaces as extremal sets). One can then give a more intrinsic proof of Theorem 1.2. This also led him to rather a complete isoperimetric calculus in Gauss space ([Eh1], [Eh2], [Eh3]). In particular, he obtained in this way an inequality of the Brunn-Minkowski type: namely for A, B convex sets in $\mathbb{R}^N$ and $\lambda \in [0,1]$,

$$(1.1) \quad \Phi^{-1}\big(\gamma_N\big(\lambda A + (1-\lambda)B\big)\big) \geq \lambda\Phi^{-1}\big(\gamma_N(A)\big) + (1-\lambda)\Phi^{-1}\big(\gamma_N(B)\big)$$

where the sum $\lambda A + (1-\lambda)B$ is understood in the sense of Minkowski as $\{x \in \mathbb{R}^N;\ x = \lambda a + (1-\lambda)b,\ a \in A, b \in B\}$. Taking B to be the Euclidean ball with center the origin and radius $r/(1-\lambda)$ and letting λ tend to 1, it is easily seen how (1.1) actually implies the isoperimetric inequality of Theorem 1.2 (for A convex). However, at the present time, (1.1) is only known for convex sets. An inequality on which (1.1) appears as an improvement (for convex sets), but which holds for arbitrary Borel sets A, B, is the so-called log-concavity of Gaussian measures:

$$(1.2) \quad \log\gamma_N\big(\lambda A + (1-\lambda)B\big) \geq \lambda\log\gamma_N(A)) + (1-\lambda)\log\gamma_N(B)\,.$$

A proof of (1.2) may be given using again the Poincaré limit but this time on the classical Brunn-Minkowski inequality on $\mathbb{R}^N$ (see [B-Z], [Pi18]...) which states that

$$\mathrm{vol}_N\big(\lambda A + (1-\lambda)B\big) \geq \big(\mathrm{vol}_N(A)\big)^{\lambda}\big(\mathrm{vol}_N(B)\big)^{1-\lambda}$$

for λ in [0,1], A, B bounded in $\mathbb{R}^N$ and where vol_N is the N-dimensional volume. Let, for $k \geq N$, $P_{k,N}$ be the projection from the ball with center the origin and radius $\sqrt{k}$ in $\mathbb{R}^k$ onto $\mathbb{R}^N$. If A, B are Borel sets in $\mathbb{R}^N$ and $0 \leq \lambda \leq 1$, it is clear using convexity that

$$P_{k,N}^{-1}\big(\lambda A + (1-\lambda)B\big) \supset \lambda P_{k,N}^{-1}(A) + (1-\lambda)P_{k,N}^{-1}(B)\,.$$

Hence, by the Brunn-Minkowski inequality (in $\mathbb{R}^k$),

$$\mathrm{vol}_k\left(P_{k,N}^{-1}(\lambda A + (1-\lambda)B)\right) \geq \left(\mathrm{vol}_k\left(P_{k,N}^{-1}(A)\right)\right)^\lambda \left(\mathrm{vol}_k\left(P_{k,N}^{-1}(B)\right)\right)^{1-\lambda}.$$

Then, one should only multiply both terms of this inequality by an appropriate normalizing sequence and, by using a variation on the Poincaré limit (which turns out to be actually simpler in this case), we obtain (1.2) for every Borel sets A and B in $\mathbb{R}^N$. One further measures on this proof the sharpness of (1.1).

As announced, and due to the properties of Gaussian distributions, the preceding inequalities and Theorem 1.2 easily extend to general finite or infinite dimensional Gaussian measures. These extensions will usually be described in their applications below. However, let us briefly indicate one infinite dimensional result which it will be useful to record here. Consider the measure $\gamma = \gamma_\infty$ on $\mathbb{R}^{\mathbb{N}}$ which is the infinite product of the canonical one-dimensional Gaussian distribution on each coordinate. The isoperimetric inequality indicates similarly that for each Borel set A in $\mathbb{R}^{\mathbb{N}}$ and each $r > 0$,

$$(1.3) \qquad \Phi^{-1}\left(\gamma_*(A_r)\right) \geq \Phi^{-1}\left(\gamma(A)\right) + r$$

where A_r is here the Euclidean or rather the Hilbertian neighborhood of order r of A in $\mathbb{R}^{\mathbb{N}}$, i.e. $A_r = A + rB_2 = \{x = a + rh; a \in A, h \in \mathbb{R}^{\mathbb{N}}, |h| \leq 1\}$ where B_2 is the unit ball of ℓ_2 (A_r is not necessarily measurable). This of course simply follows from Theorem 1.2 and a cylindrical approximation. Note that $\gamma(\ell_2) = 0$.

As a corollary to Theorem 1.2, we now express the concentration of measure phenomenon for functions on $(\mathbb{R}^N, d, \gamma_N)$. This formulation of isoperimetry will turn out to be a convenient tool in the applications. Let f be Lipschitz on $\mathbb{R}^N$ with Lipschitz norm given by

$$\|f\|_{\mathrm{Lip}} = \sup\left\{\frac{|f(x) - f(y)|}{|x - y|}; x, y \in \mathbb{R}^N\right\}.$$

Let us denote further by M_f a median of f for γ_N, i.e. M_f is a number such that $\gamma_N(f \geq M_f)$ and $\gamma_N(f \leq M_f)$ are both bigger than $1/2$. Applying the second conclusion of Theorem 1.2 to those two sets of measure larger than or equal to $1/2$ and noticing that for $t > 0$,

$$\left(\{f \geq M_f\} \cap \{f \leq M_f\}\right)_t \subset \{|f - M_f| \leq t\|f\|_{\mathrm{Lip}}\},$$

we get, for all $t > 0$,

$$(1.4) \qquad \gamma_N\left(|f - M_f| > t\right) \leq 2\Psi\left(t/\|f\|_{\mathrm{Lip}}\right) \leq \exp\left(-t^2/2\|f\|_{\mathrm{Lip}}^2\right).$$

Hence, with very high probability, f is concentrated around its median M_f. As we will see in Chapter 3, this inequality can be used to investigate the integrability properties of Gaussian random vectors. Let us note also that the preceding argument applied only to $\{f \leq M_f\}$ shows similarly that

$$\gamma_N(f > M_f + t) \le \tfrac{1}{2}\exp\left(-t^2/2\|f\|_{\text{Lip}}^2\right).$$

This inequality however appears more as a *deviation* inequality only as opposed to (1.4) which indicates a *concentration*. In the sequel, we shall come back to this distinction in various contexts.

While (1.4) appears as a direct consequence of Theorem 1.2, it should be noted that inequalities of the same type can actually be established by simple direct arguments and considerations. One such approach is the following. Let f be, as before, Lipschitz on $\mathbb{R}^N$, so that f is almost everywhere differentiable and its gradient ∇f satisfies $|\nabla f| \le \|f\|_{\text{Lip}}$. Assume now moreover that $\int f d\gamma_N = 0$. Then, for any t and $\lambda > 0$, we can write by Chebyshev's inequality,

$$\gamma_N(f > t) \le \exp(-\lambda t)\int \exp(\lambda f) d\gamma_N$$

$$\le \exp(-\lambda t)\iint \exp\left[\lambda\big(f(x) - f(y)\big)\right] d\gamma_N(x)d\gamma_N(y)$$

where in the second inequality we have used Jensen's inequality (in y) and the mean zero assumption. Now, x, y being fixed in $\mathbb{R}^N$, set, for any θ in $[0, 2\pi]$,

$$x(\theta) = x\sin\theta + y\cos\theta\,,\quad x'(\theta) = x\cos\theta - y\sin\theta\,.$$

We have

$$f(x) - f(y) = \int_0^{\pi/2} \frac{d}{d\theta}f\big(x(\theta)\big)d\theta$$

$$= \int_0^{\pi/2} \langle\nabla f\big(x(\theta)\big)\,,\ x'(\theta)\rangle d\theta\,.$$

Hence, using Jensen's inequality once more (here with respect to θ), $\gamma_N(f > t)$ is majorized by

$$\exp(-\lambda t)\frac{2}{\pi}\int_0^{\pi/2}\left(\iint \exp\left[\frac{\lambda\pi}{2}\langle\nabla f\big(x(\theta)\big)\,,\ x'(\theta)\rangle\right]d\gamma_N(x)d\gamma_N(y)\right)d\theta\,.$$

The fundamental rotational invariance of Gaussian measures indicates that, for any θ, the couple $(x(\theta), x'(\theta))$ has the same distribution as the original one (x, y) under $\gamma_N \otimes \gamma_N$. Therefore, by Fubini's theorem,

$$\gamma_N(f > t) \le \exp(-\lambda t)\iint \exp\left[\frac{\lambda\pi}{2}\langle\nabla f(x),\ y\rangle\right]d\gamma_N(x)d\gamma_N(y)\,.$$

Performing integration in y,

$$\gamma_N(f > t) \le \exp(-\lambda t)\int \exp\left(\frac{\lambda^2\pi^2}{8}|\nabla f|^2\right)d\gamma_N$$

$$\le \exp\left(-\lambda t + \frac{\lambda^2\pi^2}{8}\|f\|_{\text{Lip}}^2\right)\,.$$

If we then minimize in λ ($\lambda = 4t/\pi^2\|f\|_{\text{Lip}}^2$), we finally get

$$\gamma_N(f > t) \leq \exp\left(-2t^2/\pi^2\|f\|_{\text{Lip}}^2\right)$$

for every $t > 0$. For f not necessarily of mean zero, and applying the result to $-f$ as well, it follows that, for $t > 0$,

(1.5) $$\gamma_N\left(|f - E_f| > t\right) \leq 2\exp\left(-2t^2/\pi^2\|f\|_{\text{Lip}}^2\right)$$

where $E_f = \int f d\gamma_N$ (finite since f is Lipschitz). This inequality is of course very close in spirit to (1.4). Although (1.5) has a worse constant in the exponent, it can actually be shown, using a similar argument but with stochastic differentials with respect to Brownian motion, that we also have, for every $t > 0$,

(1.6) $$\gamma_N\left(|f - E_f| > t\right) \leq 2\exp\left(-t^2/2\|f\|_{\text{Lip}}^2\right).$$

However, the preceding argument leading to (1.5) presents the advantage to apply to more general situations such as vector valued functions (cf. [Pi16]). In any case, we retain that concentration inequalities of the type (1.4)–(1.6) are usually easier to obtain than the rather delicate isoperimetric theorems.

(1.4) and (1.6) describe the same concentration property around a median M_f or the expectation $E_f = \int f d\gamma_N$ of a Lipschitz function f. Of course, it is often easier to work with expectations rather than with medians. Actually, here these are essentially of the same order. Indeed, integrating for example (1.4) yields

$$|E_f - M_f| \leq (\pi/2)^{1/2}\|f\|_{\text{Lip}},$$

while, given (1.6), if t is chosen such that

$$2\exp\left(-t^2/2\|f\|_{\text{Lip}}^2\right) < \tfrac{1}{2},$$

for example $t = 2\|f\|_{\text{Lip}}$, we get from (1.6)

$$E_f - t \leq M_f \leq t + E_f.$$

However, it is not known whether it is possible to deduce *exactly* (1.4) and (1.6) from each other. Moreover, note that (1.4) actually shows that a median M_f is necessarily *unique*. Indeed, if $M_f < M_f'$ are two distinct medians of f, letting $t = (M_f' - M_f)/2 > 0$ gives

$$\tfrac{1}{2} \leq \gamma_N(f \geq M_f') \leq \gamma_N(f > M_f + t) \leq \Psi(t/\|f\|_{\text{Lip}}) < \tfrac{1}{2}$$

which is impossible.

Uniform measures on spheres and Gaussian measures thus satisfy isoperimetric inequalities and concentration phenomena. For our purposes, there is a useful observation which allows us to deduce from the Gaussian inequalities further inequalities for a rather large class of measures by a simple "contraction" argument. Denote by φ a Lipschitz map on $\mathbb{R}^N$ with values in $\mathbb{R}^N$ such that for some $c = c_\varphi > 0$

$$|\varphi(x) - \varphi(y)| \leq c|x - y| \quad \text{for all } x, y \text{ in } \mathbb{R}^N.$$

Denote by λ the image measure of γ_N by φ, i.e. $\lambda(A) = \gamma_N(\varphi^{-1}(A))$ for every Borel set A in $\mathbb{R}^N$. Then, there is an isoperimetric inequality of the Gaussian type for the measure λ, namely, for any measurable set A in $\mathbb{R}^N$ and any $r > 0$,

(1.7) $$\Phi^{-1}(\lambda(A_{cr})) \geq \Phi^{-1}(\lambda(A)) + r.$$

Similarly, the corresponding inequality (1.4) for λ also holds with $c\|f\|_{\text{Lip}}$ instead of $\|f\|_{\text{Lip}}$ in the right hand side. For a proof of (1.7), simply note that by Theorem 1.2,

$$\Phi^{-1}\left(\gamma_N\left(\left(\varphi^{-1}(A)\right)_r\right)\right) \geq \Phi^{-1}(\lambda(A)) + r$$

and, by the Lipschitz property of φ, clearly $(\varphi^{-1}(A))_r \subset \varphi^{-1}(A_{cr})$ from which the result follows. Inequality (1.5) and its simple proof may be extended similarly (see [Pi16]).

Then, it is of course of some interest to try to describe the class of probability measures λ which can be obtained as the image of γ_N by a contraction. While a complete description is still missing, the next examples are noteworthy (and useful for the sequel). For a further main example, see [Ta19]. Let λ be uniformly distributed on the cube $[0,1]^N$. Then λ is the image of γ_N by the map $\varphi = \Phi^{\otimes N}$, i.e. $\varphi(x) = \varphi(x_1, \ldots, x_N) = \Phi(x_1) \cdots \Phi(x_N)$, $x = (x_1, \ldots, x_N) \in \mathbb{R}^N$, for which it is easily seen that $c_\varphi = (2\pi)^{-1/2}$. If, for symmetry reasons, one is rather interested by the uniform measure on $[-1/2, +1/2]^N$, then choose $\varphi = (2\Phi - 1)^{\otimes N}$ for which $c_\varphi = (2/\pi)^{1/2}$.

Unfortunately, the preceding approach does not allow to investigate the important case of the Haar measure on $\{0,1\}^N$ (the extreme points of $[0,1]^N$), or rather on $\{-1, +1\}^N$ which we use preferably for symmetry reasons. A different way has to be taken. Denote by $\mu_N = (\frac{1}{2}\delta_{-1} + \frac{1}{2}\delta_{+1})^{\otimes N}$ the canonical probability measure (Haar measure) on the (Cantor) group $\{-1, +1\}^N$. Consider the normalized Hamming metric d on $\{-1, +1\}^N$ given by

$$d(x,y) = \frac{1}{N}\text{Card}\{i \leq N \, ; \, x_i \neq y_i\} = \frac{1}{2N}\sum_{i=1}^{N}|x_i - y_i|$$

for $x, y \in \{-1, +1\}^N$. An isoperimetric theorem for the triple $(\{-1, +1\}^N, d, \mu_N)$ is known [Har] and states in particular that if $\mu_N(A) \geq 1/2$ for some $A \subset \{-1, +1\}^N$, for $r > 0$,

$$1 - \mu_N(A_r) \leq \tfrac{1}{2} \exp(-2Nr^2)$$

where, as usual, $A_r = \{x \in \{-1, +1\}^N; d(x, A) < r\}$. Unfortunately, this result is not strong enough to yield a concentration inequality for μ_N similar to (1.4) or (1.6) since it depends on the dimension. In order to accomplish this program, we will need a stronger result (independent of N) which is the following. For a non-empty subset A of $\{-1, +1\}^N$, set, for $x \in \{-1, +1\}^N$,

$$d_A(x) = \inf\{|x - y| \; ; y \in \operatorname{Conv} A\}$$

where $\operatorname{Conv} A$ is the convex hull (in $[-1, +1]^N$) of A.

Theorem 1.3. *For any non-empty subset A of $\{-1, +1\}^N$,*

$$\int \exp(d_A^2/8) d\mu_N \leq \frac{1}{\mu_N(A)} \; .$$

Proof. We first consider the case where $\operatorname{Card} A = 1$. Then

$$\int \exp(d_A^2/8) d\mu_N = 2^{-N} \sum_{i=0}^{N} \binom{N}{i} e^{i/2}$$

$$= \left(\frac{1 + e^{1/2}}{2}\right)^N \leq 2^N = \frac{1}{\mu_N(A)}$$

since $e^{1/2} < e < 3$. This already proves the theorem when $N = 1$ since then the only case left is $A = \{-1, +1\}$ for which $d_A \equiv 0$ and the result holds.

Now, we prove Theorem 1.3 by induction over N. Assuming it holds for N, we prove it for $N + 1$. By the preceding, it is enough to consider the case where A has at least two points. Assuming without loss of generality that these points differ on the last coordinate and identifying $\{-1, +1\}^{N+1}$ with $\{-1, +1\}^N \times \{-1, +1\}$, we can then suppose that $A = A_{-1} \times \{-1\} \cup A_{+1} \times \{+1\}$ where A_{-1}, A_{+1} are non-empty subsets of $\{-1, +1\}^N$. For example, we can assume that $\mu_N(A_{-1}) \leq \mu_N(A_{+1})$ and we observe moreover that $d_A((x, 1)) \leq d_{A_{+1}}(x)$. The crucial point of the proof is contained in the following observation: for any x in $\{-1, +1\}^N$ and $0 \leq \alpha \leq 1$,

(1.8) $$d_A^2((x, -1)) \leq 4\alpha^2 + \alpha d_{A_{+1}}^2(x) + (1 - \alpha) d_{A_{-1}}^2(x) \; .$$

Indeed, for $i = -1, +1$, let $z_i \in \operatorname{Conv} A_i$ such that $|x - z_i| = d_{A_i}(x)$. We notice that $(z_i, i) \in \operatorname{Conv} A$ so that $z = (\alpha z_{+1} + (1 - \alpha) z_{-1}, -1 + 2\alpha) \in \operatorname{Conv} A$. Now

$$|(x, -1) - z|^2 = 4\alpha^2 + |x - (\alpha z_{+1} + (1 - \alpha) z_{-1})|^2$$

$$= 4\alpha^2 + |\alpha(x - z_{+1}) + (1 - \alpha)(x - z_{-1})|^2$$

$$\leq 4\alpha^2 + \alpha|x - z_{+1}|^2 + (1 - \alpha)|x - z_{-1}|^2$$

by the triangle inequality and the convexity of the square function. This proves (1.8).

For $i = -1, +1$, we set $u_i = \int \exp(d_{A_i}^2/8) d\mu_N$ and $v_i = 1/\mu_N(A_i)$ so that $u_i \leq v_i$ by the induction hypothesis. From $d_A((x, 1)) \leq d_{A_{+1}}(x)$ and (1.8) we have, for every $0 \leq \alpha \leq 1$,

$$\int \exp(d_A^2/8)d\mu_{N+1} \le \frac{1}{2}\int \exp(d_{A+1}^2/8)d\mu_N$$

$$+ \frac{1}{2}\int \exp(\alpha^2/2 + \alpha d_{A+1}^2/8 + (1-\alpha)d_{A-1}^2/8)\, d\mu_N$$

$$\le \frac{1}{2}u_{+1} + \frac{1}{2}e^{\alpha^2/2}u_{+1}^\alpha u_{-1}^{1-\alpha}$$

$$\le \frac{1}{2}v_{+1}\left[1 + e^{\alpha^2/2}\left(\frac{v_{-1}}{v_{+1}}\right)^{1-\alpha}\right]$$

by Hölder's inequality using also $u_i \le v_i$. The value of α which minimizes the preceding expression is $\alpha = -\log(v_{+1}/v_{-1})$ but, in order not to have to consider the case where $\alpha \ge 1$, let us take $\alpha = 1 - v_{+1}/v_{-1}$ (recall that we assume that $v_{+1} = 1/\mu_N(A_{+1}) \le 1/\mu_N(A_{-1}) = v_{-1}$) which gives

$$\int \exp(d_A^2/8)d\mu_{N+1} \le \frac{1}{2}v_{+1}\left[1 + e^{\alpha^2/2}(1-\alpha)^{\alpha-1}\right].$$

It is elementary to see that for $0 \le \alpha < 1$,

$$1 + e^{\alpha^2/2}(1-\alpha)^{\alpha-1} \le \frac{4}{2-\alpha}.$$

Moreover, for this value α,

$$\frac{1}{2}v_{+1}\left(\frac{4}{2-\alpha}\right) = \frac{2}{\frac{1}{v_{+1}} + \frac{1}{v_{-1}}} = \frac{2}{\mu_N(A_{+1}) + \mu_N(A_{-1})} = \frac{1}{\mu_{N+1}(A)}.$$

The proof of Theorem 1.3 is complete.

As announced, Theorem 1.3 contains a concentration estimate for Lipschitz functions which is similar to the estimate we described before for Gaussian measures. This application is one of the main interests of Theorem 1.3. However, with respect to the preceding inequalities, one main additional assumption is that the inequality will only concern *convex* Lipschitz functions f. Let f (on $\mathbb{R}^N$) be convex and Lipschitz with Lipschitz constant $\|f\|_{\mathrm{Lip}}$. Let M_f denote a median of f with respect to μ_N. Then, for every $t > 0$,

$$(1.9) \qquad \mu_N(|f - M_f| > t) \le 4\exp(-t^2/8\|f\|_{\mathrm{Lip}}^2).$$

To prove this inequality, let first $A = \{f \le M_f\}$. Since f is convex, $f \le M_f$ on $\mathrm{Conv}A$. Further, by the Lipschitz property, if $d_A(x) \le t/\|f\|_{\mathrm{Lip}}$, then $f(x) \le M_f + t$. Hence, by Chebyshev's inequality and Theorem 1.3,

$$\mu_N(f > M_f + t) \le \mu_N(d_A > t/\|f\|_{\mathrm{Lip}})$$

$$\le \frac{1}{\mu_N(A)}\exp(-t^2/8\|f\|_{\mathrm{Lip}}^2)$$

$$\le 2\exp(-t^2/8\|f\|_{\mathrm{Lip}}^2).$$

On the other hand, let $B = \{f < M_f - t\}$. As above, we see that $d_B(x) \leq t/\|f\|_{\mathrm{Lip}}$ implies $f(x) < M_f$; thus

$$\mu_N\big(d_B > t/\|f\|_{\mathrm{Lip}}\big) \geq \tfrac{1}{2}$$

by the definition of the median. Now, again by Chebyshev's inequality and Theorem 1.3, we get

$$\mu_N(B) \leq 2\exp\big(-t^2/8\|f\|_{\mathrm{Lip}}^2\big) .$$

These two inequalities together imply (1.9).

Actually, Theorem 1.3 and the subsequent concentration inequality (1.9) do not depend on N and easily extend to the case of the Haar measure $\mu = (\tfrac{1}{2}\delta_{-1} + \tfrac{1}{2}\delta_{+1})^{\otimes\mathbb{N}}$ on $\{-1,+1\}^{\mathbb{N}}$. For example, concerning (1.9), if f on $\mathbb{R}^{\mathbb{N}}$ is convex and Lipschitz in the sense that

$$|f(\alpha) - f(\beta)| \leq \|f\|_{\mathrm{Lip}}|\alpha - \beta|$$

for $\alpha - \beta \in \ell_2$, then, for M_f a median of f for μ, we have similarly that

$$(1.10) \qquad \mu\big(|f - M_f| > t\big) \leq 4\exp\big(-t^2/8\|f\|_{\mathrm{Lip}}^2\big)$$

for all $t > 0$.

Compared with the corresponding inequalities (1.4) and (1.6), the coefficient 8 in (1.9) does not seem best possible. It is not known whether 2 can be reached, something which the argument of the proof of Theorem 1.3 cannot accomplish. Let us note further that the convexity assumption on f in (1.9) cannot be dropped. This is made clear by the following example. Let $A = \{x \in \{-1,+1\}^N; \sum_{i=1}^{N} x_i \leq 0\}$ and define $f(x) = \inf\{|x - y|; \ y \in A\}$. Clearly $\|f\|_{\mathrm{Lip}} = 1$ and 0 is a median of f. Assume that N is an *even* integer. Then, as is easy to see, $f^2(x) = 2\big(\sum_{i=1}^{N} x_i\big)^+$; but then, from the central limit theorem, $\mu_N(f > cN^{1/4}) \geq 1/4$ for some $c > 0$ independent of N from which it is clear that the non-convex Lipschitz function f cannot verify an inequality as (1.9).

Despite these somewhat negative observations, Theorem 1.3 and the concentration inequality (1.9) will be used in Chapter 4 in the study of the tail behaviors and integrability properties of vector valued Rademacher series as efficiently as the Gaussian inequalities in Chapter 3.

1.2 An Isoperimetric Inequality for Product Measures

The preceding isoperimetric inequalities and concentration phenomena will be applied in the next chapters to the study of the integrability properties and tail behaviors of Gaussian and Rademacher series with vector valued coefficients. In this section, we present an isoperimetric theorem for product measures which will be our key tool in the study of general sums of independent random variables (Chapter 6). Its discovery was actually motivated by these questions.

The statement of the result is somewhat abstract but we will try to clarify its powerful meaning by making some comments and by discussing some ideas about the proof.

Given a probability space (E, Σ, μ) and a fixed, but arbitrary, integer $N \geq 1$, denote by P the product measure $\mu^{\otimes N}$ on E^N. A point x in E^N has coordinates $x = (x_1, \ldots, x_N)$, $x_i \in E$. To a subset A of E^N, we associate

$$H(A, q, k) = \{x \in E^N; \ \exists x^1, \ldots, x^q \in A$$
$$\text{such that Card} \left\{i \leq N; \ x_i \notin \{x_i^1, \ldots, x_i^q\}\right\} \leq k\} \ .$$

The set $H(A, q, k)$ can be thought of, in an isoperimetric way, as some neighborhood of A whose elements are determined by a fixed number q of points in A with at most k free coordinates. This can be made somewhat more precise in the terminology of the beginning of this chapter. For an element x in $(E^N)^q$, denote its coordinates by $x = (x^1, \ldots, x^q) = (x^\ell)_{\ell \leq q}$ where $x^\ell \in E^N$ and, as before, $x^\ell = (x_i^\ell)_{i \leq N}$. Between elements x, y in $(E^N)^q$ introduce

$$d(x, y) = \sum_{i=1}^{N} I_{\{\forall \ell = 1, \ldots, q: \ x_i^\ell \neq y_i^\ell\}}$$
$$= \text{Card} \left\{i \leq N; \ \forall \ell = 1, \ldots, q, \ x_i^\ell \neq y_i^\ell\right\} \ .$$

Then, for A in E^N, $H(A, q, k)$ can simply be interpreted as the neighborhood of order k with respect to d (although d is not a metric) in the sense that

$$H(A, q, k) = \{x \in E^N; \ d(\widetilde{x}, A^q) \leq k\}$$

where, for x in E^N, $\widetilde{x}$ is the element of $(E^N)^q$ with coordinates $\widetilde{x} = (x, \ldots, x)$.

The isoperimetric theorem estimates the size of $H(A, q, k)$ under P in terms of $P(A)$, q and k. The main conclusion is an *exponential decay* in terms of k of the measure of the complement of $H(A, q, k)$.

Theorem 1.4. *For some universal positive constant K,*

$$P_*\big(H(A, q, k)\big) \geq 1 - \left[K\left(\frac{\log\big(1/P(A)\big)}{k} + \frac{1}{q}\right)\right]^k$$

where P_* denotes inner probability.

The proof of Theorem 1.4 is isoperimetric in nature and relies on several reductions based on symmetrization (rearrangement) procedures. Below, we illustrate one typical argument in a particular case. However, it does not seem to allow for an exact solution of the isoperimetric problem which would be the determination, for any $a > 0$, of

$$\inf\big\{P_*\big(H(A, q, k)\big); \ P(A) > a\big\} \ .$$

Note further that the use of the inner probability is necessary since $H(A, q, k)$ need not be measurable when A is.

Theorem 1.4 is mostly used in application only in the typical case $P(A) \geq 1/2$ and $k \geq q$ for which we get

$$(1.11) \qquad P_*\big(H(A,q,k)\big) \geq 1 - \left(\frac{K_0}{q}\right)^k$$

where $K_0 > 0$ is a numerical constant. It will be convenient in the sequel to assume this constant K_0 to be an *integer*; we do this and K_0 will moreover always have the meaning of (1.11) throughout the book.

It has been remarked in [Tal1] that, in the case $P(A) \geq 1/2$ for example, (1.11) can actually be improved into

$$(1.12) \qquad P_*\big(H(A,q,k)\big) \geq 1 - \left[K\left(\frac{1}{k} + \frac{1}{q \log q}\right)\right]^k.$$

The gain of the factor $\log q$ is irrelevant as far as the applications presented in this work are concerned. It should be noted however that this estimate is sharp. Indeed, consider the case where $E = \{0,1\}$ and $\mu = \left(1 - \frac{r}{N}\right)\delta_0 + \frac{r}{N}\delta_1$ where r is an integer less than N and N is assumed to be large. Let A in E^N be defined as

$$A = \left\{x \in E^N;\ \sum_{i=1}^{N} x_i \leq r\right\}.$$

Then $P(A)$ is of the order of $1/2$ and, clearly,

$$H(A,q,k) = \left\{x \in E^N;\ \sum_{i=1}^{N} x_i \leq rq + k\right\}.$$

Now

$$\begin{aligned}
P\big(H(A,q,k)^c\big) &\geq \left(\frac{r}{N}\right)^{rq+k+1}\left(1 - \frac{r}{N}\right)^{N-rq-k}\binom{N}{rq+k+1} \\
&\geq \left(\frac{r}{N}\right)^{rq+k+1} e^{-2r}\left(\frac{N}{2(rq+k+1)}\right)^{rq+k+1} \\
&\geq \left[\frac{1}{K}\left(\frac{1}{q + \frac{k}{r}}\right)\right]^{rq+k+1}
\end{aligned}$$

If $k \geq q$ are fixed (large enough) and if we take r to be of the order of $k/(q \log q)$, we see that we have obtained an example for which the bound (1.12) is optimal.

As announced, we will only use (1.11) in applications. These will be mainly studied in Chapter 6 and in corollaries in Chapters 7 and 8 on strong limit theorems for sums of independent random variables. Let us therefore briefly indicate the trivial translation from the preceding abstract product measures to the setting of independent random variables. Let $\mathbf{X} = (X_i)_{i \leq N}$ be a sample of independent random variables with values in a measurable space E. By independence, they can be constructed on some product probability space

Ω^N in such a way that, for $\omega = (\omega_i)_{i \leq N}$ in Ω^N, $X_i(\omega)$ only depends on ω_i. Then (1.11) is simply that, when $k \geq q$ and $\mathbb{P}\{\mathbf{X} \in A\} \geq 1/2$ for some measurable set A in E^N, then

$$(1.13) \qquad \mathbb{P}_*\{\mathbf{X} \in H(A, q, k)\} \geq 1 - \left(\frac{K_0}{q}\right)^k .$$

Hence, when k or q is large, the sample $\mathbf{X}$ falls with high probability into $H(A, q, k)$. On this set, $\mathbf{X}$ is entirely controlled by a finite number q of points in A provided k elements of the sample are neglected. In the applications, especially to the study of sums of independent random variables, these neglected terms can essentially be thought of as the largest values of the sample. Hence, once an appropriate bound on large values has been found, a good choice of A and some relations between the various parameters q and k determine sharp bounds on the tails of sums of independent random variables. This will be one of the objects of study in Chapter 6. Further, let us note that this intuition about large values of the sample is justified in the special case of the proof of Theorem 1.4 that we give below; the final binomial arguments exactly handle the situation as we just described.

We refer to [Tal1] for a detailed proof of Theorem 1.4. However, we would like to give a proof in the simpler case where A is symmetric, i.e. invariant under the permutations of the coordinates. The rearrangement part of the proof is an inequality introduced in [Ta6], in the same spirit but easier than the rearrangement arguments needed for Theorem 1.4. The explicit computations after appropriate rearrangements are then identical to those required to prove Theorem 1.4, and rely on classical binomial estimates. This method also yields a version of Theorem 1.4 in the case of symmetric A's for $q > 1$ *not necessarily an integer*. This is another motivation for the details we are now giving. More precisely let, as before, $N \geq 1$ be an integer and let now $q > 1$; denote by N' the integer part of qN. Consider C symmetric (invariant under the permutations of the coordinates) in $E^{N'}$ such that $P'(C) > 0$ where $P' = \mu^{\otimes N'}$. For each integer k, set

$$G(C, k) = \big\{x \in E^N; \ \exists y \in C$$
$$\text{such that } \operatorname{Card}\big\{i \leq N; \ x_i \notin \{y_1, \ldots, y_{N'}\}\big\} \leq k\big\} .$$

For a comparison with $H(A, q, k)$ when q is an integer, let A in E^N and denote by $C \subset E^{qN}$ the set of all sequences $y = (y_i)_{i \leq qN}$ such that $\{y_1, \ldots, y_{qN}\}$ can be covered by q sets of the form $\{x_1, \ldots, x_N\}$ where $(x_i)_{i \leq N} \in A$. C is clearly invariant under the permutations of the coordinates and $H(A, q, k) \subset G(C, k)$. On the other hand, it is not difficult to see that when A is symmetric, the converse inclusion $G(C, k) \subset H(A, q, k)$ is also satisfied at least on the subset of E^N consisting of those $x = (x_i)_{i \leq N}$ such that $x_i \neq x_j$ whenever $i \neq j$.

Using this notation, there exist $K(q) < 1$ and $k(q, P'(C))$ large enough such that for all $k \geq k(q, P'(C))$,

$$(1.14) \qquad P_*\big(G(C, k)\big) \geq 1 - k\big(q, P'(C)\big)\big[K(q)\big]^k .$$

For simplicity, we do not indicate the explicit dependence of $K(q)$ and $k(q, P'(C))$ in function of q and $P'(C)$ but, as will be clear from the proof, these are similar to those explicited in Theorem 1.4.

In order to establish (1.14), we take upon the framework of [Ta11] and note in particular, that since the result is measure theoretic we might as well assume that $E = [0, 1]$ and μ is the Lebesgue measure λ on $[0, 1]$. The main point in the proof of (1.14) is the use of Theorem 11 in [Ta6] which ensures the existence, for any symmetric C, of a left-hereditary subset $\widetilde{C}$ of $]0, 1[^{N'}$ such that $\lambda^{N'}(\widetilde{C}) = \lambda^{N'}(C)$ and for which, for every k,

$$\lambda_*^N\big(G(C, k)\big) \geq \lambda^N\big(G(\widetilde{C}, k)\big) \ .$$

$\widetilde{C}$ is left-hereditary in the sense that whenever $(y_i)_{i \leq n} \in \widetilde{C}$ and, for $1 \leq i \leq N'$, $0 < z_i < y_i$, then $(z_i)_{i \leq N'} \in \widetilde{C}$. (When $\widetilde{C}$ is left-hereditary, so is $G(\widetilde{C}, k)$ which is therefore measurable, but in general $G(C, k)$ need not be measurable.) The conclusion now follows from an appropriate lower bound on $\lambda(G(\widetilde{C}, k))$ which will be obtained using binomial estimates. Here, the following inequality is convenient:

$$(1.15) \qquad \mathbb{P}\big\{B(n, \tau) \leq tn\big\} \leq \left[\left(\frac{\tau}{t}\right)^t \left(\frac{1 - \tau}{1 - t}\right)^{1-t}\right]^n$$

where $B(n, \tau)$ is the number of successes in a run of n Bernoulli trials with probability τ of success and $0 < t < \tau$ (see e.g. [Ho]).

We let $q = (1 + \varepsilon)^2$, $\varepsilon > 0$, and first show that for some $\alpha = \alpha(\varepsilon, \lambda^{N'}(C))$ large enough, there exists $(y_i)_{i \leq N'}$ in $\widetilde{C}$ such that, for every $1 \leq r \leq N'$,

$$\mathrm{Card}\left\{i \leq N'; \ y_i > 1 - \frac{(1 + \varepsilon)(r + \alpha)}{N'}\right\} \geq r \ .$$

Indeed,

$$\left(\frac{1 - \tau}{1 - t}\right)^{1-t} = \left(1 - \frac{\tau - t}{1 - t}\right)^{1-t} \leq \exp\big(-(\tau - t)\big)$$

so that, by (1.15),

$$\mathbb{P}\big\{B(n, \tau) \leq tn\big\} \leq \exp\left(-tn\left(\frac{\tau}{t} - 1 - \log\frac{\tau}{t}\right)\right) \ .$$

If $u \geq 1+\varepsilon$, then $u-1-\log u \geq \delta^2 u$ where $\delta = \frac{1}{4}\min(\varepsilon, \frac{1}{2})$; hence, if $\tau/t \geq 1+\varepsilon$,

$$\mathbb{P}\big\{B(n, \tau) \leq tn\big\} \leq \exp(-\delta^2 \tau n) \ .$$

Therefore, by this inequality, for every $1 \leq r \leq N$,

$$\lambda^{N'}\left((y_i)_{i \leq N'} \, ; \ \mathrm{Card}\left\{i \leq N'; \ y_i > 1 - \frac{(1 + \varepsilon)(r + \alpha)}{N'}\right\} \leq r - 1\right)$$
$$\leq \exp\big(-\delta^2(1 + \varepsilon)(r + \alpha)\big)$$

and

$$\sum_{r\geq 1} \exp\bigl(-\delta^2(1+\varepsilon)(r+\alpha)\bigr) < \lambda^{N'}(C)$$

whenever $\alpha = \alpha(\varepsilon, \lambda^{N'}(C))$ is chosen large enough. This proves the preceding claim.

Now, since $\widetilde{C}$ is left-hereditary, each sequence $(x_i)_{i\leq N}$ such that for every $r > k$,

$$\mathrm{Card}\left\{i \leq N;\ x_i \geq 1 - \frac{(1+\varepsilon)(r-k+\alpha)}{N'}\right\} \leq r - 1$$

belongs to $G(\widetilde{C}, k)$; indeed, the r-th largest element of $(x_i)_{i\leq N}$ is less than $1 - (1+\varepsilon)(r-k+\alpha)/N'$ and is therefore smaller than the $(r-k)$-th largest element of $(y_i)_{i\leq N'}$. Thus, by the left-hereditary property of $\widetilde{C}, (x_i)_{i\leq N} \in G(\widetilde{C}, k)$. The proof of (1.14) will therefore be complete if we can show that

$$\lambda^N\left((x_i)_{i\leq N};\ \forall r > k,\ \mathrm{Card}\left\{i \leq N;\ x_i > 1 - \frac{(1+\varepsilon)(r-k+\alpha)}{N'}\right\} \leq r-1\right)$$
$$\geq 1 - k\bigl(\varepsilon, \lambda^{N'}(C)\bigr)\bigl[K(\varepsilon)\bigr]^k$$

for some $K(\varepsilon) < 1$ and every large enough $k \geq k(\varepsilon, \lambda^{N'}(C))$. To this aim, note that

$$\left(\frac{\tau}{t}\right)^t = \left(1 + \frac{\tau - t}{t}\right)^t \leq \exp(\tau - t)$$

and thus, by (1.15),

$$(1.16) \quad \mathbb{P}\{B(n, \tau) \leq tn\} \leq \exp\left(-(1-t)n\left(\frac{1-\tau}{1-t} - 1 - \log\frac{1-\tau}{1-t}\right)\right).$$

If $0 < u < (1+\varepsilon)^{-1}$, then $u - 1 - \log u \geq \delta^2$ (where we recall that $\delta = \frac{1}{4}\min\left(\varepsilon, \frac{1}{2}\right)$). Hence, if $(1-\tau)/(1-t) \leq 1/1+\varepsilon$,

$$\mathbb{P}\{B(n, \tau) \leq tn\} \leq \exp\bigl(-\delta^2(1-t)n\bigr).$$

Using this inequality, if $r > k$ and $k \geq k(\varepsilon, \lambda^{N'}(C))$ are large enough, we get

$$\lambda^N\left((x_i)_{i\leq N};\ \mathrm{Card}\left\{i \leq N;\ x_i \geq 1 - \frac{(1+\varepsilon)(r-k+\alpha)}{N'}\right\} \geq r\right) \leq \exp(-\delta^2 r).$$

As announced, this is exactly what was required to conclude the proof of (1.14).

1.3 Martingale Inequalities

Martingale methods prove useful in order to establish various concentration results. These results complement the preceding isoperimetric inequalities and will prove useful in various places throughout this work. The inequalities that we state, at least some of them, are rather classical and we present them in the general spirit of the concentration properties.

Recall that $L_1 = L_1(\Omega, \mathcal{A}, \mathbb{P})$ denotes the space of all measurable functions f on Ω such that $\mathbb{E}|f| = \int |f| d\mathbb{P} < \infty$. Assume that we are given a filtration

$$\{\emptyset, \Omega\} = \mathcal{A}_0 \subset \mathcal{A}_1 \subset \cdots \subset \mathcal{A}_N = \mathcal{A}$$

of sub-σ-algebras of $\mathcal{A}$. $\mathbb{E}^{\mathcal{A}_i}$ denotes the conditional operator with respect to $\mathcal{A}_i$. Given f in L_1, set, for each $i = 1, \ldots, N$,

$$d_i = \mathbb{E}^{\mathcal{A}_i} f - \mathbb{E}^{\mathcal{A}_{i-1}} f$$

so that $f - \mathbb{E}f = \sum_{i=1}^N d_i$. $(d_i)_{i \leq N}$ defines a so-called *martingale difference sequence* characterized by the property $\mathbb{E}^{\mathcal{A}_{i-1}} d_i = 0$, $i \leq N$.

One of the typical examples of a martingale difference sequence that we have in mind is a sequence $(X_i)_{i \leq N}$ of *independent mean zero* random variables. Indeed, if $\mathcal{A}_i$ denotes the σ-algebra generated by the variables $X_1, \ldots, X_i$, by independence and the mean zero property, it is clear that $\mathbb{E}^{\mathcal{A}_{i-1}} X_i = \mathbb{E} X_i = 0$. Hence, all the results we will present for $f - \mathbb{E}f = \sum_{i=1}^N d_i$ as before apply to the sum $\sum_{i=1}^N X_i$.

The first lemma is a kind of analog in this context of the concentration property for Lipschitz functions. It expresses, in the preceding notation, the high concentration of f around its expectation in terms of the size of the differences d_i.

Lemma 1.5. *Let f in L_1 and let $f - \mathbb{E}f = \sum_{i=1}^N d_i$ be as above the sum of martingale differences with respect to $(\mathcal{A}_i)_{i \leq N}$. Assume that $\|d_i\|_\infty < \infty$ and set $a = (\sum_{i=1}^N \|d_i\|_\infty^2)^{1/2}$. Then, for every $t > 0$,*

$$\mathbb{P}\{|f - \mathbb{E}f| > t\} \leq 2 \exp(-t^2/2a^2) .$$

Proof. We first note that when φ is a random variable such that $|\varphi| \leq 1$ almost surely and $\mathbb{E}\varphi = 0$, then, for any real number λ, $\mathbb{E} \exp(\lambda \varphi) \leq \exp(\lambda^2/2)$. Indeed, simply note from the convexity of the function $x \to \exp(\lambda x)$ and from $\lambda x = \lambda(1 + x)/2 - \lambda(1 - x)/2$ that, for any $|x| \leq 1$,

$$\exp(\lambda x) \leq \operatorname{ch} \lambda + x \operatorname{sh} \lambda \leq \exp(\lambda^2/2) + x \operatorname{sh} \lambda .$$

An integration yields the claim. It clearly follows that, for any $i = 1, \ldots, N$,

$$\mathbb{E}^{\mathcal{A}_{i-1}} \exp(\lambda d_i) \leq \exp(\lambda^2 \|d_i\|_\infty^2/2) .$$

Iterating this inequality and using the properties of conditional expectation,

$$\mathbb{E}\exp\big[\lambda(f - \mathbb{E}f)\big] = \mathbb{E}\exp\Big(\lambda\sum_{i=1}^{N}d_i\Big)$$

$$= \mathbb{E}\Big(\exp\Big(\lambda\sum_{i=1}^{N-1}d_i\Big)\mathbb{E}^{\mathcal{A}_{N-1}}\exp\big(\lambda d_N\big)\Big)$$

$$\leq \mathbb{E}\exp\Big(\lambda\sum_{i=1}^{N-1}d_i\Big)\exp\big(\lambda^2\|d_N\|_\infty^2/2\big)$$

$$\cdots$$

$$\leq \exp\big(\lambda^2 a^2/2\big)\ .$$

Then, we obtain from Chebyshev's inequality that, for $t > 0$,

$$\mathbb{P}\{f - \mathbb{E}f > t\} \leq \exp\big(-\lambda t + \lambda^2 a^2/2\big)$$
$$\leq \exp\big(-t^2/2a^2\big)$$

for the optimal choice of λ. Applying this inequality to $-f$ also yields the conclusion of the lemma.

At this stage, we should point out, once and for all, that in a statement such as Lemma 1.5, it is understood that if we are interested in $f - \mathbb{E}f$ only rather than $|f - \mathbb{E}f|$, we also have

$$\mathbb{P}\{f - \mathbb{E}f > t\} \leq \exp(-t^2/2a^2)\ .$$

(This was actually explicitly established in the proof!) This general comment about a coefficient 2 in front of the exponential bound in order to take into account absolute values (f and $-f$) applies in many similar situations (we already mentioned it about the concentration inequalities of Section 1.1) and will be used in the sequel without any further comment.

When, in addition to the bounds on d_i, some information on $\mathbb{E}^{\mathcal{A}_{i-1}}d_i^2$ is available, the preceding proof basically yields the following refinement.

Lemma 1.6. *Let $f - \mathbb{E}f = \sum_{i=1}^{N}d_i$ be as before. Set $a = \max_{i\leq N}\|d_i\|_\infty$ and let $b \geq (\sum_{i=1}^{N}\|\mathbb{E}^{\mathcal{A}_{i-1}}d_i^2\|_\infty)^{1/2}$. Then, for every $t > 0$,*

$$\mathbb{P}\{|f - \mathbb{E}f| > t\} \leq 2\,\exp\Big[-\frac{t^2}{2b^2}\Big(2 - \exp\Big(\frac{at}{b^2}\Big)\Big)\Big]\ .$$

The proof is similar to that of Lemma 1.5. It uses simply that

$$\mathbb{E}^{\mathcal{A}_{i-1}}\exp(\lambda d_i) = 1 + \frac{\lambda^2}{2!}\mathbb{E}^{\mathcal{A}_{i-1}}d_i^2 + \frac{\lambda^3}{3!}\mathbb{E}^{\mathcal{A}_{i-1}}d_i^3 + \cdots$$

$$\leq 1 + \frac{\lambda^2}{2}\|\mathbb{E}^{\mathcal{A}_{i-1}}d_i^2\|_\infty\Big(1 + \frac{\lambda\|d_i\|_\infty}{3} + \frac{\lambda^2\|d_i\|_\infty^2}{3\cdot 4} + \cdots\Big)$$

$$\leq \exp\Big[\frac{\lambda^2}{2}\|\mathbb{E}^{\mathcal{A}_{i-1}}d_i^2\|_\infty e^{\lambda a}\Big]\ .$$

Turning back to Lemma 1.5, it is clear that we always have

$$|f - \mathbb{E}f| \leq \sum_{i=1}^{N} \|d_i\|_\infty .$$

Of course, this simple observation suggests the possibility of some "interpolation" between the sum of the squares $(\sum_{i=1}^{N} \|d_i\|_\infty^2)^{1/2}$ which steps in in Lemma 1.5 and this trivial bound. This kind of result is described in the next two lemmas.

Lemma 1.7. *Let $1 < p < 2$ and let $q = p/p-1$ denote the conjugate of p. Let further f be as before with $f - \mathbb{E}f = \sum_{i=1}^{N} d_i$ and set $a = \max_{i \leq N} i^{1/p} \|d_i\|_\infty$. Then, for every $t > 0$,*

$$\mathbb{P}\{|f - \mathbb{E}f| > t\} \leq 2 \exp(-t^q/C_q a^q)$$

where $C_q > 0$ only depends on q.

Proof. By homogeneity we may and do assume that $a = 1$. For any integer m we can write

$$|f - \mathbb{E}f| \leq \sum_{i=1}^{m} |d_i| + \left| \sum_{i>m} d_i \right|$$

$$\leq \sum_{i=1}^{m} i^{-1/p} + \left| \sum_{i>m} d_i \right| \leq q m^{1/q} + \left| \sum_{i>m} d_i \right| .$$

Assume first that $t > 2q$ and then denote by m the largest integer such that $t > 2q m^{1/q}$. We can apply Lemma 1.5 to $\sum_{i>m} d_i$ and thus obtain in this way, together with the preceding "interpolation" inequality,

$$\mathbb{P}\{|f - \mathbb{E}f| > t\} \leq \mathbb{P}\left\{ \left| \sum_{i>m} d_i \right| > q m^{1/q} \right\}$$

$$\leq 2 \exp\left[-q^2 m^{2/q}/2 \sum_{i>m} \|d_i\|_\infty^2 \right] .$$

Now, since $a \leq 1$,

$$\sum_{i>m} \|d_i\|_\infty^2 \leq \sum_{i>m} i^{-2/p} \leq \frac{q}{q-2} m^{1-2/p}$$

so that

$$\mathbb{P}\{|f - \mathbb{E}f| > t\} \leq 2 \exp(-q(q-2)m/2)$$
$$\leq 2 \exp(-t^q/C_q^1)$$

where $C_q^1 = 4(2q)^q/q(q-2)$. When $t \leq 2q$,

$$\mathbb{P}\{|f - \mathbb{E}f| > t\} \leq 1 \leq 2 \exp(-(2q)^q/C_q^2) \leq 2 \exp(-t^q/C_q^2)$$

where $C_q^2 = (2q)^q/\log 2$. The lemma follows with $C_q = \max(C_q^1, C_q^2)$.

Lemma 1.8. *Let $f - \mathbb{E}f = \sum_{i=1}^{N} d_i$ be as above and set $a = \max_{i \leq N} i \|d_i\|_{\infty}$. Then, for every $t > 0$,*

$$\mathbb{P}\{|f - \mathbb{E}f| > t\} \leq 16 \exp[-\exp(t/4a)] \ .$$

Proof. It is similar to the preceding one. We again assume by homogeneity that $a = 1$. When $t \leq 4$,

$$\mathbb{P}\{|f - \mathbb{E}f| > t\} \leq 16e^{-e} \leq 16 \exp[-\exp(t/4)] \ .$$

When $t > 4$, let m be the largest integer such that $t \geq 2 + \log m$. We have as before

$$|f - \mathbb{E}f| \leq \sum_{i=1}^{m} |d_i| + \left|\sum_{i>m} d_i\right| \leq (1 + \log m) + \left|\sum_{i>m} d_i\right| \ .$$

Hence,

$$\mathbb{P}\{|f - \mathbb{E}f| > t\} \leq \mathbb{P}\left\{\left|\sum_{i>m} d_i\right| > 1\right\} \leq 2 \exp\left(-1 \Big/ 2 \sum_{i>m} i^{-2}\right) \leq 2 \exp(-m/2)$$

where we have used Lemma 1.5. Since $t < 2 + \log(m + 1)$, we get

$$\mathbb{P}\{|f - \mathbb{E}f| > t\} \leq 4 \exp[-\exp(t/4)]$$

and the conclusion follows.

Notes and References

The description of the concentration of measure phenomenon is taken from the paper [G-M] by M. Gromov and V. D. Milman where further interesting examples of "Lévy families" are discussed (see also [Mi-S], [Mi3]). The use of isoperimetric concentration properties in Dvoretzky's theorem on almost spherical sections of convex bodies (see Chapter 9) was initiated by V. D. Milman [Mi1], and amplified later in [F-L-M]. Further applications to the local theory of Banach spaces are presented in [Mi-S], [Pi18], [TJ2]. The isoperimetric inequality on the sphere (Theorem 1.1) is due to P. Lévy [Lé2] and E. Schmidt [Schm]. Lévy's proof, which has not been understood for a long time, has been revived and generalized by M. Gromov [Gr1], [Mi-S]. Schmidt's proof is based on deep isoperimetric symmetrizations (rearrangements in the sense of Steiner). Accounts on symmetrizations and rearrangements, geometric inequalities and isoperimetry may be found in [B-Z], [Os]. For a short proof of Theorem 1.1, we refer to [F-L-M], or [Ba-T], [Beny] (for the two point symmetrization method). Poincaré's lemma is *not* to be found in [Po] according to [D-F]; see this paper for the history of the result. Poincaré's lemma is nicely revisited in [MK]. The Gaussian isoperimetric Theorem 1.2 is due independently to C. Borell [Bo2] and V. N. Sudakov and B. S. Tsirel'son [S-T] with the proof sketched here. C. Borell [Bo2] carefully describes the infinite dimensional extensions. A. Ehrhard introduced Gaussian symmetrization in [Eh1] and established there inequality (1.1). See further [Eh2] and also [Eh3] where the extremality of half-spaces is investigated. We refer the reader to [Ta19] where a new isoperimetric inequality is presented, that improves upon certain aspects of the Gaussian isoperimetric theorem. Log-concavity of Gaussian Radon measures in locally convex spaces has been shown by C. Borell [Bo1]. The simple proof that we suggest has been shown to us by J. Saint-Raymond. Inequality (1.5) is due to G. Pisier with the simple proof of B. Maurey [Pi16]. They actually deal with vector valued functions and the method of proof indeed ensures in the same way that if $f : E \to G$ is locally Lipschitz between two Banach spaces E and G, if γ is a Gaussian Radon measure on E and if $F : G \to \mathbb{R}$ is measurable and convex, then

$$\int F(f - \int f d\gamma) d\gamma \le \int \int F \left(\frac{\pi}{2} f'(x) \cdot y \right) d\gamma(x) d\gamma(y).$$

(1.6) comes from B. Maurey (cf. [Pi16], [Led7]). A proof of (1.6) using Yurinskii's observation (see Chapter 6 below) and a central limit theorem for martingales has been noticed by A. de Acosta and J. Zinn (oral communication). (1.7) and the subsequent examples were observed by G. Pisier [Pi16]. The isoperimetric theorem for the Haar measure on $\{-1, +1\}^N$ with respect to the Hamming metric was established by L. H. Harper [Har]. Theorem 1.3 may be found in [Ta9] where the comparison with [Har] is discussed. Some extensions to measures on $\{-1, +1\}^N$ with non-symmetric weights are described in [J-S2].

The isoperimetric theorem for subsets of a product of probability spaces, Theorem 1.4, is due to the second author [Ta11]. Inequality (1.14) for q not necessarily an integer is new, and is also due to the second author. The binomial computations closely follow the last step in [Ta11]. (1.15) comes from [Che] (see also [Ho]).

Very recently, an abstract extension of Theorem 1.3 to arbitrary product measures has been discovered by the second author [Ta22]. Much simpler than Theorem 1.4, it yields similar consequences to the bounds on the tails of sums of independent random variables studied in Chapter 6 and may thus be considered as a significant simplification of the isoperimetric argument in the study of these bounds.

The first two inequalities of Section 1.3 are rather classical. In this form, Lemma 1.5 is apparently due to [Azu]. Lemma 1.6 is the martingale analog of the classical exponential inequality of A. N. Kolmogorov [Ko] (see [Sto]), in a form put forward in [Ac7]. For sums of independent random variables, and starting with Bernstein's inequality (cf. [Ho]), this type of inequalities has been extensively studied leading to sharp versions in, for example, [Ben], [Ho] (see also Chapter 6). For simplicity, we use Lemma 1.6 in this work but the preceding references can basically be used equivalently in our applications. Lemmas 1.6 and 1.7 are taken from [Pi16]. For applications of all these inequalities to the Banach space theory, see, besides others, [Mau3], [Sch1], [Sch2], [Sch3], [J-S1], [Pi12], [B-L-M], [Mi-S], [Pi16], etc. For applications of the martingale method to rather different problems, see [R-T1], [R-T2], [R-T3].

2. Generalities on Banach Space Valued Random Variables and Random Processes

This chapter collects in rather an informal way some basic facts about processes and infinite dimensional random variables. The material that we present actually only appears as the necessary background for the subsequent analysis developed in the next chapters. Only a few proofs are given and many important results are only just mentioned or even omitted. It is therefore recommended to complement, if necessary, these partial bases with the classical references, some of which are given at the end of the chapter.

The first section describes Radon (or separable) vector valued random variables while the second makes precise some terminology and definitions about random processes and general vector valued random variables. The third section presents some important facts about symmetric random variables, especially Lévy's inequalities and Itô-Nisio's theorem. In the last paragraph, we mention some classical and useful inequalities.

Throughout this book, we deal with abstract probability spaces $(\Omega, \mathcal{A}, \mathbb{P})$ which are always assumed to be large enough to support all the random variables we will work with; this is legitimate by Kolmogorov's extension theorem. For convenience, we also assume that $(\Omega, \mathcal{A}, \mathbb{P})$ is complete, that is the σ-algebra $\mathcal{A}$ contains the negligible sets for $\mathbb{P}$.

Throughout this book also, B denotes a Banach space, that is a vector space over $\mathbb{R}$ or $\mathbb{C}$ with norm $\|\cdot\|$ and complete with respect to it. For simplicity, we shall always consider *real* Banach spaces, but actually almost everything we will present carries over to the complex case. B' denotes the topological dual of B and $f(x) = \langle f, x \rangle$ ($\in \mathbb{R}$), $f \in B'$, $x \in B$, the duality. The norm on B' is also denoted $\|f\|$, $f \in B'$.

2.1 Banach Space Valued Radon Random Variables

A Borel random variable or vector with values in a Banach space B is a measurable map X from some probability space $(\Omega, \mathcal{A}, \mathbb{P})$ into B equipped with its Borel σ-algebra $\mathcal{B}$ generated by the open sets of B. In fact, this definition of random variable is somewhat too general for our purposes since for example the sum of two random variables is not trivially a random variable. Furthermore, if B is equipped with a different σ-algebra, such as the coarsest one for which the linear functionals are measurable (cylindrical σ-algebra), the two definitions might not agree in general. We do not wish here to deal at length with measurability questions. One way to handle them is the concept of Radon or regular random variables, which amounts to some separability of the range.

A Borel random variable X with values in B is said to be regular with respect to compact sets, or *Radon*, or yet *tight*, if, for each $\varepsilon > 0$, there is a compact set $K = K(\varepsilon)$ in B such that

$$(2.1) \qquad \mathbb{P}\{X \in K\} \geq 1 - \varepsilon .$$

In other words, the image of the probability $\mathbb{P}$ by X (see below) is a Radon measure on $(B, \mathcal{B})$. Equivalently, X takes almost all its values in some separable (i.e. countably generated) closed linear subspace E of B. Indeed, under (2.1), there exists a sequence (K_n) of compact sets in B such that $\mathbb{P}\{X \in \bigcup_n K_n\} = 1$ so that X takes almost surely its values in some separable subspace of B. Conversely, let $\{x_i \, ; \, i \in \mathbb{N}\}$ be dense in E and let $\varepsilon > 0$ be fixed. By density, for each $n \geq 1$ there exists an integer N_n such that

$$\mathbb{P}\left\{X \in \bigcup_{i \leq N_n} B(x_i, 2^{-n})\right\} \geq 1 - \varepsilon \cdot 2^{-n}$$

where $B(x_i, 2^{-n})$ denotes the closed ball with center x_i and radius 2^{-n} in B. Then set

$$K = K(\varepsilon) = \bigcap_{n \geq 1} \bigcup_{i \leq N_n} B(x_i, 2^{-n}) .$$

K is closed and $\mathbb{P}\{X \in K\} \geq 1 - \varepsilon$; further, K is compact since from each sequence in K one can extract a subsequence contained, for every n, in a single ball $B(x_i, 2^{-n})$; this subsequence is, therefore, a Cauchy sequence hence converges by completness of B.

We call thus *Radon*, or *separable*, a Borel random variable X that satisfies (2.1). The preceding argument shows equivalently that X is the almost sure limit of step random variables of the form $\sum_{\text{finite}} x_i I_{A_i}$ where $x_i \in B$ and $A_i \in \mathcal{A}$. Note also that (2.1) is extended into

$$\mathbb{P}\{X \in A\} = \sup\{\mathbb{P}\{X \in K\} \, ; \, K \text{ compact}, K \subset A\}$$

for every Borel set A in B. This follows from (2.1) together with the analogous property for closed sets which holds from the very definition of the Borel σ-algebra.

Since Radon random variables have separable range, it is sometimes convenient to assume the Banach space itself to be separable. We are mostly interested in this work in results concerning *sequences* of random variables. When dealing with sequences of Radon random variables, we will therefore usually assume for convenience and without any loss in generality that the Banach space under consideration is *separable*. Further note that when B is separable all the "reasonable" definitions of σ-algebras on B coincide, in particular the Borel and cylindrical σ-algebras. Note also, and this observation will motivate parts of the next section, that if B is separable the norm can be expressed as a supremum

$$\|x\| = \sup_{f \in D} |f(x)|, \quad x \in B ,$$

over a *countable* set D of linear functionals of norm 1 (or less than or equal to 1).

For a random variable X with values in B, the probability measure $\mu = \mu_X$ which is the image of $\mathbb{P}$ by X is called the *distribution* (or *law*) of X; for any real valued bounded measurable function φ on B,

$$\mathbb{E}\varphi(X) = \int_B \varphi(x)d\mu(x) .$$

The distribution of a Radon random variable is completely determined by its finite dimensional projections. More precisely, if X and Y are Radon random variables such that for every f in B', $f(X)$ and $f(Y)$ have the same distribution (as real valued random variables), then $\mu_X = \mu_Y$. Indeed, we can assume that B is separable; the Borel σ-algebra is therefore generated by the algebra of cylinder sets. Since $\mu_{f(X)} = \mu_{f(Y)}$, μ_X and μ_Y agree on this algebra and the result follows. Note that it suffices to know that $f(X)$ and $f(Y)$ have the same law on some weakly dense subset of B'. As a consequence of the preceding, and according to the uniqueness theorem in the scalar case, the characteristic functionals on B'

$$\mathbb{E}\exp\left(if(X)\right) = \int_B \exp\left(if(x)\right)d\mu(x), \quad f \in B' ,$$

completely determines the distribution of X.

Denote by $\mathcal{P}(B)$ the space of all Radon probability measures on B. For each μ in $\mathcal{P}(B)$ consider the neighborhood

$$\left\{\nu \in \mathcal{P}(B); \left|\int_B \varphi_i d\mu - \int_B \varphi_i d\nu\right| < \varepsilon , \ i \leq N\right\}$$

where $\varepsilon > 0$ and φ_i, $i \leq N$, are real valued bounded continuous functions on B. The topology generated by these neighborhoods is called the *weak topology* and a sequence (μ_n) in $\mathcal{P}(B)$ that converges with respect to this topology is said to converge *weakly*. Observe that $\mu_n \to \mu$ weakly if and only if

$$\lim_{n \to \infty} \int \varphi d\mu_n = \int \varphi d\mu$$

for every bounded continuous φ on B. It can be shown further that this holds if and only if

$$\limsup_{n \to \infty} \mu_n(F) \leq \mu(F)$$

for each closed set F in B, or, equivalently,

$$\liminf_{n\to\infty} \mu_n(G) \geq \mu(G)$$

for each open set G.

The space $\mathcal{P}(B)$ equipped with the weak topology is known to be a complete metric space (which is separable if B is separable). Thus, in particular, in order to check that a sequence (μ_n) in $\mathcal{P}(B)$ converges weakly, it suffices to show that (μ_n) is relatively compact in the weak topology and that all possible limits are the same. The latter can be verified along linear functionals. For the former, a very useful criterion of Y. V. Prokhorov characterizes relatively compact sets of $\mathcal{P}(B)$ as those which are *uniformly tight* with respect to compact sets.

Theorem 2.1. *A family $(\mu_i)_{i\in I}$ in $\mathcal{P}(B)$ is relatively compact for the weak topology if and only if for each $\varepsilon > 0$ there is a compact set K in B such that*

$$(2.2) \qquad \mu_i(K) \geq 1 - \varepsilon \quad \text{for all } i \in I \,.$$

This compactness criterion may be expressed in various manners depending on the context. For example, (2.2) holds if and only if for each $\varepsilon > 0$ there is a *finite* set A in B such that

$$(2.3) \qquad \mu_i\big(x \in B \,;\, d(x, A) < \varepsilon\big) \geq 1 - \varepsilon \quad \text{for all } i \in I$$

where $d(x, A)$ denotes the distance (in B) from the point x to the set A.

Another equivalent formulation of Theorem 2.1 is based on the idea of finite dimensional approximation and is most useful in applications. It is sometimes refered to as "flat concentration". The idea is simply that bounded sets in finite dimension are relatively compact and, therefore, if a set of measures is concentrated near a finite dimensional subspace, then it should be close to be relatively compact. The following simple functional analytic lemma makes this fact clear. If F is a closed subspace of B, denote by $T = T_F$ the canonical quotient map $T : B \to B/F$. Then $\|T(x)\| = d(x, F)$, $x \in B$. (We denote in the same way the norm of B and the norm of B/F.)

Lemma 2.2. *A subset K of B is relatively compact if and only if it is bounded and for each $\varepsilon > 0$ there is a finite dimensional subspace F of B such that if $T = T_F$, $\|T(x)\| < \varepsilon$ for every x in K (i.e. $d(x, F) < \varepsilon$ for $x \in K$).*

According to this result and to Theorem 2.1, if, for each $\varepsilon > 0$, there is a bounded set L in B such that $\mu_i(L) \geq 1 - \varepsilon$ for every $i \in I$ and a finite dimensional subspace F of B such that

$$(2.4) \qquad \mu_i\big(x \in B \,;\, d(x, F) < \varepsilon\big) \geq 1 - \varepsilon \quad \text{for all } i \in I \,,$$

then the family $(\mu_i)_{i\in I}$ is relatively compact.

Actually, when (2.4) holds, the assumption of the existence of L is too strong. It is enough to assume that for every f in B', $(\mu_i \circ f^{-1})_{i \in I}$ is a weakly relatively compact family of probability measures on the line. If μ is a Borel measure on B and f a linear functional, $\mu \circ f^{-1}$ denotes the measure on $\mathbb{R}$ which is the image of μ by f. To check the preceding claim, let F be of dimension N and such that (2.4) is satisfied. By the Hahn-Banach theorem, there exist linear functionals $f_1, \ldots, f_n$ in the unit ball of B' such that, whenever $x \in F$ and $a > 0$, if $\max_{i \leq n} |f_i(x)| \leq a$, then $\|x\| \leq 2a$. Therefore, if $(\mu_i \circ f^{-1})_{i \in I}$ is relatively compact for every f in B' (actually a weakly dense subset of B' would suffice), and if (2.4) holds, the family $(\mu_i)_{i \in I}$ is uniformly almost concentrated on a bounded set and Prokhorov's criterion is then fulfilled.

A sequence (X_n) of Radon random variables with values in B converges *weakly* to a Radon random variable X if the sequence of distributions (μ_{X_n}) converges weakly to μ_X. For *real valued* random variables, a celebrated theorem of P. Lévy indicates that (X_n) converges weakly if and only if the corresponding sequence of characteristic functions (Fourier transforms) converges pointwise (to a continuous limit). In the vector valued case, by Theorem 2.1, (X_n) converges weakly to X as soon as $(f(X_n))$ converges weakly (as a sequence of real valued random variables) to $f(X)$ for every f in B' (or only in a weakly dense subset) *and* the sequence (X_n) is tight in the sense that, for each $\varepsilon > 0$, there exists a compact set K in B with

$$\mathbb{P}\{X_n \in K\} \geq 1 - \varepsilon$$

for all n's (or only all large enough n's since the X_n's are themselves tight). By what preceeds, this can be established by (2.3) or (2.4).

The sequence (X_n) is said to converge in *probability* (in measure) to X if, for each $\varepsilon > 0$,

$$\lim_{n \to \infty} \mathbb{P}\{\|X_n - X\| > \varepsilon\} = 0 \,.$$

It is said to be bounded in probability (or stochastically bounded) if, for each $\varepsilon > 0$, one can find $M > 0$ such that

$$\sup_n \mathbb{P}\{\|X_n\| > M\} < \varepsilon \,.$$

The topology of the convergence in probability is metrizable and a possible metric can be given by $\mathbb{E} \min(1, \|X - Y\|)$. Denote by $L_0(B) = L_0(\Omega, \mathcal{A}, \mathbb{P}; B)$ the vector space of all random variables (on $(\Omega, \mathcal{A}, \mathbb{P})$) with values in B equipped with the topology of the convergence in probability. If (X_n) converges in $L_0(B)$, it also converges weakly and the converse holds true if the limiting distribution is concentrated on one point. Thus, if one has to check for example that (X_n) converges to 0 in probability, it suffices to show that the sequence (X_n) is tight and that all possible limits are 0. The L_0 and weak topologies are thus close in this sense and may be considered as *weak* statements as opposed to the *strong* almost sure properties (defined below).

If $0 < p \le \infty$, denote by $L_p(B) = L_p(\Omega, \mathcal{A}, \mathbb{P}; B)$ the space of all (Radon) random variables X (on $(\Omega, \mathcal{A}, \mathbb{P})$) with values in B such that $\|X\|^p$ is integrable:

$$\mathbb{E}\|X\|^p = \int \|X\|^p d\mathbb{P} < \infty , \quad p < \infty$$

and $\|X\|_\infty = \text{ess sup } \|X\| < \infty$ if $p = \infty$. If $B = \mathbb{R}$, we simply set $L_p = L_p(\mathbb{R})$ $(0 \le p \le \infty)$. We denote moreover, both in the scalar and vector valued cases (and without any confusion), by $\|X\|_p$ the quantity $(\mathbb{E}\|X\|^p)^{1/p}$. The spaces $L_p(B)$ are Banach spaces for $1 \le p \le \infty$ (metric vector spaces for $0 \le p < 1$). If (X_n) converges to X in $L_p(B)$, it converges to X in $L_0(B)$, that is in probability, and a fortiori weakly.

Finally, a sequence (X_n) converges *almost surely* (almost everywhere) to X if

$$\mathbb{P}\Big\{ \lim_{n \to \infty} X_n = X \Big\} = 1 .$$

The sequence (X_n) is almost surely bounded if

$$\mathbb{P}\Big\{ \sup_n \|X_n\| < \infty \Big\} = 1 .$$

Almost sure convergence is not metrizable. It clearly implies convergence in probability which in turn implies weak convergence. Conversely, an important theorem of A.V. Skorokhod [Sk1] asserts that if (X_n) converges weakly to X, then there exist, on a possibly richer probability space, random variables X'_n and X' such that $\mu_{X'_n} = \mu_{X_n}$ for every n and $\mu_X = \mu_{X'}$ and such that $X'_n \to X'$ almost surely. This property is useful in particular in dealing with the convergence of moments, for example in central limit theorems.

We conclude this section with some remarks concerning integrability properties. As we have already seen, a Radon random variable X on $(\Omega, \mathcal{A}, \mathbb{P})$ with values in B belongs to $L_1(B)$, or is *strongly* or *Bochner integrable*, if the real valued random variable $\|X\|$ is integrable $(\mathbb{E}\|X\| < \infty)$. Suppose now that we are given X such that for each f in B' the real valued random variable $f(X)$ is integrable. If we consider the operator

$$T : B' \to L_1 = L_1(\Omega, \mathcal{A}, \mathbb{P})$$

defined by $Tf = f(X)$, T has clearly a closed graph. T is therefore a bounded operator; hence, $f \to \mathbb{E}f(X)$ defines a continuous linear map on B', that is an element, let us call it z, of the bidual B'' of B. The Radon random variable X is said to be *weakly* or *Pettis integrable* if, for each f in B', $f(X)$ is integrable, and the element z of B'' just constructed actually belongs to B. If this is the case, z is then denoted by $\mathbb{E}X$.

It is not difficult to see that, if the Radon random variable X is strongly integrable, then it is weakly integrable and $\|\mathbb{E}X\| \le \mathbb{E}\|X\|$. Indeed, we can choose, for each $\varepsilon > 0$, a compact set K in B such that $\mathbb{E}(\|X\|I_{\{X \notin K\}}) < \varepsilon$. Let $(A_i)_{i \le N}$ be a finite partition of K with sets of diameter less than ε and fix for each i a point x_i in A_i. Then set

$$Y(\varepsilon) = \sum_{i=1}^{N} x_i I_{\{X \in A_i\}} .$$

It is plain, by construction, that $\mathbb{E}\|X - Y(\varepsilon)\| < 2\varepsilon$ and that $Y(\varepsilon)$ is weakly integrable with expectation $\mathbb{E}Y(\varepsilon) = \sum_{i=1}^{N} x_i \mathbb{P}\{X \in A_i\}$. The conclusion then follows from the fact that $(\mathbb{E}Y(1/n))$ is a Cauchy sequence in B which therefore converges to an element $\mathbb{E}X$ in B satisfying $f(\mathbb{E}X) = \mathbb{E}f(X)$ for every f in B'.

In the same way, the conditional expectation of vector valued random variables can be constructed. Let X be a Radon random variable in $L_1(\Omega, \mathcal{A}, \mathbb{P}; B)$ and let $\mathcal{F}$ be a sub-σ-algebra of $\mathcal{A}$. Then one can define $\mathbb{E}^{\mathcal{F}}X$ as the Radon random variable in $L_1(\Omega, \mathcal{F}, \mathbb{P}; B)$ such that, for any F in $\mathcal{F}$,

$$\int_F \mathbb{E}^{\mathcal{F}}X \, d\mathbb{P} = \int_F X \, d\mathbb{P} .$$

It satisfies $\|\mathbb{E}^{\mathcal{F}}X\| \leq \mathbb{E}^{\mathcal{F}}\|X\|$ almost surely and $f(\mathbb{E}^{\mathcal{F}}X) = \mathbb{E}^{\mathcal{F}}f(X)$ almost surely for every f in B'. Note further that, by separability and extension of a classical martingale theorem, if X is in $L_1(\Omega, \mathcal{A}, \mathbb{P}; B)$, there exists an increasing sequence $(\mathcal{A}_N)$ of *finite* sub-σ-algebras of $\mathcal{A}$ such that if $X_N = \mathbb{E}^{\mathcal{A}_N}X$, (X_N) converges almost surely and in $L_1(B)$ to X. If X is in $L_p(B)$, $1 < p < \infty$, the convergence also takes place in $L_p(B)$.

We would like to mention for the sequel (in particular Chapter 8) that, by analogy with the preceding, if X is a Radon random variable with values in B such that, for every f in B', $\mathbb{E}f^2(X) < \infty$, the operator $Tf = f(X)$ is also bounded from B' into L_2. Furthermore, it can be shown as before that, for every ξ in L_2, ξX is weakly integrable so that $\mathbb{E}(\xi X)$ is well defined as an element of B. In particular, for f, g in B', $g(\mathbb{E}(f(X)X)) = \mathbb{E}(f(X)g(X))$, which defines the so-called "covariance structure" of a random variable X which is weakly in L_2.

2.2 Random Processes and Vector Valued Random Variables

The concept of a Radon or separable random variable is a convenient concept when we deal with weak convergence or tightness properties. Indeed, we will use it in the typical weak convergence theorem which is the central limit theorem, and also in some related questions about the law of large numbers and the law of the iterated logarithm. This concept is also a way of taking easily into account various measurability problems.

However, Radon random variables form a somewhat too restrictive setting for other types of questions. For example, if we are given a sequence (X_n) of real valued random variables such that $\sup_n |X_n| < \infty$ almost surely, and if we ask (for example) for the integrability properties or the tail behavior of this supremum, we are clearly faced with a random element of infinite dimension but we need not (and in general do not) have a Radon random vector. In other words, it would be convenient to have a notion of random variable with values in ℓ_∞. The space c_0 of all real valued sequences tending to 0 is a separable subspace of ℓ_∞ but ℓ_∞ is not separable. Recall further that every separable Banach space is isometric to a closed subspace of ℓ_∞.

On the other hand, another category of infinite dimensional random elements are the random functions or stochastic processes. Let T be a (infinite) index set which will be usually assumed to be a metric space (T, d). A *random function* or *process* $X = (X_t)_{t \in T}$ indexed by T is a collection of real valued random variables X_t, $t \in T$. By the distribution or law of X we mean the distribution on $\mathbb{R}^T$, equipped with the *cylindrical σ-algebra* generated by the cylinder sets, determined by the collection of all marginal distributions of the finite dimensional random vectors $(X_{t_1}, \ldots, X_{t_N})$, $t_i \in T$.

Throughout this book, we often study whether a given random process is almost surely bounded and/or continuous, and, when this is the case, we ask for possible integrability properties or for the tail behavior of $\sup_{t \in T} |X_t|$ whenever this makes sense. Of course, these considerations raise some nontrivial measurability questions as soon as T is no longer countable. A priori, a random process $X = (X_t)_{t \in T}$ is almost surely bounded or continuous, or has almost all its trajectories or sample paths bounded or continuous, if, for almost all ω, the path $t \to X_t(\omega)$ is bounded or continuous. However, in order to prove that a random process is almost surely bounded or continuous, and to deal with it, it is preferable and convenient to know that the sets involved in these definitions are properly measurable. It is not the focus of this work to enter into these complications, but rather to try to reduce the discussion to a simple setting in order not to hide the main ideas of the theory. Let us therefore briefly indicate in this section some possible and classical arguments used to handle these annoying measurability questions. These will then mostly be used without any further comments in the sequel.

Let $X = (X_t)_{t \in T}$ be a random process. When T is not countable, the pointwise supremum $\sup_{t \in T} |X_t(\omega)|$ is usually not well defined since one has to take into account an uncountable family of negligible sets. Therefore, it is necessary to consider a handy notion of measurable supremum of the collection $(X_t)_{t \in T}$. One possible way is to understand quantities such as $\sup_{t \in T} |X_t|$ (or similar ones, $\sup_{t \in T} X_t$, $\sup_{s, t \in T} |X_s - X_t|...$) as the essential (or lattice) supremum in L_0 of the collection of random variables $|X_t|$, $t \in T$. Even simpler, if the process X is in L_p, $0 < p < \infty$, that is, if $\mathbb{E}|X_t|^p < \infty$ for every t in T, we can simply set

$$\mathbb{E}\sup_{t \in T} |X_t|^p = \sup\left\{ \mathbb{E}\sup_{t \in F} |X_t|^p ;\ F \text{ finite in } T \right\}.$$

This lattice supremum also works in more general Orlicz spaces than L_p-spaces and will mainly be used in Chapters 11 and 12 when we will have to show that a given process is bounded, reducing basically the various estimates to the case where T is finite.

Another possibility is the probabilistic concept of separable version which allows us to deal similarly with other properties than boundedness, such as continuity. Let (T, d) be a metric space. A random process $X = (X_t)_{t \in T}$ defined on $(\Omega, \mathcal{A}, \mathbb{P})$ is said to be *separable* if there exist a negligible set $N \subset \Omega$ and a countable set S in T such that, for every $\omega \notin N$, every $t \in T$ and $\varepsilon > 0$,

$$X_t(\omega) \in \overline{\{X_s(\omega) ; \; s \in S, \; d(s,t) < \varepsilon\}}$$

where the closure is taken in $\mathbb{R} \cup \{\infty\}$. If X is separable, then, in particular, $\sup_{t \in T} |X_t(\omega)| = \sup_{t \in S} |X_t(\omega)|$ for every $\omega \notin N$, and since S is countable there is, of course, no difficulty in dealing with this type of supremum. Note that if there exists a separable random process on (T, d), then (T, d) is separable as a metric space. If (T, d) is separable and X is almost surely continuous, then X is separable.

Hence, when a random process is separable, there is no difficulty in dealing with the almost sure boundedness or continuity of the trajectories since these properties are reduced along some countable parameter set. In general however, a given random process $X = (X_t)_{t \in T}$ need not be separable. In a rather general setting, it admits a version which is separable. A random process $Y = (Y_t)_{t \in T}$ is said to be a *version* of X if, for every $t \in T$, $Y_t = X_t$ with probability one; in particular, Y has the same distribution as X. It is known that when (T, d) is separable and when $X = (X_t)_{t \in T}$ is continuous in probability, that is, for every $t_0 \in T$ and every $\varepsilon > 0$,

$$\lim_{t \to t_0} \mathbb{P}\{|X_t - X_{t_0}| > \varepsilon\} = 0 \,,$$

then X admits a separable version. Moreover, every dense sequence S in T can be chosen as a separable set. The preceding hypotheses will always be satisfied when we will need such a result so that we freely use it below.

Summarizing, the study of the almost sure boundedness and continuity of random processes can essentially be reduced with the tools of essential supremum or separable version to the setting of a *countable* index set for which no measurability question occurs. In our first part, we will therefore basically study integrability properties and tail behaviors of supremum of bounded processes indexed by a countable set. In the second part, we examine when a given process is almost surely bounded or continuous and we use separable versions.

The purposes of the first part motivate the introduction of a slightly more general notion of random variable with vector values in order to possibly unify results on Radon random variables and on ℓ_∞-valued random variables or bounded processes. One possible definition is the following. Assume that we are given a Banach space B (not necessarily separable!) such that there exists a *countable* subset D of the unit ball or sphere of the dual space B' such that

$$\|x\| = \sup_{f \in D} |f(x)| \,, \quad x \in B \,.$$

The typical example that we have in mind is the space ℓ_∞. Recall that separable Banach spaces possess this property. Given such B and D, we can say that X is a random variable with values in B if X is a map from some probability space $(\Omega, \mathcal{A}, \mathbb{P})$ into B such that $f(X)$ is Borel measurable for every f in D. We can then work freely with the measurable function $\|X\|$.

This definition includes Radon random variables. It also includes almost surely bounded processes $X = (X_t)_{t \in T}$ indexed by a countable set T; take simply $B = \ell_\infty(T)$ and $D = T$ identified with the evaluation maps. As a remark, note that when $X = (X_t)_{t \in T}$ is an almost surely continuous process on (T, d) assumed to be compact, it defines a Radon random variable in the separable Banach space $C(T)$ of all continuous functions on T.

When X and B are as above, we simply say that X is a *random variable* (or *vector*) with values in B, as opposed to a *Radon* random variable. We will try however to recall each time it will be necessary the exact setting in which we are working, not trying to avoid repetitions in this regard. When we are dealing with a separable Banach space B however, we do not distinguish and simply speak of a random variable (or a Borel random variable) with values in B.

To conclude this section, let us note that for this generalized notion of random variable with values in a Banach space B, we can also speak of the spaces $L_p(B)$, $0 \leq p \leq \infty$, as the spaces of random variables X such that $\|X\|_p = (\mathbb{E}\|X\|^p)^{1/p} < \infty$ for $0 < p < \infty$, and the corresponding concepts for $p = 0$ or ∞. Almost sure convergence of a sequence (X_n) makes sense similarly, and if we have to deal with the distribution of such a random variable X, we simply mean the set of distributions of the finite dimensional random vectors $(f_1(X), \ldots, f_N(X))$ where $f_1, \ldots, f_N \in D$. Again, in the case of a Radon random variable, this coincides with the usual definition (choose D to be weakly dense in the unit ball of B').

Finally in this section, let us mention a trivial but useful observation based on independence and Jensen's inequality. If X is a random variable with values in B in the general sense just described, let us simply say that X has mean zero if $\mathbb{E}f(X) = 0$ for all f in D (we then sometimes write with some abuse of notation that $\mathbb{E}X = 0$). Let then F be a convex function on $\mathbb{R}_+$ and let X and Y be independent random variables in B such that $\mathbb{E}F(\|X\|) < \infty$ and $\mathbb{E}F(\|Y\|) < \infty$. Then, if Y has mean zero,

(2.5) $$\mathbb{E}F(\|X + Y\|) \geq \mathbb{E}F(\|X\|) .$$

Indeed, this simply follows by convexity of $F(\|\cdot\|)$ and partial integration with respect to Y using Fubini's theorem.

2.3 Symmetric Random Variables and Lévy's Inequalities

In this paragraph, B denotes a Banach space such that for some countable set D in the unit ball of B', $\|x\| = \sup_{f \in D} |f(x)|$ for every $x \in B$.

A random variable X with values in B is called symmetric if X and $-X$ have the same distribution. Equivalently, X has the same distribution as εX where ε denotes a symmetric Bernoulli or Rademacher random variable taking values ± 1 with probability $1/2$ which is independent of X. (Although the name of Bernoulli is historically more appropriate, we will speak of Rademacher variables since this is the most commonly used terminology in the field.) This simple observation is at the basis of *randomization* (or *symmetrization*, in the probabilistic sense) which is one most powerful tool in Probability in Banach spaces.

Note that for a random variable X, there is a canonical way of generating a symmetric random variable which is not too "far" from X: consider indeed $\widetilde{X} = X - X'$ where X' in an independent copy of X, i.e., with the same distribution as X. In these notations, we will usually assume that X and X' are constructed on different probability spaces $(\Omega, \mathcal{A}, \mathbb{P})$ and $(\Omega', \mathcal{A}', \mathbb{P}')$.

We call a Rademacher sequence (or Bernoulli sequence) a sequence $(\varepsilon_i)_{i \in \mathbb{N}}$ of *independent Rademacher random variables* taking thus the values $+1$ and -1 with equal probability. A sequence (X_i) of random variables with values in B is called a *symmetric sequence* if, for every choice of signs ± 1, $(\pm X_i)$ has the same distribution as (X_i) (i.e. for every N, $(\pm X_1, \ldots, \pm X_N)$ has the same law as $(X_1, \ldots, X_N)$ in B^N). Equivalently, (X_i) has the same distribution as $(\varepsilon_i X_i)$ where (ε_i) is a Rademacher sequence which is independent of (X_i). The typical example of a symmetric sequence consists in a sequence of *independent and symmetric* random variables. In this setting of symmetric sequences, it will be convenient to denote, using Fubini's theorem, by $\mathbb{P}_\varepsilon$, $\mathbb{E}_\varepsilon$ (resp. $\mathbb{P}_X$, $\mathbb{E}_X$) the conditional probability and expectation with respect to the sequence (X_i) (resp. (ε_i)). We hope that the slight abuse of notation, ε representing (ε_i) and X, (X_i), will not get confusing in the sequel.

Partial sums of a symmetric sequence of random variables satisfy some very important inequalities known as Lévy's inequalities. They can be stated as follows. Recall that they apply to the important case of independent and symmetric random variables.

Proposition 2.3. *Let (X_i) be a symmetric sequence of random variables with values in B. For every k, set $S_k = \sum_{i=1}^{k} X_i$. Then, for every integer N and every $t > 0$,*

$$(2.6) \qquad \mathbb{P}\left\{\max_{k \leq N} \|S_k\| > t\right\} \leq 2\mathbb{P}\left\{\|S_N\| > t\right\}$$

and

$$(2.7) \qquad \mathbb{P}\left\{\max_{i \leq N} \|X_i\| > t\right\} \leq 2\mathbb{P}\left\{\|S_N\| > t\right\}.$$

If (S_k) converges in probability to S, the inequalities extend to the limit as

$$\mathbb{P}\{\sup_k \|S_k\| > t\} \le 2\mathbb{P}\{\|S\| > t\}$$

and similarly for (2.7). As a consequence of Proposition 2.3, note also that by integration by parts, for every $0 < p < \infty$,

$$\mathbb{E}\max_{k \le N} \|S_k\|^p \le 2\mathbb{E}\|S_N\|^p$$

and similarly with X_k instead of S_k.

Proof. We only detail (2.6), (2.7) being established exactly in the same way. Let $\tau = \inf\{k \le N\,;\, \|S_k\| > t\}$. We have

$$\mathbb{P}\{\|S_N\| > t\} = \sum_{k=1}^{N} \mathbb{P}\{\|S_N\| > t,\, \tau = k\}\,.$$

Now, since, for every k, $(X_1, \ldots, X_k, -X_{k+1}, \ldots, -X_N)$ has the same distribution as $(X_1, \ldots, X_N)$, and $\{\tau = k\}$ only depends on $X_1, \ldots, X_k$, we also have that

$$\mathbb{P}\{\|S_N\| > t\} = \sum_{k=1}^{N} \mathbb{P}\{\|S_k - R_k\| > t,\, \tau = k\}$$

where $R_k = S_N - S_k$, $k \le N$. Using the triangle inequality

$$2\|S_k\| \le \|S_k + R_k\| + \|S_k - R_k\| = \|S_N\| + \|S_k - R_k\|\,.$$

Then, summing the two preceding probabilities yields

$$2\mathbb{P}\{\|S_N\| > t\} \ge \sum_{k=1}^{N} \mathbb{P}\{\tau = k\} = \mathbb{P}\{\max_{k \ge N} \|S_k\| > t\}\,.$$

The proof of Proposition 2.3 is complete.

Among the consequences of Lévy's inequalities is a useful result on the convergence of series of symmetric sequences. This is known as the Lévy-Itô-Nisio theorem which we present in the context of Radon random variables.

Theorem 2.4. *Let (X_i) be a symmetric sequence of Borel random variables with values in a separable Banach space B. For each n, denote by μ_n the distribution of the n-th partial sum $S_n = \sum_{i=1}^{n} X_i$. The following are equivalent:*

(i) the sequence (S_n) converges almost surely;

(ii) (S_n) converges in probability;

(iii) (μ_n) converges weakly;

(iv) there exists a probability measure μ in $\mathcal{P}(B)$ such that $\mu_n \circ f^{-1} \to \mu \circ f^{-1}$ weakly for every f in B'.

By a simple symmetrization argument, the equivalences (i) – (iii) can also be shown to hold for sums of independent (not necessarily symmetric) random variables. We shall come back to this in Chapter 6. Furthermore, observe from the proof that the equivalence between (i) and (ii) is not restricted to the Radon setting.

Proof. (iii) ⇒ (ii). We first show that $X_i \to 0$ in probability. By difference, (X_i) is weakly relatively compact. Hence, from every subsequence, one can extract a further one, denote it by i', such that $X_{i'}$ converges weakly to some X. Thus, along every linear functional f, $f(X_{i'}) \to f(X)$ weakly. Now, $(f(S_n))$ converges in distribution as a sequence of real valued random variables so that, for all $\delta > 0$, there is an $M > 0$ such that

$$\sup_n \mathbb{P}\{|f(S_n)| > M\} < \delta^2 .$$

Now recall the symmetry assumption and the preceding notation:

$$\sup_n \mathbb{P}_X \mathbb{P}_\varepsilon \left\{ \left| \sum_{i=1}^n \varepsilon_i f(X_i) \right| > M \right\} < \delta^2$$

where (ε_i) is a Rademacher sequence which is independent of (X_i). For every n, let

$$A = \left\{ \omega ; \ \mathbb{P}_\varepsilon \left\{ \left| \sum_{i=1}^n \varepsilon_i f(X_i(\omega)) \right| > M \right\} < \delta \right\} .$$

By Fubini's theorem, $\mathbb{P}(A) = \mathbb{P}_X(A) > 1 - \delta$. If $\omega \in A$, we can apply Khintchine's inequalities (Lemma 4.1) to the sum $\sum_{i=1}^n \varepsilon_i f(X_i(\omega))$ together with Lemma 4.2 and (4.3) below to see that, if $\delta \leq 1/8$,

$$\sum_{i=1}^n f^2(X_i(\omega)) \leq 8M^2 .$$

It follows that

$$\sup_n \mathbb{P}\left\{ \sum_{i=1}^n f^2(X_i) > 8M^2 \right\} < \delta$$

from which we get $\sum_i f^2(X_i) < \infty$ almost surely. Thus, $f(X_i) \to 0$ almost surely. Hence, (X_i) is a tight sequence with only 0 as a possible limit point. This shows that $X_i \to 0$ in probability.

We then deduce (ii). Indeed, if this is not the case, (S_n) is not a Cauchy sequence in probability and there exists a strictly increasing sequence of integers (n_k) such that $T_k = S_{n_{k+1}} - S_{n_k}$ does not converge in probability to 0. Since $\sum_k T_k = \sum_i X_i$ converges weakly, we may apply the preceding step to get a contradiction.

(ii) ⇒ (i). If $S_n \to S$ in probability, there exists a sequence (n_k) of integers such that

$$\sum_k \mathbb{P}\{\|S_{n_k} - S\| > 2^{-k}\} < \infty .$$

By Lévy's inequalities,

$$\mathbb{P}\Big\{\max_{n_{k-1}<n\leq n_k}\|S_n-S_{n_{k-1}}\|>2^{-k+1}\Big\}$$

$$\leq 2\mathbb{P}\big\{\|S_{n_k}-S_{n_{k-1}}\|>2^{-k+1}\big\}$$

$$\leq 2\big(\mathbb{P}\big\{\|S_{n_k}-S\|>2^{-k}\big\}+\mathbb{P}\big\{\|S_{n_{k-1}}-S\|>2^{-k-1}\big\}\big)\ .$$

By the Borel-Cantelli lemma, (S_n) is almost surely a Cauchy sequence and thus (i) holds.

We are left with the proof of (iv) $\Rightarrow$ (iii) since the other implications are obvious. By Prokhorov's criterion and (2.4), it is enough to show that for every $\varepsilon>0$ there exists a finite dimensional subspace F of B such that, for every n,

$$\mathbb{P}\big\{d(S_n,F)>\varepsilon\big\}<\varepsilon\ .$$

Since μ is a Radon measure, it suffices to show that

$$\mathbb{P}\big\{d(S_n,F)>\varepsilon\big\}\leq 2\mu\big(x;\ d(x,F)>\varepsilon\big)$$

for any n, $\varepsilon>0$ and any closed subspace F in B. Now, since B is separable, for every closed subspace F in B, there is a countable subset $D=\{f_m\}$ of the unit ball of B' such that $d(x,F)=\sup_{f\in D}|f(x)|$ for every x. For every m, $(f_1(S_n),\ldots,f_m(S_n))$ is weakly convergent in $\mathbb{R}^m$ to the corresponding marginal of μ. Hence, by Lévy's inequality (2.6) in the limit, for every n,

$$\mathbb{P}\big\{\max_{i\leq m}|f_i(S_m)|>\varepsilon\big\}\leq 2\mu\big(x;\ \max_{i\leq m}|f_i(x)|>\varepsilon\big)\ .$$

The conclusion is easy:

$$\mathbb{P}\big\{d(S_n,F)>\varepsilon\big\}=\mathbb{P}\big\{\sup_{f\in D}|f(S_n)|>\varepsilon\big\}$$

$$\leq\sup_m\mathbb{P}\big\{\max_{i\leq m}|f_i(S_n)|>\varepsilon\big\}$$

$$\leq 2\sup_m\mu\big(x;\ \max_{i\leq m}|f_i(x)|>\varepsilon\big)$$

$$\leq 2\mu\big(x;\ \sup_{f\in D}|f(x)|>\varepsilon\big)$$

$$\leq 2\mu\big(x;\ d(x,F)>\varepsilon\big)\ .$$

Theorem 2.4 is thus established.

2.4 Some Inequalities
for Real Valued Random Variables

We conclude this chapter with two elementary and classical inequalities for real valued random variables which it will be useful to record at this stage. The first one is a version of the binomial inequalities (compare with (1.15),

(1.16)) while the second is the inequality which stands at the basis of the Borel-Cantelli lemma.

Lemma 2.5. *Let (A_i) be a sequence of independent sets such that $a = \sum_i \mathbb{P}(A_i) < \infty$. Then, for every integer n,*

$$\mathbb{P}\left\{\sum_i I_{A_i} \geq n\right\} \leq \frac{a^n}{n!} \leq \left(\frac{ea}{n}\right)^n .$$

Proof. We have

$$\mathbb{P}\left\{\sum_i I_{A_i} \geq n\right\} \leq \sum \prod_{j=1}^{n} \mathbb{P}(A_{i_j})$$

where the summation is over all choices of indexes $i_1 < \cdots < i_n$. Now

$$\sum \prod_{j=1}^{n} \mathbb{P}(A_{i_j}) = \frac{1}{n!} \sum_{\substack{\text{distinct} \\ i_1,\ldots,i_n}} \prod_{j=1}^{n} \mathbb{P}(A_{i_j})$$

$$\leq \frac{1}{n!} \sum_{\substack{\text{all} \\ i_1,\ldots,i_n}} \prod_{j=1}^{n} \mathbb{P}(A_{i_j}) = \frac{a^n}{n!} .$$

Lemma 2.6. *Let $(Z_i)_{i \leq N}$ be independent positive random variables. Then, for every $t > 0$,*

$$\mathbb{P}\left\{\max_{i \leq N} Z_i > t\right\} \geq \sum_{i=1}^{N} \mathbb{P}\{Z_i > t\} \Big/ \left(1 + \sum_{i=1}^{N} \mathbb{P}\{Z_i > t\}\right) .$$

In particular, if $\mathbb{P}\left\{\max_{i \leq N} Z_i > t\right\} \leq \frac{1}{2}$,

$$\sum_{i=1}^{N} \mathbb{P}\{Z_i > t\} \leq 2\mathbb{P}\left\{\max_{i \leq N} Z_i > t\right\} .$$

Proof. For $x \geq 0$, $1 - x \leq \exp(-x)$ and $1 - \exp(-x) \geq x/(1 + x)$. Thus, by independence,

$$\mathbb{P}\left\{\max_{i \leq N} Z_i > t\right\} = 1 - \prod_{i=1}^{N}(1 - \mathbb{P}\{Z_i > t\})$$

$$\geq 1 - \exp\left(-\sum_{i=1}^{N} \mathbb{P}\{Z_i > t\}\right)$$

$$\geq \sum_{i=1}^{N} \mathbb{P}\{Z_i > t\} \Big/ 1 + \sum_{i=1}^{N} \mathbb{P}\{Z_i > t\} .$$

If $\mathbb{P}\{\max_{i \leq N} Z_i > t\} \leq \frac{1}{2}$, the preceding inequality ensures that $\sum_{i=1}^{N} \mathbb{P}\{Z_i > t\} \leq 1$ so that the second conclusion of the lemma follows from the first one.

Notes and References

The following references will hopefully complete appropriately this survey chapter.

Basics on metric spaces, infinite dimensional vector spaces, Banach spaces, etc. can be found in all classical treatises on Functional Analysis such as [Dun-S]. Informations on Banach spaces more in the spirit of this book are given in [Da], [Bea], [Li-T1], [Li-T2] as well as in the references therein.

Probability distributions on metric spaces and weak convergence are presented in [Par], [Bi]. Various accounts on random variables with values in Banach spaces may be found in [Ka1], [Schw2], [HJ3], [Ar-G2], [Li], [V-T-C]. Prokhorov's criterion comes from [Pro1]; the terminology "flat concentration" is used in [Ac1]. For Skorokhod's theorem [Sk1], see also [Du3]. The necessary elements on vector valued martingales and their convergence may be found in [Ne3].

Generalities on random processes and separability are given in [Doo], [Me], [Ne1]. More in the context of Probability in Banach spaces, see [Ba], [J-M3], [Ar-G2].

Symmetric sequences and randomization techniques were first considered by J.-P. Kahane in [Ka1] who gave a proof of Lévy's inequalities [Lé1] in this setting of vector valued random variables. See also [HJ1], [HJ2], [HJ3]. Theorem 2.4 is due to P. Lévy [Lé1] on the line and to K. Itô and M. Nisio [I-N] for independent Banach space valued random variables. For symmetric sequences, see J. Hoffmann-Jørgensen [HJ3]. Our proof follows [Ar-G2].

The inequalities of Section 4 can be found in all classical treatises on Probability Theory (e.g. [Fe1]).

Part I

Banach Space Valued Random Variables and Their Strong Limiting Properties

3. Gaussian Random Variables

With this chapter, we really enter into the subject of Probability in Banach spaces. The study of Gaussian random vectors and processes may indeed be considered as one of the fundamental topics of the theory since it inspires many other parts of the field both in the results themselves and in the techniques of investigation. Historically, the developments also followed this line of progress.

In this chapter, we shall be interested in the integrability properties and the tail behaviors of norms of Gaussian random vectors or of bounded processes as well as in the basic comparison properties of Gaussian processes. The question of when a Gaussian process is almost surely bounded (or continuous) will be addressed and completely solved in Chapters 11 and 12. The study of the tail behavior of the norm of a Gaussian random vector is based on the isoperimetric tools introduced in the first chapter. This study will be a reference for the corresponding results for other types of random vectors like Rademacher series, stable random variables, and even sums of independent random variables which will be treated in the next chapters. This will be the subject of the first section of this chapter. The second examines the corresponding results for chaos. The last paragraph is devoted to the important comparison properties of Gaussian random variables. These now appear at the basis of the rather deep present knowledge about the regularity properties of the sample paths of Gaussian processes (cf. Chapter 12).

First, we recall the basic definitions and some classical properties of Gaussian variables. A real valued mean zero random variable X in $L_2(\Omega, \mathcal{A}, \mathbb{P})$ is said to be Gaussian (or normal) if its Fourier transform satisfies

$$\mathbb{E} \exp(itX) = \exp(-\sigma^2 t^2 / 2), \quad t \in \mathbb{R},$$

where $\sigma = \|X\|_2 = (\mathbb{E}X^2)^{1/2}$. When we speak of a Gaussian variable, we always mean a *centered* Gaussian (or equivalently *symmetric*) variable. X is said to be standard if $\sigma = 1$. Throughout this work, the sequence denoted by $(g_i)_{i \in \mathbb{N}}$ will always mean a sequence of *independent standard Gaussian random variables*; we sometimes call it *orthogaussian* or *orthonormal sequence* (because orthogonality and independence are equivalent for Gaussian variables). In other words, for each N, the vector $g = (g_1, \ldots, g_N)$ is distributed as the canonical Gaussian distribution γ_N on $\mathbb{R}^N$ with density

$$(2\pi)^{-N/2} \exp(-|x|^2 / 2).$$

A random vector $X = (X_1, \ldots, X_N)$ in $\mathbb{R}^N$ is Gaussian if for all real numbers $\alpha_1, \ldots, \alpha_N$, $\sum_{i=1}^{N} \alpha_i X_i$ is a real valued Gaussian random variable. Such a Gaussian vector can always be diagonalized and regarded under the canonical distribution γ_N. Indeed, if $\Gamma = AA^t = (\mathbb{E} X_i X_j)_{1 \leq i,j \leq N}$ denotes the symmetric (semi-) positive definite covariance matrix, Γ completely determines the distribution of X which is the same as that of Ag where $g = (g_1, \ldots, g_N)$.

One fundamental property of the Gaussian distributions is their *rotational* invariance which may be expressed in various manners. For example, if $g = (g_1, \ldots, g_N)$ is distributed according to γ_N in $\mathbb{R}^N$ and if U is an orthogonal matrix in $\mathbb{R}^N$, then Ug is also distributed according to γ_N. As a simple consequence, if (α_i) is a finite sequence of real numbers, $\sum_i g_i \alpha_i$ has the same distribution as $g_1 \left(\sum_i \alpha_i^2 \right)^{1/2}$. In particular, since g_1 has moments of all orders,

$$\left\| \sum_i g_i \alpha_i \right\|_p = \|g_1\|_p \left(\sum_i \alpha_i^2 \right)^{1/2}$$

for any $0 < p < \infty$ so that the span of (g_i) in L_p is isometric to ℓ_2. Another description of this invariance by rotation (which was used in the proof of (1.5) in Chapter 1) is the following: if X is Gaussian (in $\mathbb{R}^N$) and if Y is an independent copy of X, then for every θ, the rotation of angle θ of the vector (X, Y), i.e.

$$(X \sin \theta + Y \cos \theta, \; X \cos \theta - Y \sin \theta),$$

has the same distribution as (X, Y). These properties are trivially verified on the covariances.

A Radon random variable X with values in a Banach space B is Gaussian if for any continuous linear functional f on B, $f(X)$ is a real valued Gaussian variable. A typical example is given by a convergent series $\sum_i g_i x_i$ (which converges in any sense by Theorem 2.4) where (x_i) is a sequence in B. Actually, we shall see later on in this chapter that every Radon Gaussian vector may be represented in distribution in this way. Finally, a process $X = (X_t)_{t \in T}$ indexed by a set T is called Gaussian if each finite linear combination $\sum_i \alpha_i X_{t_i}$, $\alpha_i \in \mathbb{R}$, $t_i \in T$, is Gaussian. The covariance structure $\Gamma(s,t) = \mathbb{E} X_s X_t$, $s,t \in T$, completely determines the distribution of the Gaussian process X. Since the distributions of these infinite dimensional Gaussian variables are determined by the finite dimensional projections, the rotational invariance trivially extends. For example, if X is a Gaussian process or variable (in B), and if (X_i) is a sequence of independent copies of X, then, for any finite sequence (α_i) of real numbers, $\sum_i \alpha_i X_i$ has the same distribution as $\left(\sum_i \alpha_i^2 \right)^{1/2} X$.

3.1 Integrability and Tail Behavior

By their very definition, (norms of) real or finite dimensional Gaussian variables admit high integrability properties. One of the purposes of this section will be to show how these extend to an infinite dimensional setting. We thus investigate the integrability and tail behavior of the norm $\|X\|$ of a Gaussian Radon variable or of the supremum of an almost surely bounded process. In order to state the results in a somewhat unified way, it is convenient to adopt the terminology introduced in Section 2.2. Unless otherwise specified, we thus deal in this section with a Banach space B such that, for some countable subset D of the unit ball of B', $\|x\| = \sup_{f \in D} |f(x)|$ for all x in B. We say that X is a Gaussian random variable in B if $f(X)$ is measurable for every f in D and if every finite linear combination $\sum_i \alpha_i f_i(X)$, $\alpha_i \in \mathbb{R}$, $f_i \in D$, is Gaussian.

Let therefore X be Gaussian in B. *Two* main parameters of the distribution of X will determine the behavior of $\mathbb{P}\{\|X\| > t\}$ (when $t \to \infty$): some quantity in the (*strong*) topology of the norm, median or exceptation of $\|X\|$ for example (or some other quantity in $L_p(B)$, $0 \leq p < \infty$ – see below), and *weak* variances, $\mathbb{E}f^2(X)$, $f \in D$. More precisely, let $M = M(X)$ be a median of $\|X\|$, that is a number satisfying both

$$\mathbb{P}\{\|X\| \leq M\} \geq \tfrac{1}{2}, \quad \mathbb{P}\{\|X\| \geq M\} \geq \tfrac{1}{2}.$$

Actually, for the purposes of tail behaviors, integrability properties or moment equivalences, it would be enough to consider M such that $\mathbb{P}\{\|X\| \leq M\} > \varepsilon > 0$ as we will see later; the concept of median is however crucial for concentration results as opposed to the preceding ones which arise more with deviation inequalities. Besides M, consider

$$\sigma = \sigma(X) = \sup_{f \in D} \left(\mathbb{E}f^2(X)\right)^{1/2}.$$

Note that this supremum is finite and is actually controlled by M. Indeed, for every f in D, $\mathbb{P}\{|f(X)| \leq M\} \geq \tfrac{1}{2}$; now $f(X)$ is a real valued Gaussian variable with variance $\mathbb{E}f^2(X)$ and this inequality implies that $(\mathbb{E}f^2(X))^{1/2} \leq 2M$ since $\mathbb{P}\{|g| \leq \tfrac{1}{2}\} < 0.4 < \tfrac{1}{2}$ where g is a standard normal variable. Hence, $\sigma \leq 2M < \infty$. Let us mention that if X is a Gaussian Radon variable with values in B, then σ is really meant to be $\sup_{\|f\| \leq 1}(\mathbb{E}f^2(X))^{1/2}$. This is always understood in the sequel.

The main conclusions about the behavior of $\mathbb{P}\{\|X\| > t\}$ are obtained from the Gaussian isoperimetric inequality (Theorem 1.2) and the subsequent concentration results described in Section 1.1. In order to conveniently use the isoperimetric inequality, we reduce, as is usual, the distribution of X to the canonical Gaussian distribution $\gamma = \gamma_\infty$ on $\mathbb{R}^{\mathbb{N}}$ (i.e. γ is the distribution of the orthonormal sequence (g_i)). The procedure is classical. Set $D = \{f_n, \, n \geq 1\}$. By the Gram-Schmidt orthonormalization procedure applied to the sequence $(f_n(X))_{n \geq 1}$ in L_2, we can write (in distribution)

$$f_n(X) = \sum_{i=1}^{n} a_i^n g_i \,, \quad n \geq 1 \,.$$

In other words, if $x = (x_i)$ is a generic point in $\mathbb{R}^{\mathbb{N}}$, the sequence $(f_n(X))$ has the same distribution as the sequence $(\sum_{i=1}^{n} a_i^n x_i)$ under γ. For notational convenience, we extend the meaning of f_n by letting $f_n(x) = \sum_{i=1}^{n} a_i^n x_i$ for $n \geq 1$ and $x = (x_i)$ in $\mathbb{R}^{\mathbb{N}}$. Furthermore, if we set, for $x \in \mathbb{R}^{\mathbb{N}}$, $\|x\| = \sup_{n \geq 1} |f_n(x)|$, the probabilistic study of $\|X\|$ then amounts to the study of $\|x\|$ under γ. Note that, in this notation,

(3.1) $$\sigma = \sigma(X) = \sup_{|h| \leq 1} \|h\|$$

where, as usual, $|\cdot|$ is the ℓ_2-norm. We use this simple reduction throughout the various proofs below.

As announced, the next lemma describes the isoperimetric concentration of $\|X\|$ around its median M measured in terms of σ. Recall Φ, the distribution function of a standard normal variable, Φ^{-1}, $\Psi = 1 - \Phi$ and the estimate $\Psi(t) \leq \frac{1}{2} \exp(-t^2/2)$, $t \geq 0$.

Lemma 3.1. *Let X be a Gaussian random variable in B with median $M = M(X)$ and supremum of weak variances $\sigma = \sigma(X)$. Then, for every $t > 0$,*

$$\mathbb{P}\{|\|X\| - M| > t\} \leq 2\Psi(t/\sigma) \leq \exp(-t^2/2\sigma^2) \,.$$

Proof. We use Theorem 1.2 with the preceding reduction. Let $A = \{x \in \mathbb{R}^{\mathbb{N}} \,; \|x\| \leq M\}$. Then $\gamma(A) \geq \frac{1}{2}$. By Theorem 1.2, or rather (1.3) here, if A_t is the Hilbertian neighborhood of order $t > 0$ of A, $\gamma_*(A_t) \geq \Phi(t)$. Now, if $x \in A_t$, $x = a + th$ where $a \in A$ and $|h| \leq 1$; hence, by (3.1),

$$\|x\| \leq M + t\|h\| \leq M + t\sigma$$

and therefore $A_t \subset \{x \,; \|x\| \leq M + \sigma t\}$. Applying the same argument to $A = \{x \,; \|x\| \geq M\}$ clearly concludes the proof of Lemma 3.1.

The proof of Lemma 3.1 of course just repeats the concentration property (1.4) for Lipschitz functions as is clear from the fact that $\|x\|$ (on $\mathbb{R}^{\mathbb{N}}$) is Lipschitz with constant σ. This observation tells us that, following (1.5), we have similarly (and at some cheaper price) a concentration of $\|X\|$ around its expectation, that is, for any $t > 0$,

(3.2) $$\mathbb{P}\{|\|X\| - \mathbb{E}\|X\|| > t\} \leq 2\exp(-2t^2/\pi^2\sigma^2) \,.$$

It is clear that $\mathbb{E}\|X\| < \infty$ from Lemma 3.1; actually, this can also be deduced from a finite dimensional version of (3.2) together with an approximation argument. ((1.6) of course yields (3.2) with the best constant $-t^2/2\sigma^2$ in the exponential.) As usual, (3.2) is interesting since it is often easier to work with expectations rather than with medians. Repeating in this framework some of the comments of Section 1.1, note that a median $M = M(X)$ of $\|X\|$ is unique and that, by integrating for example the inequality of Lemma 3.1,

$$|\mathbb{E}\|X\| - M| \leq (\pi/2)^{1/2}\sigma.$$

As we already mentioned, the integrability theorems which we will deduce from Lemma 3.1 (or (3.2)) actually only use half of it, that is only the deviation inequality

$$\mathbb{P}\{\|X\| > M + \sigma t\} \leq \Psi(t), \quad t > 0,$$

(and similarly with $\mathbb{E}\|X\|$ instead of M). The concentration around M (or $\mathbb{E}\|X\|$) will however be crucial for some other questions dealt with, for example, in Chapter 9 where this result and the relative weights of M and σ can be used in Geometry of Banach spaces. Actually, for the integrability theorems even only the knowledge of s such that $\mathbb{P}\{\|X\| \leq s\} \geq \varepsilon > 0$ is sufficient. Indeed, if, in the proof of Lemma 3.1, we apply the isoperimetric inequality to $A = \{x;\ \|x\| \leq s\}$, then, we get, for every $t > 0$,

$$\mathbb{P}\{\|X\| > s + \sigma t\} \leq \Psi(\Phi^{-1}(\varepsilon) + t).$$

It follows for example that, for $t > 0$,

$$\mathbb{P}\{\|X\| > s + \sigma t\} \leq \exp(\Phi^{-1}(\varepsilon)^2/2) \exp(-t^2/8).$$

As we have seen, the information σ on weak moments is weaker than the one, M or $\mathbb{E}\|X\|$, on the strong topology. We already noted that $\sigma \leq 2M$ and we can add the trivial estimate $\sigma \leq (\mathbb{E}\|X\|^2)^{1/2}$ (which is finite). In general, σ is much smaller than $2M$. For the canonical distribution γ_N on $\mathbb{R}^N$, we already have that $\sigma = 1$ and that M is of the order of $\sqrt{N}$. In the preceding inequality, σ can be replaced by one of the strong parameters yielding weaker but still interesting bounds. For example, in the context of the preceding inequality, we observe that $\sigma \leq s/\Phi^{-1}\left(\frac{1+\varepsilon}{2}\right)$ from which it follows that, for $t > 0$,

$$(3.3) \quad \begin{aligned} &\mathbb{P}\{\|X\| > t\} \\ &\leq \exp\left(\frac{1}{2}\Phi^{-1}(\varepsilon)^2 + \frac{1}{8}\Phi^{-1}\left(\frac{1+\varepsilon}{2}\right)\right)\exp\left(-\frac{t^2}{32s^2}\Phi^{-1}\left(\frac{1+\varepsilon}{2}\right)^2\right). \end{aligned}$$

This inequality seems complicated but it however gives as before an exponential squared tail for $\mathbb{P}\{\|X\| > t\}$. Note further that $\Phi^{-1}\left(\frac{1+\varepsilon}{2}\right)$ becomes large when ε goes to 1. While we deduce (3.3) from isoperimetric methods, it should be noticed that such an inequality can actually be established by a direct argument which we would like to briefly sketch. Let Y denote an independent copy of X. By the rotational invariance of Gaussian measures, $(X + Y)/\sqrt{2}$ and $(X - Y)/\sqrt{2}$ are independent with the same distribution as X. Now, if for $s < t$, $\|X + Y\| \leq s\sqrt{2}$ and $\|X - Y\| > t\sqrt{2}$, we have from the triangle inequality that both $\|X\|$ and $\|Y\|$ are larger than $(t - s)/\sqrt{2}$. Hence, by independence and identical distribution,

$$\mathbb{P}\{\|X\| \leq s\}\mathbb{P}\{\|X\| > t\} = \mathbb{P}\{\|X + Y\| \leq s\sqrt{2}, \ \|X - Y\| > t\sqrt{2}\}$$
$$\leq \mathbb{P}\{\|X\| > (t - s)/\sqrt{2}, \ \|Y\| > (t - s)/\sqrt{2}\}$$
$$\leq \left(\mathbb{P}\{\|X\| > (t - s)/\sqrt{2}\}\right)^2.$$

Iterating this inequality with $\mathbb{P}\{\|X\| \leq s\} = \varepsilon > \frac{1}{2}$ and

$$t = t_n = \left(\sqrt{2^{n+1}} - 1\right)\left(\sqrt{2} + 1\right)s$$

easily yields that, for each $t \geq s$,

$$(3.4) \qquad \mathbb{P}\{\|X\| > t\} \leq \exp\left(-\frac{t^2}{24s^2}\log\frac{\varepsilon}{1 - \varepsilon}\right),$$

an inequality indeed similar in nature to (3.3).

Let us record at this stage an inequality of the preceding type which will be convenient in the sequel and in which only the strong parameter steps in. From Lemma 3.1 (for instance!), and $\sigma^2 \leq \mathbb{E}\|X\|^2$, $M^2 \leq 2\mathbb{E}\|X\|^2$, we have, for every $t > 0$,

$$(3.5) \qquad \mathbb{P}\{\|X\| > t\} \leq 4\exp\left(-t^2/8\mathbb{E}\|X\|^2\right).$$

To conclude these comments, let us also describe a bound for the maximum of a finite number of norms of vector valued Gaussian variables X_i, $i \leq N$. Assume that $\max_{i \leq N} \sigma(X_i) \leq 1$. For any $\delta \geq 0$, we have by an integration by parts and Lemma 3.1 that,

$$\mathbb{E}\max_{i \leq N}|\|X_i\| - M(X_i)| \leq \delta + \sum_{i=1}^{N}\int_{\delta}^{\infty}\mathbb{P}\{|\|X_i\| - M(X_i)| > t\}\,dt$$
$$\leq \delta + N\int_{\delta}^{\infty}\exp(-t^2/2)\,dt$$
$$\leq \delta + N\sqrt{\frac{\pi}{2}}\exp(-\delta^2/2).$$

Then, simply let $\delta = (2\log N)^{1/2}$ so that we have obtained, by homogeneity,

$$(3.6) \qquad \mathbb{E}\max_{i \leq N}\|X_i\| \leq 2\max_{i \leq N}\mathbb{E}\|X_i\| + 3(\log N)^{1/2}\max_{i \leq N}\sigma(X_i).$$

The next corollary describes some applications of Lemma 3.1 and of the preceding inequalities to the tail behavior and the integrability properties of the norm of a Gaussian random vector.

Corollary 3.2. *Let X be a Gaussian random variable in B with corresponding $\sigma = \sigma(X)$. Then*

$$\lim_{t \to \infty}\frac{1}{t^2}\log\mathbb{P}\{\|X\| > t\} = -\frac{1}{2\sigma^2}$$

or, equivalently,

$$\mathbb{E}\,\exp\Big(\frac{1}{2\alpha^2}\|X\|^2\Big) < \infty \quad \text{if and only if } \alpha > \sigma.$$

Furthermore, all the moments of $\|X\|$ are equivalent (and equivalent to $M = M(X)$, the median of $\|X\|$): for any $0 < p, q < \infty$, there exists a constant $K_{p,q}$ depending on p and q only such that for any Gaussian vector X,

$$\|X\|_p \leq K_{p,q}\|X\|_q.$$

In particular, $K_{p,2} = K\sqrt{p}$ ($p \geq 2$) where K is a numerical constant.

Proof. The fact that the limit is less than or equal to $-1/2\sigma^2$ easily follows from Lemma 3.1 while the minoration simply uses $\mathbb{P}\{\|X\| > t\} \geq \mathbb{P}\{|f(X)| > t\}$ for every f in D. The equivalence with the exponential integrability is easy thanks to Chebyshev's inequality and an integration by parts. Concerning the equivalence of moments, let M be the median of $\|X\|$. If we integrate the inequality of Lemma 3.1,

$$\mathbb{E}|\|X\| - M|^p = \int_0^\infty \mathbb{P}\{|\|X\| - M| > t\}\,dt^p$$
$$\leq \int_0^\infty \exp(-t^2/2\sigma^2)\,dt^p \leq (K\sqrt{p}\,\sigma)^p$$

for some numerical constant K. Now, this inequality is stronger than what we need since $\sigma \leq 2M$ and since M can be majorized by $(2\mathbb{E}\|X\|^q)^{1/q}$ for every $q > 0$. The proof is complete.

As we already mentioned, we use Lemma 3.1 in this proof but the inequalities discussed prior to Corollary 3.2 can be used similarly (to some extent).

Let (X_n) be a sequence of Gaussian random variables which is bounded in probability, that is, for each $\varepsilon > 0$ there exists $A > 0$ such that

$$\sup_n \mathbb{P}\{\|X_n\| > A\} < \varepsilon.$$

Then (X_n) is bounded in all the L_p-spaces. Indeed, if $M(X_n)$ is the median of $\|X_n\|$, certainly $\sup_n M(X_n) < \infty$ and the preceding equivalence of moments confirms the claim. In particular, if (X_n) is a Gaussian sequence (in the sense that $(X_{n_1}, \ldots, X_{n_N})$ is Gaussian in B^N for all $n_1, \ldots, n_N$) which converges in probability, this convergence takes place in L_p for every p.

Although rather sharp, the previous corollary can be refined. The refinement which we describe relies on a more elaborate use of the Gaussian isoperimetric inequality and confirms the rôle of the two parameters, weak and strong, used to measure the size of the distribution of the norm of a Gaussian vector. Let X be as above a Gaussian random variable in B and recall that $\sigma = \sigma(X) = \sup_{f \in D}(\mathbb{E}f^2(X))^{1/2}$. Consider

$$\tau = \tau(X) = \inf\{\lambda \geq 0;\ \mathbb{P}\{\|X\| \leq \lambda\} > 0\},$$

that is the first jump of the distribution of $\|X\|$. This jump can actually be shown to be unique [Ts]. In the case X is Radon, $\tau = 0$. One way to prove this fact is to first observe that for every $\varepsilon > 0$ and x in B

$$\left(\mathbb{P}\{\|X - x\| \leq \varepsilon\}\right)^2 \leq \mathbb{P}\{\|X\| \leq \varepsilon\sqrt{2}\}.$$

Indeed, if Y is an independent copy of X, by symmetry and independence,

$$\begin{aligned}
\left(\mathbb{P}\{\|X - x\| \leq \varepsilon\}\right)^2 &= \mathbb{P}\{\|X - x\| \leq \varepsilon\}\mathbb{P}\{\|Y + x\| \leq \varepsilon\} \\
&\leq \mathbb{P}\{\|(X - x) + (Y + x)\| \leq 2\varepsilon\} \\
&\leq \mathbb{P}\{\|X\| \leq \varepsilon\sqrt{2}\}
\end{aligned}$$

since $X + Y$ has the law of $\sqrt{2}X$. Note that this inequality can actually be improved to

$$\mathbb{P}\{\|X - x\| \leq \varepsilon\} \leq \mathbb{P}\{\|X\| \leq \varepsilon\}$$

by the Φ^{-1}- or log-concavity ((1.1) and (1.2)) but is sufficient for our modest purpose here. If X is Radon, and if we assume that $P\{\|X\| \leq \varepsilon_0\} = 0$ for some $\varepsilon_0 > 0$, then there is a sequence (x_n) in B such that, by separability,

$$\mathbb{P}\{\exists n; \|X - x_n\| \leq \varepsilon_0/\sqrt{2}\} = 1.$$

Then

$$1 \leq \sum_n \mathbb{P}\{\|X - x_n\| \leq \varepsilon_0/\sqrt{2}\} \leq \sum_n \left(\mathbb{P}\{\|X\| \leq \varepsilon_0\}\right)^{1/2} = 0$$

and thus, necessarily $\tau = \tau(X) = 0$ in this case.

On the other hand, let us recall the typical example in which $\tau > 0$. Consider X in ℓ_∞ given by the sequence of its coordinates $(g_n/(2\log(n+1))^{1/2})$ where (g_n) is the orthogaussian sequence. Then, by a simple use of the Borel-Cantelli lemma, it is easily seen that

$$(3.7) \qquad \limsup_{n \to \infty} \frac{|g_n|}{(2\log(n+1))^{1/2}} = 1 \quad \text{almost surely}$$

so that $\tau = 1$ in this case.

As announced, the following theorem refines Corollary 3.2 and involves both σ and τ in the integrability result.

Theorem 3.3. *Let X be a Gaussian random variable with values in B and let $\sigma = \sigma(X)$ and $\tau = \tau(X)$. Then, for any $\tau' > \tau$,*

$$\mathbb{E}\exp\left(\frac{1}{2\sigma^2}(\|X\| - \tau')^2\right) < \infty.$$

Before we prove this theorem, let us interpret this result as a tail behavior. We have seen in Corollary 3.2 that

$$\lim_{t\to\infty} \frac{1}{t^2}\log \mathbb{P}\{\|X\| > t\} = -\frac{1}{2\sigma^2}$$

which can be rewritten as

$$\lim_{t\to\infty} \frac{1}{t}\Phi^{-1}(\mathbb{P}\{\|X\| \le t\}) = \frac{1}{\sigma}$$

($\Phi^{-1}(1-u)$ is equivalent to $(2\log\frac{1}{u})^{1/2}$ when $u \to 0$). From (1.1), we know that the function $\Phi^{-1}(\mathbb{P}\{\|X\| \le t\})$ is concave on $]\tau, \infty[$. Theorem 3.3 can therefore be described equivalently as

$$0 \ge \lim_{t\to\infty} \left[\Phi^{-1}(\mathbb{P}\{\|X\| \le t\}) - \sigma t\right] \ge -\frac{\tau}{\sigma}.$$

That is, the concave function $\Phi^{-1}(\mathbb{P}\{\|X\| \le t\})$ on $]\tau, \infty[$ has an asymptote $(t/\sigma) + \ell$ with $-\tau/\sigma \le \ell \le 0$. Note that $\tau = 0$ and hence $\ell = 0$ if X is Radon.

The proof of Theorem 3.3 appears as a consequence of the following (deviation) lemma which is of independent interest and which may be compared to Lemma 3.1.

Lemma 3.4. *For every $\tau' > \tau$, there is an integer N such that for every $t > 0$,*

$$\mathbb{P}\{\|X\| > \tau' + \sigma t\} \le \mathbb{P}\left\{\left(\sum_{i=1}^{N} g_i^2\right)^{1/2} > t\right\}.$$

In particular,

$$\mathbb{P}\{\|X\| > \tau' + \sigma t\} \le K(N)(1+t)^N \exp(-t^2/2)$$

where $K(N)$ is a positive constant depending on N only.

To see why the theorem follows from the lemma, let $\tau' > \tau'' > \tau$ and $\varepsilon = (\tau' - \tau'')/\sigma > 0$. Applying Lemma 3.4 to $\tau'' > \tau$, we get, for some integer N,

$$\int_{\{\|X\|>\tau'\}} \exp\left(\frac{1}{2\sigma^2}(\|X\| - \tau')^2\right) d\mathbb{P}$$

$$\le 1 + \int_0^\infty \mathbb{P}\{\|X\| > \tau' + \sigma t\}\, t\, \exp\left(\frac{t^2}{2}\right) dt$$

$$\le 1 + K(N)\int_0^\infty (1 + \varepsilon + t)^{N+1}\exp\left(\frac{t^2}{2} - \frac{(t+\varepsilon)^2}{2}\right) dt$$

$$\le 1 + K(N)\int_0^\infty (1 + \varepsilon + t)^{N+1}\exp(-\varepsilon t)\, dt < \infty.$$

Proof of Lemma 3.4. It is based again on the Gaussian isoperimetric inequality in the form of the following lemma for which we introduce the following notation. Given an integer N, a point x in $\mathbb{R}^{\mathbb{N}}$ can be decomposed in (y, z) with $y \in \mathbb{R}^N$ and $z \in \mathbb{R}^{]N,\infty[}$. By Fubini's theorem, $d\gamma(x) = d\gamma_N(y)d\gamma_{]N,\infty[}(z)$. If

A is a Borel set in $\mathbb{R}^{\mathbb{N}}$, set $B = \{z \in \mathbb{R}^{]N,\infty[} ; \ (0,z) \in A\}$. Recall that for $t \geq 0$, A_t denotes the Euclidean or Hilbertian neighborhood of A.

Lemma 3.5. *With this notation, if* $\gamma_{]N,\infty[}(B) \geq 1/2$, *then, for any* $t > 0$,

$$\gamma^*\big(x \in \mathbb{R}^{\mathbb{N}} ; \ x \notin A_t\big) \leq \gamma_{N+1}\big(y' \in \mathbb{R}^{N+1} ; \ |y'| > t\big).$$

Proof. Since $\gamma_{]N,\infty[}(B) \geq 1/2$, Theorem 1.2 implies that $(\gamma_{]N,\infty[})_*(B_s) \geq \Phi(s)$ where B_s is of course the Hilbertian neighborhood of order $s \geq 0$ of B in $\mathbb{R}^{]N,\infty[}$. Let $y \in \mathbb{R}^N$, $t \geq |y|$ and $s = (t^2 - |y|^2)^{1/2}$. If $z \in B_s$, then, by definition, $z = b + sk$ where $b \in B$, $k \in \mathbb{R}^{]N,\infty[}$, $|k| \leq 1$; thus

$$x = (y, z) = (0, b) + (|y|^2 + s^2)^{1/2}h = (0, b) + th$$

where $h \in \mathbb{R}^{\mathbb{N}}$ and $|h| \leq 1$. Hence $x \in A_t$. From this observation, we get, via Fubini's theorem, that, for $t > 0$,

$$\gamma^*(x; x \notin A_t) \leq \gamma_N(y; |y| > t)$$
$$+ \gamma_N \otimes (\gamma_{]N,\infty[})^*\big((y,z); |y| \leq t, \ z \notin B_{(t^2-|y|^2)^{1/2}}\big)$$
$$\leq \gamma_N(y; |y| > t)$$
$$+ \int_{|y| \leq t} \int_{s > (t^2 - |y|^2)^{1/2}} \exp\Big(-\frac{s^2}{2}\Big) \frac{ds}{\sqrt{2\pi}} d\gamma_N(y)$$
$$\leq \int_{s^2 + |y|^2 > t^2} \exp\Big(-\frac{s^2}{2}\Big) \frac{ds}{\sqrt{2\pi}} d\gamma_N(y)$$

which is the announced result. Lemma 3.5 is thus established.

Now, we are in a position to prove Lemma 3.4. Of course, we use the reduction to $(\mathbb{R}^{\mathbb{N}}, \gamma)$ as in Lemma 3.3. Let $\tau' > \tau$ and $A = \{x; \|x\| \leq \tau'\}$. We first show that there exists an integer N such that with the notation introducing Lemma 3.5, if $B = \{z \in \mathbb{R}^{]N,\infty[} ; \ (0,z) \in A\}$, then $\gamma_{]N,\infty[}(B) \geq 1/2$. By the hypothesis, $\gamma(A) > 0$. Since $\|x\| = \sup_{n \geq 1} |f_n(x)|$ and since $f_n(x)$ only depends on the first n coordinates of x, there exists N such that

$$\gamma_N\big(y; \max_{n \leq N} |f_n(y)| \leq \tau'\big) \leq 4\gamma(A)/3.$$

Then, by Fubini's theorem, there exists y in $\mathbb{R}^N$ such that

$$\gamma_{]N,\infty[}(z; \ (y,z) \in A) \geq 3/4.$$

By symmetry,

$$\gamma_{]N,\infty[}(z; \ (y,-z) \in A) \geq 3/4$$

so that the intersection of the two preceding sets has a measure larger than or equal to $1/2$. If z belongs to this intersection, $\|(y,z)\| \leq \tau'$ and $\|(y,-z)\| = \|(-y,z)\| \leq \tau'$ and therefore $\|(0,z)\| \leq \tau'$. This shows that $\gamma_{]N,\infty[}(B) \geq 1/2$.

From Lemma 3.5 we then get an upper bound for the complement of A_t, $t > 0$. Since

$$A_t \subset \left\{ x \, ; \, \|x\| \leq \tau' + t\sigma \right\}$$

the first inequality in Lemma 3.4 follows. The second is an easy and classical consequence of the first inequality. The proof of Lemma 3.4 is complete.

To conclude this paragraph we briefly describe the series representation of Gaussian Radon Banach space valued random variables which is easily deduced from the integrability properties. Recall that (g_i) denotes an orthogaussian sequence.

Proposition 3.6. *Let X be a Gaussian Radon random variable on $(\Omega, \mathcal{A}, \mathbb{P})$ with values in a Banach space B. Then X has the same distribution as $\sum_i g_i x_i$ for some sequence (x_i) in B where this series converges almost surely and in all L_p's.*

Proof. Let H be the closure in $L_2 = L_2(\Omega, \mathcal{A}, \mathbb{P})$ of the variables of the form $f(X)$ with f in B'. Then H may be assumed to be separable since X is Radon and H consists of Gaussian variables. Let (g_i) denote an orthonormal basis of H and denote by $\mathcal{A}_N$ the σ-algebra generated by $g_1, \ldots, g_N$. It is easily seen that

$$\mathbb{E}^{\mathcal{A}_N} X = \sum_{i=1}^{N} g_i x_i$$

with $x_i = \mathbb{E}(g_i X)$. Now, recall that $\mathbb{E}\|X\| < \infty$ (for example) so that, by the martingale convergence theorem (cf. Section 2.1) the series $\sum_i g_i x_i$ converges almost surely to X. Since $\mathbb{E}\|X\|^p < \infty$, $0 < p < \infty$, it converges also in $L_p(B)$ for every p.

3.2 Integrability of Gaussian Chaos

In this section, the integrability and the tail behavior properties of real and vector valued Gaussian chaos are investigated as a natural continuation of the preceding section. The Gaussian variables studied so far may indeed be considered as chaos of the first order. Let us first briefly (and partially) describe the concept of chaos.

Consider the Hermite polynomials $\{h_k, k \in \mathbb{N}\}$ on $\mathbb{R}$ defined by the series expansion

$$\exp\left(\lambda x - \frac{\lambda^2}{2} \right) = \sum_{k=0}^{\infty} \frac{\lambda^k}{\sqrt{k!}} h_k(x), \quad \lambda, x \in \mathbb{R}.$$

The Hermite polynomials form an orthonormal basis of $L_2(\mathbb{R}, \gamma_1)$. Similarly, if $\underline{k} \in \mathbb{N}^{(\mathbb{N})}$, i.e. $\underline{k} = (k_1, k_2, \ldots)$, $k_i \in \mathbb{N}$, with $|\underline{k}| = \sum_i k_i < \infty$, set, for $x = (x_i) \in \mathbb{R}^{\mathbb{N}}$,

$$H_{\underline{k}}(x) = h_{k_1}(x_1)h_{k_2}(x_2)\cdots .$$

Then $\{H_{\underline{k}}, \ \underline{k} \in \mathbb{N}^{(\mathbb{N})}\}$ forms an orthonormal basis of $L_2(\mathbb{R}^{\mathbb{N}}, \gamma)$ where we recall that $\gamma = \gamma_\infty$ is the canonical Gaussian product measure on $\mathbb{R}^{\mathbb{N}}$.

For $0 \leq \varepsilon \leq 1$, introduce the bounded linear operator $T(\varepsilon) : L_2(\gamma) \to L_2(\gamma)$ defined by

$$T(\varepsilon)H_{\underline{k}} = \varepsilon^{|\underline{k}|}H_{\underline{k}}$$

for any $\underline{k}$ with $|\underline{k}| < \infty$. It is not difficult to check that $T(\varepsilon)$ extends to a positive contraction on every $L_p(\gamma)$, $1 \leq p < \infty$. Actually, this is clear from the following integral representation of $T(\varepsilon)$: if f is in $L_2(\gamma)$ and x in $\mathbb{R}^{\mathbb{N}}$,

$$T(\varepsilon)f(x) = \int f\big(\varepsilon x + (1 - \varepsilon^2)^{1/2}y\big)d\gamma(y) .$$

If $t \geq 0$, $T_t = T(e^{-t})$ is known as the Hermite or Ornstein-Uhlenbeck semigroup.

The operators $T(\varepsilon)$, $0 \leq \varepsilon \leq 1$, satisfy a very important *hypercontractivity* property which is in relation with the integrability properties of Gaussian variables and chaos. This property indicates that for $1 < p < q < \infty$ and ε such that $|\varepsilon| \leq [(p-1)/(q-1)]^{1/2}$, $T(\varepsilon)$ maps $L_p(\gamma)$ into $L_q(\gamma)$ *with norm 1*, i.e. for any f in $L_p(\gamma)$,

(3.8) $$\|T(\varepsilon)f\|_q \leq \|f\|_p .$$

A function f in $L_2(\gamma)$ can be written as

$$f = \sum_{\underline{k}} H_{\underline{k}}f_{\underline{k}}$$

where $f_{\underline{k}} = \int f H_{\underline{k}}d\gamma$ and the sum runs over all $\underline{k}$'s in $\mathbb{N}^{(\mathbb{N})}$. We can also write

$$f = \sum_{d=0}^{\infty}\Big(\sum_{|\underline{k}|=d} H_{\underline{k}}f_{\underline{k}}\Big) = \sum_{d=0}^{\infty} Q_d f .$$

$Q_d f$ is named the chaos of degree d of f. Since $h_0 = 1$, $Q_0 f$ is simply the mean of f; $h_1(x) = x$, so chaos of degree 1 are Gaussian series $\sum_i g_i \alpha_i$. Chaos of degree 2 are of the type

$$\sum_{i \neq j} g_i g_j \alpha_{ij} + \sum_i (g_i^2 - 1)\alpha_i ,$$

etc. Now, the very definition of the Hermite operators $T(\varepsilon)$ shows that the action of $T(\varepsilon)$ on a chaos of order d is simply the multiplication by ε^d, that is $T(\varepsilon)Q_d f = \varepsilon^d Q_d f$. This observation together with (3.8) has some interesting consequence. If we let, in (3.8), $p = 2$, $q > 2$ and $\varepsilon = (q-1)^{-1/2}$, then we see that

$$\|Q_d f\|_q \leq (q-1)^{d/2}\|Q_d f\|_2 .$$

Of course, these inequalities imply some strong exponential integrability properties of $Q_d f$. This follows for example from the next easy lemma which is obtained via the expansion of the exponential function.

Lemma 3.7. *Let d be an integer and let Z be a positive random variable. The following are equivalent:*

(i) there is a constant K such that for any $p \geq 2$

$$\|Z\|_p \leq K p^{d/2} \|Z\|_2 \, ;$$

(ii) for some $\alpha > 0$,

$$\mathbb{E} \exp(\alpha Z^{2/d}) < \infty \, .$$

The integrability properties of Gaussian chaos which we obtain in this way extend to chaos with coefficients in a Banach space. In particular, in the case of series, this provides an alternate approach to the integrability of Gaussian variables presented in Section 3.1; note the right order of magnitude of $K_{p,2} = K\sqrt{p}$. In this paragraph, we shall actually be concerned with a more precise description of the tail behavior of Gaussian chaos similar to those obtained previously for chaos of the first order. This will be accomplished again with the tool of the Gaussian isoperimetric inequality. For simplicity, we only treat the case of chaos of order 2. (Let us mention that this reduction simplifies, by elementary symmetry considerations, several non-trivial polarization arguments that are necessary in general and which are thus somewhat hidden in our treatment; we refer to [Bo4], [Bo7] for details on these aspects.) With respect to the preceding description of chaos, we will study more precisely *homogeneous* Gaussian *polynomials* which basically correspond, for the degree 2, to convergent series of the type $\sum_{i,j} g_i g_j x_{ij}$ where (x_{ij}) is a sequence in a Banach space. Following the work of C. Borell [Bo4], [Bo7], since the constant 1 belongs to the closure of the homogeneous Gaussian polynomials of degree 2 (at least if the underlying Gaussian measure is infinite dimensional), the chaos described previously (as well as and their vector valued version) are limits, in probability of homogeneous Gaussian polynomials of the corresponding degree. This framework therefore includes, for the results which we will describe, the usual meaning of Gaussian or Wiener chaos. (For the second order, some aspects of this comparison can also be made apparent with simple symmetrization arguments.) We still use the terminology of chaos in this setting.

Now, let us describe the framework in which we will work. As in Section 3.1, the case of convergent quadratic sums is somewhat too restrictive. Let again B be a Banach space with $D = \{f_n, \, n \geq 1\}$ in the unit ball of B' such that $\|x\| = \sup_{f \in D} |f(x)|$ for all x in B. Following the reduction to the canonical Gaussian distribution on $\mathbb{R}^{\mathbb{N}}$, we say that a random variable X with values in B is a Gaussian chaos of order 2 if, for each n, there exists a bilinear form

$$Q_n(x, x') = \sum_{i,j} x_i x'_j a^n_{ij}$$

on $\mathbb{R}^{\mathbb{N}} \times \mathbb{R}^{\mathbb{N}}$ such that the sequence $(f_n(X))$ has the same distribution as $(\sum_{i,j} g_i g_j a^n_{ij})$ where we recall that (g_i) is the canonical orthogaussian sequence. Therefore, for each n, $\sum_{i,j} g_i g_j a^n_{ij}$ converges almost surely (or only in probability). Furthermore, if we set, by analogy, $\|Q(x, x)\| = \sup_{n \geq 1} |Q_n(x, x)|$, $x \in \mathbb{R}^{\mathbb{N}}$, we are reduced as usual to the study of the tail behavior and integrability properties of $\|Q\|$ under the canonical Gaussian product measure γ on $\mathbb{R}^{\mathbb{N}}$.

To measure the size of the tail $\mathbb{P}\{\|X\| > t\}$ we use, as in the preceding section, several parameters of the distribution of X. First, consider the "decoupled" symmetric chaos Y associated to X defined in distribution by

$$f_n(Y) = \sum_{i,j} g_i g'_j b^n_{ij}, \quad n \geq 1$$

where $b^n_{ij} = a^n_{ij} + a^n_{ji}$ and where (g'_j) is an independent copy of (g_i). According to our usual notation, we denote below by $\mathbb{E}'$, $\mathbb{P}'$ partial expectation and probability with respect to (g'_j). Let then M and m be such that

$$\mathbb{P}\{\|X\| \leq M\} \geq 3/4 \quad \text{and} \quad \mathbb{P}\{\sup_{f \in D} (\mathbb{E}' f^2(Y))^{1/2} \leq m\} \geq 3/4.$$

Moreover, set

$$\sigma = \sigma(X) = \sup_{|h| \leq 1} \|Q(h, h)\|.$$

If we recall the situation for Gaussian variables in the previous section, then we see that σ and M correspond to the weak and strong parameters respectively. The new and important parameter m appears as some intermediate quantity involving both the weak and strong topologies. Let us show that these parameters are actually well defined. It will suffice to show that the decoupled chaos Y exists. The key observation is that, for every $t > 0$,

$$(3.9) \qquad \mathbb{P}\{\|Y\| > t\} \leq 2\mathbb{P}\{\|X\| > t/2\sqrt{2}\}.$$

We establish this inequality for finite sums and a norm given by a finite supremum; then, a limiting argument shows that the quadratic sums defining Y converge as soon as the corresponding ones for X do, and, by increasing the finite supremum to the norm, that (3.9) is satisfied. We reduce to $(\mathbb{R}^{\mathbb{N}}, \gamma)$ and recall that $\|Q(x, x)\| = \sup_{n \geq 1} |Q_n(x, x)|$. Let, on $\mathbb{R}^{\mathbb{N}} \times \mathbb{R}^{\mathbb{N}}$,

$$\widehat{Q}_n(x, x') = \sum_{i,j} x_i x'_j b^n_{ij}$$

and set, with some further abuse of notation, $\|\widehat{Q}(x, x')\| = \sup_{n \geq 1} |\widehat{Q}_n(x, x')|$. Then, simply note that for x, x' in $\mathbb{R}^{\mathbb{N}}$,

$$2\widehat{Q}_n(x, x') = Q_n(x + x', x + x') - Q_n(x - x', x - x').$$

Furthermore, $x + x'$ and $x - x'$ have both the same distribution under $d\gamma(x)d\gamma(x')$ than $\sqrt{2}x$ under $d\gamma(x)$. From this and the preceding comments, it follows that Y is well defined and that (3.9) is satisfied.

It might be noteworthy to remark that (3.9) can essentially be reversed. Since the couple $((x+x')/\sqrt{2}, (x-x')/\sqrt{2})$ has the same distribution as (x, x'), one easily verifies that for all $t > 0$,

$$\mathbb{P}\{\|X - X'\| > t\} \leq 3\mathbb{P}\{\|Y\| > t/3\}$$

where X' is an independent copy of X. If the diagonal terms of X are all zero, we see by Jensen's inequality that all moments of X and Y are equivalent. In particular, it follows from this observation and the subsequent arguments that if X is a *real valued* chaos (i.e. D is reduced to one point) with all its diagonal terms equal to zero, then the parameters M and m are equivalent (at least if M is large enough, see below), and are equivalent to $\|X\|_2$ and $\|Y\|_2$. We will use this observation later on in Chapter 11, Section 11.3.

Let $\widehat{M}$ be such that $\gamma \otimes \gamma((x, x'); \|\widehat{Q}(x, x')\| \leq \widehat{M}) \geq 7/8$. By Fubini's theorem, $\gamma(A) \geq 3/4$ where

$$A = \left\{x; \gamma(x'; \|\widehat{Q}(x, x')\| \leq \widehat{M}) \geq 1/2\right\}.$$

Conditionally on x, $\|\widehat{Q}(x, x')\|$ is the norm of a Gaussian variable in x'. If $x \in A$, $\widehat{M}$ is larger than the median of $\|\widehat{Q}(x, x')\|$ and thus, by what was observed early in the preceding section, the supremum of the weak variances is less than $2\widehat{M}$. But this supremum is simply

$$\sup_{n \geq 1}\left(\int \widehat{Q}_n(x, x')^2 d\gamma(x')\right)^{1/2}.$$

Hence, we may simply take $m = 2\widehat{M}$ which is therefore well defined and finite. (Notice, for later purposes, that if M is chosen to satisfy $\mathbb{P}\{\|X\| \leq M\} \geq 15/16$, we can take, by (3.9), $\widehat{M} = 2\sqrt{2}M$, and thus also $m = 4\sqrt{2}M$.)

Concerning σ, for every $k \in \ell_2$, $|k| \leq 1$, $\gamma(x; \|\widehat{Q}(x, k)\| \leq m) \geq 3/4 \geq 1/2$. Hence, by the same reasoning as before,

$$\sup_{|h| \leq 1}\|\widehat{Q}(h, k)\| = \sup_{n \geq 1}\left(\int \widehat{Q}_n(x, k)^2 d\gamma(x)\right)^{1/2} \leq 2m.$$

Therefore,

$$\sup_{|h|, |k| \leq 1}\|\widehat{Q}(h, k)\| \leq 2m.$$

But this supremum is easily seen to be bigger than 2σ so that $\sigma \leq m < \infty$.

As for the series in the previous paragraph, notice that it is usually easier to work with expectations and moments rather than with medians or quantiles. Actually, by independence and the results of Section 3.1 it is not difficult to see that $\mathbb{E}\|Y\|^2 < \infty$ (actually $\mathbb{E}\|Y\|^p < \infty$ for every p) and that we have the following hierarchy in the parameters:

$$\left(\mathbb{E}\|Y\|^2\right)^{1/2} \geq \left(\mathbb{E}\sup_{f\in D}\mathbb{E}'f^2(Y)\right)^{1/2} \geq 2\sigma\,.$$

After these long preliminaries to properly describe the various parameters which we are using, we now state and prove the lemma which describes the tail $\mathbb{P}\{\|X\| > t\}$ of a Gaussian chaos X in terms of these parameters. We shall not be interested here in concentration properties; some, however, can be obtained at the expense of some complications. Recall, also that for *real valued* chaos, the parameters M and m are equivalent so that the inequality of the lemma may be formulated only in terms of M and σ. This will be used in Section 11.3.

Lemma 3.8. *Let X be a Gaussian chaos of order 2 as just defined with corresponding parameters M, m and σ. Then, for every $t > 0$,*

$$\mathbb{P}\{\|X\| > M + mt + \sigma t^2\} \leq \exp(-t^2/2)\,.$$

Proof. It is a simple consequence of Theorem 1.2. We use the preceding notation. Let

$$A_1 = \{x;\ \|Q(x,x)\| \leq M\}\,,$$
$$A_2 = \{x;\ \sup_{|k|\leq 1}\|\widehat{Q}(x,k)\| \leq m\}$$

and set $A = A_1 \cap A_2$ so that, by the definitions of M and m, $\gamma(A) \geq 1/2$. By Theorem 1.2, more precisely (1.3), for any $t > 0$, $\gamma_*(A_t) \geq \Phi(t)$. Now, if $x \in A_t$, $x = a + th$ where $a \in A$, $|h| \leq 1$. Thus, for every n,

$$Q_n(x,x) = Q_n(a,a) + t\widehat{Q}_n(a,h) + t^2 Q_n(h,h)$$

and therefore

$$A_t \subset \{x;\ \|Q(x,x)\| \leq M + tm + t^2\sigma\}\,.$$

The Lemma 3.8 is established.

The next corollary is the qualitative result drawn from the preceding lemma concerning the integrability and tail behavior of a Gaussian chaos. It corresponds to Corollary 3.2 in the first section.

Corollary 3.9. *Let X be a Gaussian chaos of order 2 with corresponding $\sigma = \sigma(X)$. Then*

$$\lim_{t\to\infty}\frac{1}{t}\log\mathbb{P}\{\|X\| > t\} = -\frac{1}{2\sigma}$$

or, equivalently,

$$\mathbb{E}\exp\left(\frac{1}{2\alpha}\|X\|\right) < \infty \quad \textit{if and only if}\quad \alpha > \sigma\,.$$

Furthermore, all the moments of X are equivalent.

Proof. That the limit is $\leq -1/2\sigma$ follows from Lemma 3.8; this also implies that $\mathbb{E}\exp(\|X\|/2\alpha) < \infty$ for $\alpha > \sigma$. To prove the converse assertion, let $\varepsilon > 0$ and choose $|h| \leq 1$ such that $\|Q(h,h)\| \geq \sigma - \varepsilon$. Given this $h = (h_i)_{i\geq 1}$, there exists an orthonormal basis $(h^i)_{i\geq 1}$ of ℓ_2 such that $h_1^i = h_i$ for every $i \geq 1$. By the rotational invariance of Gaussian measures, the distribution of $y = (\langle x, h^i\rangle)$ under γ is the same as the distribution of x. If we then set $\widetilde{y} = (\langle \widetilde{x}, h^i\rangle)$ where $\widetilde{x} = (0, x_2, x_3, \ldots)$, we can write $y = x_1 h + \widetilde{y}$ and x_1 and $\widetilde{y}$ are independent. Since, for each n,

$$Q_n(y,y) = x_1^2 Q_n(h,h) + x_1 \widehat{Q}_n(\widetilde{y}, h) + Q_n(\widetilde{y}, \widetilde{y})$$

(where $\widehat{Q}_n$, $\widehat{Q}$ were introduced prior to Lemma 3.8) and since $x_1 h - \widetilde{y}$ is distributed as y, a simple symmetry argument shows that one can find M such that

$$\gamma\big(x;\, \|Q(\widetilde{y}, \widetilde{y})\| \leq M\,,\ \|\widehat{Q}(\widetilde{y}, h)\| \leq M\big) \geq 1/2\,.$$

Then, we deduce from Fubini's theorem that for every $t > 0$

$$\begin{aligned}
\gamma\big(x;\, \|Q(x,x)\| > t\big) &= \gamma\big(x;\, \|Q(y,y)\| > t\big) \\
&\geq \tfrac{1}{2}\gamma\big(x;\, x_1^2(\sigma - \varepsilon) > t + |x_1|M + M\big)\,.
\end{aligned}$$

The proof of the first claim in Corollary 3.9 is then easily completed. That $\mathbb{E}\exp(\|X\|/2\sigma) = \infty$ can be established in the same way.

We have seen before Lemma 3.8 that if M is chosen such that $\mathbb{P}\{\|X\| \leq M\} \geq 15/16$, then we can take $m \leq 4\sqrt{2}M$ and σ satisfies $\sigma \leq 4\sqrt{2}M$. Hence, if $p > 0$, integrating the inequality of Lemma 3.8 immediately yields

$$\|X\|_p \leq K_p M$$

for some constant K_p depending on p only. Given $q > 0$, simply take $M = (16\mathbb{E}\|X\|^q)^{1/q}$ from which the equivalence of the moments of X follows. Note that K_p is of the order of p when $p \to \infty$, in accordance with what we observed at the beginning with the hypercontractive estimates. The proof of Corollary 3.9 is thus complete.

We conclude this section with a refinement of the previous corollary along the lines of what we obtained in Theorem 3.3. Let X be, as before, a Gaussian chaos variable of order 2 with values in B and recall the symmetric decoupled chaos Y associated to X. Then set

$$\tau = \tau(X) = \inf\big\{\lambda > 0;\, \mathbb{P}\big\{\sup_{f\in D} \big(\mathbb{E}' f^2(Y)\big)^{1/2} \leq \lambda\big\} > 0\big\}\,.$$

We can state

Theorem 3.10. *Let X be a Gaussian chaos with corresponding $\sigma = \sigma(X)$ and $\tau = \tau(X)$. Then, for every $\tau' > \tau$,*

$$\mathbb{E}\,\exp\Big(\frac{1}{2}\Big(\Big(\frac{\|X\|}{\sigma}\Big)^{1/2} - \frac{\tau'}{2\sigma}\Big)^2\Big) < \infty\,.$$

Proof. We show that for every $\tau' > \tau$ there exist an integer N and a positive number t_0 such that for every $t \geq t_0$

$$(3.10) \qquad \mathbb{P}\{\|X\| > \tau't + \sigma t^2\} \leq \mathbb{P}\left\{\left(\sum_{i=1}^{N} g_i^2\right)^{1/2} > t\right\}.$$

This suffices to establish the theorem. Indeed, if $\tau' > \tau'' > \tau$, let $\varepsilon = (\tau' - \tau'')/2\sigma > 0$. Then, applying (3.10) to τ'', for some integer N,

$$\int_{\{(\frac{\|X\|}{\sigma})^{1/2} > \frac{\tau'}{2\sigma}\}} \exp\left(\frac{1}{2}\left(\left(\frac{\|X\|}{\sigma}\right)^{1/2} - \frac{\tau'}{2\sigma}\right)^2\right) d\mathbb{P}$$

$$\leq \exp\left(\frac{t_0^2}{2}\right) + \int_{t_0}^{\infty} \mathbb{P}\left\{\left(\frac{\|X\|}{\sigma}\right)^{1/2} > \frac{\tau'}{2\sigma} + t\right\} t \exp\left(\frac{t^2}{2}\right) dt$$

$$\leq \exp\left(\frac{t_0^2}{2}\right) + \int_{t_0}^{\infty} \mathbb{P}\left\{\left(\frac{\|X\|}{\sigma}\right)^{1/2} > \frac{\tau''}{2\sigma} + (t+\varepsilon)\right\} t \exp\left(\frac{t^2}{2}\right) dt$$

$$\leq \exp\left(\frac{t_0^2}{2}\right) + \int_{t_0}^{\infty} \mathbb{P}\{\|X\| > \tau''(t+\varepsilon) + \sigma(t+\varepsilon)^2\} t \exp\left(\frac{t^2}{2}\right) dt$$

$$\leq \exp\left(\frac{t_0^2}{2}\right) + K(N) \int_0^{\infty} (1+\varepsilon+t)^{N+1} \exp\left(\frac{t^2}{2} - \frac{(t+\varepsilon)^2}{2}\right) dt < \infty.$$

Let us therefore prove (3.10). Recall the notation in the proof of Lemma 3.4. Given an integer N, a point x in $\mathbb{R}^{\mathbb{N}}$ is decomposed in (y,z) with $y \in \mathbb{R}^N$ and $z \in \mathbb{R}^{]N,\infty[}$. Furthermore, $\gamma = \gamma_N \otimes \gamma_{]N,\infty[}$ and if A is a Borel set in $\mathbb{R}^{\mathbb{N}}$, we set $B = \{z \in \mathbb{R}^{]N,\infty[}; (0,z) \in A\}$.

If $\tau' > \tau$, $\gamma(A_2) > 0$ where

$$A_2 = \left\{x; \sup_{|k| \leq 1} \|\widehat{Q}(x,k)\| \leq \tau'\right\}$$

and where $\widehat{Q}$ is the quadratic form associated to the decoupled chaos Y (cf. the proof of Lemma 3.8). Choose M large enough such that if

$$A_1 = \{x; \|Q(x,x)\| \leq M\},$$

and $A_3 = A_1 \cap A_2$, then $\gamma(A_3) > 0$. There exists an integer ℓ such that if

$$A_3' = \left\{x; \max_{n \leq \ell} |Q_n(x,x)| \leq M, \max_{n \leq \ell} \sup_{|k| \leq 1} |\widehat{Q}_n(x,k)| \leq \tau'\right\},$$

then $\gamma(A_3') < 4\gamma(A_3)/3$. By a simple approximation, replacing M and τ' by $M+\varepsilon$ and $\tau'+\varepsilon$ if necessary, we may assume that the bilinear forms $Q_n, \widehat{Q}_n$, $n \leq \ell$, only depend on a finite number of coordinates. Hence, for some integer N, $A_3' = L \times \mathbb{R}^{]N,\infty[}$ with $L \subset \mathbb{R}^N$. Then, by Fubini's theorem, there exists y in $\mathbb{R}^N$ such that

$$\gamma_{]N,\infty[}\big(z; (y,z) \in A_3\big) \geq 3/4.$$

By symmetry, we also have

$$\eta_{N,\infty[}\big(z;\,(y,-z)\in A_3\big)\ge 3/4\,.$$

The intersection of these two sets of measure larger than $3/4$ has therefore measure larger than or equal to $1/2$. Let z belong to this intersection. By convexity,

$$\sup_{|k|\le 1}\|\widehat{Q}((0,z),k)\|\le \tau'\,.$$

Moreover, since

$$\|Q((y,z),(y,z))\|\,,\ \|Q((y,-z),(y,-z))\|\le M\,,$$

we get, by summing,

$$\|Q((0,z),(0,z))\|\le M+\sup_{n\ge 1}\Big|\sum_{i,j=1}^{N} y_i y_j a_{ij}^n\Big|\le M+\sigma|y|^2\,.$$

Hence, if we set $M'=M+\sigma|y|^2$ and

$$A=\big\{x;\ \|Q(x,x)\|\le M',\ \sup_{|k|\le 1}\|\widehat{Q}(x,k)\|\le \tau'\big\}\,,$$

it follows that $B=\{z\in\mathbb{R}^{]N,\infty[};\ (0,z)\in A\}$ satisfies $\eta_{N,\infty[}(B)\ge 1/2$. We are then in a position to apply Lemma 3.5 from which we get that, for every $t>0$,

$$\gamma^*(x;\ x\notin A_t)\le \gamma_{N+1}(y';\ |y'|>t)\,.$$

Now it has simply to be observed that

$$A_t\subset\big\{x;\ \|Q(x,x)\|\le M'+t\tau'+t^2\sigma\big\}\,.$$

Indeed, if $x\in A_t$, $x=a+th$ with $a\in A$, $|h|\le 1$, then

$$\|Q(x,x)\|\le \|Q(a,a)\|+t\|\widehat{Q}(a,h)\|+t^2\|Q(h,h)\|\,.$$

(3.10) then easily follows from the preceding, working with $\tau'>\tau$ and t large enough. The proof of Theorem 3.10 is complete.

To conclude, let us briefly mention the corresponding results for the chaos of degree $d>2$. If X is such a chaos, and if σ is defined analogously, Corollary 3.9 reads:

$$\lim_{t\to\infty}\frac{1}{t^{2/d}}\log\mathbb{P}\{\|X\|>t\}=-\frac{1}{2\sigma^{2/d}}\,.$$

Theorem 3.10 is somewhat more difficult to translate. τ is defined appropriately from the associated d-decoupled symmetric chaos for which one takes weak moments on $d-1$ coordinates and the strong parameter on the remaining one. We then get that, for all $\tau'>\tau$,

$$\mathbb{E}\exp\left(\frac{1}{2}\left(\Big(\frac{\|X\|}{\sigma}\Big)^{1/d}-\frac{\tau'}{d\sigma}\right)^2\right)<\infty\,.$$

One might possibly imagine further refinements involving $d+1$ parameters.

3.3 Comparison Theorems

In the last part of this chapter, we investigate the Gaussian comparison theorems which, together with integrability, are very important and useful tools in Probability in Banach spaces. These results, which may be considered as geometrical, first step in with the so-called Slepian's lemma of which the further results are variations. Here, we present some of these statements in the scope of their further applications.

Assume that we are given two Gaussian random vectors $X = (X_1, \ldots, X_N)$ and $Y = (Y_1, \ldots, Y_N)$ in $\mathbb{R}^N$. In order to describe the question which we would like to study, let us assume first, as an example, that the covariance structure of X dominates that of Y in the (strong) sense that for every α in $\mathbb{R}^N$

$$\mathbb{E}\langle \alpha, Y \rangle^2 \leq \mathbb{E}\langle \alpha, X \rangle^2 \, .$$

Then, for any convex set C in $\mathbb{R}^N$

(3.11) $$\mathbb{P}\{Y \notin C\} \leq 2\mathbb{P}\{X \notin C\} \, .$$

Indeed, we may of course assume for the proof that X and Y are independent. If Z is a Gaussian variable independent from Y with covariance $\mathbb{E}Z_i Z_j = \mathbb{E}X_i X_j - \mathbb{E}Y_i Y_j$ (which is positive definite from the assumption), then X has the same distribution as $Y + Z$. By independence and symmetry, X has also the same distribution as $Y - Z$. Hence, by convexity,

$$\mathbb{P}\{Y \notin C\} = \mathbb{P}\{\tfrac{1}{2}[(Y + Z) + (Y - Z)] \notin C\}$$
$$\leq \mathbb{P}\{Y + Z \notin C\} + \mathbb{P}\{Y - Z \notin C\}$$

which is (3.11).

It should be noted that deeper tools can actually yield (3.11) with the numerical constant 1 instead of 2 for all convex and *symmetric* (with respect to the origin) sets C. Indeed, by Fubini's theorem,

$$\mathbb{P}\{X \in C\} = \mathbb{P}\{Y + Z \in C\} = \int \mathbb{P}\{Y + z \in C\} \, d\mathbb{P}_Z(z) \, .$$

Now, the concavity inequalities (1.1) or (1.2) and the symmetry ensure that, for every z in $\mathbb{R}^N$,

$$\mathbb{P}\{Y \in C - z\} \leq \mathbb{P}\{Y \in C\} \, ,$$

hence $\mathbb{P}\{X \in C\} \leq \mathbb{P}\{Y \in C\}$.

Typically (3.11) (or the preceding) is used to show a property such as the following one: if (X_n) is a sequence of Gaussian Radon random variables with values in a Banach space B such that $\mathbb{E}f^2(X_n) \leq \mathbb{E}f^2(X)$ for all f in B', all n and some Gaussian Radon variable X in B, then the sequence (X_n) is tight. Indeed, since X is Radon, for every $\varepsilon > 0$, there exists a compact set C which may be chosen to be convex such that $\mathbb{P}\{X \in C\} \geq 1 - \varepsilon$. Since B may be assumed to be separable, there is a sequence (f_k) in B such that $x \in K$

whenever $f_k(x) \leq 1$ for each k. The conclusion is then immediate from (3.11) which implies that $\mathbb{P}\{X_n \in C\} \geq 1 - 2\varepsilon$ for every n.

The first comparison property which we now present is the abstract statement from which we will deduce many interesting consequences. Its quite easy proof clearly describes the Gaussian properties which enter into these questions. The result is similar in nature to (3.11) but under weaker conditions on the covariance structures.

Theorem 3.11. *Let* $X = (X_1, \ldots, X_N)$ *and* $Y = (Y_1, \ldots, Y_N)$ *be Gaussian random variables in* $\mathbb{R}^N$. *Assume that*

$$\mathbb{E}X_i X_j \leq \mathbb{E}Y_i Y_j \quad \text{if} \quad (i,j) \in A,$$
$$\mathbb{E}X_i X_j \geq \mathbb{E}Y_i Y_j \quad \text{if} \quad (i,j) \in B,$$
$$\mathbb{E}X_i X_j = \mathbb{E}Y_i Y_j \quad \text{if} \quad (i,j) \notin A \cup B$$

where A *and* B *are subsets of* $\{1, \ldots, N\} \times \{1, \ldots, N\}$. *Let* f *be a function on* $\mathbb{R}^N$ *such that its second derivatives in the sense of distributions satisfy*

$$D_{ij}f \geq 0 \quad \text{if} \quad (i,j) \in A,$$
$$D_{ij}f \leq 0 \quad \text{if} \quad (i,j) \in B.$$

Then

$$\mathbb{E}f(X) \leq \mathbb{E}f(Y).$$

Proof. As before, we may assume that X and Y are independent. Set, for $t \in [0,1]$, $X(t) = (1-t)^{1/2}X + t^{1/2}Y$ and $\varphi(t) = \mathbb{E}f(X(t))$. We have

$$\varphi'(t) = \sum_{i=1}^{N} \mathbb{E}\big(D_i f\big(X(t)\big)X_i'(t)\big).$$

Now, let t and i be fixed. It is easily seen that, for every j,

$$\mathbb{E}X_j(t)X_i'(t) = \tfrac{1}{2}\mathbb{E}(Y_j Y_i - X_j X_i).$$

The hypotheses of the theorem then indicate that we can write

$$X_j(t) = \alpha_j X_i'(t) + Z_j$$

where Z_j is orthogonal to $X_i'(t)$ and $\alpha_j \geq 0$ if $(i,j) \in A$, $\alpha_j \leq 0$ if $(i,j) \in B$, $\alpha_j = 0$ if $(i,j) \notin A \cup B$. If we now examine $\mathbb{E}(D_i f(X(t))X_i'(t))$ as a function of the α_j's (for $(i,j) \in A \cup B$), the hypotheses on f show that this function is increasing in terms of those α_j's such that $(i,j) \in B$. But, by the orthogonality and therefore independence properties, this function vanishes when all the α_j's are 0, since

$$\mathbb{E}\big(D_i f(Z)X_i'(t)\big) = \mathbb{E}D_i f(Z)\mathbb{E}X_i'(t) = 0.$$

Hence $\mathbb{E}(D_i f(X(t))X_i'(t)) \geq 0$, $\varphi'(t) \geq 0$ and therefore $\varphi(0) \leq \varphi(1)$ which is the conclusion of the theorem.

As a first corollary, we state Slepian's lemma. It is simply obtained by taking in the theorem $A = \{(i,j);\ i \neq j\}$, $B = \emptyset$ and $f = I_G$ where G is a product of half-lines $] -\infty, \lambda_i]$ for which the hypotheses about the second derivatives are immediately verified. The claim concerning expectations of maxima in the next statement simply follows from the integration by parts formula

$$\mathbb{E}X = \int_0^\infty \mathbb{P}\{X > t\}\, dt - \int_0^\infty \mathbb{P}\{X < -t\}\, dt\,.$$

Corollary 3.12. *Let X and Y be Gaussian in $\mathbb{R}^N$ such that*

$$\begin{cases} \mathbb{E}X_i X_j \leq \mathbb{E}Y_i Y_j & \text{for all } i \neq j\,, \\ \mathbb{E}X_i^2 = \mathbb{E}Y_i^2 & \text{for all } i\,. \end{cases}$$

Then, for all real numbers λ_i, $i \leq N$,

$$\mathbb{P}\left\{ \bigcup_{i=1}^N (Y_i > \lambda_i) \right\} \leq \mathbb{P}\left\{ \bigcup_{i=1}^N (X_i > \lambda_i) \right\}.$$

In particular, by integration by parts,

$$\mathbb{E}\max_{i\leq N} Y_i \leq \mathbb{E}\max_{i\leq N} X_i\,.$$

The next corollary relies on a more elaborate use of Theorem 3.11. It will be useful in various applications both in this chapter and then in Chapters 9 and 15.

Corollary 3.13. *Let $X = (X_{i,j})$ and $Y = (Y_{i,j})$, $1 \leq i \leq n$, $1 \leq j \leq m$, be Gaussian random vectors such that*

$$\begin{cases} \mathbb{E}X_{i,j} X_{i,k} \leq \mathbb{E}Y_{i,j} Y_{i,k} & \text{for all } i,j,k\,, \\ \mathbb{E}X_{i,j} X_{\ell,k} \geq \mathbb{E}Y_{i,j} Y_{\ell,k} & \text{for all } i \neq \ell \text{ and } j,k\,, \\ \mathbb{E}X_{i,j}^2 = \mathbb{E}Y_{i,j}^2 & \text{for all } i,j\,. \end{cases}$$

Then, for all real numbers $\lambda_{i,j}$,

$$\mathbb{P}\left\{ \bigcap_{i=1}^n \bigcup_{j=1}^m (Y_{i,j} > \lambda_{i,j}) \right\} \leq \mathbb{P}\left\{ \bigcap_{i=1}^n \bigcup_{j=1}^m (X_{i,j} > \lambda_{i,j}) \right\}.$$

In particular,

$$\mathbb{E}\min_{i\leq n}\max_{j\leq m} Y_{i,j} \leq \mathbb{E}\min_{i\leq n}\max_{j\leq m} X_{i,j}\,.$$

Proof. Let $N = mn$. For $I \in \{1, \ldots, N\}$ let $i = i(I)$, $j = j(I)$ be the unique $1 \leq i \leq n$, $1 \leq j \leq m$ such that $I = m(i-1) + j$. Consider then X and Y as random vectors in $\mathbb{R}^N$ indexed in this way, i.e. $X_I = X_{i(I),j(I)}$. Let

$$A = \{(I, J); \ i(I) = i(J)\},$$
$$B = \{(I, J); \ i(I) \neq i(J)\}.$$

Then, the first set of hypotheses of Theorem 3.11 is fulfilled. Taking further f to be the indicator function of the set

$$\bigcup_{i=1}^{n} \bigcap_{I/i(I)=i} \{x \in \mathbb{R}^N; \ x_I > \lambda_{i,j(I)}\},$$

Theorem 3.11 implies the conclusion by taking complements.

In the preceding results the comparison was made possible by some assumptions on the respective covariances of the Gaussian vectors with, especially, conditions of equality on the "diagonal". In practice, it is often more convenient to deal with the corresponding L_2-metrics $\|X_i - X_j\|_2$ which do not require those special conditions on the diagonal. The next statement is a simple consequence of Corollary 3.12 in this direction.

Corollary 3.14. *Let* $X = (X_1, \ldots, X_N)$ *and* $Y = (Y_1, \ldots, Y_N)$ *be Gaussian variables in* $\mathbb{R}^N$ *such that for every* i, j

$$\mathbb{E}|Y_i - Y_j|^2 \leq \mathbb{E}|X_i - X_j|^2.$$

Then

$$\mathbb{E} \max_{i \leq N} Y_i \leq 2\mathbb{E} \max_{i \leq N} X_i.$$

Proof. Replacing $X = (X_i)_{i \leq N}$ by $(X_i - X_1)_{i \leq N}$ we may and do assume that $X_1 = 0$ and similarly that $Y_1 = 0$. Let $\sigma = \max_{i \leq N}(\mathbb{E}X_i^2)^{1/2}$ and consider the Gaussian random variables $\overline{X}$ and $\overline{Y}$ in $\mathbb{R}^N$ defined by

$$\overline{X}_i = X_i + g(\sigma^2 + \mathbb{E}Y_i^2 - \mathbb{E}X_i^2)^{1/2},$$
$$\overline{Y}_i = Y_i + g\sigma$$

where $i \leq N$ and where g is standard normal variable which is independent from X and Y. It is easily seen that

$$\mathbb{E}\overline{Y}_i^2 = \mathbb{E}\overline{X}_i^2 = \sigma^2 + \mathbb{E}Y_i^2$$

while

$$\mathbb{E}|\overline{Y}_i - \overline{Y}_j|^2 = \mathbb{E}|Y_i - Y_j|^2 \leq \mathbb{E}|X_i - X_j|^2 \leq \mathbb{E}|\overline{X}_i - \overline{X}_j|^2$$

so that $\mathbb{E}\overline{X}_i\overline{X}_j \leq \mathbb{E}\overline{Y}_i\overline{Y}_j$ for all $i \neq j$. Thus, the hypotheses of Corollary 3.12 are fulfilled and therefore

$$\mathbb{E}\max_{i\leq N}\overline{Y}_i \leq \mathbb{E}\max_{i\leq N}\overline{X}_i.$$

Now, clearly, $\mathbb{E}\max_{i\leq N}\overline{Y}_i = \mathbb{E}\max_{i\leq N}Y_i$ while

$$\mathbb{E}\max_{i\leq N}\overline{X}_i \leq \mathbb{E}\max_{i\leq N}X_i + \sigma\mathbb{E}g^+$$

where we have used that $\mathbb{E}Y_i^2 \leq \mathbb{E}X_i^2$ (since $X_1 = Y_1 = 0$). Now,

$$\begin{aligned}
\sigma = \max_{i\leq N}(\mathbb{E}X_i^2)^{1/2} &= \frac{1}{\mathbb{E}|g|}\max_{i\leq N}\mathbb{E}|X_i| \\
&\leq \frac{1}{\mathbb{E}|g|}\mathbb{E}\max_{i,j\leq N}|X_i - X_j| \\
&= \frac{2}{\mathbb{E}|g|}\mathbb{E}\max_{i\leq N}X_i = \frac{1}{\mathbb{E}g^+}\mathbb{E}\max_{i\leq N}X_i
\end{aligned}$$

where we have used again, in the first inequality, that $X_1 = 0$. This bound on σ and the preceding arguments finish the proof.

If $X = (X_1, \ldots, X_N)$ is Gaussian in $\mathbb{R}^N$, then, by symmetry, we have

$$\mathbb{E}\max_{i,j}|X_i - X_j| = \mathbb{E}\max_{i,j}(X_i - X_j) = 2\mathbb{E}\max_i X_i.$$

The comparison theorems usually deal with $\max_i X_i$ or $\max_{i,j}|X_i - X_j| = \max_{i,j}(X_i - X_j)$ rather than with $\max_i|X_i|$. Of course, for every $i_0 \leq N$,

$$\mathbb{E}\max_i X_i \leq \mathbb{E}\max_i|X_i| \leq \mathbb{E}|X_{i_0}| + \mathbb{E}\max_{i,j}|X_i - X_j|$$

$$\leq \mathbb{E}|X_{i_0}| + 2\mathbb{E}\max_i X_i.$$

In general however, the comparison results do not apply directly to $\mathbb{E}\max_i|X_i|$ (see however [Si1], [Si2]); take for example $Y_i = X_i + cg$ where g is a standard normal random variable which is independent of X in Corollary 3.14 and let c tend to infinity. Actually, one convenient feature of $\mathbb{E}\max_i X_i$ is that for any real valued mean zero random variable Z, $\mathbb{E}\max_i(X_i + Z) = \mathbb{E}\max_i X_i$.

The numerical constant 2 in the preceding corollary is not the best possible and can be improved to 1 with, however, a somewhat more complicated proof. On the other hand, under the hypotheses of Corollary 3.12, we also have that, for every $\lambda > 0$,

$$\mathbb{P}\Big\{\max_{i,j}|Y_i - Y_j| > \lambda\Big\} \leq 2\mathbb{P}\Big\{\max_i X_i > \frac{\lambda}{2}\Big\}.$$

Following the proof of Corollary 3.14, we can then obtain that if X and Y are Gaussian vectors in $\mathbb{R}^N$ such that for all i, j, $\|Y_i - Y_j\|_2 \leq \|X_i - X_j\|_2$, then, for $\lambda > 0$,

$$\mathbb{P}\Big\{\max_{i,j}|Y_i - Y_j| > \lambda\Big\} \leq 2\mathbb{P}\Big\{\max_{i,j}|X_i - X_j| > \frac{\lambda}{4}\Big\}$$

(3.12)

$$+ 2\mathbb{P}\Big\{\max_{i,j}(\mathbb{E}|X_i - X_j|^2)^{1/2}g^+ > \frac{\lambda}{4}\Big\}.$$

Of course, this inequality can be integrated by parts. This observation suggests that the functional $\max_{i,j}|X_i - X_j|$ is perhaps more natural in comparison theorems. The next result (due to X. Fernique [Fer4]), which we state without proof, completely answers these questions.

Theorem 3.15. *Let X and Y be Gaussian random vectors in $\mathbb{R}^N$ such that for every i, j*

$$\mathbb{E}|Y_i - Y_j|^2 \leq \mathbb{E}|X_i - X_j|^2.$$

Then, for every non-negative convex increasing function F on $\mathbb{R}_+$

$$\mathbb{E}F\big(\max_{i,j}|Y_i - Y_j|\big) \leq \mathbb{E}F\big(\max_{i,j}|X_i - X_j|\big).$$

There is also a version of Corollary 3.13 with conditions on L_2-distances. Again, the proof is more involved so that we only state the result. It is due to Y. Gordon [Gor1], [Gor3].

Theorem 3.16. *Let $X = (X_{i,j})$ and $Y = (Y_{i,j})$, $1 \leq i \leq n$, $1 \leq j \leq m$, be Gaussian random vectors such that*

$$\begin{cases} \mathbb{E}|Y_{i,j} - Y_{i,k}|^2 \leq \mathbb{E}|X_{i,j} - X_{i,k}|^2 & \text{for all } i, j, k, \\ \mathbb{E}|Y_{i,j} - Y_{\ell,k}|^2 \geq \mathbb{E}|Y_{i,j} - Y_{\ell,k}|^2 & \text{for all } i \neq \ell \text{ and } j, k. \end{cases}$$

Then

$$\mathbb{E}\min_{i\leq n}\max_{j\leq m} Y_{i,j} \leq \mathbb{E}\min_{i\leq n}\max_{j\leq m} X_{i,j}.$$

Among the various consequences of the preceding comparison properties, let us start with an elementary one which we only present in the scope of the next chapter where a similar result for Rademacher averages is obtained. We use Theorem 3.15 for convenience but a somewhat weaker result can be obtained with (3.12).

Corollary 3.17. *Let T be a bounded set in $\mathbb{R}^N$ and consider the Gaussian process indexed by T defined as $\sum_{i=1}^N g_i t_i$, $t = (t_1, \ldots, t_N) \in T \subset \mathbb{R}^N$. Further, let $\varphi_i : \mathbb{R} \to \mathbb{R}$, $i \leq N$, be contractions with $\varphi_i(0) = 0$. Then, for any non-negative convex increasing function F on $\mathbb{R}_+$,*

$$\mathbb{E}F\Big(\frac{1}{2}\sup_{t\in T}\Big|\sum_{i=1}^N g_i\varphi_i(t_i)\Big|\Big) \leq \mathbb{E}F\Big(2\sup_{t\in T}\Big|\sum_{i=1}^N g_i t_i\Big|\Big).$$

Proof. Let $u \in T$. We can write by convexity:

$$\mathbb{E}F\left(\frac{1}{2}\sup_{t \in T}\left|\sum_{i=1}^{N} g_i \varphi_i(t_i)\right|\right)$$

$$\leq \frac{1}{2}\mathbb{E}F\left(\sup_{t \in T}\left|\sum_{i=1}^{N} g_i(\varphi_i(t_i) - \varphi_i(u_i))\right|\right) + \frac{1}{2}\mathbb{E}F\left(\left|\sum_{i=1}^{N} g_i \varphi_i(u_i)\right|\right)$$

$$\leq \frac{1}{2}\mathbb{E}F\left(\sup_{s,t}\left|\sum_{i=1}^{N} g_i(\varphi_i(s_i) - \varphi_i(t_i))\right|\right) + \frac{1}{2}\mathbb{E}F\left(\left|\sum_{i=1}^{N} g_i u_i\right|\right)$$

where we have used that $|\varphi_i(u_i)| \leq |u_i|$ since $\varphi_i(0) = 0$. Now, by Theorem 3.15 (and a trivial approximation reducing to a finite supremum), the preceding quantity is further majorized by

$$\frac{1}{2}\mathbb{E}F\left(\sup_{s,t}\left|\sum_{i=1}^{N} g_i(s_i - t_i)\right|\right) + \frac{1}{2}\mathbb{E}F\left(\left|\sum_{i=1}^{N} g_i u_i\right|\right)$$

since by contraction, for every s, t,

$$\sum_{i=1}^{N} |\varphi_i(s_i) - \varphi_i(t_i)|^2 \leq \sum_{i=1}^{N} |s_i - t_i|^2 .$$

Corollary 3.17 is therefore established.

A most important consequence of the comparison theorems is the so-called Sudakov minoration. We shall come back to it in Chapter 12 when we investigate the regularity of Gaussian processes but it is fruitful to already record it at this stage. To introduce it, let us first observe the following easy facts. Let $X = (X_1, \ldots, X_N)$ be Gaussian in $\mathbb{R}^N$. Then

$$(3.13) \qquad \mathbb{E}\max_{i \leq N} X_i \leq 3(\log N)^{1/2} \max_{i \leq N}(\mathbb{E}X_i^2)^{1/2} .$$

Indeed, assume by homogeneity that $\max_{i \leq N} \mathbb{E}X_i^2 \leq 1$; then, for every $\delta \geq 0$, we obtain by an integration by parts,

$$\mathbb{E}\max_{i \leq N} X_i \leq \mathbb{E}\max_{i \leq N}|X_i| \leq \delta + N \int_{\delta}^{\infty} \mathbb{P}\{|g| > t\}\, dt$$

$$\leq \delta + N\sqrt{\frac{\pi}{2}}\exp(-\delta^2/2)$$

where g is a standard normal variable. Then simply choose $\delta = (2\log N)^{1/2}$. Note the comparison with (3.6).

The preceding inequality is *two-sided* for the *canonical* Gaussian random vector $(g_1, \ldots, g_N)$ where we recall that g_i are independent standard normal variables. Namely, for some positive numerical constant K,

(3.14) $$K^{-1}(\log N)^{1/2} \leq \mathbb{E}\max_{i\leq N} g_i \leq K(\log N)^{1/2}.$$

Indeed, since $\mathbb{E}\max(g_1, g_2) \geq 1/3$ (for example), we may assume N to be large enough. Note that, by the independence and the identical distribution properties, for every $\delta > 0$,

$$\mathbb{E}\max_{i\leq N} |g_i| \geq \int_0^\delta \left[1 - (1 - \mathbb{P}\{|g| > t\})^N\right] dt$$

$$\geq \delta\left[1 - (1 - \mathbb{P}\{|g| > \delta\})^N\right].$$

Now,

$$\mathbb{P}\{|g| > \delta\} = \sqrt{\frac{2}{\pi}} \int_\delta^\infty \exp(-t^2/2)\, dt \geq \sqrt{\frac{2}{\pi}} \exp(-(\delta+1)^2/2).$$

Choose then for example $\delta = (\log N)^{1/2}$ (N large) so that $\mathbb{P}\{|g| > \delta\} \geq 1/N$ and hence

$$\mathbb{E}\max_{i\leq N} |g_i| \geq \delta\left[1 - \left(1 - \frac{1}{N}\right)^N\right] \geq \delta\left(1 - \frac{1}{e}\right).$$

Since $\mathbb{E}\max_{i\leq N} |g_i| \leq \mathbb{E}|g| + 2\mathbb{E}\max_{i\leq N} g_i$, this proves the lower bound in (3.14) since N is assumed to be large enough. The upper bound has been established before.

If (T, d) is a metric or pseudo-metric space (d need not separate the points of T), denote by $N(T, d; \varepsilon)$ the minimal number of open balls of radius $\varepsilon > 0$ in the metric d which are necessary to cover T (the results which we present are actually also valid with closed balls). Of course, $N(T, d; \varepsilon)$ need not be finite in which case we agree that $N(T, d; \varepsilon) = \infty$. $N(T, d; \varepsilon)$ is finite for each $\varepsilon > 0$ if and only if (T, d) is totally bounded. We refer to the numbers $N(T, d; \varepsilon)$ as the *entropy numbers*.

Let $X = (X_t)_{t\in T}$ be a Gaussian process indexed by a set T. As will be further and deeply investigated in Chapter 12, one fruitful way to analyze the regularity properties of the process X is to study the "geometric" properties of T with respect to the L_2-pseudo-metric d_X induced by X which is defined as

$$d_X(s, t) = \|X_s - X_t\|_2, \quad s, t \in T.$$

The next theorem is an estimate of the size of the entropy numbers $N(T, d_X; \varepsilon)$ for each $\varepsilon > 0$ in terms of the supremum of X. It is known as Sudakov's minoration. In the statement, we simply let

$$\mathbb{E}\sup_{t\in T} X_t = \sup\left\{\mathbb{E}\sup_{t\in F} X_t \,;\, F \text{ finite in } T\right\}.$$

Theorem 3.18. *Let $X = (X_t)_{t\in T}$ be a Gaussian process with L_2-metric d_X. Then, for each $\varepsilon > 0$,*

$$\varepsilon\left(\log N(T, d_X; \varepsilon)\right)^{1/2} \leq K\mathbb{E}\sup_{t\in T} X_t$$

where $K > 0$ is a numerical constant. In particular, if X is almost surely bounded, then (T, d_X) is totally bounded.

Proof. Let N be such that $N(T, d_X; \varepsilon) \geq N$. There exists $U \subset T$ with $\mathrm{Card}\, U = N$ and such that $d_X(u, v) > \varepsilon$ for all $u \neq v$ in U. Let $(g_u)_{u \in U}$ be standard normal independent variables and consider $X'_u = \frac{\varepsilon}{\sqrt{2}} g_u$, $u \in U$. Then, clearly, for all $u \neq v$,

$$\|X'_u - X'_v\|_2 = \varepsilon < d_X(u, v).$$

Therefore, by Corollary 3.14,

$$\mathbb{E} \sup_{u \in U} X'_u \leq 2 \mathbb{E} \sup_{u \in U} X_u.$$

If we now recall (3.14), then we see that

$$\mathbb{E} \sup_{u \in U} X'_u \geq K^{-1} \frac{\varepsilon}{\sqrt{2}} (\log \mathrm{Card}\, U)^{1/2} = K^{-1} \frac{\varepsilon}{\sqrt{2}} (\log N)^{1/2}.$$

The conclusion follows.

Theorem 3.18 admits a slight strengthening when the process is continuous.

Corollary 3.19. *If the Gaussian process $X = (X_t)_{t \in T}$ has a version with almost all bounded and continuous sample paths on (T, d_X), then*

$$\lim_{\varepsilon \to 0} \varepsilon \big(\log N(T, d_X; \varepsilon)\big)^{1/2} = 0.$$

Proof. We denote by X itself the bounded continuous (and therefore separable) version. By the integrability properties of Gaussian vectors and the compactness of (T, d_X) (since X is also bounded, Theorem 3.18),

$$\lim_{\delta \to 0} \mathbb{E} \sup_{d_X(s,t) < \delta} |X_s - X_t| = 0.$$

For every $\eta > 0$, let $\delta > 0$ be small enough such that

$$\mathbb{E} \sup_{d_X(s,t) < \delta} |X_s - X_t| < \eta.$$

Let A be finite in T such that the balls of radius δ with centers in A cover T (such an A exists by Theorem 3.18). Let $\varepsilon > 0$. By Theorem 3.18, for every s in A, there exists $A_s \subset T$ satisfying

$$\varepsilon (\log \mathrm{Card}\, A_s)^{1/2} \leq K\eta$$

and such that if $t \in T$ and $d_X(s, t) < \delta$ there exists u in A_s with $d_X(u, t) < \varepsilon$. Then let $B = \bigcup_{s \in A} A_s$. Each point of T is within distance ε of an element of B; hence

$$N(T, d_X; \varepsilon) \le \operatorname{Card} B \le \operatorname{Card} A \max_s \operatorname{Card} A_s.$$

Therefore

$$\varepsilon \left(\log N(T, d_X; \varepsilon)\right)^{1/2} \le \varepsilon (\log \operatorname{Card} A)^{1/2} + K\eta.$$

Letting ε, and then η, tend to 0 concludes the proof.

Sudakov's minoration is the occasion of a short digression on some *dual* formulation. Let T be a convex body in $\mathbb{R}^N$, i.e. T is compact convex symmetric about the origin with non-empty interior in $\mathbb{R}^N$ (T is a Banach ball). Consider the Gaussian process $X = (X_t)_{t \in T}$ defined as

$$X_t = \sum_{i=1}^{N} g_i t_i, \quad t = (t_1, \ldots, t_N) \in T \subset \mathbb{R}^N.$$

Set

$$\ell(T) = \mathbb{E} \sup_{t \in T} |X_t| = \mathbb{E} \sup_{t \in T} \left| \sum_{i=1}^{N} g_i t_i \right| = \int \sup_{t \in T} |\langle x, t \rangle| d\gamma_N(x).$$

Of course, the L_2-metric of X is simply the Euclidean metric in $\mathbb{R}^N$. If A, B are sets in $\mathbb{R}^N$, denote by $N(A, B)$ the minimal number of translates of B by elements of A which are necessary to cover A. For example, $N(T, d_X; \varepsilon) = N(T, \varepsilon B_2)$ where B_2 is the Euclidean (open, for consistency) unit ball of $\mathbb{R}^N$. Sudakov's minoration indicates that the rate of growth of $N(T, \varepsilon B_2)$ when $\varepsilon \to 0$ is controlled by $\ell(T)$. It may be interesting to point out here that the dual version of this result is also true. Namely, the sup-norm of X controls similarly $N(B_2, \varepsilon T^\circ)$ where $T^\circ = \{x \in \mathbb{R}^N ; \langle x, y \rangle \le 1 \text{ for all } y \text{ in } T\}$ is the *polar* set of T; more precisely, for some numerical constant $K > 0$,

(3.15)
$$\sup_{\varepsilon > 0} \varepsilon \left(\log N(B_2, \varepsilon T^\circ)\right)^{1/2} \le K\ell(T).$$

The proof of (3.15) is rather simple, in fact simpler than the proof of Sudakov's minoration. Let $a = 2\ell(T)$. Then

$$\gamma_N(aT^\circ) = \mathbb{P}\{\sup_{t \in T} |X_t| \le a\} \ge 1/2$$

where γ_N is the canonical Gaussian measure on $\mathbb{R}^N$. Now let $\varepsilon > 0$ and n be such that

$$N(B_2, \varepsilon T^\circ) = N\left(\frac{a}{\varepsilon} B_2, aT^\circ b\right) \ge n.$$

There exist $z_1, \ldots, z_n$ in $\frac{a}{\varepsilon} B_2$ such that, for all $i \ne j$, $(z_i + aT^\circ) \cap (z_j + aT^\circ) = \emptyset$. Hence,

$$1 \ge \gamma_N \left(\bigcup_{i=1}^{n} (z_i + aT^\circ)\right) = \sum_{i=1}^{n} \gamma_N(z_i + aT^\circ).$$

For any z in $\mathbb{R}^N$, a change of variables indicates that

$$\gamma_N\left(z + aT^\circ\right) = \exp(-|z|^2/2) \int_{aT^\circ} \exp\langle z, x\rangle d\gamma_N(x),$$

and thus, by Jensen's inequality and the symmetry of T°,

$$\gamma_N\left(z + aT^\circ\right) \geq \exp(-|z|^2/2)\gamma_N\left(aT^\circ\right).$$

Therefore, since $z_i \in \frac{a}{\varepsilon}B_2$, $i \in N$, and since $\gamma_N(aT^\circ) \geq 1/2$, we finally get $2 \geq n\exp(-a^2/2\varepsilon^2)$ which is exactly (3.15).

It is noteworthy that Theorem 3.18 and its dual version (3.15) can easily be shown to be equivalent. Let us sketch how Sudakov's minoration can be deduced from (3.15) using a simple duality argument. First observe that for every $\varepsilon > 0$, $2T \cap (\frac{\varepsilon^2}{2}T^\circ) \subset \varepsilon B_2$. Indeed, if $t \in 2T$ and $t \in \frac{\varepsilon^2}{2}T^\circ$,

$$|t|^2 = \langle t, t\rangle \leq \|t\|_T\|t\|_{T^\circ} \leq 2 \cdot \frac{\varepsilon^2}{2} = \varepsilon^2,$$

where $\|\cdot\|_T$ (respectively $\|\cdot\|_{T^\circ}$) is the Banach norm (gauge) induced by T (T°). It follows that

$$N(T, \varepsilon B_2) \leq N\left(T, 2T \cap \left(\frac{\varepsilon^2}{2}T^\circ\right)\right) = N\left(T, \frac{\varepsilon^2}{2}T^\circ\right).$$

By homogeneity and elementary properties of entropy numbers,

$$N\left(T, \frac{\varepsilon^2}{2}T^\circ\right) \leq N(T, 2\varepsilon B_2)N\left(2\varepsilon B_2, \frac{\varepsilon^2}{2}T^\circ\right)$$

$$\leq N(T, 2\varepsilon B_2)N\left(B_2, \frac{\varepsilon}{4}T^\circ\right).$$

Thus, for every $\varepsilon > 0$,

$$\varepsilon\left(\log N(T, \varepsilon B_2)\right)^{1/2} \leq \varepsilon\left(\log N(T, 2\varepsilon B_2)\right)^{1/2} + 4M$$

where $M = \sup_{\varepsilon > 0} \varepsilon(\log N(B_2, \varepsilon T^\circ))^{1/2}$. Then one easily deduces that

$$(3.16) \qquad \sup_{\varepsilon > 0} \varepsilon\left(\log N(T, \varepsilon B_2)\right)^{1/2} \leq 8M.$$

The converse inequality may be shown similarly. By duality, we have $B_2 \subset \text{Conv}(\frac{\varepsilon}{2}T^\circ, \frac{2}{\varepsilon}T) \subset \frac{\varepsilon}{2}T^\circ + \frac{2}{\varepsilon}T$. Then

$$N(B_2, \varepsilon T^\circ) \leq N\left(\frac{\varepsilon}{2}T^\circ + \frac{2}{\varepsilon}T, \varepsilon T^\circ\right)$$

$$= N\left(\frac{2}{\varepsilon}T, \varepsilon T^\circ\right)$$

$$\leq N\left(\frac{2}{\varepsilon}T, \frac{1}{4}B_2\right)N\left(\frac{1}{4}B_2, \frac{\varepsilon}{2}T^\circ\right)$$

and we can conclude as above that

(3.17) $$\sup_{\varepsilon > 0} \varepsilon \left(\log N(B_2, \varepsilon T^\circ) \right)^{1/2} \le 4M'$$

where $M' = \sup_{\varepsilon > 0} \varepsilon (\log N(T, \varepsilon B_2))^{1/2}$.

The last application of the comparison theorems that we present concerns tensor products of Gaussian measures. For simplicity, we only deal with Radon random variables. Let E and F be two Banach spaces. If $x \in E$ and $y \in F$, $x \otimes y$ is the bilinear form on $E' \times F'$ which maps $(f, h) \in E' \times F'$ into $f(x)h(y)$. The linear tensor product $E \otimes F$ consists of all finite sums $u = \sum_i x_i \otimes y_i$ with $x_i \in E$, $y_i \in F$. On $E \otimes F$, we consider the injective tensor product norm

$$\|u\|_\vee = \sup \left\{ \left| \sum_i f(x_i)h(y_i) \right| ; \ \|f\| \le 1, \ \|h\| \le 1 \right\},$$

i.e. the norm of u as a bounded bilinear form on $E' \times F'$. The completion of $E \otimes F$ with respect to this norm is called the injective tensor product of E and F and is denoted by $E \check{\otimes} F$.

Consider now $X = \sum_i g_i x_i$ (resp. $Y = \sum_j g_j y_j$) a Gaussian random variable with values in E (resp. F). Here (g_i) denotes as usual an orthogaussian sequence. Let further (g_{ij}) be a doubly-indexed orthogaussian sequence. Given such convergent series X and Y with values in E and F respectively, one might wonder whether

$$\sum_{i,j} g_{ij} x_i \otimes y_j$$

converges almost surely in the injective tensor product space $E \check{\otimes} F$. This question has a positive answer and this is the conclusion of the next theorem. Recall that $\sigma(X) = \sup_{\|f\| \le 1} (\mathbb{E} f^2(X))^{1/2} = \sup_{\|f\| \le 1} (\sum_i f^2(x_i))^{1/2}$, $\sigma(Y)$ being defined similarly.

Theorem 3.20. *Let $X = \sum_i g_i x_i$ and $Y = \sum_j g_j y_j$ be convergent Gaussian series with values in E and F respectively and with corresponding $\sigma(X)$ and $\sigma(Y)$. Then $G = \sum_{i,j} g_{ij} x_i \otimes y_j$ converges almost surely in the injective tensor product $E \check{\otimes} F$ and the following inequality holds:*

$$\max \left(\sigma(X) \mathbb{E} \|Y\|, \ \sigma(Y) \mathbb{E} \|X\| \right) \le \mathbb{E} \|G\|_\vee \le \sigma(X) \mathbb{E} \|Y\| + \sigma(Y) \mathbb{E} \|X\|.$$

Proof. To prove that G converges, it is enough to establish the right hand side inequality of the theorem for finite sums and to use a limiting argument. By definition of the tensor product norm, the left hand side is easy; note that it indicates in the same way that the convergence of G implies that of X and Y. In the sequel, we therefore only deal with finite sequences (x_i) and (y_j). The idea of the proof is to compare G, considered as a Gaussian process indexed by the product of the unit balls of E' and F', to another process built as a kind of (independent) "sum" of X and Y. Consider namely the Gaussian process $\widetilde{G}$ indexed by $E' \times F'$:

$$\widetilde{G}(f,h) = \sigma(X)\sum_j g'_j h(y_j) + \left(\sum_j h^2(y_j)\right)^{1/2} \sum_i g_i f(x_i), \quad f \in E', \ h \in F'$$

where (g_i), (g'_j) are independent orthogaussian sequences. Actually, rather than to compare G to $\widetilde{G}$, it is convenient to replace G by

$$\widehat{G}(f,h) = \sum_{i,j} g_{ij} f(x_i) h(y_j) + \sigma(X)\left(\sum_j h^2(y_j)\right)^{1/2} g, \quad f \in E', \ h \in F'$$

where g is a standafd normal variable which is independent of (g_{ij}). Clearly, by Jensen's inequality and independence,

$$\mathbb{E}\|G\|_{\vee} \leq \mathbb{E} \sup_{\|f\|=1} \sup_{\|h\|=1} \widehat{G}(f,h).$$

The reason for the introduction of $\widehat{G}$ is that we will use Corollary 3.12 where we need some special information on the diagonal of the covariance structures, something which is given by $\widehat{G}$. Indeed, it is easily verified that for f, $f' \in E'$, h, $h' \in F'$,

$$\mathbb{E}\widehat{G}(f,h)\widehat{G}(f',h') - \mathbb{E}\widetilde{G}(f,h)\widetilde{G}(f',h')$$
$$= \left(\sigma(X)^2 - \sum_i f(x_i)f'(x_i)\right)$$
$$\times \left[\left(\sum_j h^2(y_j)\right)^{1/2}\left(\sum_j h'^2(y_j)\right)^{1/2} - \sum_j h(y_j)h'(y_j)\right].$$

Hence, this difference is always positive and is equal to 0 when $h = h'$. We are thus in a position to apply Corollary 3.12 to the Gaussian processes $\widehat{G}$ and $\widetilde{G}$. After an approximation and compactness argument, it implies

$$\mathbb{E} \sup_{\|h\|=1} \sup_{\|f\|=1} \widehat{G}(f,h) \leq \mathbb{E} \sup_{\|h\|=1} \sup_{\|f\|=1} \widetilde{G}(f,h).$$

To get the inequality of the theorem, we need simply note that

$$\mathbb{E} \sup_{\|h\|=1} \sup_{\|f\|=1} \widetilde{G}(f,h) \leq \sigma(X)\mathbb{E}\|Y\| + \sigma(Y)\mathbb{E}\|X\|.$$

Therefore, the proof of Theorem 3.20 is complete. Notice that Theorem 3.15 yields a somewhat simplified proof.

As a consequence of Corollary 3.13, we also have that

$$\mathbb{E} \inf_{\|h\|=1} \sup_{\|f\|=1} \widetilde{G}(f,h) \leq \mathbb{E} \inf_{\|h\|=1} \sup_{\|f\|=1} \widehat{G}(f,h).$$

Now, notice that

$$\mathbb{E} \inf_{\|h\|=1} \sup_{\|f\|=1} \tilde{G}(f,h) = \sigma(X)\mathbb{E} \inf_{\|h\|=1} h(Y) + \inf_{\|h\|=1}\left(\sum_j h^2(y_j)\right)^{1/2} \mathbb{E}\|X\|$$

so that we have the following lower bound:

(3.18)
$$\mathbb{E} \inf_{\|h\|=1} \sup_{\|f\|=1} \widehat{G}(f,h)$$
$$\geq \sigma(X)\mathbb{E} \inf_{\|h\|=1} h(Y) + \inf_{\|h\|=1}\left(\sum_j h^2(y_j)\right)^{1/2} \mathbb{E}\|X\| \, .$$

This inequality has some interesting consequences together with that in Theorem 3.20. One application is the following corollary which will be of interest in Chapter 9.

Corollary 3.21. *Let X be a Gaussian Radon random variable with values in a Banach space B. Let also (X_i) be independent copies of X. Then, for every N, if $\alpha = (\alpha_1, \ldots, \alpha_N)$ is a generic point in $\mathbb{R}^N$,*

$$\mathbb{E} \sup_{|\alpha|=1} \left\|\sum_{i=1}^N \alpha_i X_i\right\| \leq \mathbb{E}\|X\| + \sigma(X)\sqrt{N}$$

and

$$\mathbb{E} \inf_{|\alpha|=1} \left\|\sum_{i=1}^N \alpha_i X_i\right\| \geq \mathbb{E}\|X\| - \sigma(X)\sqrt{N} \, .$$

Proof. X can be represented as $X = \sum_i g_i x_i$ for some sequence (x_i) in B. Consider

$$Y = \sum_{j=1}^N g_j e_j$$

where (e_j) is the canonical basis of ℓ_2^N. Then Theorem 3.20 in the tensor space $B \bar{\otimes} \ell_2^N$ immediately yields

$$\mathbb{E} \sup_{|\alpha|=1} \left\|\sum_{i=1}^N \alpha_i X_i\right\| \leq \mathbb{E}\|X\| + \sigma(X)\mathbb{E}\left(\left(\sum_{j=1}^N g_j^2\right)^{1/2}\right)$$

and thus the first inequality of the corollary follows since obviously $\mathbb{E}\left((\sum_{j=1}^N g_j^2)^{1/2}\right) \leq \sqrt{N}$. For the second, use (3.18); indeed, in this case,

$$\mathbb{E} \inf_{|\alpha|=1} \sup_{\|f\|=1} \left(\sum_{i,j} g_{ij}\alpha_j f(x_i) + \sigma(X)\left(\sum_j \alpha_j^2\right)^{1/2} g\right)$$

$$= \mathbb{E} \inf_{|\alpha|=1} \sup_{\|f\|=1} \left(\sum_{j=1}^N \alpha_j f(X_j) + \sigma(X)g\right)$$

$$= \mathbb{E} \inf_{|\alpha|=1} \left\|\sum_{j=1}^N \alpha_j X_j\right\|$$

and thus

$$\mathbb{E} \inf_{|\alpha|=1} \left\| \sum_{j=1}^{N} \alpha_j X_j \right\| \geq \mathbb{E} \|X\| - \sigma(X) \mathbb{E} \left(\left(\sum_{j=1}^{N} g_j^2 \right)^{1/2} \right).$$

The proof is therefore complete.

Notes and References

Some general references about Gaussian processes and measures (for both this chapter and Chapters 11 and 12) are [Ne2], [B-C], [Fer4], [Ku], [Su4], [J-M3], [Fer9], [V-T-C], [Ad]. The interested reader may find there (completed with the papers [Bo3], [Ta1]) various topics about Gaussian measures on abstract spaces, such as zero-one laws, reproducing kernel Hilbert space, etc., not developed in this book.

The history of the integrability properties of norms of infinite dimensional Gaussian random vectors starts with the papers [L-S] and [Fer2]. Fernique's simple argument is the one leading to (3.4) and applies to the rather general setting of measurable seminorms. The proof by H. J. Landau and L. A. Shepp is isoperimetric and led eventually to the Gaussian isoperimetric inequality. A. V. Skorokhod [Sk2] had an argument to show that $\mathbb{E} \exp(\alpha\|X\|) < \infty$ (using the strong Markov property of Brownian motion); J. Hoffmann-Jørgensen indicates in [HJ3] a way to get from this partial conclusion the usual exponential squared integrability. Corollary 3.2 is due to M. B. Marcus and L. A. Shepp [M-S2]. Our description of the integrability properties and tail behavior is isoperimetric and follows C. Borell [Bo2] (and the exposition of [Eh4]). The concentration result in Lemma 3.1 is issued from Chapter 1. Theorem 3.3 was established in [Ta2] where examples describing its optimality are given. In particular, $\Phi^{-1}(\mathbb{P}\{\|X\| \leq t\})$ approaches its asymptote as slowly as one may wish. Lemma 3.5, in a slight improved form, was used in [Go]; see also [G-K1], [G-K2], [G-K3]. Theorem 3.3 has some interpretation in large deviations; indeed, the limit in Corollary 3.2

$$\lim_{t\to\infty} \frac{1}{t^2} \log \mathbb{P}\{\|X\| > t\} = -\frac{1}{2\sigma^2}$$

is a large deviation result for complements of balls centered at the origin. Theorem 3.3 improves this limit into

$$\lim_{t\to\infty} \left(\frac{1}{t} \log \mathbb{P}\{\|X\| > t\} + \frac{t}{2\sigma^2} \right) = 0$$

(for Radon variables) which appears as a "normal" deviation result for complements of balls; similar results for different sets night hold as well (on large deviations, see e.g. [Az], [Ja3], [De-St]). That $\tau = \tau(X) = 0$ for a Gaussian Radon measure was recorded in [D-HJ-S] and that τ is the unique jump of the distribution of $\|X\|$ is due to B. S. Tsirelson [Ts].

Homogeneous chaos were introduced by N. Wiener [Wie] and are presented, e.g. in [Ne2]. Their order of integrability was first investigated in [Schr] and [Var]. Hypercontractivity of the Hermite semigroup has been discovered by E. Nelson [Nel]. L. Gross [Gro] translated this property into a logarithmic Sobolev inequality and uses a two point inequality and the central limit theorem to provide an alternate proof (see also Chapter 4 and [Bec] for further deep results in Fourier Analysis along these lines). The relevance of hypercontractivity to the integrability of the Gaussian chaos (and its extension to the vector valued case) was noticed by C. Borell [Bo4], [Bo6]; however, the deep work [Bo4] develops the isoperimetric approach which we closely follow here (and that is further developed in [Bo7]). The introduction of decoupled chaos is motivated by [Kw4] (following [Bo6], [Bo7]). Theorem 3.10 is perhaps new.

Inequality (3.11) with its best constant 1 (for C symmetric) is due to T. W. Anderson [An]. Slepian's lemma appeared in [Sl]. Its geometric meaning makes it probably more ancient as was noted by several authors [Su1], [D-St], [Gr2]. In relation with this lemma, let us mention its "two-sided" analog studied by Z. Šidák [Si1], [Si2] which expresses that if $X = (X_1, \ldots, X_N)$ is a Gaussian vector in $\mathbb{R}^N$, for any positive numbers λ_i, $i \leq N$,

$$\mathbb{P}\left\{\bigcup_{i=1}^{N}(|X_i| \leq \lambda_i)\right\} \geq \prod_{i=1}^{N}\mathbb{P}\{|X_i| \leq \lambda_i\}.$$

We refer to [To] for more inequalities on Gaussian distributions in finite dimension. Slepian's lemma was first used in the study of Gaussian processes in [Su1], [M-S1] and [M-S2] where Corollary 3.14 is established. Theorem 3.15 was announced by V. N. Sudakov in [Su2] (see also [Su4]) and established in this form by X. Fernique [Fer4]; credit is also due to S. Chevet (unpublished). Y. Gordon [Gor1], [Gor2] discovered Corollary 3.13 and Theorem 3.16 motivated by Dvoretzky's theorem (cf. Chapter 9). [Gor3] contains a more general and simplified proof of Theorem 3.16 and some applications. Our exposition of the inequalities by Slepian and Gordon is based on Theorem 3.11 of J.-P. Kahane [Ka2]. Sudakov's minoration was observed in [Su1], [Su3]. Its dual version (3.15) appeared in the context of the local theory of Banach spaces and the duality of entropy numbers [P-TJ]. The consideration of $\ell(T)$ (with this notation) goes back to [L-P]. The simple proof of (3.15) presented here is due to the second author. [This proof was communicated in particular to the author of [Go] (where a probabilistic application is obtained) who gives a strickingly creative acknowledgement of the fact.] Further applications of the method are presented in [Ta19]. The equivalence between Sudakov's minoration and its dual version ((3.16) and (3.17)) is due to N. Tomczak-Jaegermann [TJ1] (and her argument actually shows a closer relationship between the entropy numbers and the dual entropy numbers, cf. [TJ1], [Ca]; this will partially be used in Section 15.5 below). Tensor products of Gaussian measures were initiated by S. Chevet [Ch1], [Ch2]; see also [Car]. The best constants in Theorem 3.20 and the inequality (3.18) follow from [Gor1], [Gor2] from where Corollary 3.21 is also taken.

4. Rademacher Averages

This chapter is devoted to Rademacher averages $\sum_i \varepsilon_i x_i$ with vector valued coefficients as a natural analog of the Gaussian averages $\sum_i g_i x_i$. The properties we examine are entirely similar to those investigated in the Gaussian case. In this way, we will see how isoperimetric methods can be used to yield strong integrability properties of convergent Rademacher series and chaos. This is studied in Sections 4.3 and 4.4. Some comparison results are also available in the form, for example, of a version of Sudakov's minoration presented in Section 4.5. However, we start in the first two sections with some basic facts about Rademacher averages with real coefficients as well as with the so-called contraction principle, a most valuable tool in Probability in Banach spaces.

Thus, we assume that we are given, on some probability space $(\Omega, \mathcal{A}, \mathbb{P})$, a sequence (ε_i) of independent random variables taking the values ± 1 with probability $1/2$, that is symmetric Bernoulli or Rademacher random variables. We usually call (ε_i) a *Rademacher sequence*. If (ε_i) is considered alone, one might take, as a concrete example, Ω to be the Cantor group $\{-1, +1\}^{\mathbb{N}}$, $\mathbb{P}$ its canonical product probability measure (Haar measure) $\mu = \left(\frac{1}{2}\delta_{-1} + \frac{1}{2}\delta_{+1}\right)^{\otimes \mathbb{N}}$ and ε_i the coordinate maps. We thus investigate finite or convergent sums $\sum_i \varepsilon_i x_i$ with vector valued coefficients x_i. As announced, the first paragraph is devoted to some preliminaries in the real case.

4.1 Real Rademacher Averages

If (α_i) is a sequence of real numbers, a trivial application of the three series theorem (or Lemma 4.2 below) indicates that the series $\sum_i \varepsilon_i \alpha_i$ converges almost surely (or in probability) if and only if $\sum_i \alpha_i^2 < \infty$. Actually, the sum $\sum_i \varepsilon_i \alpha_i$ has remarkable properties in connection with the sum of the squares $\sum_i \alpha_i^2$ and it is the purpose of this paragraph to recall some of these.

Since we will only be interested in estimates which easily extend to infinite sums, there is no loss in generality to assume, as is usual, that we deal with finite sequences (α_i), i.e. finitely many α_i's only are non-zero.

The first main observation is the classical *subgaussian* estimate which draws its name from the Gaussian type tail. We can obtain it as a consequence of Lemma 1.5 since $\sum_i \varepsilon_i \alpha_i$ clearly defines a mean zero sum of martingale differences, or directly by the same argument: indeed, given thus a finite sequence (α_i) of real numbers, for every $\lambda > 0$,

$$\mathbb{E}\exp\left(\lambda \sum_i \varepsilon_i \alpha_i\right) = \prod_i \mathbb{E}\exp(\lambda \varepsilon_i \alpha_i) \le \prod_i \exp\left(\frac{\lambda^2}{2}\alpha_i^2\right) = \exp\left(\frac{\lambda^2}{2}\sum_i \alpha_i^2\right).$$

Hence, by Chebyshev's inequality, for every $t > 0$,

$$(4.1) \qquad \mathbb{P}\left\{\left|\sum_i \varepsilon_i \alpha_i\right| > t\right\} \le 2\exp\left(-t^2/2\sum_i \alpha_i^2\right).$$

In particular, a convergent Rademacher series $\sum_i \varepsilon_i \alpha_i$ satisfies exponential squared integrability properties exactly as Gaussian variables.

This simple inequality (4.1) is extremely useful. Moreover, it is sharp in various instances and we have in particular the following converse which we record at this stage for further purposes: there is a numerical constant $K \ge 1$ such that if (α_i) and t satisfy $t \ge K(\sum_i \alpha_i^2)^{1/2}$ and $t\max_i |\alpha_i| \le K^{-1}\sum_i \alpha_i^2$, then

$$(4.2) \qquad \mathbb{P}\left\{\sum_i \varepsilon_i \alpha_i > t\right\} \ge \exp\left(-Kt^2/\sum_i \alpha_i^2\right).$$

This inequality, actually in a more precise form concerning the choice of the constants, can be deduced for example from the more general Kolmogorov minoration inequality given below as Lemma 8.1. However, let us give a direct proof of (4.2). Assume that $(\alpha_i)_{i\ge 1}$ is such that, by homogeneity, $\sum_i \alpha_i^2 = 1$ and that $t \ge 2$ and $|\alpha_i| \le 1/16t$ for each i. Define n_j, $j \le k$ ($n_0 = 0$), by

$$n_j = \inf\left\{n > n_{j-1};\ \sum_{i=n_{j-1}+1}^{n} \alpha_i^2 > \frac{1}{16^2 t^2}\right\}.$$

Since $\sum_i \alpha_i^2 = 1$, $k \le 16^2 t^2$. On the other hand, for each $j \le k$,

$$\sum_{i=n_{j-1}+1}^{n_j} \alpha_i^2 \le \sum_{i=n_{j-1}+1}^{n_j-1} \alpha_i^2 + \frac{1}{16^2 t^2} \le \frac{2}{16^2 t^2}$$

so that $k \ge 4 \cdot 16t^2$ since

$$\sum_{i>n_k} \alpha_i^2 \le \frac{1}{16^2 t^2} \le \frac{1}{2}.$$

Set $I_j = \{n_{j-1}+1,\ldots,n_j\}$, $1 \le j \le k$. By independence, we can write

$$\mathbb{P}\left\{\sum_i \varepsilon_i \alpha_i > t\right\} \ge \prod_{j\le k} \mathbb{P}\left\{\sum_{i\in I_j} \varepsilon_i \alpha_i > \frac{1}{4\cdot 16t}\right\}$$

$$\ge \prod_{j\le k} \mathbb{P}\left\{\sum_{i\in I_j} \varepsilon_i \alpha_i > \frac{1}{4}\left(\sum_{i\in I_j} \alpha_i^2\right)^{1/2}\right\}.$$

Now, (4.3) and Lemma 4.2 below indicate together that

$$\mathbb{P}\left\{ \sum_i \varepsilon_i \beta_i > \frac{1}{4} \left(\sum_i \beta_i^2 \right)^{1/2} \right\} \geq \frac{1}{16}.$$

It follows that

$$\mathbb{P}\left\{ \sum_i \varepsilon_i \alpha_i > t \right\} \geq \left(\frac{1}{16} \right)^k \geq \exp(-16^2 t^2 \log 16)$$

which is the desired result (with e.g. $K = 16^2 \log 16$).

The subgaussian inequality (4.1) can be used to yield a simple proof of the classical Khintchine inequalities.

Lemma 4.1. *For any* $0 < p < \infty$*, there exist positive finite constants* A_p *and* B_p *depending on* p *only such that for any finite sequence* (α_i) *of real numbers,*

$$A_p \left(\sum_i \alpha_i^2 \right)^{1/2} \leq \left\| \sum_i \varepsilon_i \alpha_i \right\|_p \leq B_p \left(\sum_i \alpha_i^2 \right)^{1/2}.$$

Proof. By homogeneity, assume that $\sum_i \alpha_i^2 = 1$. Then, by the integration by parts formula and (4.1),

$$\mathbb{E} \left| \sum_i \varepsilon_i \alpha_i \right|^p = \int_0^\infty \mathbb{P} \left\{ \left| \sum_i \varepsilon_i \alpha_i \right| > t \right\} dt^p$$

$$\leq 2 \int_0^\infty \exp(-t^2/2)\, dt^p = B_p^p.$$

For the left hand side inequality, it is enough, by Jensen's inequality, to consider the case $p < 2$. By means of Hölder's inequality, we get

$$1 = \mathbb{E} \left| \sum_i \varepsilon_i \alpha_i \right|^2 = \mathbb{E} \left(\left| \sum_i \varepsilon_i \alpha_i \right|^{2p/3} \left| \sum_i \varepsilon_i \alpha_i \right|^{2-2p/3} \right)$$

$$\leq \left(\mathbb{E} \left| \sum_i \varepsilon_i \alpha_i \right|^p \right)^{2/3} \left(\mathbb{E} \left| \sum_i \varepsilon_i \alpha_i \right|^{6-2p} \right)^{1/3}$$

$$\leq \left(\mathbb{E} \left| \sum_i \varepsilon_i \alpha_i \right|^p \right)^{2/3} B_{6-2p}^{2-2p/3}$$

from which the conclusion follows.

The best possible constants A_p and B_p in Khintchine's inequalities are known [Ha]. We retain from the preceding proof that $B_p \leq K\sqrt{p}$ $(p \geq 1)$ for some numerical constant K. We will use also the known fact [Sz] that $A_1 = 2^{-1/2}$ (in order to deal with a specific value), i.e.

(4.3) $$\left(\sum_i \alpha_i^2 \right)^{1/2} = \left\| \sum_i \varepsilon_i \alpha_i \right\|_2 \leq \sqrt{2} \left\| \sum_i \varepsilon_i \alpha_i \right\|_1.$$

Khintchine's inequalities show how the Rademacher sequence defines a basic unconditional sequence and spans ℓ_2 in the spaces L_p, $0 < p < \infty$. We could also add $p = 0$ in this claim as is shown by the simple (but useful) following lemma. The interest of this lemma goes beyond this application and it will be mentioned many times throughout this book.

Lemma 4.2. *Let Z be a positive random variable such that for some $q > p > 0$ and some constant C,*

$$\|Z\|_q \leq C\|Z\|_p\,.$$

Then, if $t > 0$ is such that $\mathbb{P}\{Z > t\} \leq (2C^p)^{q/(p-q)}$, we have

$$\|Z\|_p \leq 2^{1/p}t \quad and \quad \|Z\|_q \leq 2^{1/p}Ct\,.$$

Proof. By Hölder's inequality,

$$\mathbb{E}Z^p \leq t^p + \int_{\{Z>t\}} Z^p d\mathbb{P} \leq t^p + \|Z\|_q^p \big(\mathbb{P}\{Z > t\}\big)^{1-p/q} \leq 2t^p$$

where the last inequality is obtained from the choice of t.

Note, as a consequence of this lemma, that if (X_n) is a sequence of random variables (real or vector valued) such that for some $q > p > 0$ and some $C > 0$, $\|X_n\|_q \leq C\|X_n\|_p$ for every n, and if (X_n) converges in probability to some variable X, then, since $\sup_n \|X_n\|_q < \infty$ by Lemma 4.2, (X_n) also converges to X in $L_{q'}$ for all $q' < q$.

While the subspace generated by (ε_i) in L_p for $0 \leq p < \infty$ is ℓ_2, in L_∞ however, this subspace is isometric to ℓ_1. Indeed, for any finite sequence (α_i) of real numbers, there exists a choice of signs $\varepsilon_i = \pm 1$ such that $\varepsilon_i \alpha_i = |\alpha_i|$ for every i. Hence,

$$\left\|\sum_i \varepsilon_i \alpha_i\right\|_\infty = \sum_i |\alpha_i|\,.$$

Therefore, it might be of some interest to try to have an idea of the span of the Rademacher sequence in spaces "between" L_p ($p < \infty$) and L_∞. Among the possible intermediary spaces, we may consider Orlicz spaces with exponential rates. A Young function $\psi : \mathbb{R}_+ \to \mathbb{R}_+$ is convex, increasing with $\lim_{x\to\infty} \psi(x) = \infty$ and $\psi(0) = 0$. Denote by $L_\psi = L_\psi(\Omega, \mathcal{A}, \mathbb{P})$ the Orlicz space of all real valued random variables X (defined on $(\Omega, \mathcal{A}, \mathbb{P})$) such that $\mathbb{E}\psi(|X|/c) < \infty$ some $c > 0$. Equipped with the norm

$$\|X\|_\psi = \inf\{c > 0;\ \mathbb{E}\psi(|X|/c) \leq 1\}\,,$$

L_ψ defines a Banach space. If for example $\psi(x) = x^p$, $1 \leq p < \infty$, then $L_\psi = L_p$. We shall be interested in the exponential functions

$$\psi_q(x) = \exp(x^q) - 1, \quad 1 \leq q < \infty\,.$$

To handle the small convexity problem when $0 < q < 1$, we can let $\psi_q(x) = \exp(x^q) - 1$ for $x \geq x(q)$ large enough and take ψ_q to be linear on $[0, x(q)]$.

The first observation is that (ε_i) still spans a subspace isomorphic to ℓ_2 in L_{ψ_q} whenever $q \leq 2$. We assume that $1 \leq q \leq 2$ for simplicity, but the proof is similar when $0 < q < 1$. Letting $X = \sum_i \varepsilon_i \alpha_i$ where (α_i) is a finite sequence of real numbers such that, by homogeneity, $\sum_i \alpha_i^2 = 1$, we have, by the integration by parts formula and (4.1),

$$\mathbb{E}\,\psi_q\left(\frac{|X|}{c}\right) = \int_0^\infty \mathbb{P}\{|X| > c\,t\}\,d(e^{t^q} - 1)$$

$$\leq 2q \int_0^\infty \exp\left(-\frac{c^2 t^2}{2} + t^q\right) t^{q-1} dt\,.$$

Hence, when $q \leq 2$ and $c \geq B'_q$ is large enough, $\mathbb{E}\,\psi_q(|X|/c) \leq 1$ so that $\|X\|_{\psi_q} \leq B'_q$. On the other hand, since $e^x - 1 \geq x$,

$$\mathbb{E}\,\psi_q\left(\frac{|X|}{c}\right) \geq \mathbb{E}\left(\frac{|X|^q}{c^q}\right) \geq \left(\frac{A_q}{c}\right)^q > 1$$

for $c \leq A'_q$ small enough so that $\|X\|_{\psi_q} \geq A'_q$. Hence, as claimed, when $q \leq 2$,

$$A'_q\left(\sum_i \alpha_i^2\right)^{1/2} \leq \left\|\sum_i \varepsilon_i \alpha_i\right\|_{\psi_q} \leq B'_q\left(\sum_i \alpha_i^2\right)^{1/2}$$

for any sequence (α_i) of real numbers.

For $q > 2$, the span of (ε_i) in L_{ψ_q} is no longer isomorphic to ℓ_2. This span actually appears as some interpolation space between ℓ_2 and ℓ_1. To see this, we simply follow the observation which led to Lemma 1.7. Recall that for $0 < p < \infty$ we denote by $\ell_{p,\infty}$ the space of all real sequences $(\alpha_i)_{i\geq 1}$ such that

$$\|(\alpha_i)\|_{p,\infty} = \left(\sup_{t>0} t^p \mathrm{Card}\{i\,;\ |\alpha_i| > t\}\right)^{1/p} < \infty\,.$$

Equivalently, $\|(\alpha_i)\|_{p,\infty} = \sup_{i\geq 1} i^{1/p}\alpha_i^*$ where (α_i^*) is the non-increasing rearrangement of the sequence $(|\alpha_i|)_{i\geq 1}$. These spaces are known to show up in interpolation of ℓ_p-spaces. The functional $\|\cdot\|_{p,\infty}$, which is equivalent to a norm when $p > 1$, may be compared to the ℓ_p-norms as follows: for $r < p$,

$$\|(\alpha_i)\|_{p,\infty} \leq \left(\sum_i |\alpha_i|^p\right)^{1/p} \leq \left(\frac{p}{p-r}\right)^{1/p}\|(\alpha_i)\|_{r,\infty}\,.$$

Now, let $2 < q < \infty$ and denote by p the conjugate of $q : 1/p + 1/q = 1$, $1 < p < \infty$. The next lemma describes how in this case the span of (ε_i) in L_{ψ_q} is isomorphic to $\ell_{p,\infty}$.

Lemma 4.3. Let $2 < q < \infty$ and $p = q/q - 1$. There exist positive finite constants A'_q, B'_q depending on q (p) only such that for any finite sequence

(α_i) *of real numbers*

$$A_q'\|(\alpha_i)\|_{p,\infty} \leq \left\|\sum_i \varepsilon_i \alpha_i\right\|_{\psi_q} \leq B_q'\|(\alpha_i)\|_{p,\infty}.$$

Proof. By symmetry, $(\varepsilon_i \alpha_i)$ has the same distribution as $(\varepsilon_i|\alpha_i|)$. Moreover, by identical distribution and the definition of $\|(\alpha_i)\|_{p,\infty}$, we may assume that $|\alpha_1| \geq \cdots \geq |\alpha_i| \geq \cdots$. The martingale inequality of Lemma 1.7 then yields, for every $t > 0$,

$$(4.4) \qquad \mathbb{P}\left\{\left|\sum_{i \geq 1} \varepsilon_i \alpha_i\right| > t\right\} \leq 2\exp\left(-t^q/C_q\|(\alpha_i)\|_{p,\infty}^q\right).$$

As above, one then deduces the right hand side of the inequality of Lemma 4.3 from a simple integration by parts. Turning to the left hand side inequality, we use the contraction principle in the form of Theorem 4.4 below. It implies indeed, by symmetry and monotonicity of $(|\alpha_i|)$, that, for every m

$$\mathbb{E}\exp\left|\sum_{i \geq 1} \varepsilon_i \alpha_i\right|^q \geq \mathbb{E}\exp\left(|\alpha_m|^q\left|\sum_{i=1}^m \varepsilon_i\right|^q\right).$$

Hence,

$$\left\|\sum_{i \geq 1} \varepsilon_i \alpha_i\right\|_{\psi_q} \geq |\alpha_m|\left\|\sum_{i=1}^m \varepsilon_i\right\|_{\psi_q}.$$

Now, since $\sum_{i=1}^m \varepsilon_i = m$ with probability 2^{-m}, it is easily seen that

$$\left\|\sum_{i=1}^m \varepsilon_i\right\|_{\psi_q} \geq (1 + \log 2)^{-1/q} m^{1/p}.$$

Summarizing, we have obtained

$$\left\|\sum_{i \geq 1} \varepsilon_i \alpha_i\right\|_{\psi_q} \geq (1 + \log 2)^{-1/q} \sup_{m \geq 1} m^{1/p}|\alpha_m|$$

which is the result.

Furthermore, we should point out that the inequality corresponding to Lemma 1.8 states in this setting that for any finite sequence (α_i) and any $t > 0$,

$$(4.5) \qquad \mathbb{P}\left\{\left|\sum_i \varepsilon_i \alpha_i\right| > t\right\} \leq 16\exp\left[-\exp\left(t/4\|(\alpha_i)\|_{1,\infty}\right)\right].$$

The rest of this chapter is mainly devoted to extensions of the previous classical results to Rademacher averages with Banach space valued coefficients. Of course, the vector valued setting is characterized by *the lack of the orthogonality property*

$$\mathbb{E}\left|\sum_i \varepsilon_i \alpha_i\right|^2 = \sum_i \alpha_i^2 \,.$$

Various substitutes have therefore to be investigated, but the extension program will basically be fulfilled. A classification of Banach spaces according to the preceding orthogonality property is at the origin of the notions of type and cotype of Banach spaces which will be discussed later on in Chapter 9. Here, we study integrability and comparison theorems for Rademacher averages with vector valued coefficients.

4.2 The Contraction Principle

It is plain from Khintchine's inequalities that if (α_i) and (β_i) are two sequences of real numbers such that $|\alpha_i| \leq |\beta_i|$ for every i, one can compare $\|\sum_i \varepsilon_i \alpha_i\|_p$ and $\|\sum_i \varepsilon_i \beta_i\|_p$ for all p's. This comparison is thus based on the sum of the squares and orthogonality. However, it extends to vector valued coefficients, even in an improved form. This property is known as the *contraction principle* to which this paragraph is devoted. The main result is expressed in the following fundamental theorem.

Theorem 4.4. *Let $F : \mathbb{R}_+ \to \mathbb{R}_+$ be convex. For any finite sequence (x_i) in a Banach space B and any real numbers (α_i) such that $|\alpha_i| \leq 1$ for every i, we have*

$$(4.6) \qquad \mathbb{E}F\left(\left\|\sum_i \alpha_i \varepsilon_i x_i\right\|\right) \leq \mathbb{E}F\left(\left\|\sum_i \varepsilon_i x_i\right\|\right) .$$

Furthermore, for any $t > 0$,

$$(4.7) \qquad \mathbb{P}\left\{\left\|\sum_i \alpha_i \varepsilon_i x_i\right\| > t\right\} \leq 2\mathbb{P}\left\{\left\|\sum_i \varepsilon_i x_i\right\| > t\right\} .$$

Proof. The function on $\mathbb{R}^N$

$$(\alpha_1, \ldots, \alpha_N) \to \mathbb{E}F\left(\left\|\sum_{i=1}^N \alpha_i \varepsilon_i x_i\right\|\right)$$

is convex. Therefore, on the compact convex set $[-1, +1]^N$, it attains its maximum at an extreme point, that is a point $(\alpha_i)_{i \leq N}$ such that $\alpha_i = \pm 1$. For such values of α_i, by symmetry, both terms in (4.6) are equal. This proves (4.6). Concerning (4.7), replacing α_i by $|\alpha_i|$, we may assume by symmetry that $\alpha_i \geq 0$. Moreover, by identical distribution, we suppose that $\alpha_1 \geq \cdots \geq \alpha_N \geq \alpha_{N+1} = 0$. Set $S_k = \sum_{i=1}^k \varepsilon_i x_i$. Then

$$\sum_{i=1}^N \alpha_i \varepsilon_i x_i = \sum_{k=1}^N \alpha_k(S_k - S_{k-1}) = \sum_{k=1}^N (\alpha_k - \alpha_{k+1})S_k \,.$$

It follows that

$$\left\|\sum_{i=1}^{N}\alpha_i\varepsilon_i x_i\right\| \leq \max_{k\leq N}\|S_k\|.$$

We conclude by Lévy's inequalities (Proposition 2.3).

As a simple consequence of inequality (4.7), notice the following fact. Let (α_i) and (β_i) be two sequences of real numbers such that $|\beta_i| \leq |\alpha_i|$ for every i; then, if the series with vector coefficients $\sum_i \alpha_i\varepsilon_i x_i$ converges almost surely or equivalently in probability, then the same holds for the series $\sum_i \beta_i\varepsilon_i x_i$.

Theorem 4.4 admits several easy generalizations which will be used mostly without further comments in the sequel. Let us briefly indicate some of these generalizations. Recall that a sequence (η_i) of real valued random variables is called a symmetric sequence when (η_i) has the same distribution, as a sequence, as $(\varepsilon_i\eta_i)$ where (ε_i) is independent from (η_i). It is then clear, by independence and Fubini's theorem, that Theorem 4.4 also applies to (η_i) in place of (ε_i). Furthermore, as is easy also, if the α_i's are now random variables independent of (ε_i) (or (η_i)), such that $\|\alpha_i\|_\infty \leq 1$ for all i, the conclusion of Theorem 4.4 still holds true. Moreover, as we will see in Chapter 6, the fixed points x_i can also be replaced by vector valued random variables which are independent of the Rademacher sequence (ε_i).

The next lemma is another extension and formulation of the contraction principle.

Lemma 4.5. *Let $F : \mathbb{R}_+ \to \mathbb{R}_+$ be convex and let (η_i) be a symmetric sequence of real valued random variables such that $\mathbb{E}|\eta_i| < \infty$ for every i. Then, for any finite sequence (x_i) in a Banach space,*

$$\mathbb{E}F\left(\inf_i \mathbb{E}|\eta_i|\left\|\sum_i \varepsilon_i x_i\right\|\right) \leq \mathbb{E}F\left(\left\|\sum_i \eta_i x_i\right\|\right).$$

Proof. By the symmetry assumption, (η_i) has the same distribution as $(\varepsilon_i|\eta_i|)$ where, as usual, (ε_i) is independent from (η_i). Using first Jensen's inequality and partial integration, and then the contraction principle (4.6), we get

$$\mathbb{E}F\left(\left\|\sum_i \eta_i x_i\right\|\right) = \mathbb{E}F\left(\left\|\sum_i \varepsilon_i|\eta_i|x_i\right\|\right)$$

$$\geq \mathbb{E}F\left(\left\|\sum_i \varepsilon_i\mathbb{E}|\eta_i|x_i\right\|\right)$$

$$\geq \mathbb{E}F\left(\inf_i \mathbb{E}|\eta_i|\left\|\sum_i \varepsilon_i x_i\right\|\right).$$

Note that in case the η_i's have a common distribution, the inequality reduces to the application of Jensen's inequality.

An example of a sequence (η_i) of particular interest is given by the or-thogaussian sequence (g_i) consisting of independent standard normal vari-ables. Since $\mathbb{E}|g_i| = (2/\pi)^{1/2}$, we have from the previous lemma and its nota-tion, that

$$(4.8) \qquad \mathbb{E}F\left(\left\|\sum_i \varepsilon_i x_i\right\|\right) \leq \mathbb{E}F\left(\left(\frac{\pi}{2}\right)^{1/2}\left\|\sum_i g_i x_i\right\|\right).$$

Hence, Gaussian averages always "dominate" the corresponding Rademacher averages. In particular, from the integrability properties of Gaussian series (Corollary 3.2), if the series $\sum_i g_i x_i$ converges, so does $\sum_i \varepsilon_i x_i$. One might wonder whether a converse inequality or implication holds true. Letting, for simplicity, $F(t) = t$, the contraction principle applied conditionally on (g_i), which is assumed to be independent from (ε_i), yields

$$\mathbb{E}_\varepsilon \left\|\sum_{i=1}^N \varepsilon_i g_i x_i\right\| \leq \max_{i \leq N} |g_i| \mathbb{E}_\varepsilon \left\|\sum_{i=1}^N \varepsilon_i x_i\right\|$$

for any finite sequence $(x_i)_{i \leq N}$ in a Banach space B where we recall that $\mathbb{E}_\varepsilon$ indicates partial integration with respect to the Rademacher sequence (ε_i). If we now recall from (3.13) that $\mathbb{E}\max_{i \leq N}|g_i| \leq K(\log(N+1))^{1/2}$ for some numerical constant K, we see, by integrating the previous inequality, that

$$(4.9) \qquad \mathbb{E}\left\|\sum_{i=1}^N g_i x_i\right\| \leq K\left(\log(N+1)\right)^{1/2}\mathbb{E}\left\|\sum_{i=1}^N \varepsilon_i x_i\right\|.$$

This is *not* the converse of inequality (4.8) since the constant depends on the number of elements x_i which are considered. (4.8) is actually best possible in general Banach spaces as is shown by the example of the canonical basis of ℓ_∞ together with the left hand side of (3.14). This example is however extremal in the sense that if a Banach space does not contain subspaces isomorphic to finite dimensional subspaces of ℓ_∞, then (4.9) holds with a constant independent of N (but depending on the Banach space). We shall come back to this later on in Chapter 9 (see (9.12)). So far, we retain that Gaussian averages dominate Rademacher averages and that the converse is not true in general or only holds in the form (4.9).

The next lemma is yet another form of the contraction principle.

Lemma 4.6. *Let $F : \mathbb{R}_+ \to \mathbb{R}_+$ be convex. Let further (η_i) and (ξ_i) be two symmetric sequences of real valued random variables such that for some constant $K \geq 1$ and every i and $t > 0$*

$$\mathbb{P}\{|\eta_i| > t\} \leq K\mathbb{P}\{|\xi_i| > t\}.$$

Then, for any finite sequence (x_i) in a Banach space,

$$\mathbb{E}F\left(\left\|\sum_i \eta_i x_i\right\|\right) \leq \mathbb{E}F\left(K\left\|\sum_i \xi_i x_i\right\|\right).$$

Proof. Let (δ_i) be independent of (η_i) such that $\mathbb{P}\{\delta_i - 1\} = 1 - \mathbb{P}\{\delta_i = 0\} = 1/K$ for every i; then, for every $t > 0$,

$$\mathbb{P}\{|\delta_i \eta_i| > t\} \le \mathbb{P}\{|\xi_i| > t\}.$$

Taking inverses of the distribution functions, it is easily seen (and classical) that the sequences $(\delta_i \eta_i)$ and (ξ_i) can be constructed on some rich enough probability space in such a way that, almost surely,

$$|\delta_i \eta_i| \le |\xi_i| \quad \text{for every } i.$$

From the contraction principle and the symmetry assumption, it follows that

$$\mathbb{E}F\left(\left\|\sum_i \delta_i \eta_i x_i\right\|\right) \le \mathbb{E}F\left(\left\|\sum_i \xi_i x_i\right\|\right).$$

The proof is then completed via Jensen's inequality applied to the sequence (δ_i) since $\mathbb{E}\delta_i = 1/K$.

Notice that if, in the preceding lemma,

$$\mathbb{P}\{|\eta_i| > t\} \le K\mathbb{P}\{|\xi_i| > t\}$$

only for all $t \ge t_0 > 0$, then the conclusion would be somewhat weakened: we have

$$\mathbb{E}F\left(\left\|\sum_i \eta_i x_i\right\|\right) \le \frac{1}{2}\mathbb{E}F\left(2Kt_0\left\|\sum_i \varepsilon_i x_i\right\|\right) + \frac{1}{2}\mathbb{E}F\left(2K\left\|\sum_i \xi_i x_i\right\|\right).$$

Indeed, simply note that if

$$\xi_i' = t_0 \varepsilon_i I_{\{|\xi_i| \le t_0\}} + \xi_i I_{\{|\xi_i| > t_0\}}$$

where (ε_i) is a Rademacher sequence which is independent of the sequences (η_i) and (ξ_i), then the couple (η_i, ξ_i') satisfies the hypothesis of Lemma 4.6. Then use the convexity and the contraction principle to get rid of the indicator functions.

4.3 Integrability and Tail Behavior of Rademacher Series

On the basis of the scalar results described in the first paragraph, we now investigate integrability properties of Rademacher series with vector valued coefficients. The typical object of study is a convergent series $\sum_i \varepsilon_i x_i$ where (x_i) is a sequence in a Banach space B. This defines a Radon random variable in B. Motivated by the Gaussian study of the previous chapter, there is a somewhat larger setting corresponding to what could be called almost surely bounded *Rademacher processes*. That is, for some set T assumed to be countable for simplicity, let (x_i) be a sequence of functions on T such that for each t, $\sum_i \varepsilon_i x_i(t)$ converges almost surely (or in probability). In other words, $\sum_i x_i(t)^2 < \infty$ for each t. Assuming that

$$\sup_{t \in T} \left| \sum_i \varepsilon_i x_i(t) \right| < \infty \quad \text{almost surely},$$

we are interested in the integrability properties and tail behavior of this almost surely finite supremum.

As in the previous chapter, in order to unify the exposition, we assume that we are given a Banach space B such that for some countable subset D in the unit ball of B', $\|x\| = \sup_{f \in D} |f(x)|$. We deal with a random variable X with values in B such that there exists a sequence (x_i) of points in B such that $\sum_i f^2(x_i) < \infty$ for every f in D and for which $(f_1(X), \ldots, f_N(X))$ has the same distribution as $(\sum_i \varepsilon_i f_1(x_i), \ldots, \sum_i \varepsilon_i f_N(x_i))$ for every finite subset $\{f_1, \ldots, f_N\}$ of D. We then speak of X as a vector valued Rademacher series (although this terminology is somewhat improper) or as an almost surely bounded Rademacher process. For such an X, we investigate the integrability and tail behavior of $\|X\|$. The size of the tail $\mathbb{P}\{\|X\| > t\}$ will be measured in terms of *two* parameters similar to those used in the Gaussian case. As for Gaussian random vectors, we indeed consider

$$\sigma = \sigma(X) = \sup_{f \in D} \left(\mathbb{E} f^2(X) \right)^{1/2} = \sup_{f \in D} \left(\sum_i f^2(x_i) \right)^{1/2} = \sup_{|h| \leq 1} \sup_{f \in D} \left| \sum_i h_i f(x_i) \right|$$

(where $h = (h_i) \in \ell_2$). Recall that if $X = \sum_i \varepsilon_i x_i$ converges almost surely in a (arbitrary) Banach space B, thus defining a Radon random variable, we can simply let

$$\sigma(X) = \sup_{\|f\| \leq 1} \left(\mathbb{E} f^2(X) \right)^{1/2}.$$

It is easy to see that σ is finite. In fact, σ is controlled by some quantity associated to the L_0-topology of the norm $\|X\|$: if M is for example such that $\mathbb{P}\{\|X\| > M\} \leq 1/8$, then we have $\sigma \leq 2\sqrt{2}M$. To see this, first recall that by (4.3), for any f in D,

$$\left(\mathbb{E} \left| \sum_i \varepsilon_i f(x_i) \right|^2 \right)^{1/2} \leq \sqrt{2}\, \mathbb{E} \left| \sum_i \varepsilon_i f(x_i) \right|.$$

It then follows from Lemma 4.2 and the definition of M that

$$\left(\sum_i f^2(x_i) \right)^{1/2} = \left(\mathbb{E} \left| \sum_i \varepsilon_i f(x_i) \right|^2 \right)^{1/2} \leq 2\sqrt{2}M$$

and, thus, our preceding claim follows. The tail behavior and the integrability properties of $\|X\|$ are measured in terms of this number σ, supremum of weak moments, and some quantity, median or expectation, related to the L_0-topology of the norm (strong moments). The main ingredient is the isoperimetric Theorem 1.3 and the related concentration inequality. The next statement summarizes more or less the various results about this question.

Theorem 4.7. *Let X be a Rademacher series in B as defined above with corresponding $\sigma = \sigma(X)$. Let moreover $M = M(X)$ denote a median of $\|X\|$. Then, for every $t > 0$,*

$$(4.10) \qquad \mathbb{P}\{|\|X\| - M| > t\} \leq 4\exp(-t^2/8\sigma^2).$$

In particular, there exists $\alpha > 0$ such that $\mathbb{E}\exp(\alpha\|X\|^2) < \infty$ and all moments of X are equivalent: that is, for any $0 < p, q < \infty$, there is a constant $K_{p,q}$ depending on p, q only such that

$$\|X\|_p \leq K_{p,q}\|X\|_q.$$

Proof. Recall the Haar measure μ on $\{-1, +1\}^{\mathbb{N}}$. The function φ on $\mathbb{R}^{\mathbb{N}}$ defined by $\varphi(\alpha) = \sup_{f \in D}|\sum_i \alpha_i f(x_i)|$, $\alpha = (\alpha_i)$, is μ-almost everywhere finite by definition of X. It is convex and Lipschitz on ℓ_2 with Lipschitz constant σ since

$$|\varphi(\alpha) - \varphi(\beta)| \leq \sup_{f \in D}\left|\sum_i (\alpha_i - \beta_i)f(x_i)\right| \leq \sigma|\alpha - \beta|.$$

Then, inequality (4.10) is simply the concentration inequality (1.10) issued from Theorem 1.3 and applied to this convex Lipschitz function on $(\mathbb{R}^{\mathbb{N}}, \mu)$. Alternatively, one may obtain (4.10) directly from Theorem 1.3 by a simple finite dimensional approximation, first with a finite supremum in f, then with finite sums. Recall, as is usual in this setting, that we also have, for every $t > 0$,

$$(4.11) \qquad \mathbb{P}\{\|X\| > M + t\} \leq 2\exp(-t^2/8\sigma^2).$$

An integration by parts on the basis of (4.10) already ensures that $\mathbb{E}\exp(\alpha\|X\|^2) < \infty$ for every $\alpha > 0$ small enough, namely less than $1/8\sigma^2$. Concerning the equivalence of moments, if M' is chosen such that $\mathbb{P}\{\|X\| > M'\} \leq 1/8$, we know that $\sigma \leq 2\sqrt{2}M'$, and, from (4.11) and $M \leq M'$, that

$$\mathbb{P}\{\|X\| > M' + t\} \leq 2\exp(-t^2/8\sigma^2)$$

for every $t > 0$. By an integration by parts, for any $0 < p < \infty$,

$$\mathbb{E}|\|X\| - M'|^p \leq M'^p + \int_0^\infty \mathbb{P}\{\|X\| > M' + t\}\, dt^p$$
$$\leq M'^p + K_p\sigma^p \leq K_p'M'^p.$$

Since $M' \leq (8\mathbb{E}\|X\|^q)^{1/q}$ for every $0 < q < \infty$, the claim of the theorem is established. Note that we can take $K_{p,2} = K\sqrt{p}$ $(p \geq 2)$ for some numerical constant K.

It is noteworthy that the equivalence of moments in Theorem 4.7, due to J.-P. Kahane [Ka1], provides an alternate proof of Khintchine's inequalities (with the right order of magnitude of the constants when $p \to \infty$). They are therefore sometimes referred to in the literature as the *Khintchine-Kahane inequalities*. Note also that (4.10) (or rather (4.11)) implies the weaker but sometimes convenient inequality which corresponds perhaps more directly to the subgaussian inequality (4.1): for every $t > 0$,

$$(4.12) \qquad \mathbb{P}\{\|X\| > t\} \leq 2 \exp\left(-t^2/32\mathbb{E}\|X\|^2\right).$$

For the proof, use (4.11) together with the fact that $M \leq (2\mathbb{E}\|X\|^2)^{1/2}$ and $\sigma^2 \leq \mathbb{E}\|X\|^2$.

For an almost surely convergent series $X = \sum_i \varepsilon_i x_i$ in B, it is very easy to see that the exponential square integrability result of Theorem 4.7 can be refined into $\mathbb{E}\exp(\alpha\|X\|^2) < \infty$ *for every* $\alpha > 0$. Set indeed $X_N = \sum_{i=1}^N \varepsilon_i x_i$ for each N. X_N converges to X almost surely and in $L_2(B)$ by Theorem 4.7. In particular, $\sigma(X_N - X) \to 0$. Then, let $\alpha > 0$ and choose an integer N such that the median of $\|X - X_N\|$ is less than 1 (say) and such that $8\sigma(X - X_N)^2\alpha < 1$. It follows from (4.11) that for every $t > 0$

$$\mathbb{P}\{\|X - X_N\| > t + 1\} \leq 2 \exp\left(-t^2/8\sigma(X - X_N)^2\right).$$

Hence, by the choice of N, $\mathbb{E}(\exp\alpha\|X - X_N\|^2) < \infty$ from which the claim follows since $\|X_N\|$ is bounded.

It is still true that the preceding observation holds in the more general setting of almost surely bounded Rademacher processes. However, the proof is somewhat more complicated and uses a variation about the converse subgaussian inequality (4.2).

Theorem 4.8. *Let X be a vector valued Rademacher series as in Theorem 4.7. Then*

$$\mathbb{E}\exp(\alpha\|X\|^2) < \infty \quad \text{for every} \ \alpha > 0.$$

Proof. It relies on the following lemma.

Lemma 4.9. *There is a numerical constant K with the following property. Let $(\alpha_i)_{i \geq 1}$ be a decreasing sequence of positive numbers such that $\sum_i \alpha_i^2 < \infty$. If $t \geq K(\sum_i \alpha_i^2)^{1/2}$, define n to be the smallest integer such that $\sum_{i=1}^n \alpha_i > t$. Then*

$$\mathbb{P}\left\{\left|\sum_i \varepsilon_i \alpha_i\right| > t\right\} \geq \frac{1}{2}\exp\left(-Kt^2/\sum_{i \geq n}\alpha_i^2\right).$$

Proof. Let $\sigma = (\sum_{i \geq n}\alpha_i^2)^{1/2}$ and let $K \geq 1$ be the numerical constant of (4.2). We distinguish between two cases. If $n \leq 2Kt^2/\sigma^2$, by definition of n,

$$\mathbb{P}\left\{\sum_{i=1}^{n}\varepsilon_i\alpha_i > t\right\} \geq 2^{-n} \geq \exp(-2Kt^2/\sigma^2).$$

If $n > 2Kt^2/\sigma^2$, by definition of n and since $\alpha_i \leq t$ for every i,

$$\max_{i\geq n}\alpha_i = \alpha_n \leq \frac{1}{n}\sum_{i=1}^{n}\alpha_i \leq \frac{2t}{n} \leq \frac{K\sigma^2}{t}.$$

Therefore, since K is the numerical constant of (4.2),

$$\mathbb{P}\left\{\sum_{i\geq n}\varepsilon_i\alpha_i > t\right\} \geq \exp(-Kt^2/\sigma^2).$$

Lemma 4.9 follows by Lévy's inequality (2.6) and by changing K into $2K$.

To establish Theorem 4.8, we show a quantitative version of the result. Namely, for every $\alpha > 0$, there exists $\varepsilon = \varepsilon(\alpha) > 0$ such that if M satisfies $\mathbb{P}\{\|X\| > M\} < \varepsilon$, then, for all $t > 0$,

$$(4.13) \qquad \mathbb{P}\{\|X\| > KM(t+1)\} \leq 2\exp(-\alpha t^2)$$

for some numerical constant K. If $\mathbb{P}\{\|X\| > M\} \leq 1/8$, then we know that $\sigma = \sigma(X) \leq 2\sqrt{2}M$. Let then $M' = 2\sqrt{2}KM$ where $K \geq 1$ is the constant of Lemma 4.9. We thus assume that $\varepsilon < 1/8$. By this lemma (applied to the non-increasing rearrangement of $(|f(x_i)|)$ with $t = M'$), there exist, for each f in D, sequences $(u_i(f))$ and $(v_i(f))$ such that, for every i, $f(x_i) = u_i(f) + v_i(f)$ with the following properties:

$$\sum_i |u_i(f)| \leq M', \quad \exp\left(-KM'^2/\sum_i v_i(f)^2\right) \leq 2\varepsilon.$$

In particular,

$$\|X\| \leq M' + \sup_{f\in D}\left|\sum_i \varepsilon_i v_i(f)\right|$$

and the Rademacher process $(\sum_i \varepsilon_i v_i(f))_{f\in D}$ satisfies the following properties:

$$\mathbb{P}\left\{\sup_{f\in D}\left|\sum_i \varepsilon_i v_i(f)\right| > 2M'\right\} < \varepsilon < \frac{1}{2}$$

and

$$\sup_{f\in D}\sum_i v_i(f)^2 \leq KM'^2\left(\log\frac{1}{2\varepsilon}\right)^{-1}.$$

Given $\alpha > 0$, choose $\varepsilon = \varepsilon(\alpha) > 0$ small enough in order that $K(\log\frac{1}{2\varepsilon})^{-1}$ is smaller than $1/8\alpha$. Then, from (4.11), for every $t > 0$,

$$\mathbb{P}\left\{\sup_{f\in D}\left|\sum_i \varepsilon_i v_i(f)\right| > M'(t+2)\right\} \leq 2\exp(-\alpha t^2).$$

Hence,

$$\mathbb{P}\{\|X\| > M'(t+3)\} \le 2\exp(-\alpha t^2)$$

which gives (4.13).

Now, we can easily conclude the proof of Theorem 4.8. For each N, set $Z_N = \sup_{f \in D} |\sum_{i>N} \varepsilon_i f(x_i)|$. (Z_N) defines a reverse submartingale with respect to the family of σ-algebras generated by $\varepsilon_{N+1}, \varepsilon_{N+2}, \ldots$, $N \in \mathbb{N}$. By Theorem 4.7,

$$\sup_N \mathbb{E}Z_N \le \mathbb{E}\|X\| < \infty.$$

Therefore, (Z_N) converges almost surely. Its limit is measurable with respect to the tail algebra, and is therefore degenerate. Hence, there exists $M < \infty$ such that, for every $\varepsilon > 0$, one can find N large enough such that $\mathbb{P}\{Z_N > M\} < \varepsilon$. Apply then (4.13) to Z_N to easily conclude the proof of Theorem 4.8 since $\|\sum_{i=1}^N \varepsilon_i x_i\|$ is bounded.

Going back to the equivalence of moments of Theorem 4.7, it is interesting to point out that these inequalities contain the corresponding inequalities for Gaussian averages. This can be seen from a simple argument involving the central limit theorem. Denote by (ε_{ij}) a doubly indexed Rademacher sequence; for any finite sequence (x_i) in B and any n, $0 < p, q < \infty$,

$$\left\|\sum_i \left(\sum_{j=1}^n \frac{\varepsilon_{ij}}{\sqrt{n}}\right) x_i\right\|_p \le K_{p,q} \left\|\sum_i \left(\sum_{j=1}^n \frac{\varepsilon_{ij}}{\sqrt{n}}\right) x_i\right\|_q.$$

When n tends to infinity, $(\sum_{j=1}^n \varepsilon_{ij}/\sqrt{n})$ converges in distribution to a standard normal random variable. It follows that

$$\left\|\sum_i g_i x_i\right\|_p \le K_{p,q} \left\|\sum_i g_i x_i\right\|_q.$$

From the right order of magnitude of the constants $K_{p,q}$ in function of p and q, the exponential squared integrability of Gaussian random vectors (Lemma 3.7) also follows.

As yet another remark, we would like to outline a different approach to the conclusions of Theorem 4.7 based on the Gaussian isoperimetric inequality and contractions of Gaussian measures (see (1.7) in Chapter 1). Let (u_i) denote a sequence of independent random variables uniformly distributed on $[-1, +1]$. As described by (1.7), there is an isoperimetric inequality of Gaussian type for these measures. As in the previous chapter (cf. (3.2)), we can obtain, for example, that for any (finite) sequence (x_i) in B and any $t > 0$

$$(4.14) \qquad \mathbb{P}\left\{\left\|\sum_i u_i x_i\right\| > 2\mathbb{E}\left\|\sum_i u_i x_i\right\| + t\right\} \le \exp(-t^2/\pi\sigma^2)$$

where σ is as above, i.e. $\sigma = \sup_{f \in D} (\sum_i f^2(x_i))^{1/2}$. Now, we would like to apply the contraction principle to replace the sequence (u_i) in (4.14) by the Rademacher sequence. To perform this, we need to translate the inequality

(4.14) in some moment inequalities in order to be able to apply Theorem 4.4 and Lemma 4.5. We make use of the following easy equivalence (in the spirit of Lemma 3.7).

Lemma 4.10. *Let Z be a positive random variable and α, β be positive numbers. The following are equivalent:*

(i) there is a constant $K > 0$ such that for every $t > 0$,

$$\mathbb{P}\{Z > K(\beta + \alpha t)\} \leq K \exp(-t^2/K);$$

(ii) there is a constant $K > 0$ such that for every $p \geq 1$

$$\|Z\|_p \leq K(\beta + \alpha\sqrt{p}).$$

Furthermore, the constants in (i) and (ii) only depend on each other.

From (4.14) and this lemma, it follows that for some numerical constant K,

$$\left\|\sum_i u_i x_i\right\|_p \leq K\left(\left\|\sum_i u_i x_i\right\|_1 + \sigma\sqrt{p}\right)$$

for every $p \geq 1$. The contraction principle (Theorem 4.4 and Lemma 4.5) applies to give

$$\left\|\sum_i \varepsilon_i x_i\right\|_p \leq 2K\left(\left\|\sum_i \varepsilon_i x_i\right\|_1 + \sigma\sqrt{p}\right)$$

$(\mathbb{E}|u_i| = 1/2)$. Going back to a tail estimate via Lemma 4.10 we get

$$(4.15) \quad \mathbb{P}\left\{\left\|\sum_i \varepsilon_i x_i\right\| > K\left(\mathbb{E}\left\|\sum_i \varepsilon_i x_i\right\| + t\right)\right\} \leq K \exp(-t^2/K\sigma^2)$$

for some numerical constant $K > 0$ and every $t > 0$.

This inequality easily extends to infinite sums and to bounded Rademacher processes (starting with a norm given by a finite supremum and passing to the limit). It is possible to obtain from it all the conclusions of Theorem 4.7. The most important difference is that (4.10) expresses a concentration property. However, each time (4.10) is only used as a deviation inequality (i.e. in the form of (4.11)), then (4.15) can be used equivalently.

4.4 Integrability of Rademacher Chaos

Let us consider the canonical representation of the Rademacher functions (ε_i) as the coordinate maps on $\Omega = \{-1, +1\}^{\mathbb{N}}$ equipped with the natural product probability measure $\mu = (\frac{1}{2}\delta_{-1} + \frac{1}{2}\delta_{+1})^{\otimes\mathbb{N}}$. For any finite subset A of $\mathbb{N}$, define $w_A = \prod_{i \in A} \varepsilon_i$ $(w_\emptyset = 1)$. It is known that the Walsh system $\{w_A; A \subset \mathbb{N}, \text{Card } A < \infty\}$ defines an orthonormal basis of $L_2(\mu)$.

For $0 \leq \varepsilon \leq 1$, introduce the operator $T(\varepsilon) : L_2(\mu) \to L_2(\mu)$ defined by

$$T(\varepsilon)w_A = \varepsilon^{\operatorname{Card} A} w_A$$

for $A \subset \mathbb{N}$, $\operatorname{Card} A < \infty$. Since, as in the Gaussian case, $T(\varepsilon)$ is a convolution operator, it extends to a positive contraction on all $L_p(\mu)$'s, $1 \leq p < \infty$. One striking property of the operator $T(\varepsilon)$ is the hypercontractivity property similar to that observed for the Hermite semigroup in the preceding chapter. Namely, for $1 < p < q < \infty$ and $\varepsilon \leq [(p-1)/(q-1)]^{1/2}$, $T(\varepsilon)$ maps $L_p(\mu)$ into $L_q(\mu)$ *with norm* 1, i.e. for each f in $L_p(\mu)$

(4.16) $$\|T(\varepsilon)f\|_q \leq \|f\|_p.$$

This property can be deduced from a sharp two point inequality together with a convolution argument. Moreover, it implies the corresponding Gaussian hypercontractivity property using the central limit theorem ([Gro], [Bec]).

A function f in $L_2(\mu)$ can be written as $f = \sum_A w_A f_A$ where $f_A = \int f w_A \, d\mu$ and the sum runs over all $A \subset \mathbb{N}$, $\operatorname{Card} A < \infty$. We can write

$$f = \sum_{d=0}^{\infty} \left(\sum_{\operatorname{Card} A = d} w_A f_A \right) = \sum_{d=0}^{\infty} Q_d f.$$

$Q_d f$ is called the chaos of degree or order d of f. Chaos of degree 1 are simply Rademacher series $\sum_i \varepsilon_i \alpha_i$, chaos of degree 2 are quadratic series of the type $\sum_{i \neq j} \varepsilon_i \varepsilon_j \alpha_{ij}$, etc. Chaos of degree d are characterized by the fact that the action of the operator $T(\varepsilon)$ is the multiplication by ε^d, that is

$$T(\varepsilon)Q_d f = \varepsilon^d Q_d f.$$

Using (4.16) with $p = 2$, $q > 2$ and $\varepsilon = (q-1)^{-1/2}$, we have

$$\|Q_d f\|_q \leq (q-1)^{d/2} \|Q_d f\|_2.$$

As we know, this type of inequalities implies strong integrability properties of $Q_d f$; recalling Lemma 3.7, it follows indeed, as in the Gaussian case, that

$$\mathbb{E} \exp(\alpha |Q_d f|^{2/d}) < \infty$$

for some (actually all) $\alpha > 0$.

This approach extends to chaos with values in a Banach space providing in particular a different proof of some of the results of the preceding section for chaos of the first order. As for Gaussian chaos, we however would like to complete these integrability results with some more precise tail estimates. This is the object of what follows where, once more, we use isoperimetric methods. For simplicity, we only deal with the chaos of degree, or order, 2, in a setting similar to that developed in the Gaussian case.

We keep the setting of the preceding paragraph with a Banach space B for which there exists a countable subset D of the unit ball of B' such that $\|x\| = \sup_{f \in D} |f(x)|$ for all x in B. We say that a random variable X

with values in B is a Rademacher chaos, of the second order thus, if there is a sequence (x_{ij}) in B such that $\sum_{i,j} \varepsilon_i \varepsilon_j f(x_{ij})$ converges almost surely (or only in probability) for every f in D and such that $(f(X))_{f \in D}$ has the same distribution as $(\sum_{i,j} \varepsilon_i \varepsilon_j f(x_{ij}))_{f \in D}$. We assume for simplicity that the diagonal terms x_{ii} are zero, in which case the results are more complete. We briefly discuss the general case at the end of the section.

The idea of this study will be to follow the approach to Gaussian chaos using isoperimetric methods in the form of Theorem 1.3. However, with respect to the Gaussian case, some convexity and decoupling results appear to be slightly more complicated here and we will try to detail some of these difficulties.

Thus, let X be a Rademacher chaos (of order 2) as just defined. Our aim will be to try to find estimates of the tail $\mathbb{P}\{\|X\| > t\}$ in terms of some parameters of the distribution of X. These are similar to those used in the Gaussian setting. Let us consider first the "decoupled" chaos $Y = (\sum_{i,j} \varepsilon_i \varepsilon_j' f(y_{ij}))_{f \in D}$ where (ε_j') is an independent copy of (ε_i) and where $y_{ij} = x_{ij} + x_{ji}$. Moreover, let M be a number such that $\mathbb{P}\{\|X\| \leq M\}$ is large enough, for example

$$\mathbb{P}\{\|X\| \leq M\} > 63/64.$$

Let also m be such that

$$\mathbb{P}\left\{ \sup_{|h| \leq 1} \sup_{f \in D} \left| \sum_{i,j} \varepsilon_i h_j f(y_{ij}) \right| \leq m \right\} > 15/16$$

and set

$$\sigma = \sigma(X) = \sup_{|h| \leq 1} \sup_{f \in D} \left| \sum_{i,j} h_i h_j f(x_{ij}) \right|.$$

It might be worthwhile to mention at this stage that these parameters, as well as the decoupled chaos Y, are well defined. Towards this aim, we use a *decoupling* argument which will also be useful in the proof of the main result below.

First, let us assume that we deal with a norm $\| \cdot \|$ given by a *finite* supremum. Once the right estimates are established, we will simply need to increase these supremum to the norm. For each N, let us set further $X_N = \sum_{i,j=1}^{N} \varepsilon_i \varepsilon_j x_{ij}$. If M' is such that $\mathbb{P}\{\|X\| \leq M'\} > 127/128$, for all large enough N's (recall the norm is a finite supremum), $\mathbb{P}\{\|X_N\| \leq M'\} \geq 127/128$. Let $I \subset \{1, \ldots, N\}$ and let (η_i) be defined as $\eta_i = -1$ if $i \in I$, $\eta_i = +1$ if $i \notin I$. By symmetry,

$$\mathbb{P}\left\{ \|X_N\| \leq M, \left\| \sum_{i,j=1}^{N} \varepsilon_i \eta_i \varepsilon_j \eta_j x_{ij} \right\| \leq M \right\} \geq 63/64,$$

and thus, by difference,

$$\mathbb{P}\left\{ \left\| \sum_{i \in I, j \notin I} \varepsilon_i \varepsilon_j y_{ij} \right\| \leq M \right\} \geq 63/64.$$

Of course, we are now in a position to "decouple" in the sense that $\sum_{i \in I, j \notin I} \varepsilon_i \varepsilon_j y_{ij}$ has the same distribution as $\sum_{i \in I, j \notin I} \varepsilon_i \varepsilon'_j y_{ij}$. Let us assume for clarity that (ε_i) and (ε'_j) are constructed on two different probability spaces $(\Omega, \mathcal{A}, \mathbb{P})$ and $(\Omega', \mathcal{A}', \mathbb{P}')$ respectively. We claim that, for some numerical constant K,

$$(4.17) \qquad \mathbb{E} \left\| \sum_{i \in I, j \notin I} \varepsilon_i \varepsilon'_j y_{ij} \right\|^2 \leq K M'^2.$$

This simply follows from Fubini's theorem and the integrability results for one-dimensional chaos, i.e. series. Indeed, let

$$A = \left\{ \omega; \ \mathbb{P}' \left\{ \left\| \sum_{i \in I, j \notin J} \varepsilon_i(\omega) \varepsilon'_j y_{ij} \right\| \leq M' \right\} \geq 7/8 \right\}.$$

Then $\mathbb{P}(A) \geq 7/8$. For ω in A, Theorem 4.7 applied to the sum in ε'_j implies

$$\mathbb{E}' \left\| \sum_{i \in I, j \notin J} \varepsilon_i(\omega) \varepsilon'_j y_{ij} \right\|^2 \leq K M'^2$$

for some numerical constant K. Now, the same result applied to the sum in ε_i but in $L_2(\Omega', \mathbb{P}'; B)$ implies the announced claim (4.17) since $\mathbb{P}(A) \geq 7/8$. Then, we note that

$$\sum_{i,j=1}^N \varepsilon_i \varepsilon'_j y_{ij} = 4 \cdot \frac{1}{2^N} \sum_{I \subset \{1, \ldots, N\}} \left(\sum_{i \in I, j \notin J} \varepsilon_i \varepsilon'_j y_{ij} \right).$$

Therefore, from (4.17), and for some possibly different constant K,

$$\mathbb{E} \left\| \sum_{i,j=1}^N \varepsilon_i \varepsilon'_j y_{ij} \right\|^2 \leq K M'^2.$$

By a Cauchy sequence argument, it then easily follows that for each f in D, $\sum_{i,j} \varepsilon_i \varepsilon'_j f(y_{ij})$ converges in L_2 (and almost surely by Fubini's theorem). Hence, Y is well defined and, *increasing the finite supremum to the norm*, also satisfies $\mathbb{E} \|Y\|^2 \leq K M'^2$. To control m and σ, observe that, by independence,

$$\mathbb{E} \|Y\|^2 \geq \mathbb{E} \sup_{|h| \leq 1} \sup_{f \in D} \left| \sum_{i,j} \varepsilon_i h_j f(y_{ij}) \right|^2$$

$$\geq \sup_{|h|, |k| \leq 1} \sup_{f \in D} \left| \sum_{i,j} h_i k_j f(y_{ij}) \right|^2 \geq 4 \sigma^2.$$

Furthermore, by Theorem 4.7 again (although for a different norm), we may take m to be equivalent to

$$\left(\mathbb{E} \sup_{|h| \leq 1} \sup_{f \in D} \left| \sum_{i,j} \varepsilon_i h_j f(y_{ij}) \right|^2 \right)^{1/2}$$

and thus $\sigma \leq Km \leq K^2M'$ for some numerical constant K. In particular moreover, we will be allowed to deal with finite sums in the estimates we are looking for since the preceding scheme and these inequalities justify all the necessary approximations.

After describing the parameters, the decoupling and the approximation arguments which we will use, we are now ready to state and to prove our result about the tail behavior of $\mathbb{P}\{\|X\| > t\}$ where X is as above a chaos of the second order.

Theorem 4.11. *Let* $X = (\sum_{i,j} \varepsilon_i \varepsilon_j f(x_{ij}))_{f \in D}$ *be a Rademacher chaos with* $x_{ii} = 0$ *for every* i*, and let* M, m, σ *be its parameters as just described. Then, for every* $t > 0$*,*

$$\mathbb{P}\{\|X\| > 2(M + mt + \sigma t^2)\} \leq 20 \exp(-t^2/144)\,.$$

Moreover, $\mathbb{E} \exp(\alpha\|X\|) < \infty$ *for every* $\alpha > 0$*.*

Proof. As announced, for the proof of the tail estimate, we can assume that we deal with a finite sum $X = \sum_{i,j=1}^{N} \varepsilon_i \varepsilon_j x_{ij}$. Recall that

$$X = 2 \cdot \frac{1}{2^N} \sum_{I \subset \{1,\dots,N\}} \left(\sum_{i \in I, j \notin I} \varepsilon_i \varepsilon_j y_{ij} \right).$$

First, we estimate, for every $I \subset \{1, \dots, N\}$, the tail probability of $\|\sum_{i \in I, j \notin I} \varepsilon_i \varepsilon_j y_{ij}\|$. For simplicity in the notation, we thus assume for this step that $y_{ij} = 0$ if $i \notin I$ or $j \in I$. Recall that, by decoupling and difference,

$$\mathbb{P}\left\{ \left\| \sum_{i,j} \varepsilon_i \varepsilon_j y_{ij} \right\| \leq M \right\} \geq 31/32\,.$$

In the same way, we have

$$\mathbb{P}\left\{ \sup_{|h| \leq 1} \left\| \sum_{i,j} \varepsilon_i h_j y_{ij} \right\| \leq m \right\} \geq 7/8\,.$$

Let

$$B = \left\{ \omega;\ \mathbb{P}'\left\{ \left\| \sum_{i,j} \varepsilon_i(\omega)\varepsilon_j' y_{ij} \right\| \leq M \right\} \geq 3/4,\ \sup_{|h| \leq 1} \left\| \sum_{i,j} \varepsilon_i(\omega) h_j y_{ij} \right\| \leq m \right\}.$$

Thus, $\mathbb{P}(B) \geq 3/4$. If $\omega \in B$, we see that we control the one-dimensional parameters in the summation with respect to ε_j'. Therefore, it follows from Theorem 1.3 that, for $t > 0$,

$$\mathbb{P}(A_1) \geq \mathbb{P}(B)\big(1 - 2\exp(-t^2/8)\big)$$

where

$$A_1 = \left\{ (\omega, \omega'); \; \omega \in B, \; \left\| \sum_{i,j} \varepsilon_i(\omega) \varepsilon'_j(\omega') y_{ij} \right\| \leq M + mt \right\}.$$

Also from Theorem 1.3, we have, for $t > 0$,

$$\mathbb{P}' \left\{ \sup_{|h| \leq 1} \left\| \sum_{i,j} h_i \varepsilon'_j y_{ij} \right\| > m + 6\sigma t \right\} \leq 2 \exp(-t^2/8)$$

where we have used the simple fact that

$$\sup_{|h|, |k| \leq 1} \left\| \sum_{i,j} h_i k_j y_{ij} \right\| \leq 6\sigma.$$

Hence, if we let, for every $t > 0$

$$A = \left\{ (\omega, \omega'); \; \omega \in B, \; \left\| \sum_{i,j} \varepsilon_i(\omega) \varepsilon'_j(\omega') y_{ij} \right\| \leq M + mt \right.$$

$$\left. \text{and} \; \sup_{|h| \leq 1} \left\| \sum_{i,j} h_i \varepsilon'_j y_{ij} \right\| \leq m + 4\sigma t \right\}$$

we have

$$\mathbb{P}(A) \geq \mathbb{P}(B) \big(1 - 4 \exp(-t^2/18) \big).$$

Then, let

$$B' = \left\{ \omega'; \; \mathbb{P} \left\{ \omega \in B; \; \left\| \sum_{i,j} \varepsilon_i(\omega) \varepsilon'_j(\omega') y_{ij} \right\| \leq M + mt \right\} \geq 1/2, \right.$$

$$\left. \sup_{|h| \leq 1} \left\| \sum_{i,j} h_i \varepsilon'_j(\omega') y_{ij} \right\| \leq m + 4\sigma t \right\}.$$

By Fubini's theorem, $\mathbb{P}'(B') \geq 1 - 8 \exp(-t^2/18)$. If ω' is in B', then we are in a position to apply the one-dimensional integrability results for the sum in ε_i since we control the corresponding parameters; this gives:

$$\mathbb{P} \left\{ \left\| \sum_{i,j} \varepsilon_i \varepsilon'_j y_{ij} \right\| \leq M + 2mt + 4\sigma t^2 \right\}$$

$$= \mathbb{P} \left\{ \left\| \sum_{i,j} \varepsilon_i \varepsilon'_j y_{ij} \right\| \leq M + mt + (m + 4\sigma t)t \right\}$$

$$\geq 1 - 10 \exp(-t^2/18).$$

Summarizing, we have obtained that for every $I \subset \{1, \ldots, N\}$ and every $t > 0$,

$$(4.18) \quad \mathbb{P} \left\{ \left\| \sum_{i \in I, j \notin I} \varepsilon_i \varepsilon_j y_{ij} \right\| > M + tm + \sigma t^2 \right\} \leq 10 \exp(-t^2/72).$$

Now, we recall that

$$\sum_{i,j=1}^{N} \varepsilon_i \varepsilon_j x_{ij} = 2 \cdot \frac{1}{2^N} \sum_{I \subset \{1,\dots,N\}} \left(\sum_{i \in I, j \notin I} \varepsilon_i \varepsilon_j y_{ij} \right).$$

Thus, we are basically left to show that the preceding tail estimate is stable by convex combination. However, this is easily proved. Indeed, let $u = u(t) = \sigma t^2 + mt$ and denote by $t = t(u)$ the inverse function (on $\mathbb{R}_+$). The function $\psi(u) = \exp(t(u)^2/144) - 1$ is convex and increasing with $\psi(0) = 0$ (elementary computation). By (4.18) and integration by parts, we have

$$\mathbb{E}\,\psi\left(\left(\left\| \sum_{i \in I, j \notin I} \varepsilon_i \varepsilon_j y_{ij} \right\| - M \right)^+ \right) \leq 10.$$

Hence, by convexity

$$\mathbb{E}\,\psi\left(\left(\frac{1}{2} \|X\| - M \right)^+ \right) \leq 10.$$

Thus, for every $t > 0$,

$$\mathbb{P}\left\{ \left(\frac{1}{2}\|X\| - M \right)^+ > mt + \sigma t^2 \right\} \leq \frac{10}{\exp(t^2/144) - 1}$$

from which the tail estimate of the theorem follows. Note that the somewhat unsatisfactory factor 2 came in only at the very end from the decoupling formula.

By the preceding tail estimate, we already know that $\mathbb{E}\exp(\alpha\|X\|) < \infty$ for some $\alpha > 0$. To establish that this holds for every $\alpha > 0$, it clearly suffices, by the decoupling argument developed above (and Fatou's lemma) to show this integrability result for the decoupled chaos Y. We make use of the proof of Theorem 4.8 and of (4.13). Let $Z_N = \|\sum_{i,j>N} \varepsilon_i \varepsilon_j' y_{ij}\|$. By the reverse submartingale theorem, Z_N converges almost surely to some degenerate distribution. Using then (4.13) and Fubini's theorem as in the previous proof of the tail estimate, we find that there exists $M > 0$ such that, for every $\alpha > 0$, one can find N large enough with

$$\mathbb{P}\{Z_N > KM(t+1)\} \leq K \exp(-\alpha t)$$

for every $t > 0$, K numerical constant. By the one-dimensional integrability properties, the proof of Theorem 4.11 is easily completed.

We conclude this part on Rademacher chaos with a few words about the case where the diagonal elements are non-zero. Assume that we are given a finite sequence (x_{ij}) in a Banach space B. We just learned that there is a numerical constant K such that for every $p \geq 1$

$$\left\| \sum_{i \neq j} \varepsilon_i \varepsilon_j x_{ij} \right\|_p \leq Kp \left\| \sum_{i \neq j} \varepsilon_i \varepsilon_j x_{ij} \right\|_1$$

(cf. Lemma 3.7). Denote by (ε_i') a Rademacher sequence which is independent of (ε_i). Then, by independence and Jensen's inequality,

$$\left\|\sum_{i\neq j}\varepsilon_i\varepsilon_j x_{ij}\right\|_1 \leq \left\|\sum_{i\neq j}(\varepsilon_i\varepsilon_j - \varepsilon_i'\varepsilon_j')x_{ij}\right\|_1$$

$$= \left\|\sum_{i,j}(\varepsilon_i\varepsilon_j - \varepsilon_i'\varepsilon_j')x_{ij}\right\|_1$$

$$\leq 2\left\|\sum_{i,j}\varepsilon_i\varepsilon_j x_{ij}\right\|_1.$$

Hence, by difference,

$$\left\|\sum_i x_{ii}\right\| \leq 3\left\|\sum_{i,j}\varepsilon_i\varepsilon_j x_{ij}\right\|_1.$$

Therefore, for every $p \geq 1$,

$$\left\|\sum_{i,j}\varepsilon_i\varepsilon_j x_{ij}\right\|_p \leq (3+2Kp)\left\|\sum_{i,j}\varepsilon_i\varepsilon_j x_{ij}\right\|_1,$$

from which we deduce similar integrability properties for chaos with non-zero diagonal terms. However, we do not know whether the tail estimate of Theorem 4.11 extends to this setting.

4.5 Comparison Theorems

The norm of a (finite) Rademacher sum $\|\sum_{i=1}^N \varepsilon_i x_i\|$ with coefficients in a Banach space B is the supremum of a Rademacher process. For the purposes of this study, one convenient representation is

$$\left\|\sum_{i=1}^N \varepsilon_i x_i\right\| = \sup_{t\in T}\left|\sum_{i=1}^N \varepsilon_i t_i\right|$$

where T is the (compact) subset of $\mathbb{R}^N$ defined by $T = \{t = (f(x_i))_{i\leq N} ;$ $f \in B'$, $\|f\| \leq 1\}$. Therefore, we present the results about the comparison of Rademacher processes of this type, i.e. $\sum_{i=1}^N \varepsilon_i t_i$, $t = (t_1,\ldots,t_N) \in T$, T (compact) subset of $\mathbb{R}^N$.

We learned in the preceding chapter how Gaussian processes can be compared by means of their L_2-metrics. While this is not completely possible for Rademacher processes, one can however investigate analogs of some of the usual consequences of the Gaussian comparison theorems. More precisely, we establish a comparison theorem for Rademacher averages when coordinates are contracted and we prove a version of Sudakov's minoration in this context.

We start with the comparison theorem, which is analogous to Corollary 3.17. A map $\varphi : \mathbb{R} \to \mathbb{R}$ is called a contraction if $|\varphi(s) - \varphi(t)| \leq |s - t|$ for all $s, t \in \mathbb{R}$. If h is a map on some set T, we set for simplicity (and with some abuse of notation) $\|h(t)\|_T = \|h\|_T = \sup_{t \in T} |h(t)|$.

Theorem 4.12. *Let* $F : \mathbb{R}_+ \to \mathbb{R}_+$ *be convex and increasing. Let further* $\varphi_i : \mathbb{R} \to \mathbb{R}$, $i \leq N$, *be contractions such that* $\varphi_i(0) = 0$. *Then, for any bounded subset* T *in* $\mathbb{R}^N$

$$\mathbb{E}F\left(\frac{1}{2}\left\|\sum_{i=1}^N \varepsilon_i \varphi_i(t_i)\right\|_T\right) \leq \mathbb{E}F\left(\left\|\sum_{i=1}^N \varepsilon_i t_i\right\|_T\right).$$

Before we turn to the proof, note the following. The numerical constant $\frac{1}{2}$ is optimal as can be seen from the example of the subset T of $\mathbb{R}^2$ consisting of the points $(1,1)$ and $(-1,-1)$ with $\varphi_1(x) = x$, $\varphi_2(x) = -|x|$ and $F(x) = x$. One typical application of Theorem 4.12 is of course given when $\varphi_i(x) = |x|$ for each i. Another application, which will be useful in the sequel, is the following: if $(x_i)_{i \leq N}$ are points in a Banach space, then

$$(4.19) \qquad \mathbb{E}\left(\sup_{\|f\| \leq 1}\left|\sum_{i=1}^N \varepsilon_i f^2(x_i)\right|\right) \leq 4\mathbb{E}\left\|\sum_{i=1}^N \varepsilon_i \|x_i\| x_i\right\|.$$

(Recall that by the contraction principle, we can replace the right hand side by $4\max_{i \leq N} \|x_i\| \mathbb{E} \|\sum_{i=1}^N \varepsilon_i x_i\|$.) To deduce (4.19) from the theorem, simply let T be as before, i.e. $T = \{t = (f(x_i))_{i \leq N}; \|f\| \leq 1\}$, and take

$$\varphi_i(s) = \min\left(\frac{s^2}{2\|x_i\|^2}, \frac{\|x_i\|^2}{2}\right), \quad s \in \mathbb{R}, \quad i \leq N.$$

As we mentioned before, theorems like Theorem 4.12 for Gaussian averages follow from the Gaussian comparison properties. The Rademacher case involves independent (and conceptually simpler) proofs to which we now turn.

Proof of Theorem 4.12. We first show that if $G : \mathbb{R} \to \mathbb{R}$ is convex and increasing, then

$$(4.20) \qquad \mathbb{E}G\left(\sup_{t \in T}\sum_{i=1}^N \varepsilon_i \varphi_i(t_i)\right) \leq \mathbb{E}G\left(\sup_{t \in T}\sum_{i=1}^N \varepsilon_i t_i\right).$$

By conditioning and iteration, it suffices to show that if T is a subset of $\mathbb{R}^2$ and φ a contraction on $\mathbb{R}$ such that $\varphi(0) = 0$, then

$$\mathbb{E}G\left(\sup_{t \in T}(t_1 + \varepsilon_2 \varphi(t_2))\right) \leq \mathbb{E}G\left(\sup_{t \in T}(t_1 + \varepsilon_2 t_2)\right)$$

$(t = (t_1, t_2))$. We show that for all t and s in T, the right hand side is always larger than or equal to

$$I = \tfrac{1}{2}G\big(t_1 + \varphi(t_2)\big) + \tfrac{1}{2}G\big(s_1 - \varphi(s_2)\big).$$

We may assume that

(∗) $$t_1 + \varphi(t_2) \geq s_1 + \varphi(s_2)$$

and

(∗∗) $$s_1 - \varphi(s_2) \geq t_1 - \varphi(t_2).$$

We distinguish between the following cases.

1st *case.* $t_2 \geq 0$, $s_2 \geq 0$. Assume to begin with that $s_2 \leq t_2$. We show

$$2I \leq G(t_1 + t_2) + G(s_1 - s_2).$$

Set $a = s_1 - \varphi(s_2)$, $b = s_1 - s_2$, $a' = t_1 + t_2$, $b' = t_1 + \varphi(t_2)$ so that we would like to prove that

(4.21) $$G(a) - G(b) \leq G(a') - G(b').$$

Since φ is a contraction with $\varphi(0) = 0$, and $s_2 \geq 0$, $|\varphi(s_2)| \leq s_2$. Hence, $a \geq b$ and, by (∗), $b' \geq b$. Furthermore, again by contraction and $s_2 \leq t_2$,

$$a - b = s_2 - \varphi(s_2) \leq t_2 - \varphi(t_2) = a' - b'.$$

Since G is convex and increasing, for every positive x, the map $G(\cdot + x) - G(\cdot)$ is increasing. Thus, for $x = a - b \geq 0$ and with $b \leq b'$, we get

$$G(a) - G(b) \leq G\big(b' + (a - b)\big) - G(b').$$

Using that $b' + a - b \leq a'$ then yields the announced claim (4.21). When $s_2 \geq t_2$, the argument is similar changing s into t and φ into $-\varphi$.

2nd *case.* $t_2 \leq 0$, $s_2 \leq 0$. It is completely similar to the preceding case.

3rd *case.* $t_2 \geq 0$, $s_2 \leq 0$. Since $\varphi(t_2) \leq t_2$, $-\varphi(s_2) \leq -s_2$, we have

$$2I \leq G(t_1 + t_2) + G(s_1 - s_2)$$

and the result follows.

4th *case.* $t_2 \leq 0$, $s_2 \geq 0$. Similar to the third case.

This completes the proof of (4.20).

We conclude the proof of the theorem. By convexity,

$$\mathbb{E}F\left(\frac{1}{2}\left\|\sum_{i=1}^{N} \varepsilon_i \varphi_i(t_i)\right\|_T\right) \leq \frac{1}{2}\left(\mathbb{E}F\left(\sup_{t \in T}\left(\sum_{i=1}^{N} \varepsilon_i \varphi_i(t_i)\right)^+\right)\right.$$

$$+ \left.\mathbb{E}F\left(\sup_{t \in T}\left(\sum_{i=1}^{N} \varepsilon_i \varphi_i(t_i)\right)^-\right)\right)$$

$$\leq \mathbb{E}F\left(\sup_{t \in T}\left(\sum_{i=1}^{N} \varepsilon_i \varphi_i(t_i)\right)^+\right)$$

where, in the second step, we have used that, by symmetry, $(-\varepsilon_i)$ has the same distribution as (ε_i) and $(-\cdot)^- = (\cdot)^+$. Applying (4.20) to $F((\cdot)^+)$ which is convex and increasing on $\mathbb{R}$ then immediately yields the conclusion. The proof of Theorem 4.12 is complete.

After comparison properties, we describe in the last part of this chapter a version of the Sudakov minoration inequality (Theorem 3.18) for Rademacher processes.

If T is a compact subset of $\mathbb{R}^N$, set

$$r(T) = \mathbb{E} \sup_{t \in T} \left| \sum_{i=1}^{N} \varepsilon_i t_i \right|.$$

Denote by $d_2(s,t) = |s-t|$ the Euclidean metric on $\mathbb{R}^N$ and recall the entropy numbers $N(T, d_2; \varepsilon)$: $N(T, d_2; \varepsilon)$ is the minimal number of (open) balls of radius $\varepsilon > 0$ in the metric d_2 which are necessary to cover T. Equivalently $N(T, d_2; \varepsilon) = N(T, \varepsilon B_2)$ where we denote by $N(A, B)$ the minimal number of translates of B by elements of A necessary to cover A, and where B_2 is the Euclidean (open) unit ball. (We do not specify later if the balls are open or closed since this distinction clearly does not affect the various results.) The next result is the main step in the proof of Sudakov's minoration for Rademacher processes.

Proposition 4.13. *There is a numerical constant K such that for any $\varepsilon > 0$, if T is a subset of $\mathbb{R}^N$ such that $\max_{i \leq N} |t_i| \leq \varepsilon^2/Kr(T)$ for every $t \in T$, then*

$$\varepsilon \big(\log N(T, d_2; \varepsilon) \big)^{1/2} \leq Kr(T).$$

Proof. As an intermediary step, we first show that when $T \subset B_2$ and $\max_{i \leq N} |t_i| \leq 1/Kr(T)$ for every $t \in T$, then

(4.22) $\log N \big(T, \tfrac{1}{2} B_2 \big)^{1/2} \leq Kr(T)$

where K is a numerical constant.

Let g be a standard normal variable. For $s > 0$, set $h = g I_{\{|g| > s\}}$. The first simple observation of this proof is that, whenever $\lambda \leq s/4$,

(4.23) $\mathbb{E} \exp(\lambda h) \leq 1 + 16\lambda^2 \exp(-s^2/32) \leq \exp\big[16\lambda^2 \exp(-s^2/32)\big].$

For a proof, consider for example $f(\lambda) = \mathbb{E}\exp(\lambda h) - 1 - 16\lambda^2 \exp(-s^2/32)$ for $\lambda \geq 0$. Since $f(0) = f'(0) = 0$, it suffices to check that $f''(\lambda) \leq 0$ when $\lambda \leq s/4$. Now,

$$f''(\lambda) = \mathbb{E}(h^2 \exp \lambda h) - 32 \exp(-s^2/32).$$

By definition of h and a change of variables,

$$\mathbb{E}(h^2 \exp \lambda h) = \int_{|x|>s} x^2 \exp\Big(\lambda x - \frac{x^2}{2}\Big) \frac{dx}{\sqrt{2\pi}}$$

$$= \exp\Big(\frac{\lambda^2}{2}\Big) \int_{|x+\lambda|>s} (x+\lambda)^2 \exp\Big(-\frac{x^2}{2}\Big) \frac{dx}{\sqrt{2\pi}}.$$

When $\lambda \le s/4 \le s/2$, $s < |x + \lambda| \le |x| + s/2 \le 2|x|$ so that

$$\mathbb{E}(h^2 \exp \lambda h) \le 4 \exp\Big(\frac{\lambda^2}{2}\Big) \int_{|x|>s/2} x^2 \exp\Big(-\frac{x^2}{2}\Big) \frac{dx}{\sqrt{2\pi}}$$

$$\le 16 \exp\Big(\frac{\lambda^2}{2}\Big) \int_{|x|>s/2} \exp\Big(-\frac{x^2}{4}\Big) \frac{dx}{\sqrt{2\pi}}$$

$$\le 32 \exp\Big(\frac{\lambda^2}{2} - \frac{s^2}{16}\Big) \le 32 \exp\Big(-\frac{s^2}{32}\Big)$$

which gives the result.

Now, let us show (4.22). There is nothing to prove if $T \subset \frac{1}{2}B_2$ so that we may assume that there is an element t of T with $|t| \ge 1/2$; then, by (4.3), $r(T) \ge 1/2\sqrt{2}$. By definition of $N(T, \frac{1}{2}B_2)$ there exists a subset U of T of cardinality $N(T, \frac{1}{2}B_2)$ such that $d_2(u, v) \ge 1/2$ whenever u, v are distinct elements in U. Let us then consider the Gaussian process $(\sum_{i=1}^{N} g_i u_i)_{u \in U}$ where (g_i) is an orthogaussian sequence. As a consequence of Sudakov's minoration and the Gaussian integrability properties (Theorem 3.18 and Corollary 3.2), there is a numerical constant $K' \ge 1$ such that

$$\mathbb{P}\Big\{\sup_{u \in U}\Big|\sum_{i=1}^{N} g_i u_i\Big| > (K')^{-1}(\log \operatorname{Card} U)^{1/2}\Big\} \ge \frac{1}{2}.$$

We use a kind of *a priori* estimate argument. Let $K = (100K')^2$ and assume that $\max_{i \le N} |t_i| \le 1/Kr(T)$ for every $t \in T$. We claim that whenever $\alpha \ge 1$ satisfies $(\log \operatorname{Card} U)^{1/2} \le \alpha Kr(T)$, then

$$\mathbb{P}\Big\{\sup_{u \in U}\Big|\sum_{i=1}^{N} g_i u_i\Big| > \alpha \frac{K}{2K'} r(T)\Big\} < \frac{1}{2}$$

so that, intersecting the sets, $(\log \operatorname{Card} U)^{1/2} \le \alpha Kr(T)/2$. Of course, this ensures that $(\log \operatorname{Card} U)^{1/2} \le Kr(T)$ which is the conclusion (4.22).

For every i, set $h_i = g_i I_{\{|g_i|>s\}}$, $k_i = g_i - h_i$. By the triangle inequality and the contraction principle, since $|k_i| \le s$,

$$\mathbb{P}\Big\{\sup_{u \in U}\Big|\sum_{i=1}^{N} g_i u_i\Big| > \alpha \frac{K}{2K'} r(T)\Big\} \le \frac{4K's}{\alpha K} + \mathbb{P}\Big\{\sup_{u \in U}\Big|\sum_{i=1}^{N} h_i u_i\Big| > \alpha \frac{K}{4K'} r(T)\Big\}.$$

By (4.23), for any $\lambda \le sKr(T)/4$, and since $U \subset T \subset B_2$,

$$\mathbb{P}\left\{\sup_{u\in U}\left|\sum_{i=1}^{N}h_iu_i\right| > \alpha\frac{K}{4K'}r(T)\right\}$$

$$\leq 2\mathrm{Card}\,U\exp\left[-\lambda\alpha\frac{K}{4K'}r(T) + 16\lambda^2\exp\left(-\frac{s^2}{32}\right)\right]$$

$$\leq 2\exp\left[\alpha^2K^2r(T)^2 - \lambda\alpha\frac{K}{4K'}r(T) + 16\lambda^2\exp\left(-\frac{s^2}{32}\right)\right].$$

Let $s = \alpha K/10K'$ and $\lambda = \alpha K^2 r(T)/40K'$. Then

$$\mathbb{P}\left\{\sup_{u\in U}\left|\sum_{i=1}^{N}g_iu_i\right| > \alpha\frac{K}{4K'}r(T)\right\}$$

$$\leq \frac{2}{5} + 2\exp\left[\alpha^2K^2r(T)^2\left(1 - \frac{K}{160(K')^2} + \frac{16K^2}{(40K')^2}\exp\left(-\frac{\alpha^2K^2}{32(10K')^2}\right)\right)\right].$$

If we recall that $\alpha \geq 1$, $r(T) \geq 1/2\sqrt{2}$ and $K = (100K')^2 \geq 10^4$, it is clear that the preceding probability is made strictly less than $1/2$. This was the announced claim.

To reach the full conclusion of the proposition, we use a simple iteration procedure. For each t and $\delta > 0$, denote by $B_2(t,\delta)$ the Euclidean ball with center t and radius δ. Let $\varepsilon > 0$ and let k be an integer such that $2^{-k} < \varepsilon \leq 2^{-k+1}$. Then,

$$N(T, d_2; \varepsilon) = N(T, \varepsilon B_2) \leq N(T, 2^{-k}B_2).$$

Clearly

$$N(T, 2^{-k}B_2) \leq \prod_{\ell\leq k}\sup_{t\in T}N\big(T\cap B_2(t, 2^{-\ell+1}), 2^{-\ell}B_2\big).$$

By homogeneity, (4.22) tells us that

$$N(T\cap B_2(t, 2^{-\ell+1}), 2^{-\ell}B_2) \leq \exp(K2^{2\ell-2}r(T)^2).$$

Hence,

$$N(T, d_2; \varepsilon) \leq \exp\left(K\sum_{\ell\leq k}2^{2\ell-2}r(T)^2\right) \leq \exp\left(4K\frac{r(T)^2}{\varepsilon^2}\right).$$

Therefore, Proposition 4.13 is established.

The previous proposition yields a first version of Sudakov's minoration for Rademacher processes which however still involves a factor depending on the dimension. It can be stated as follows.

Corollary 4.14. *Let T be a subset of $\mathbb{R}^N$; for every $\varepsilon > 0$*

$$\varepsilon\big(\log N(T, d_2; \varepsilon)\big)^{1/2} \leq Kr(T)\left(\log\left(2 + \frac{\sqrt{N}}{r(T)}\right)\right)^{1/2}$$

where $K > 0$ is some numerical constant.

Proof. As before, let us first assume that $T \subset B_2$ and estimate $N(T, \frac{1}{2}B_2)$. Denote by K_1 the numerical constant in Proposition 4.13. We can write

$$N\left(T, \frac{1}{2}B_2\right) \leq N\left(T, \frac{1}{K_1 r(T)}B_\infty\right) N\left((T-T) \cap \frac{1}{K_1 r(T)}B_\infty, \frac{1}{2}B_2\right)$$

$$\leq N\left(B_2, \frac{1}{K_1 r(T)}B_\infty\right) N\left((T-T) \cap \frac{1}{K_1 r(T)}B_\infty, \frac{1}{2}B_2\right)$$

where B_∞ is the unit ball for the sup-norm in $\mathbb{R}^N$. It is known that (see [Schü, Theorem 1])

$$\left(\log N\left(B_2, \frac{1}{K_1 r(T)}B_\infty\right)\right)^{1/2} \leq K_2 r(T)\left(\log\left(2 + \frac{\sqrt{N}}{r(T)}\right)\right)^{1/2}$$

where K_2 is a numerical constant (it can be assumed that $r(T)$ is bounded below). Combining with Proposition 4.13 we get that for some numerical constant K_3,

$$\left(\log N\left(T, \frac{1}{2}B_2\right)\right)^{1/2} \leq K_3 r(T)\left(\log\left(2 + \frac{\sqrt{N}}{r(T)}\right)\right)^{1/2}.$$

Then, we can use an iteration argument similar to that used in the proof of Proposition 4.13 to obtain the inequality of the corollary. The proof is complete.

Now, we turn to another version of Sudakov's minoration. The example of T consisting of the canonical basis of $\mathbb{R}^N$ for which clearly $r(T) = 1$ and $N(T, B_2) = N$ indicates that Sudakov's minoration for Gaussian variables cannot extend litterally to Rademacher averages. On the other hand, note that if $T \subset B_1$, the ℓ_1^N-ball, then $r(T) \leq 1$. This suggests the possibility of some interpolation and of a minoration involving *both* B_2 and B_1, the unit balls of ℓ_2^N and ℓ_1^N respectively. This is the conclusion of the next statement.

Theorem 4.15. *Let T be a (compact) subset of $\mathbb{R}^N$ and let $r(T) = \mathbb{E}\sup_{t \in T} |\sum_{i=1}^{N} \varepsilon_i t_i|$. There exists a numerical constant K such that if $\varepsilon > 0$ and if $D = Kr(T)B_1 + \varepsilon B_2$, then*

$$\varepsilon\left(\log N(T, D)\right)^{1/2} \leq Kr(T)$$

where we recall that $N(T, D)$ is the minimal number of translates of D by elements of T which are necessary to cover T.

Proof. The idea is to use Proposition 4.13 by changing the balls D into ℓ_2-balls for another T. Let K_1 be the constant of the conclusion in Proposition 4.13 and set $K = 3K_1$ which we would like to fit the statement of Theorem 4.15. Set $a = \varepsilon^2/Kr(T)$ and let M be an integer larger than $r(T)/a$. Define a map φ in the following way:

$$\varphi : [-Ma, +Ma] \to \mathbb{R}^{[-M,M]}$$
$$u \to (\varphi(u)_j)_{-M \le j \le M} \, ;$$

$\varphi(u)_j$ is defined according to the following rule: if $u \in [0, Ma]$, $k = k(u)$ is the integer part of u/a; we then set

$$\varphi(u)_j = a \quad \text{for} \ 1 \le j \le k$$
$$\varphi(u)_{k+1} = u - ka$$
$$\varphi(u)_j = 0 \quad \text{for all other values of} \ j \, .$$

If $u \in [-Ma, 0]$, we let $\varphi(u)_j = \varphi(-u)_{-j}$. We mention some elementary properties of φ. First, for every u, u' in $[-Ma, Ma]$,

$$(4.24) \qquad \sum_{j=-M}^{M} |\varphi(u)_j - \varphi(u')_j| = |u - u'| \, .$$

Another elementary property is the following. Suppose that we are given u, u' in $[-Ma, Ma]$ and assume that $u' < u$. Let us define $v' \le v$ in the following way: if $u \ge 0$, let $v = k(u)a$ and $v' = (k(u') + 1)a$ or $v' = k(u')a$ according whether $u' \ge 0$ or $u' < 0$; if $u < 0$, we let $v = (k(u) - 1)a$ and $v' = k(u')a$. Then, we have

$$(4.25) \qquad \sum_{j=-M}^{M} |\varphi(u)_j - \varphi(u')_j|^2 = |u - v|^2 + |u' - v'|^2 + a|v - v'| \, .$$

Once these properties have been observed, let

$$\psi : T \to \left(\mathbb{R}^{[-M,M]}\right)^{N}$$
$$t = (t_i)_{i \le N} \to \left((\varphi(t_i)_j)_{-M \le j \le M}\right)_{i \le N} \, .$$

Of course, ψ is well defined since if $t \in T$, then $|t_i| \le r(T) \le Ma$ for every $i = 1, \ldots, N$. Consider now a doubly indexed Rademacher sequence (ε_{ij}) and another sequence (ε_i') which we assume to be independent of (ε_{ij}). Then, by symmetry,

$$r(\psi(T)) = \mathbb{E} \sup_{t \in T} \left| \sum_{i=1}^{N} \sum_{j=-M}^{M} \varepsilon_{ij} \varphi(t_i)_j \right| = \mathbb{E} \sup_{t \in T} \left| \sum_{i=1}^{N} \varepsilon_i' \left(\sum_{j=-M}^{M} \varepsilon_{ij} \varphi(t_i)_j \right) \right| .$$

Now, for every choice of (ε_{ij}), every i and every t, t' in T, by (4.24),

$$\left| \sum_{j=-M}^{M} \varepsilon_{ij} \varphi(t_i)_j - \sum_{j=-M}^{M} \varepsilon_{ij} \varphi(t_i')_j \right| \le |t_i - t_i'| \, .$$

This means that, with respect to (ε_i'), we are in a position to apply the comparison Theorem 4.12 from which we get that $r(\psi(T)) \le 3r(T)$. Now, by construction, for every t, i, j,

$$|\varphi(t_i)_j| \leq a = \frac{\varepsilon^2}{3K_1 r(T)} \leq \frac{\varepsilon^2}{K_1 r(\psi(T))}.$$

Hence, by Proposition 4.13 applied to $\psi(T)$ in $(\mathbb{R}^{[-M,M]})^N$,

$$\varepsilon\left(\log N\left(\psi(T), \varepsilon B_2\right)\right)^{1/2} \leq K_1 r\left(\psi(T)\right) \leq K r(T).$$

Now, this implies the conclusion; indeed, by (4.25), if t, t' are such that $|\psi(t) - \psi(t')| \leq \varepsilon$, then

$$t \in t' + K r(T) B_1 + \varepsilon B_2.$$

Hence $N(T, D) \leq N(\psi(T), \varepsilon B_2)$ and the proof of Theorem 4.15 is therefore complete.

We conclude this chapter with a remark on the tensorization of Rademacher series. As in the Gaussian case, (and we follow here the notations introduced in Section 3.3) we ask ourselves if, given (x_i) and (y_i) in Banach spaces E and F respectively such that $\sum_i \varepsilon_i x_i$ and $\sum_i \varepsilon_i y_i$ both converge almost surely, is this also true for $\sum_{i,j} \varepsilon_{ij} x_i \otimes y_j$ in the injective tensor product $E \check{\otimes} F$. To investigate this question, first recall (4.8). For a large class of Banach spaces (which will be described in Chapter 9 as the spaces having a finite cotype), convergence of Rademacher series $\sum_i \varepsilon_i x_i$ and of corresponding Gaussian series $\sum_i g_i x_i$ are equivalent. Therefore, according to Theorem 3.20, if E and F have this property, the answer to the preceding question is yes. What we would like to briefly point out here is that this is not the case in general.

Let $(x_i)_{i \leq N}$ (resp. $(y_i)_{i \leq N}$) be a finite sequence in E (resp. F) and set

$$X = \sum_{i=1}^{N} \varepsilon_i x_i \quad \left(\text{resp. } Y = \sum_{i=1}^{N} \varepsilon_i y_i\right).$$

Recall

$$\sigma(X) = \sup_{\|f\| \leq 1} \left(\sum_{i=1}^{N} f^2(x_i)\right)^{1/2} \quad \left(\sigma(Y) = \sup_{\|f\| \leq 1} \left(\sum_{i=1}^{N} f^2(y_i)\right)^{1/2}\right).$$

Then, for the Rademacher average $\sum_{i,j=1}^{N} \varepsilon_{ij} x_i \otimes y_j$ in $E \check{\otimes} F$ we have:

(4.26)
$$\mathbb{E}\left\|\sum_{i,j=1}^{N} \varepsilon_{ij} x_i \otimes y_j\right\|_{\vee}$$
$$\leq K\left(\log(N+1)\right)^{1/2}\left(\sigma(X)\mathbb{E}\|Y\| + \sigma(Y)\mathbb{E}\|X\|\right)$$

where K is a numerical constant. This inequality is an immediate consequence of, respectively, (4.8), Theorem 3.20 and (4.9). The point of this observation is that (4.26) is best possible in general. To see this, let $E = \ell_\infty^N$, x_i be the elements of the canonical basis, and $F = \mathbb{R}$, $y_i = N^{-1/2}$, $i = 1, \ldots, N$. Then clearly $\sigma(X) = \sigma(Y) = 1$, $\mathbb{E}\|X\| = 1$, $\mathbb{E}\|Y\| \leq 1$. However, by definition of the tensor product norm,

$$\mathbb{E}\Big\| \sum_{i,j=1}^{N} \varepsilon_{ij} x_i \otimes y_j \Big\|_\vee = \mathbb{E}\Big(\max_{i \leq N} \frac{1}{\sqrt{N}} \Big| \sum_{j=1}^{N} \varepsilon_{ij} \Big| \Big)$$

and this quantity turns out to be of the order of $(\log N)^{1/2}$. Indeed, by (4.2), for some numerical constant $K > 0$,

$$(4.27) \qquad \mathbb{P}\Big\{ \frac{1}{\sqrt{N}} \sum_{i=1}^{N} \varepsilon_i > K^{-1}(\log N)^{1/2} \Big\} \geq \frac{1}{N}$$

(at least for all large enough N's). Then

$$\mathbb{E}\Big(\max_{i \leq N} \frac{1}{\sqrt{N}} \Big| \sum_{i=1}^{N} \varepsilon_{ij} \Big| \Big) \geq \int_0^{K^{-1}(\log N)^{1/2}} \mathbb{P}\Big\{ \max_{i \leq N} \frac{1}{\sqrt{N}} \Big| \sum_{j=1}^{N} \varepsilon_{ij} \Big| > t \Big\} \, dt$$

$$\geq \int_0^{K^{-1}(\log N)^{1/2}} \Big[1 - \Big(1 - \mathbb{P}\Big\{ \frac{1}{\sqrt{N}} \Big| \sum_{i=1}^{N} \varepsilon_i \Big| > t \Big\} \Big)^N \Big] \, dt$$

$$\geq \int_0^{K^{-1}(\log N)^{1/2}} \Big[1 - \Big(1 - \frac{1}{N} \Big)^N \Big] \, dt$$

$$\geq K^{-1} \Big(1 - \frac{1}{e} \Big) (\log N)^{1/2}$$

which proves our claim.

Notes and References

The name of Bernoulli is historically more appropriate for a random variable taking the values ± 1 with equal probability. Strictly speaking, the Rademacher sequence is the sequence on $[0, 1]$ defined by $r_i(t) = \text{sin}(2\pi i t)$ $(i \geq 1)$. We decided to use the terminology of Rademacher sequence since it is commonly used in the field as well as in the Geometry of Banach spaces.

The best constants in Khintchine's inequality were obtained by U. Haagerup [Ha]. See [Sz] for the case $p = 1$ (4.3). Lemma 4.2 is in the spirit of the Paley-Zygmund inequality (cf. [Kal]). Lemma 4.3 has been observed in [R-S] and [Pi10] and used in Probability in Banach spaces in [M-P2] (cf. Chapter 13).

The contraction principle has been discovered by J.-P. Kahane [Kal]. Some further extensions have been obtained by J. Hoffmann-Jørgensen [HJ1], [HJ2], [HJ3]. Lemma 4.6 is taken from [J-M2].

In [Kal] (first edition), J.-P. Kahane showed that an almost surely convergent Rademacher series $X = \sum_i \varepsilon_i x_i$ with coefficients in a Banach space satisfies $\mathbb{E}\exp(\alpha\|X\|) < \infty$ for every $\alpha > 0$ and that all its moments are equivalent. Using Lemma 4.5, S. Kwapień [Kw3] improved this integrability to $\mathbb{E}\exp(\alpha\|X\|^2) < \infty$ for some (and also all) $\alpha > 0$ (Theorem 4.7). The proof of this result presented here is different and is based on isoperimetry. Theorem 4.8 about bounded Rademacher processes is perhaps new; its proof uses Lemma 4.9 which was noticed independently in [MS2].

The hypercontractivity inequality (4.16) was established by L. Gross (as a logarithmic Sobolev inequality) [Gro] and by W. Beckner (as a two point inequality and with the extension to the complex case) [Bec]. Its interest as for the integrability of Rademacher chaos was pointed out by C. Borell [Bo5]. Complete details may be found in [Pi4] where the early contribution of A. Bonami [Bon] is pointed out. The decoupling argument used in the proof of Theorem 4.11 is inspired from [B-T1]. General results on the decoupling tool may be found in [K-S], [Kw4], [MC-T1], [MC-T2], [Zi3] etc.

The comparison Theorem 4.12 is due to the second named author and first appeared in [L-T4] (the proof presented here being simpler). Proposition 4.13 and Theorem 4.15 are recent results of the second author while Corollary 4.14 is essentially in [C-P] (see also [Pa] for some earlier related results). (4.26) and the fact that it is best possible belong to the folklore. Theorem 4.15 is in particular applied in [Ta18].

5. Stable Random Variables

After Gaussian variables and Rademacher series, we investigate in this chapter another important class of random variables and vectors, namely stable random variables. Stable random variables are fundamental in Probability Theory and, as will be seen later, also play a rôle in structure theorems of Banach spaces. The literature is rather extensive on this topic and we only concentrate here on the parts of the theory which will be of interest and use to us in the sequel. In particular, we do not attempt to study stable measures in the natural more general setting of infinitely divisible distributions. We refer to [Ar-G2] and [Li] for such a study. We only concentrate on the aspects of stable distributions analogous to those developed in the preceding chapters on Gaussian and Rademacher variables. In particular, our study is based on a most useful representation of stable random variables detailed in the first paragraph. The second section examines integrability properties and tail behavior of norms of infinite dimensional stable random variables. Finally, the last section is devoted to some comparison theorems.

We recall that, for $0 < p < \infty$, $L_{p,\infty} = L_{p,\infty}(\Omega, \mathcal{A}, \mathbb{P})$ denotes the space of all real valued random variables X on $(\Omega, \mathcal{A}, \mathbb{P})$ such that

$$\|X\|_{p,\infty} = \left(\sup_{t>0} t^p \mathbb{P}\{|X| > t\}\right)^{1/p} < \infty.$$

$\|\cdot\|_{p,\infty}$ is only a quasi-norm but is equivalent to a norm when $p > 1$; take for example

$$(5.1) \qquad N_p(X) = \sup\left\{\mathbb{P}(A)^{-1/q} \int_A |X| d\mathbb{P} \; ; \; A \in \mathcal{A}, \; \mathbb{P}(A) > 0\right\}$$

where $q = p/p - 1$ is the conjugate of p, for which we have that, for all X, $\|X\|_{p,\infty} \leq N_p(X) \leq q\|X\|_{p,\infty}$ (integration by parts). A random variable X in L_p is of course in $L_{p,\infty}$, satisfying even $\lim_{t\to\infty} t^p \mathbb{P}\{|X| > t\} = 0$. Conversely, the space of all random variables such that this is limit 0 is the closure in the $L_{p,\infty}$-norm of the step random variables. Recall also the comparisons with the L_r-norms: for every $r > p$ and every random variable X,

$$\|X\|_{p,\infty} \leq \|X\|_p \leq \left(\frac{r}{r-p}\right)^{1/p} \|X\|_{r,\infty}.$$

Finally, if B is a Banach space, we denote by $L_{p,\infty}(B)$ the space of all random variables X in B such that $\|X\| \in L_{p,\infty}$; we simply let $\|X\|_{p,\infty} = \|\|X\|\|_{p,\infty}$.

As in the Gaussian case, we only consider *symmetric* stable random variables. A real valued (symmetric) random variable X is called p-stable, $0 < p \leq 2$, if, for some $\sigma \geq 0$, its Fourier transform is of the form

$$\mathbb{E} \exp(itX) = \exp\left(-\sigma^p |t|^p / 2\right), \quad t \in \mathbb{R}.$$

$\sigma = \sigma_p = \sigma_p(X)$ is called the *parameter* of the stable random variable X with *index p*. A 2-stable random variable with parameter σ is just Gaussian with variance σ^2. If $\sigma = 1$, X is called *standard*. As for Gaussian variables, when we speak of a *standard p-stable sequence* (θ_i), we always mean a sequence of independent standard p-stable random variables θ_i (the index p will be clear from the context).

Despite the analogy in the definition, the case $p = 2$ corresponding to Gaussian variables and the case $0 < p < 2$ present some quite important differences. For example, while stable distributions have densities, these cannot be easily expressed in general. Further, if Gaussian variables have exponential moments, a non-zero p-stable random variable X, $0 < p < 2$, is not even in L_p. It can however be shown that $\|X\|_{p,\infty} < \infty$ and actually that

$$(5.2) \qquad \lim_{t \to \infty} t^p \mathbb{P}\{|X| > t\} = c_p^p \sigma^p$$

where σ is the parameter of X and $c_p > 0$ only depends on p (cf. e.g. [Fe1]). X has therefore moments of order r for every $r < p$ and $\|X\|_r = c_{p,r}\sigma$ where $c_{p,r}$ depends on p and r only.

Stable random variables are characterized by their fundamental "stability" property (from which they draw their name): if (θ_i) is a standard p-stable sequence, for any finite sequence (α_i) of real numbers, $\sum_i \alpha_i \theta_i$ has the same distribution as $\left(\sum_i |\alpha_i|^p\right)^{1/p} \theta_1$. In particular, by what preceeds, for any $r < p$,

$$\left\|\sum_i \alpha_i \theta_i\right\|_r = c_{p,r} \left(\sum_i |\alpha_i|^p\right)^{1/p}$$

so that the span in L_r, $r < p$, of the sequence (θ_i) is isometric to ℓ_p. This property, which is analogous to what we learned in the Gaussian case but with a smaller spectrum in r, is of fundamental interest in the study of ℓ_p^n-subspaces of Banach spaces (cf. Chapter 9). In this order of ideas and among the consequences of (5.2), we would like to note further that if (θ_i) is a standard p-stable sequence with $0 < p < 2$ and (α_i) a sequence of real numbers such that $\sup_i |\alpha_i \theta_i| < \infty$ almost surely, then $\sum_i |\alpha_i|^p < \infty$. Indeed, we have from the Borel-Cantelli lemma (cf. Lemma 2.6) that for some $M' > 0$

$$\sum_i \mathbb{P}\{|\alpha_i \theta_i| > M'\} < \infty.$$

It already follows that (α_i) is bounded, i.e. $\sup_i |\alpha_i| \leq M''$ for some M''. By (5.2), there exists t_0 such that for all $t \geq t_0$

$$\mathbb{P}\{|\theta_1| > t\} \geq \frac{1}{2c_p^p t^p}.$$

Hence, letting $M = \max(M', t_0 M'')$, we have

$$\infty > \sum_i \mathbb{P}\{|\alpha_i \theta_i| > M\} \geq \frac{1}{2M^p c_p^p} \sum_i |\alpha_i|^p.$$

This kind of result is of course completely different in the case $p = 2$ for which we recall (see (3.7)), as an example and for the matter of comparison, that if (g_i) is an orthogaussian sequence,

$$\limsup_{i \to \infty} \frac{|g_i|}{\left(2 \log(i+1)\right)^{1/2}} = 1 \quad \text{almost surely}.$$

A random variable $X = (X_1, \ldots, X_N)$ with values in $\mathbb{R}^N$ is p-stable if each linear combination $\sum_{i=1}^N \alpha_i X_i$ is a real valued p-stable variable. A random process $X = (X_t)_{t \in T}$ indexed by a set T is called p-stable if for every $t_1, \ldots, t_N$ in T, $(X_{t_1}, \ldots, X_{t_N})$ is a p-stable random vector. Similarly, a Radon random variable X with values in a Banach space B is p-stable if $f(X)$ is p-stable for every f in B'. By their very definition, all these p-stable random vectors satisfy the fundamental and *characteristic* stability property of stable distributions: if, and only if, X is p-stable, if X_i are independent copies of X, then

$$\sum_i \alpha_i X_i \quad \text{has the same distribution as} \quad \left(\sum_i |\alpha_i|^p\right)^{1/p} X$$

for every finite sequence (α_i) of real numbers.

It will almost always be assumed in the sequel that $0 < p < 2$. The case $p = 2$ corresponding to Gaussian variables was investigated in Chapter 3.

5.1 Representation of Stable Random Variables

p-stable $(0 < p < 2)$ finite or infinite dimensional random variables can be given (in distribution) a series representation. This representation can be thought of as some central limit theorem with stable limits and we will actually have the opportunity to verify this observation in the sequel. This representation is a most valuable tool in the study of stable random variables; it almost allows to think of stable variables as sums of nicely behaved independent random variables for which a large variety of tools is available. We use it almost automatically each time we deal with stable distributions.

To introduce this representation we first investigate the scalar case. We need some notation. Let (λ_i) be independent random variables with common exponential distribution $\mathbb{P}\{\lambda_i > t\} = e^{-t}$, $t \geq 0$. Set $\Gamma_j = \sum_{i=1}^j \lambda_i$, $j \geq 1$, which will always have the same meaning throughout the book. The sequence

$(\Gamma_j)_{j \geq 1}$ defines the successive times of jump of a standard Poisson process (cf. [Fel]). As is easy to see,

$$\mathbb{P}\{\Gamma_j \leq t\} = \int_0^t \frac{x^{j-1}}{(j-1)!} \, e^{-x} \, dx \,, \quad t \geq 0 \,.$$

In particular,

$$\mathbb{P}\{\Gamma_j^{-1/p} > t\} \leq \frac{1}{j!} \cdot \frac{1}{t^{pj}}$$

for all $0 < p < \infty$, $j \geq 1$ and $t > 0$. It already follows that while

(5.3)
$$\lim_{t \to \infty} t^p \mathbb{P}\{\Gamma_1^{-1/p} > t\} = 1 \,,$$

for $j \geq 2$ we have

(5.4)
$$\|\Gamma_j^{-1/p}\|_p < \infty$$

(actually $\|\Gamma_j^{-1/p}\|_r < \infty$ for any $r < pj$), hence $\lim_{t \to \infty} t^p \mathbb{P}\{\Gamma_j^{-1/p} > t\} = 0$.

By the strong law of large numbers, $\Gamma_j/j \to 1$ almost surely. A powerful method will be to replace Γ_j by j which is non random. We already quote at this stage a first observation: for any $\alpha > 0$ and $j > \alpha$,

(5.5)
$$\mathbb{E}(\Gamma_j^{-\alpha}) = \frac{\Gamma(j - \alpha)}{\Gamma(j)} \sim \frac{1}{j^\alpha}$$

as can easily be seen from Stirling's formula (Γ is the gamma function).

Provided with these easy observations, the series representation of p-stable random variables can be formulated as follows.

Theorem 5.1. *Let $0 < p < 2$ and let η be a symmetric real valued random variable such that $\mathbb{E}|\eta|^p < \infty$. Denote further by (η_j) independent copies of η assumed to be independent from the sequence (Γ_j). Then, the almost surely convergent series*

$$X = \sum_{j=1}^\infty \Gamma_j^{-1/p} \eta_j$$

defines a p-stable random variable with parameter $\sigma = c_p^{-1} \|\eta\|_p$ (where c_p has been introduced in (5.2)).

Proof. Let us first convince ourselves that the sum defining X converges almost surely. To this aim, we prove a little bit more than is necessary but this will be useful later in the proof. Let us show indeed that

(5.6)
$$\lim_{j_0 \to \infty} \sup_{N \geq j_0} \left\| \sum_{j=j_0}^N \Gamma_j^{-1/p} \eta_j \right\|_{p,\infty} = 0 \,.$$

If this is satisfied, the sum $\sum_{j=1}^\infty \Gamma^{-1/p} \eta_j$ is in particular convergent in all L_r, $r < p$, and thus in probability. Since $(\Gamma_j^{-1/p} \eta_j)$ is a symmetric sequence, the series converges almost surely by Itô-Nisio's theorem. Thus, let us establish (5.6). An alternate proof making use of some of the material introduced later in this chapter is given at the end of Section 5.2. For every $t > 0$ and $2 \leq j_0 \leq N$,

$$\mathbb{P}\left\{\left|\sum_{j=j_0}^{N} \Gamma_j^{-1/p} \eta_j\right| > t\right\} \leq \mathbb{P}\{\exists j \geq j_0, \ |\eta_j| > tj^{1/p}\}$$

$$+ \frac{1}{t^2} \sum_{j \geq j_0} \mathbb{E}(\Gamma_j^{-2/p} \eta_j^2 I_{\{|\eta_j| \leq tj^{1/p}\}}).$$

Clearly

$$\mathbb{P}\{\exists j \geq j_0, \ |\eta_j| > tj^{1/p}\} \leq \sum_{j \geq j_0} \mathbb{P}\{|\eta| > tj^{1/p}\} \leq \frac{1}{t^p} \mathbb{E}(|\eta|^p I_{\{|\eta| > tj_0^{1/p}\}})$$

while, by (5.5), provided j_0 is large enough, for some constants C, C',

$$\frac{1}{t^2} \sum_{j \geq j_0} \mathbb{E}(\Gamma_j^{-2/p} \eta_j^2 I_{\{|\eta_j| \leq tj^{1/p}\}}) \leq \frac{C}{t^2} \mathbb{E}\left(|\eta|^2 \sum_{j \geq j_0} j^{-2/p} I_{\{j \geq |\eta|^p/t^p\}}\right)$$

$$\leq \frac{C'}{t^2} \cdot \frac{1}{j_0^{(2/p)-1}} \mathbb{E}(|\eta|^2 I_{\{|\eta| \leq tj_0^{1/p}\}})$$

$$+ \frac{C'}{t^p} \mathbb{E}(|\eta|^p I_{\{|\eta| > tj_0^{1/p}\}}).$$

The conclusion follows: for every $\varepsilon > 0$, $a > 0$ and j_0 large enough, we have obtained

$$\sup_{N \geq j_0} \left\|\sum_{j=j_0}^{N} \Gamma_j^{-1/p} \eta_j\right\|_{p,\infty}^p \leq \varepsilon^p + \frac{C'a^2}{\varepsilon^{2-p} j_0^{(2/p)-1}} + C'\mathbb{E}(|\eta|^p I_{\{|\eta| > a\}})$$

$$+ (C' + 1)\mathbb{E}(|\eta|^p I_{\{|\eta| > \varepsilon j_0^{1/p}\}}).$$

Since $\mathbb{E}|\eta|^p < \infty$, we can let j_0 tend to infinity, then a also, and then ε to 0 to get the conclusion.

In order to establish the theorem, we show that X satisfies the characteristic property of p-stable random variables, namely that if X_1 and X_2 are independent copies of X, for all real numbers α_1, α_2,

$$\alpha_1 X_1 + \alpha_2 X_2 \text{ has the same distribution as } (|\alpha_1|^p + |\alpha_2|^p)^{1/p} X.$$

Write $X_i = \sum_{j=1}^{\infty} \Gamma_{ji}^{-1/p} \eta_{ji}$, $i = 1, 2$, where $\{(\Gamma_{ji})_{j \geq 1}, (\eta_{ji})_{j \geq 1}\}$ for $i = 1, 2$ are independent copies of $\{(\Gamma_j)_{j \geq 1}, (\eta_j)_{j \geq 1}\}$. Set $a_i = |\alpha_i|^p$. Consider the non-decreasing rearrangement $\{\gamma_j^*; \ j \geq 1\}$ of the countable set $\{\Gamma_{ji}/a_i; \ j \geq 1, \ i = 1, 2\}$. The sequence $\{\Gamma_{ji}/a_i; \ j \geq 1\}$ corresponds to the successive times of the jumps of a Poisson process N^i of parameter a_i, $i = 1, 2$. It is easily seen that $\{\gamma_j^*; \ j \geq 1\}$ then corresponds to the sequence of the successive times of jump of the process $N^1 + N^2$. Now $N^1 + N^2$ is a Poisson process of parameter $a_1 + a_2$. Hence $\{\gamma_j^* \, ; \; j \geq 1\}$ has the same distribution as the sequence $\{\Gamma_j/(a_1 + a_2) \, ; \; j \geq 1\}$. Therefore, we have the following equalities in distribution:

$$\alpha_1 X_1 + \alpha_2 X_2 = \sum_{j=1}^{\infty} (\gamma_j^*)^{-1/p} \eta_j = (a_1 + a_2)^{1/p} \sum_{j=1}^{\infty} \Gamma_j^{-1/p} \eta_j$$

$$= (|\alpha_1|^p + |\alpha_2|^p)^{1/p} X \, .$$

Hence X is p-stable. The final step of the proof consists in showing that X has parameter $c_p^{-1}\|\eta\|_p$. To this aim, we identify the limit (5.2) and use the fact established in the first part of this proof. We have indeed from (5.6) and (5.4) that

$$\lim_{t \to \infty} t^p \mathbb{P} \left\{ \left| \sum_{j=2}^{\infty} \Gamma_j^{-1/p} \eta_j \right| > t \right\} = 0 \, .$$

Now, from (5.3) and independence, we see that

$$\lim_{t \to \infty} t^p \mathbb{P} \left\{ \Gamma_1^{-1/p} |\eta_1| > t \right\} = \mathbb{E}|\eta|^p \, .$$

Hence, combining these observations yields

$$\lim_{t \to \infty} t^p \mathbb{P} \left\{ |X| > t \right\} = \mathbb{E}|\eta|^p$$

and by comparison with (5.2) we indeed get that X has parameter $c_p^{-1}\|\eta\|_p$. The proof of Theorem 5.1 is complete.

After the scalar case, we now attack the case of infinite dimensional stable random vectors and processes. The key tool in this investigation is the concept of *spectral measure*. The spectral measure of a stable variable arises from the general theory of Lévy-Khintchine representations of infinitely divisible distributions. We do no follow here this approach but rather outline, for the modest purposes of our study, a somewhat weaker but simple description of spectral measures of stable distributions in infinite dimension. It will cover the applications which we have in mind and explains the flavor of the result. We refer to [Ar-G2] and [Li] for the more general infinitely divisible theory.

We state and prove the existence of a spectral measure of a stable distribution in the context of a random process $X = (X_t)_{t \in T}$ indexed by a countable set T in order to avoid measurability questions. Anyway, this is the basic result from which the useful corollaries are easily deduced.

Theorem 5.2. *Let $0 < p < 2$ and let $X = (X_t)_{t \in T}$ be a p-stable process indexed by a countable set T. There exists a positive finite measure m on $\mathbb{R}^T$ (equipped with its cylindrical σ-algebra) such that for every finite sequence (α_j) of real numbers*

$$\mathbb{E}\exp\left(i\sum_j \alpha_j X_{t_j}\right) = \exp\left(-\frac{1}{2}\int_{\mathbb{R}^T}\Big|\sum_j \alpha_j x_{t_j}\Big|^p dm(x)\right).$$

m is called a spectral measure of X (it is not necessarily unique).

Before the proof, let us just mention that in the case $p = 2$ we can simply take for m the distribution of the Gaussian process X.

Proof. In a first step, assume that T is a finite set $\{t_1,\ldots,t_N\}$. Recall that if θ is real p-stable with parameter σ, and $r < p$, then $\|\theta\|_r = c_{p,r}\sigma$. It follows that for every $\alpha = (\alpha_1,\ldots,\alpha_N)$ in $\mathbb{R}^N$,

$$\mathbb{E}\exp\left(i\sum_{j=1}^N \alpha_j X_{t_j}\right) = \exp\left(-\frac{1}{2}\sigma(\alpha)^p\right)$$

$$= \exp\left[-\frac{1}{2c_{p,r}^p}\left(\mathbb{E}\Big|\sum_{j=1}^N \alpha_j X_{t_j}\Big|^r\right)^{p/r}\right]$$

where $\sigma(\alpha)$ denotes the parameter of $\sum_{j=1}^N \alpha_j X_{t_j}$. For every $r < p$, define then a positive finite measure m_r on the unit sphere S for the sup-norm $\|\cdot\|_\infty$ on $\mathbb{R}^N$ by setting, for every bounded measurable function φ on S,

$$\int_S \varphi(y)dm_r(y) = \frac{1}{c_{p,r}^r}\int_{\mathbb{R}^N} \varphi\left(\frac{x}{\|x\|_\infty}\right)\|x\|_\infty^r d\mathbb{P}_X(x)$$

where $\mathbb{P}_X$ is the law of $X = (X_{t_1},\ldots,X_{t_N})$. Hence, for any $\alpha = (\alpha_1,\ldots,\alpha_N)$ in $\mathbb{R}^N$,

$$\mathbb{E}\exp\left(i\sum_{j=1}^N \alpha_j X_{t_j}\right) = \exp\left[-\frac{1}{2}\left(\int_S \Big|\sum_{j=1}^N \alpha_j x_j\Big|^r dm_r(x)\right)^{p/r}\right].$$

Now, the total mass $|m_r|$ of m_r is easily seen to be majorized by

$$|m_r| \le \left(\inf_{x\in S}\sum_{j=1}^N |x_j|^r\right)^{-1}\sum_{j=1}^N \sigma(e_j)^r$$

where e_j, $1 \le j \le N$, are the unit vectors of $\mathbb{R}^N$. Therefore $\sup_{r<p}|m_r| < \infty$. Let m be a cluster point (in the weak-star sense) of $(m_r)_{r<p}$; m is a positive finite measure which is clearly a spectral measure of X. This proves the theorem in the finite dimensional case.

Assume now that $T = \{t_1, t_2, \ldots\}$. It is not difficult to see that we may assume the stable process X to be almost surely bounded. Indeed, if this is not the case, by the integrability property (5.2) and the Borel-Cantelli lemma, there exists a sequence $(a_t)_{t\in T}$ of positive numbers such that if $Z_t = a_t X_t$, the p-stable process $Z = (Z_t)_{t\in T}$ satisfies $\sup_{t\in T}|Z_t| < \infty$ almost surely. Then, if we can construct a spectral measure m' on $\mathbb{R}^T$ for Z, we define m in such a way that for any bounded measurable function φ on $\mathbb{R}^T$ depending on finitely many coordinates only,

$$\int_{\mathbb{R}^T} \varphi(x)dm(x) = \int_{\mathbb{R}^T} \varphi\left(\left(\frac{x_t}{a_t}\right)\right)dm'(x)$$

where $x = (x_t) \in \mathbb{R}^T$. Then the positive finite measure m on $\mathbb{R}^T$ is a spectral measure for the p-stable process X. We therefore assume that X is almost surely bounded. For each N, the preceding finite dimensional step provides us with a spectral measure m_N concentrated on the unit sphere of ℓ_∞^N of the random vector $(X_{t_1}, \ldots, X_{t_N})$ in $\mathbb{R}^N$. Denote by (Y_j^N) independent random variables distributed as $m_N/|m_N|$. If we recall the sequence (Γ_j) of the representation and if we let (ε_j) denote a Rademacher sequence, then, the sequences (Y_j^N), (Γ_j), (ε_j) being assumed to be independent, Theorem 5.1 indicates that $(X_{t_1}, \ldots, X_{t_N})$ has the same distribution as

$$c_p|m_N|^{1/p} \sum_{j=1}^\infty \Gamma_j^{-1/p} \varepsilon_j Y_j^N .$$

Our next step is to show that $\sup_N |m_N| < \infty$. Since X is assumed to be almost surely bounded, we can choose a finite number u such that $\mathbb{P}\{\sup_{t \in T} |X_t| > u\} \leq 1/4$. By Lévy's inequality (2.7) and the preceding representations, it follows that, for every N,

$$\frac{1}{2} \geq 2\mathbb{P}\left\{\sup_{t \in T} |X_t| > u\right\} \geq 2\mathbb{P}\left\{\max_{i \leq N} |X_{t_i}| > u\right\}$$

$$\geq \mathbb{P}\left\{c_p|m_N|^{1/p}\Gamma_1^{-1/p} > u\right\}.$$

We deduce that $|m_N| \leq c_p^{-p}u^p$ and thus that $\sup_N |m_N| < \infty$ since u was chosen independently of N. As before, if m denotes then a cluster point (in the weak-star sense) of the bounded sequence (m_N) of positive measures, m is immediately seen to fulfil the conclusion of Theorem 5.2; m is a spectral measure of X and the proof is complete.

As an immediate corollary to Theorems 5.1 and 5.2 we can now state the following:

Corollary 5.3. *Let $0 < p < 2$ and let $X = (X_t)_{t \in T}$ be a p-stable random process indexed by a countable set T with spectral measure m. Let (Y_j) be a sequence of independent random variables distributed as $m/|m|$ (in $\mathbb{R}^T$). Let further (η_j) be real valued independent symmetric random variables with the same law as η where $\mathbb{E}|\eta|^p < \infty$ and assume the sequences (Y_j), (η_j), (Γ_j) to be independent. Then, the random process $X = (X_t)_{t \in T}$ has the same distribution as*

$$\left(c_p\|\eta\|_p^{-1}|m|^{1/p} \sum_{j=1}^\infty \Gamma_j^{-1/p} \eta_j Y_j(t)\right)_{t \in T} .$$

Remark 5.4. If m is a spectral measure of an *almost surely bounded* p-stable process $X = (X_t)_{t \in T}$, then necessarily

$$\int \|x\|_{\infty}^p dm(x) < \infty$$

where we have denoted by $\| \cdot \|_{\infty}$ the $\ell_{\infty}(T)$-norm. This can be seen for example from the preceding representation; indeed, by Lévy's inequalities applied conditionally on (Γ_j), we must have $\sup_{j \geq 1} \|Y_j\|_{\infty}/\Gamma_j^{1/p} < \infty$ almost surely. Now recall that from the strong law of large numbers $\Gamma_j/j \to 1$ with probability one and therefore we also have $\sup_{j \geq 1} \|Y_j\|_{\infty}/j^{1/p} < \infty$. The claim thus follows from the Borel-Cantelli lemma and the fact that the independent random variables Y_j are distributed as $m/|m|$. Actually, a close inspection of the proof of Theorem 5.2 shows that for a bounded process we directly constructed a spectral measure concentrated on the unit ball of $\ell_{\infty}(T)$. It is in fact convenient in many problems to work with a spectral measure concentrated on the unit sphere (or only ball) of $\ell_{\infty}(T)$. To this aim, if m is a spectral measure of X that satisfies $\int \|x\|_{\infty}^p dm(x) < \infty$, let m_1 be the image of the measure $\|x\|_{\infty}^p dm(x)$ by the map $x \to x/\|x\|_{\infty}$. Then m_1 is a spectral measure of X concentrated on the unit sphere of $\ell_{\infty}(T)$ with total mass

$$|m_1| = \int \|x\|_{\infty}^p dm(x).$$

It can be shown that a *symmetric* spectral measure *on the unit sphere* of $\ell_{\infty}(T)$ is *unique* and this actually follows from (5.10) below (at least in the Radon case). However, this uniqueness is rather irrelevant for our purposes. By analogy with the scalar case, we can define the *parameter* of X by

(5.7) $$\sigma_p(X) = |m_1|^{1/p} = \left(\int \|x\|_{\infty}^p dm(x) \right)^{1/p}.$$

Note that by the uniqueness of m_1, $\sigma_p(X)$ is well defined (cf. also (5.11)). The terminology of parameter extends the real case. Sometimes, this parameter plays rôles analogous to the σ's encountered in the study of Gaussian and Rademacher variables; it is however quite different in nature as will become apparent later on.

Now, we inspect the consequences of the preceding results in the case of a p-stable Radon random variable in a Banach space.

Corollary 5.5. *Let X be a p-stable $(0 < p < 2)$ Radon random variable with values in a Banach space B. Then there is a positive finite Radon measure m on B satisfying $\int \|x\|^p dm(x) < \infty$ such that for every f in B'*

$$\mathbb{E} \exp(if(X)) = \exp\left(-\frac{1}{2} \int_B |f(x)|^p dm(x)\right).$$

Furthermore, if (Y_j) is a sequence of independent random variables distributed as $m/|m|$, and (η_j) a sequence of real valued symmetric random variables with the same law as η where $\mathbb{E}|\eta|^p < \infty$, the series

$$c_p \|\eta\|_p^{-1} |m|^{1/p} \sum_{j=1}^{\infty} \Gamma_j^{-1/p} \eta_j Y_j$$

converges almost surely in B where, as usual, the sequences (Γ_j), (η_j) and (Y_j) are assumed to be independent from each other, and this series is distributed as X.

Proof. We may and do assume that B is separable. Let D be a countable weakly dense set in the unit ball of B'. By Corollary 5.3, there exists a positive finite measure m on the unit ball of $\ell_\infty(D)$ such that if $(Y_j^f)_{f \in D}$ are independent and distributed as $m/|m|$, and independent of (Γ_j) and (η_j), then $(f(X))_{f \in D}$ has the same distribution as

$$\left(c_p \|\eta\|_p^{-1} |m|^{1/p} \sum_{j=1}^{\infty} \Gamma_j^{-1/p} \eta_j Y_j^f \right)_{f \in D}.$$

Then, let (x_n) be a dense sequence in B and denote, for each n, by F_n the subspace generated by $x_1, \ldots, x_n$. By Lévy's inequality (2.7) applied to the norms $\inf_{z \in F_n} \sup_{f \in D} | \cdot - f(z)|$, and conditionally on the sequence (Γ_j), we get that for all n and $\varepsilon > 0$,

$$2\mathbb{P}\{ \inf_{z \in F_n} \|X - z\| > \varepsilon \} \geq \mathbb{P}\{ c_p \|\eta\|_p^{-1} |m|^{1/p} \inf_{z \in F_n} \sup_{f \in D} |\Gamma_1^{-1/p} \eta_1 Y_1^f - z| > \varepsilon \}.$$

Since X is a Radon variable in B, the left hand side of this inequality can be made, for every $\varepsilon > 0$, arbitrarily small for all large enough n's. It easily follows that we can define a random variable with values in B, call it Y, such that $f(Y) = Y_1^f$ almost surely for every f in D. The same argument as before via Lévy's inequalities indicates that Y is Radon. By the density of D, the law of Y is, up to a multiplicative constant, a spectral measure of X and we have, by Remark 5.4, that $\mathbb{E}\|Y\|^p < \infty$. The convergence of the series representation indeed takes place in B by the Itô-Nisio theorem (Theorem 2.4). Corollary 5.5 is therefore established.

As in (5.7) we define the parameter $\sigma_p(X) = \left(\int \|x\|^p dm(x) \right)^{1/p}$. We can choose further (following Remark 5.4) the spectral measure to be symmetrically distributed on the unit sphere of B; it is then unique (cf. (5.10) below). Of course, a typical example of a p-stable Radon random variable with values in a Banach space B is given by an almost surely convergent series $X = \sum_i \theta_i x_i$ where (θ_i) is a standard p-stable sequence and (x_i) is a sequence in B. In this case, the spectral measure is discrete and can be explicitly described. For example, since necessarily $\sup_i \|\theta_i x_i\| < \infty$ almost surely, we learned in the beginning of this chapter that, when $0 < p < 2$, $\sum_i \|x_i\|^p < \infty$. Let then m be given by

$$m = \sum_i \frac{\|x_i\|^p}{2} \left(\delta_{+\frac{x_i}{\|x_i\|}} + \delta_{-\frac{x_i}{\|x_i\|}} \right).$$

Then m is a spectral measure for X (symmetric and concentrated on the unit sphere of B). We note in this case that the parameter $\sigma_p(X)$ of X (cf. (5.7)) is simply

$$\sigma_p(X) = \left(\sum_i \|x_i\|^p \right)^{1/p} < \infty .$$

This property induces a rather deep difference with the Gaussian situation in which $\sum_i \|x_i\|^2$ is not necessarily finite if $\sum_i g_i x_i$ converges. As yet another difference, note that if convergent series $\sum_i g_i x_i$ completely describe the class of Gaussian Radon random vectors, this is no longer the case when $p < 2$ (as soon as the spectral measure is not discrete). The series representation might then be thought of as a kind of substitute for this property.

The representation theorems were described with a sequence (η_j) of independent identically distributed real valued symmetric random variables with $\mathbb{E}|\eta_j|^p < \infty$. Two choices are of particular interest. First, there is the simple choice of a Rademacher sequence. A second choice is an orthogaussian sequence. It then appears that p-stable vectors and processes may be seen as conditionally Gaussian. Various Gaussian results can then be used to yield, after integration, similar consequences for stable random variables. This main idea, and the two preceding choices, will be used extensively in the subsequent study of p-stable random variables and processes.

To conclude this section, we would like to come back to the comparison of Γ_j with j initiated in the beginning. The representation now clearly indicates what kind of properties would be desirable. First, as an easy consequence of $\Gamma_j / j \to 1$ almost surely and of the contraction principle (conditionally) in the form of Theorem 4.4, it is plain that, in the previous notation, $\sum_{j=1}^{\infty} \Gamma_j^{-1/p} \eta_j Y_j$ converges almost surely if and only if $\sum_{j=1}^{\infty} j^{-1/p} \eta_j Y_j$ does. The next two observations will be used as quantitative versions of this result. As a simple consequence of the expression of $\mathbb{P}\{\Gamma_j \leq t\}$ and Stirling's formula, we have

$$(5.8) \qquad \left\| \sup_{j \geq 1} \left(\frac{j}{\Gamma_j} \right)^{1/p} \right\|_{p,\infty} \leq K_p$$

for some $K_p < \infty$ and similarly with $(\Gamma_j / j)^{1/p}$ (for which actually all moments exist). More important perhaps is the following:

$$(5.9) \qquad \sum_{j \geq 2} \| \Gamma_j^{-1/p} - j^{-1/p} \|_p^{\min(p,1)} < \infty .$$

Note that the sum in (5.9) starts from 2 in order that $\Gamma_j^{-1/p}$ is in L_p. It suffices to show that for all large enough j's

$$\mathbb{E}|\Gamma_j^{-1/p} - j^{-1/p}|^p \leq K_p \frac{1}{j^{p/2+1}} .$$

We have

$$\mathbb{E}|\Gamma_j^{-1/p} - j^{-1/p}|^p \le \frac{1}{j}\mathbb{P}\{\Gamma_j - j > j\} + \int_{\{\Gamma_j \le 2j\}} \Gamma_j^{-1}\left|1 - \left(\frac{\Gamma_j}{j}\right)^{1/p}\right|^p d\mathbb{P}.$$

Now, if $0 \le x \le 2$, $|1 - x^{1/p}| \le K_p|1 - x|$, so that, by Chebyshev's and Hölder's inequalities the preceding is bounded by

$$\frac{1}{j^2} + K_p^p\left(\mathbb{E}\left|1 - \frac{\Gamma_j}{j}\right|^2\right)^{p/2}\left(\mathbb{E}(\Gamma_j^{2/(p-2)})\right)^{(2-p)/2}$$
$$= \frac{1}{j^2} + K_p^p\frac{1}{j^{p/2}}\left(\mathbb{E}(\Gamma_j^{2/(p-2)})\right)^{(2-p)/2}.$$

By (5.5), the claim follows.

5.2 Integrability and Tail Behavior

We investigate the integrability properties of p-stable Radon random variables and almost surely bounded processes, $0 < p < 2$. We already know from the real case some severe limitations compared to the Gaussian case $p = 2$. This study could be based entirely on the representation and the result of the next chapter on sums of independent random variables substituting, as we just learned, j to Γ_j. However, there is a first a priori simple result which it will be convenient to record. This is Proposition 5.6 below. We then use the representation, combined with some results of the next chapter, to give some precise information on tail behaviors.

As usual, in order to unify our statement on Radon random variables and bounded processes, let us assume that we are given a Banach space B with a countable subset D in the unit ball of B' such that $\|x\| = \sup_{f \in D}|f(x)|$ for all x in B. X is a random variable with values in B if $f(X)$ is measurable for every f in D. It is p-stable if each finite linear combination $\sum_i \alpha_i f_i(X)$, $\alpha_i \in \mathbb{R}$, $f_i \in D$, is a real p-stable random variable.

Proposition 5.6. *Let $0 < p < 2$ and let X be a p-stable random variable in B. Then $\|X\|_{p,\infty} < \infty$. Furthermore, all the moments of X of order $r < p$ are equivalent, and are equivalent to $\|X\|_{p,\infty}$, i.e. for every $r < p$, there exists $K_{p,r} > 0$ such that for every p-stable variable X*

$$K_{p,r}^{-1}\|X\|_r \le \|X\|_{p,\infty} \le K_{p,r}\|X\|_r.$$

Proof. As in the previous chapters, we show that the moments of X are controlled by some parameter in the $L_0(B)$-topology. Indeed, let t_0 be such that $\mathbb{P}\{\|X\| > t_0\} \le 1/4$. Let (X_i) be independent copies of X. Since, for each N,

$$\frac{1}{N^{1/p}}\sum_{i=1}^{N} X_i \text{ has the same distribution as } X,$$

we get from Lévy's inequality (2.7) that

$$\frac{1}{4} \geq \mathbb{P}\{\|X\| > t_0\} \geq \mathbb{P}\left\{\left\|\sum_{i=1}^{N} X_i\right\| > t_0 N^{1/p}\right\}$$

$$\geq \frac{1}{2}\mathbb{P}\{\max_{i \leq N} \|X_i\| > t_0 N^{1/p}\}.$$

By Lemma 2.6 and identical distribution, it follows that

$$\mathbb{P}\{\|X\| > t_0 N^{1/p}\} \leq \frac{1}{N}$$

which therefore holds for every $N \geq 1$. By a trivial interpolation, $\|X\|_{p,\infty} \leq 2^{1/p} t_0$. To show the equivalence of moments, simply note that for $0 < r < p$, if $t_0 = (4\mathbb{E}\|X\|^r)^{1/r}$, then

$$\mathbb{P}\{\|X\| > t_0\} \leq \frac{1}{t_0^r} \mathbb{E}\|X\|^r \leq \frac{1}{4}.$$

As a consequence of this proposition, we see that if (X_n) is a p-stable sequence of random variables converging almost surely to X, or only in probability, then (X_n) also converges in $L_{p,\infty}$ and therefore in L_r for every $r < p$. This follows from the preceding equivalence of moments together with Lemma 4.2, or directly from the proof of Proposition 5.6; indeed, for every $\varepsilon > 0$, $\mathbb{P}\{\|X_n - X\| > \varepsilon\}$ can be made smaller than $1/4$ for all large enough n's and then $\|X_n - X\|_{p,\infty} \leq 2^{1/p}\varepsilon$.

Thus, integrability properties of infinite dimensional p-stable random vectors are similar to the finite dimensional properties. This observation can be pushed further to extend (5.2). The proof is based on the representation and mimicks the last argument in the proof of Theorem 5.1. For notational convenience and simplicity of the exposition, we present this result in the setting of Radon random variables but everything goes through in the case of almost surely bounded stable processes.

Therefore, let X be a p-stable Radon random variable with values in a Banach space B. According to Corollary 5.5 (and Remark 5.4), let m be a spectral measure of X, which is symmetrically distributed on the unit sphere of B. Then, for every measurable set A in the unit sphere of B such that $m(\partial A) = 0$ where ∂A is the boundary of A,

$$(5.10) \qquad \lim_{t \to \infty} t^p \mathbb{P}\left\{\|X\| > t, \frac{X}{\|X\|} \in A\right\} = c_p^p m(A).$$

This shows in particular the uniqueness of such a spectral measure as announced in Remark 5.4. If we recall the parameter $\sigma_p(X) = |m|^{1/p}$ of X (cf. (5.7)), we have in particular

$$(5.11) \qquad \lim_{t \to \infty} t^p \mathbb{P}\{\|X\| > t\} = c_p^p \sigma_p(X)^p.$$

To better describe the idea of the proof of (5.10), let us first establish the particular case (5.11). We use a result of the next chapter on sums of independent random variables. Let (Y_j) be independent random variables distributed as $m/|m|$. According to Corollary 5.5, X has the same distribution as $c_p|m|^{1/p}\sum_{j=1}^{\infty}\Gamma_j^{-1/p}Y_j$. The main observation is that $\mathbb{E}\|\sum_{j=2}^{\infty}\Gamma_j^{-1/p}Y_j\|^p < \infty$. By (5.9), it is enough to have $\mathbb{E}\|\sum_{j=2}^{\infty}j^{-1/p}Y_j\|^p < \infty$. But then we are dealing with a sum of independent random variables. Anticipating on the next chapter, we invoke Theorem 6.11 to see that indeed $\mathbb{E}\|\sum_{j=2}^{\infty}j^{-1/p}Y_j\|^p < \infty$. Hence it follows that

$$(5.12) \qquad \lim_{t\to\infty} t^p \mathbb{P}\left\{\left\|\sum_{j=2}^{\infty}\Gamma_j^{-1/p}Y_j\right\| > t\right\} = 0 .$$

Since Y_1 is concentrated on the unit sphere of B, combining with (5.3) we get that

$$\lim_{t\to\infty} t^p \mathbb{P}\left\{\left\|\sum_{j=1}^{\infty}\Gamma_j^{-1/p}Y_j\right\| > t\right\} = 1 ,$$

hence the result.

We next turn to (5.10). By homogeneity, assume that $c_p|m|^{1/p} = 1$. We only establish that for every closed subset set F of the unit sphere of B

$$\limsup_{t\to\infty} t^p \mathbb{P}\left\{\|X\| > t, \ \frac{X}{\|X\|} \in F\right\} \leq \mathbb{P}\{Y_1 \in F\} .$$

The corresponding lower bound for open sets is established similarly yielding thus (5.10) since Y_1 is distributed as $m/|m|$. For each $\varepsilon > 0$, we set $F_\varepsilon = \{x \in B ; \ \exists y \in F, \ \|x - y\| \leq \varepsilon\}$. Set also $Z = \sum_{j=1}^{\infty}\Gamma_j^{-1/p}Y_j$ which is distributed as X. For every $\varepsilon, t > 0$

$$\mathbb{P}\left\{\|X\| > t, \ \frac{X}{\|X\|} \in F\right\} \leq \mathbb{P}\left\{\|Z\| > t, \ \frac{Z}{\|Z\|} \in F, \ \left\|\sum_{j=2}^{\infty}\Gamma_j^{-1/p}Y_j\right\| \leq \varepsilon t\right\}$$
$$+ \mathbb{P}\left\{\left\|\sum_{j=2}^{\infty}\Gamma_j^{-1/p}Y_j\right\| > \varepsilon t\right\} .$$

By (5.12) we will need only concentrate on the first probability on the right hand side of this inequality. Thus, assume that $\|Z\| > t$, $Z/\|Z\| \in F$ and $\|\sum_{j=2}^{\infty}\Gamma_j^{-1/p}Y_j\| \leq \varepsilon t$. Since

$$Z = \Gamma_1^{-1/p}Y_1 + \sum_{j=2}^{\infty}\Gamma_j^{-1/p}Y_j ,$$

we deduce from the triangle inequality that $\Gamma_1^{-1/p}Y_1/\|Z\| \in F_\varepsilon$. Further

$$\left\|\frac{\Gamma_1^{-1/p}Y_1}{\|Z\|} - Y_1\right\| \leq \left|\frac{\Gamma_1^{-1/p}}{\|Z\|} - 1\right|$$

and since $(1 - \varepsilon)t < \|Z\| - \varepsilon t \leq \Gamma_1^{-1/p} \leq \|Z\| + \varepsilon t$, we get that $Y_1 \in F_{2\varepsilon}$. Summarizing,

$$\mathbb{P}\left\{\|X\| > t, \ \frac{X}{\|X\|} \in F\right\} \leq \mathbb{P}\{\Gamma_1^{-1/p} > (1 - \varepsilon)t, \ Y_1 \in F_{2\varepsilon}\}$$
$$+ \mathbb{P}\left\{\left\|\sum_{j=2}^{\infty} \Gamma_j^{-1/p} Y_j\right\| > \varepsilon t\right\}.$$

By the independence of Γ_1 and Y_1, and (5.3), (5.12), it follows that for every $\varepsilon > 0$

$$\limsup_{t \to \infty} t^p \mathbb{P}\left\{\|X\| > t, \ \frac{X}{\|X\|} \in F\right\} \leq (1 + \varepsilon)\mathbb{P}\{Y_1 \in F_\varepsilon\} + \varepsilon.$$

Since $\varepsilon > 0$ is arbitrary, this proves the claim.

Along the same line of ideas, it is possible to obtain from the representation a concentration inequality of $\|X\|$ around its expectation $(p > 1)$. The argument relies on a concentration idea for sums of independent random variables presented in Section 6.3 but we already explain here the result. For simplicity, we deal as before with Radon random variables and the case $p > 1$. The case $0 < p \leq 1$ can be discussed similarly with $\mathbb{E}\|X\|^r$, $r < p$, instead of $\mathbb{E}\|X\|$.

Proposition 5.7. *Let* $1 < p < 2$ *and let* X *be a* p-*stable Radon random variable with values in a Banach space* B. *Then, with* $\sigma_p(X)$ *denoting the parameter of* X, *for all* $t > 0$,

$$\mathbb{P}\{|\|X\| - \mathbb{E}\|X\|| > t\} \leq C_p \frac{\sigma_p(X)^p}{t^p}$$

where $C_p > 0$ *only depends on* p.

Proof. Let (Y_j) be independent identically distributed random variables in the unit sphere of B such that X has the law of $c_p \sigma_p(X) Z$ where $Z = \sum_{j=1}^{\infty} \Gamma_j^{-1/p} Y_j$ is almost surely convergent in B (Corollary 5.5). By the triangle inequality,

$$|\|Z\| - \mathbb{E}\|Z\|| \leq \left|\left\|\sum_{j=1}^{\infty} j^{-1/p} Y_j\right\| - \mathbb{E}\left\|\sum_{j=1}^{\infty} j^{-1/p} Y_j\right\|\right|$$
$$+ \sum_{j=1}^{\infty} |\Gamma_j^{-1/p} - j^{-1/p}| + \sum_{j=1}^{\infty} \mathbb{E}|\Gamma_j^{-1/p} - j^{-1/p}|.$$

By (5.9), $\sum_{j=1}^{\infty} \mathbb{E}|\Gamma_j^{-1/p} - j^{-1/p}| < \infty$, and, while only $\|\Gamma_1^{-1/p} - 1\|_{p,\infty} < \infty$,

$$\left\|\sum_{j=2}^{\infty} |\Gamma_j^{-1/p} - j^{-1/p}|\right\|_p \leq \sum_{j=2}^{\infty} \|\Gamma_j^{-1/p} - j^{-1/p}\|_p < \infty.$$

To estimate $\||\sum_{j=1}^{\infty} j^{-1/p} Y_j\| - \mathbb{E}\|\sum_{j=1}^{\infty} j^{-1/p} Y_j\||$, we can use the (martingale) quadratic inequality (6.11) to see that

$$\mathbb{E}\left| \left\|\sum_{j=1}^{\infty} j^{-1/p} Y_j\right\| - \mathbb{E}\left\|\sum_{j=1}^{\infty} j^{-1/p} Y_j\right\| \right|^2 \leq \sum_{j=1}^{\infty} j^{-2/p} < \infty .$$

Combining these various estimates, the proof is easily completed.

As a parenthesis, note that if X is a p-stable random variable with parameter $\sigma_p(X)$, applying Lévy's inequalities to the representation yields

$$(5.13) \qquad\qquad \sigma_p(X) \leq K_p \|X\|_{p,\infty}$$

for some constant K_p depending on p only. (It is also a consequence of (5.11).) Thus, the "strong" norm of a p-stable variable always dominates its parameter $\sigma_p(X)$. Let us mention that (5.13) is two-sided when $0 < p < 1$. This follows again from the representation together with the fact that $\sum_{j=1}^{\infty} j^{-1/p} < \infty$ when $p < 1$. We will see later how this is no longer the case when $1 \leq p < 2$.

We conclude this paragraph with some useful inequalities for real valued independent random variables which, at first sight, do not seem to be connected with stable distributions. They will however be much helpful later on in the study of various questions involving stable distributions such as in Chapters 9 and 13. They also allow us to evaluate some interesting norms of p-stable variables.

Recall that for $0 < p < \infty$ and $(\alpha_i)_{i \geq 1}$ a sequence of real numbers, we set

$$\|(\alpha_i)\|_{p,\infty} = \left(\sup_{t>0} t^p \mathrm{Card}\{i\,;\, |\alpha_i| > t\} \right)^{1/p} = \sup_{i \geq 1} i^{1/p} \alpha_i^*$$

where $(\alpha_i^*)_{i \geq 1}$ is the non-increasing rearrangement of the sequence $(|\alpha_i|)_{i \geq 1}$. The basic inequality which we present is contained in the following lemma.

Lemma 5.8. *Let $0 < p < \infty$. Let (Z_i) be independent positive random variables. Then*

$$\sup_{t>0} t^p \mathbb{P}\{\|(Z_i)\|_{p,\infty} > t\} \leq 2e \sup_{t>0} t^p \sum_i \mathbb{P}\{Z_i > t\} .$$

Proof. By homogeneity (replacing Z_i by Z_i^p), it suffices to deal with the case $p = 1$. If $(Z_i^*)_{i \geq 1}$ denotes the non-increasing rearrangement of the sequence $(Z_i)_{i \geq 1}$, $Z_n^* > u$ if and only if $\sum_i I_{\{Z_i > u\}} \geq n$. Hence, if $a = \sum_i \mathbb{P}\{Z_i > u\}$, by Lemma 2.5, for all $n \geq 1$,

$$\mathbb{P}\{Z_n^* > u\} \leq \left(\frac{ea}{n} \right)^n .$$

Now

$$\mathbb{P}\{\|(Z_i)\|_{1,\infty} > t\} = \mathbb{P}\{\sup_{n \geq 1} n Z_n^* > t\}$$

$$\leq \sum_{n=1}^{\infty} \mathbb{P}\left\{Z_n^* > \frac{t}{n}\right\}$$

$$\leq \sum_{n=1}^{\infty} \left(\frac{e}{n} \sum_i \mathbb{P}\left\{Z_i > \frac{t}{n}\right\}\right)^n.$$

Assuming by homogeneity that $\sup_{u>0} u \sum_i \mathbb{P}\{Z_i > u\} \leq 1$, we see that if $t > 2e$

$$\mathbb{P}\{\|(Z_i)\|_{1,\infty} > t\} \leq \sum_{n=1}^{\infty} \left(\frac{e}{t}\right)^n \leq \frac{2e}{t}$$

while this inequality is trivial for $t \leq 2e$. Lemma 5.7 is therefore established.

Note that the t^{-p} tail of $\mathbb{P}\{\sup_{n\geq 1} n^{1/p} Z_n^* > t\}$ is actually given by the *largest* term Z_1^*. The next terms are smaller and, indeed, the preceding proof indicates that for all $t > 0$ and all integers $k \geq 1$,

$$(5.14) \qquad \mathbb{P}\left\{\sup_{n \geq k} n^{1/p} Z_n^* > t\right\} \leq \left(\frac{2e}{t^p}\right)^k \sup_{u>0} u^p \sum_i \mathbb{P}\{Z_i > u\}.$$

Motivated by the representation of stable variables, the preceding lemma has an interesting consequence in the case where $Z_i = Y_i/i^{1/p}$, (Y_i) being a sequence of independent identically distributed random variables.

Corollary 5.9. *Let $0 < p < \infty$ and let Y be a positive random variable such that $\mathbb{E}Y^p < \infty$. Let (Y_i) be independent copies of Y and set $Z_i = Y_i/i^{1/p}$, $i \geq 1$. Then, if (Z_i^*) is as usual the non-increasing rearrangement of the sequence (Z_i), for every $t > 0$ and every integer $n \geq 1$*

$$\mathbb{P}\{n^{1/p} Z_n^* > e^{1/p}\|Y\|_p t\} \leq \frac{1}{t^{np}}.$$

Furthermore,

$$\mathbb{P}\left\{\sup_{n \geq k} n^{1/p} Z_n^* > (2e)^{1/p}\|Y\|_p t\right\} \leq \frac{1}{t^{kp}}$$

for all $k \geq 1$ and $t > 0$.

Proof. Note that for every $u > 0$

$$\sum_{i=1}^{\infty} \mathbb{P}\{Z_i > u\} = \sum_{i=1}^{\infty} \mathbb{P}\{Y > ui^{1/p}\} \leq \frac{1}{u^p}\mathbb{E}Y^p.$$

Hence, the first inequality of the lemma simply follows from $\mathbb{P}\{Z_n^* > t\} \leq (ea/n)^n$ (Lemma 2.5) with $u = (e/n)^{1/p}\|Y\|_p t$. The second inequality follows similarly from Lemma 5.8 and (5.14).

Let us note that the previous simple statements yield an alternate proof of (5.6). Let us simply indicate how to establish that $\left\|\sum_{j=1}^{\infty} \Gamma_j^{-1/p} \eta_j\right\|_{p,\infty} < \infty$ whenever $\mathbb{E}|\eta|^p < \infty$ with these tools. For every $t > 0$,

$$\mathbb{P}\left\{\left|\sum_{j=1}^{\infty} \Gamma_j^{-1/p} \eta_j\right| > 3t\right\} \leq \mathbb{P}\left\{\left|\sum_{j=1}^{\infty} \eta_j(\Gamma_j^{-1/p} - j^{-1/p})\right| > t\right\}$$

$$+ \mathbb{P}\left\{\left|\sum_{j=1}^{\infty} j^{-1/p} \eta_j\right| > 2t\right\}.$$

By (5.3) and (5.9), the first probability on the right hand side of this inequality is of the order of t^{-p}. We are thus left with $\mathbb{P}\{|\sum_{j=1}^{\infty} j^{-1/p} \eta_j| > 2t\}$. Set $Z_j = \eta_j/j^{1/p}$, $j \geq 1$, and let (Z_j^*) be the non-increasing rearrangement of the sequence $(|Z_j|)$. Denote further by (ε_j) a Rademacher sequence which is independent of (η_j). Then, by symmetry and identical distribution,

$$\mathbb{P}\left\{\left|\sum_{j=1}^{\infty} j^{-1/p} \eta_j\right| > 2t\right\} = \mathbb{P}\left\{\left|\sum_{j=1}^{\infty} \varepsilon_j Z_j^*\right| > 2t\right\}$$

$$\leq \mathbb{P}\{Z_1^* > t\} + \mathbb{P}\left\{\left|\sum_{j=2}^{\infty} \varepsilon_j Z_j^*\right| > t\right\}$$

$$\leq \mathbb{P}\{Z_1^* > t\} + \mathbb{P}\left\{\sup_{j\geq 2} j^{1/p} Z_j^* > \sqrt{t}\right\}$$

$$+ \mathbb{P}\left\{\left|\sum_{j=2}^{\infty} \varepsilon_j Z_j^*\right| > t, \ \sup_{j\geq 2} j^{1/p} Z_j^* \leq \sqrt{t}\right\}.$$

Recall that $\mathbb{E}|\eta_j|^p < \infty$ and that $Z_j = \eta_j/j^{1/p}$. By Corollary 5.9, the first two terms in the last estimate are of the order of t^{-p}. Concerning the third one, we can use for example the subgaussian inequality (4.1) conditionally on the sequence (Z_j) and find in this way a bound of the order of $\exp(-K_p t)$. This shows that $\|\sum_{j=1}^{\infty} \Gamma_j^{-1/p} \eta_j\|_{p,\infty} < \infty$. The limit (5.6) can be established similarly.

Lemma 5.8 has some further interesting consequences concerning the evaluation of the norm of certain stable vectors. Consider a standard p-stable sequence (θ_i) $(0 < p < 2)$ and (α_i) a finite (for simplicity) sequence of real numbers. We might be interested in $\|(\sum_i |\alpha_i \theta_i|^r)^{1/r}\|_{p,\infty}$, $0 < r \leq \infty$, as a function of the α_i's. In other words, we would like to examine the p-stable random variable in ℓ_r whose coordinates are $(\alpha_i \theta_i)$. Lemma 5.8 indicates to start with that

$$(5.15) \qquad \| \|(\alpha_i \theta_i)\|_{p,\infty}\|_{p,\infty} \leq (2e)^{1/p} \|\theta_1\|_{p,\infty} \left(\sum_i |\alpha_i|^p\right)^{1/p}.$$

This inequality is in fact two-sided; we indeed have

$$(5.16) \qquad \| \sup_i |\alpha_i \theta_i| \|_{p,\infty} \sim \left(\sum_i |\alpha_i|^p \right)^{1/p}$$

where the equivalence sign means a two-sided inequality up to a constant K_p depending on p only. The right hand side follows from (5.15) while for the left hand side, we may assume by homogeneity that $\sum_i |\alpha_i|^p = 1$; then by Lemma 2.6 and (5.2), if t is large enough and satisfies $\mathbb{P}\{\sup_i |\alpha_i \theta_i| > t\} \leq 1/2$,

$$\mathbb{P}\Big\{\sup_i |\alpha_i \theta_i| > t\Big\} \geq \frac{1}{2} \sum_i \mathbb{P}\{|\alpha_i \theta_i| > t\} \geq (K_p t^p)^{-1}$$

from which (5.16) clearly follows.

Since for $r > p$

$$\sup_i |\alpha_i \theta_i| \leq \left(\sum_i |\alpha_i \theta_i|^r \right)^{1/r} \leq \left(\frac{r}{r-p} \right)^{1/r} \|(\alpha_i \theta_i)\|_{p,\infty} ,$$

it is plain that (5.16) extends for $r > p$ into

$$(5.17) \qquad \left\| \left(\sum_i |\alpha_i \theta_i|^r \right)^{1/r} \right\|_{p,\infty} \sim \left(\sum_i |\alpha_i|^p \right)^{1/p}$$

where the equivalence is up to $K_{p,r}$ depending on p, r only. When $r < p$, we can simply use Proposition 5.6 and the equivalence of moments to see that, by Fubini,

$$(5.18) \qquad \left\| \left(\sum_i |\alpha_i \theta_i|^r \right)^{1/r} \right\|_{p,\infty} \sim \left(\sum_i |\alpha_i|^r \right)^{1/r} .$$

We are thus left with the slightly more complicated case $r = p$. It states that

$$(5.19) \qquad \left\| \left(\sum_i |\alpha_i \theta_i|^p \right)^{1/p} \right\|_{p,\infty} \sim \left(\sum_i |\alpha_i|^p \left[1 + \log \left(\frac{(\sum_i |\alpha_i|^p)^{1/p}}{|\alpha_i|} \right) \right] \right)^{1/p} .$$

Assume by homogeneity that $\sum_i |\alpha_i|^p = 1$. For the upper bound, we note that for every $t > 0$,

$$\mathbb{P}\Big\{ \left(\sum_i |\alpha_i \theta_i|^p \right)^{1/p} > t \Big\} \leq \mathbb{P}\Big\{ \sum_i |\alpha_i \theta_i|^p I_{\{|\alpha_i \theta_i| \leq t\}} > t^p \Big\}$$
$$+ \mathbb{P}\Big\{ \sup_i |\alpha_i \theta_i| > t \Big\} .$$

By (5.16), we need only be concerned with the first probability on the right of this inequality. By integration by parts, it is easily seen that there exists a large enough constant K_p such that, if $t^p \geq K_p \sum_i |\alpha_i|^p (1 + \log \frac{1}{|\alpha_i|})$, then

$$\sum_i \mathbb{E}\big(|\alpha_i \theta_i|^p I_{\{|\alpha_i \theta_i| \leq t\}} \big) \leq \frac{t^p}{2} .$$

Therefore, for such t's, by Chebyshev's inequality,

$$\mathbb{P}\left\{\sum_i |\alpha_i\theta_i|^p I_{\{|\alpha_i\theta_i|\leq t\}} > t^p\right\}$$

$$\leq \mathbb{P}\left\{\sum_i |\alpha_i\theta_i|^p I_{\{|\alpha_i\theta_i|\leq t\}} - \mathbb{E}\left(|\alpha_i\theta_i|^p I_{\{|\alpha_i\theta_i|\leq t\}}\right) > \frac{t^p}{2}\right\}$$

$$\leq \frac{4}{t^{2p}}\sum_i \mathbb{E}\left(|\alpha_i\theta_i|^{2p} I_{\{|\alpha_i\theta_i|\leq t\}}\right).$$

Again, integrating by parts, this quantity is seen to be less than $8\|\theta_1\|_{p,\infty}^p t^{-p}$. If we bring together all these informations we see that these yield the upper bound in (5.19). The lower bound is proved similarly and we thus leave it to the interested reader.

5.3 Comparison Theorems

In this section, we are concerned with comparisons of stable proceses analogous to those described for Gaussian and Rademacher processes. It will appear that, in general, these cannot be extended to the stable setting. However, several interesting results are still available. All of them are based on the observation described at the end of Section 5.1, namely that stable variables can be represented as conditionally Gaussian. Gaussian techniques can then be used to yield some positive consequences for p-stable variables, $0 < p < 2$. This line of investigation will also be the key idea in Section 12.2.

We would like to start with the inequality (5.13). We noticed that this inequality is two-sided for $0 < p < 1$. Therefore, in some sense, the study of p-stable variables with $0 < p < 1$ is not really interesting since the parameter, that is the mere existence of a spectral measure m satisfying $\int \|x\|^p dm(x) < \infty$ (cf. Remark 5.4), completely describes the boundedness and the size of the variable. Things are quite different when $1 \leq p < 2$ as the following example shows.

Assume that $1 < p < 2$, although the case $p = 1$ can be treated completely similarly. Consider the p-stable random variable X in $\mathbb{R}^N$ equipped with the sup-norm given by the representation $\sum_{j=1}^{\infty} \Gamma_j^{-1/p} Y_j$ where Y_j are independent and distributed as the Haar measure μ_N on $\{-1, +1\}^N$. Then the quantity $\sigma_p(X)$ of (5.7) is 1. Let us show however that $\|X\|_{p,\infty}$ is of the order of $(\log N)^{1/q}$ (for N large) where q is the conjugate of p ($\log\log N$ when $p = 1$). We only prove the lower bound, the upper bound being similar. First note that by the contraction principle

$$\mathbb{E}\left\|\sum_{j=1}^{\infty} j^{-1/p} Y_j\right\| \leq \mathbb{E}\sup_{j\geq 1}\left(\frac{\Gamma_j}{j}\right)^{1/p} \mathbb{E}\|X\|.$$

By (5.8), we know that $\mathbb{E}\sup_{j\geq 1}(\Gamma_j/j)^{1/p} \leq K_p$ for some K_p depending on p

only. Now, let Z be the real valued random variable $\sum_{j=1}^{\infty} j^{-1/p} \varepsilon_j$ where (ε_j) is a Rademacher sequence and denote by $Z_1, \ldots, Z_N$, N independent copies of Z. By definition, and since we consider $\mathbb{R}^N$ with the sup-norm,

$$\mathbb{E}\left\| \sum_{j=1}^{\infty} j^{-1/p} Y_j \right\| = \mathbb{E} \max_{i \leq N} |Z_i|$$

so that we simply need to bound this maximum from below. For $t > 0$, let ℓ be the smallest integer such that $(\ell+1)^{1/q} > t$. By Lévy's inequality (2.6),

$$\mathbb{P}\{|Z| > t\} \geq \frac{1}{2} \mathbb{P}\left\{ \left| \sum_{j=1}^{\ell+1} j^{-1/p} \varepsilon_j \right| > t \right\}.$$

With probability $2^{-\ell}$,

$$\left| \sum_{j=1}^{\ell+1} j^{-1/p} \varepsilon_j \right| = \sum_{j=1}^{\ell+1} j^{-1/p} \geq (\ell+1)^{1/q} > t$$

so that

$$\mathbb{P}\{|Z| > t\} \geq 2^{-\ell-1} \geq \frac{1}{2} \exp(-t^q).$$

Let then $t = 2\mathbb{E} \max_{i \leq N} |Z_i|$ so that in particular $\mathbb{P}\{\max_{i \leq N} |Z_i| > t\} \leq 1/2$. By Lemma 2.6 we have $N\mathbb{P}\{|Z| > t\} \leq 1$. From the preceding lower bound, it follows that $t \geq K_p^{-1}(\log N)^{1/q}$ for some $K_p > 0$. Therefore, we have obtained that $\|X\|_{p,\infty} \geq K_p^{-1}(\log N)^{1/q}$ while $\sigma_p(X) = 1$. This clearly indicates what the differences can be between $\|X\|_{p,\infty}$ and the parameter $\sigma_p(X)$ of a infinite dimensional p-stable random variable X with $1 \leq p < 2$.

According to the previous observations, the next results on comparison theorems and Sudakov's minoration for p-stable random variables are restricted to the case $1 \leq p < 2$.

First, we address the question of comparison properties in the form of Slepian's lemma for stable random vectors. Consider two p-stable, $1 \leq p < 2$, random vectors $X = (X_1, \ldots, X_N)$ and $Y = (Y_1, \ldots, Y_N)$ in $\mathbb{R}^N$. In analogy with the Gaussian case denote by $d_X(i,j)$ (resp. $d_Y(i,j)$) the parameter of the real valued p-stable variable $X_i - X_j$ (resp. $Y_i - Y_j$), $1 \leq i, j \leq N$. One of the ideas of the Gaussian comparison theorems was that if one can compare d_X and d_Y, then one should be able to compare the distributions or averages of $\max_{i \leq N} X_i$ and $\max_{i \leq N} Y_i$ (cf. Corollary 3.14 and Theorem 3.15). In the stable case with $p < 2$, the following simple example furnishes a very negative result to begin with. Let X be as in the preceding example and take Y to be the canonical p-stable vector in $\mathbb{R}^N$ given by $Y = (\theta_1, \ldots, \theta_N)$ $((\theta_i)$ is a standard p-stable sequence). It is easily seen that $d_X(i,j) = 2^{1/q}$ while $d_Y(i,j) = 2^{1/p}$, $i \neq j$. Thus d_X and d_Y are equivalent. However, assuming for example that $p > 1$, we know that

$$\mathbb{E} \max_{i \leq N} X_i \leq K_p (\log N)^{1/q}$$

while, as a consequence of (5.16),

$$\mathbb{E} \max_{i \leq N} \theta_i \geq K_p^{-1} N^{1/p}$$

(at least for every N large enough – compare $\mathbb{E} \max_{i \leq N} \theta_i$ and $\mathbb{E} \max_{i \leq N} |\theta_i|$). Thus, one can measure on this example, the gap which may arise in comparison theorems for p-stable vectors when $p < 2$.

Nevertheless, some positive results remain. This is for example the case for Sudakov's minoration (Theorem 3.18). If $X = (X_t)_{t \in T}$ is a p-stable process $(1 \leq p < 2)$ indexed by some set T, denote as before by $d_X(s, t)$ the parameter of the p-stable random variable $X_s - X_t$, $s, t \in T$. Since $\|X_s - X_t\|_r = c_{p,r} d_X(s, t)$, $r < p$, d_X (d_X^r if $p = 1$) defines a pseudo-metric on T. Recall that $N(T, d_X; \varepsilon)$ is the smallest (possibly infinite) number of open balls of radius $\varepsilon > 0$ in the metric d_X which cover T. Then, we have the following extension of Sudakov's minoration. (This minoration will be improved in Section 12.2.) The idea of the proof is to represent X as conditionally Gaussian and then apply the Gaussian inequalities. As is usual in similar contexts, we simply let here

$$\| \sup_{t \in T} |X_t| \|_{p,\infty} = \sup \{ \| \sup_{t \in F} |X_t| \|_{p,\infty} ; \; F \text{ finite in } T \} .$$

Theorem 5.10. *Let $X = (X_t)_{t \in T}$ be a p-stable random process with $1 \leq p < 2$ and associated pseudo-metric d_X. There is a constant $K_p > 0$ depending only on $p > 1$ such that if q is the conjugate of p, for every $\varepsilon > 0$,*

$$\varepsilon \big(\log N(T, d_X; \varepsilon) \big)^{1/q} \leq K_p \| \sup_{t \in T} |X_t| \|_{p,\infty} .$$

When $p = 1$, the lower bound has to be replaced by $\varepsilon \log^+ \log N(T, d_X; \varepsilon)$. In both cases, if X is almost surely bounded, (T, d_X) is totally bounded.

Proof. We only show the result when $1 < p < 2$. The case $p = 1$ seems to require independent deeper tools; we refer to [Ta8] for this investigation. Let $N(T, d_X; \varepsilon) \geq N$; there exists $U \subset T$ with cardinality N such that $d_X(s, t) > \varepsilon$ for $s \neq t$ in U. Consider the p-stable process $(X_t)_{t \in U}$ and let m be a spectral measure of this random vector (in $\mathbb{R}^N$ thus). Let (Y_j) be independent and distributed as $m/|m|$ and let further (g_j) be an orthogaussian sequence. As usual in the representation, the sequences (Y_j), (Γ_j), (g_j) are independent. From Corollary 5.3, $(X_t)_{t \in U}$ has the same distribution as

$$c_p \|g_1\|_p^{-1} |m|^{1/p} \sum_{j=1}^{\infty} \Gamma_j^{-1/p} g_j Y_j .$$

In relation with this representation, we introduce *random* distances (on U). Denote by ω the randomness in the sequences (Y_j) and (Γ_j) and set, for each such ω and s, t in U,

$$d_\omega(s,t) = \left(\mathbb{E}_g\left|c_p\|g_1\|_p^{-1}|m|^{1/p}\sum_{j=1}^{\infty}\Gamma_j(\omega)^{-1/p}g_j\left(Y_j(\omega,s)-Y_j(\omega,t)\right)\right|^2\right)^{1/2}$$

$$= c_p\|g_1\|_p^{-1}|m|^{1/p}\left(\sum_{j=1}^{\infty}\Gamma_j(\omega)^{-2/p}|Y_j(\omega,s)-Y_j(\omega,t)|^2\right)^{1/2}$$

where $\mathbb{E}_g$ denotes partial integration with respect to the Gaussian sequence (g_i). Accordingly, note that for all $\lambda \in \mathbb{R}$ and s,t

$$\mathbb{E}\exp i\lambda(X_s - X_t) = \exp\left(-\frac{1}{2}|\lambda|^p d_X^p(s,t)\right) = \mathbb{E}\exp\left(-\frac{1}{2}\lambda^2 d_\omega^2(s,t)\right).$$

It follows that for every u and $\lambda > 0$ and s,t in U,

$$\mathbb{P}\{d_\omega(s,t) \le u d_X(s,t)\} = \mathbb{P}\left\{\exp\left(-\frac{\lambda^2}{2}d_\omega^2(s,t)\right) \ge \exp\left(-\frac{\lambda^2}{2}u^2 d_X^2(s,t)\right)\right\}$$

$$\le \exp\left(\frac{\lambda^2}{2}u^2 d_X^2(s,t) - \frac{\lambda^p}{2}d_X^p(s,t)\right).$$

Minimizing over $\lambda > 0$, namely taking $\lambda = [(2u^2)^{\alpha/2p}d_X(s,t)]^{-1}$ where $\frac{1}{\alpha} = \frac{1}{p} - \frac{1}{2}$ yields, for all $u > 0$,

$$(5.20) \qquad \mathbb{P}\{d_\omega(s,t) \le u d_X(s,t)\} \le \exp(-c_\alpha u^{-\alpha})$$

where $c_\alpha = (4 \cdot 2^{\alpha/2})^{-1}$.

Now recall that $d_X(s,t) > \varepsilon$ for $s \ne t$ in U. From (5.20), we thus get that for all $u > 0$

$$\mathbb{P}\{\exists s \ne t \text{ in } U, \ d_\omega(s,t) \le \varepsilon u\} \le (\operatorname{Card} U)^2 \exp(-c_\alpha u^{-\alpha}).$$

Choose $u > 0$ such that this probability is less than $1/2$ (say), more precisely take

$$u = c_\alpha^{1/\alpha}\left(\log(2N^2)\right)^{-1/\alpha}$$

where $N = \operatorname{Card} U$. Hence, on a set Ω_0 of ω's of probability bigger than $1/2$, $d_\omega(s,t) > \varepsilon u$ for all $s \ne t$ in U. By Sudakov's minoration for Gaussian processes (Theorem 3.18), for some numerical constant $K > 0$ and all ω's in Ω_0

$$c_p\|g_1\|_p^{-1}|m|^{1/p}\mathbb{E}_g\sup_{t\in U}\left|\sum_{j=1}^{\infty}\Gamma_j(\omega)^{-1/p}g_jY_j(\omega,t)\right| \ge \frac{1}{K}\cdot\frac{\varepsilon u}{2}(\log N)^{1/2}$$

(since $N(U, d_\omega; \varepsilon u/2) \ge N$). Now, by partial integration,

$$\mathbb{E}\sup_{t\in U}|X_t| \ge c_p\|g_1\|_p^{-1}|m|^{1/p}\int_{\Omega_0}\mathbb{E}_g\sup_{t\in U}\left|\sum_{j=1}^{\infty}\Gamma_j^{-1/p}g_jY_j(t)\right|d\mathbb{P}$$

$$\ge \frac{1}{4K}\varepsilon u(\log N)^{1/2}.$$

By the choice of u and Proposition 5.6, $\|\sup_{t\in U}|X_t|\|_{p,\infty} \geq K_p^{-1}\varepsilon(\log N)^{1/q}$. If we now recall that $N \leq N(T,d_X;\varepsilon)$ was arbitrary, the proof is seen to be complete.

There is also an extension of Corollary 3.19 whose proof is completely similar. That is, if $X = (X_t)_{t\in T}$ is a p-stable random process, $1 \leq p < 2$, with almost all trajectories bounded and continuous on (T,d_X) (or having a version with these properties), then

$$(5.21) \qquad \lim_{\varepsilon\to 0} \varepsilon\big(\log N(T,d_X;\varepsilon)\big)^{1/q} = 0, \quad p > 1$$

(and

$$\lim_{\varepsilon\to 0} \varepsilon\log^+\log N(T,d_X;\varepsilon) = 0, \quad p = 1).$$

In the last part of this chapter, we briefly investigate tensorization of stable random variables which are analogous to those studied for Gaussian and Rademacher series with vector valued coefficients. One question is the following: if (x_i) is a sequence in a Banach space E and (y_j) is a sequence in a Banach space F such that $\sum_i \theta_i x_i$ and $\sum_j \theta_j y_j$ are both almost surely convergent, where (θ_i) is a standard p-stable sequence, is it the same for $\sum_{i,j} \theta_{ij} x_i \otimes y_j$, where (θ_{ij}) is a doubly indexed standard p-stable sequence, in the injective tensor product $E\check{\otimes}F$ of the Banach spaces E and F? Theorem 3.20 has provided a positive answer in the case $p = 2$ and the object of what follows will be to show that this remains valid when $p < 2$. However, we have to somewhat widen this study; we indeed know that, contrary to the Gaussian case, not all p-stable Radon random variables in a Banach space can be represented as a convergent series of the type $\sum_i \theta_i x_i$. We use instead spectral measures. Let thus $0 < p < 2$ and let U and V be p-stable Radon random variables with values in E and F respectively. Denote by m_U (resp. m_V) the symmetric spectral measure of U (resp. V) concentrated on the unit sphere of E (resp. F). One can then define naturally the symmetric measure $m_U \otimes m_V$ on the unit sphere of $E\check{\otimes}F$. Is this measure the spectral measure of some p-stable random variable with values in $E\check{\otimes}F$? The next theorem describes a positive answer to this question.

Theorem 5.11. *Let $0 < p < 2$ and let U and V be p-stable Radon random variables with values in Banach spaces E and F respectively. Let m_U and m_V be the respective symmetric spectral measures on the unit spheres of E and F. Then, there exists a p-stable Radon random variable W with values in $E\check{\otimes}F$ with spectral measure $m_U \otimes m_V$. Moreover, for some constant K_p depending on p only,*

$$\|W\|_{p,\infty} \leq K_p\big(\sigma_p(U)\|V\|_{p,\infty} + \sigma_p(V)\|U\|_{p,\infty}\big).$$

Proof. The idea is again to use Gaussian randomization and to benefit in a conditional way of the Gaussian comparison theorems. Let (Y_j) (resp. (Z_j)) be independent random variables with values in E (resp. F) and distributed as $m_U/|m_U|$ (resp. $m_V/|m_V|$). From Corollary 5.5,

$$U' = \sum_{j=1}^{\infty} \Gamma_j^{-1/p} g_j Y_j \quad \text{and} \quad V' = \sum_{j=1}^{\infty} \Gamma_j^{-1/p} g_j Z_j$$

converge almost surely in E and F respectively where (g_j) is an orthogaussian sequence and, as usual, (Γ_j), (g_j), (Y_j) and (Z_j) are independent. Our aim is to show that

$$W' = \sum_{j=1}^{\infty} \Gamma_j^{-1/p} g_j Y_j \otimes Z_j$$

is almost surely convergent in $E \check{\otimes} F$ and satisfies

(5.22) $\|W'\|_{p,\infty} \leq K_p (\|U'\|_{p,\infty} + \|V'\|_{p,\infty})$.

W' induces a p-stable Radon random variable W with values in $E \check{\otimes} F$ and spectral measure $m_U \otimes m_V$. Since $\sigma_p(U) = |m_U|^{1/p}$, $\sigma_p(V) = |m_V|^{1/p}$, $\sigma_p(W) = |m_U|^{1/p} |m_V|^{1/p}$, homogeneity and the normalizations in Corollary 5.5 easily lead to the conclusion.

We establish inequality (5.22) for sums U', V', W' as before but only for finitely many terms in the summations, simply indicated by $\sum_j$. Convergence will follow from (5.22) by a simple limiting argument.

Let f, f' be in the unit ball of E', h, h' in the unit ball of F'. Since $\|Y_j\| = \|Z_j\| = 1$, for every j, with probability one,

$$|f(Y_j)h(Z_j) - f'(Y_j)h'(Z_j)| \leq |f(Y_j) - f'(Y_j)| + |h(Z_j) - h'(Z_j)| .$$

Let (g'_j) be another orthogaussian sequence independent from (g_j) and denote by $\mathbb{E}_g$ conditional integration with respect to those sequences. By the preceding inequality

$$\left(\mathbb{E}_g \Big| \sum_j \Gamma_j^{-1/p} g_j [f \otimes h(Y_j \otimes Z_j) - f' \otimes h'(Y_j \otimes Z_j)] \Big|^2 \right)^{1/2}$$

$$\leq \left(2\mathbb{E}_g \Big| \sum_j \Gamma_j^{-1/p} g_j (f(Y_j) - f'(Y'_j)) + \sum_j \Gamma_j^{-1/p} g'_j (h(Z_j) - h'(Z_j)) \Big|^2 \right)^{1/2} .$$

Therefore, it follows from the Gaussian comparison theorems in the form of Corollary 3.14 (or Theorem 3.15) that, almost surely in (Γ_j), (Y_j), (Z_j),

$$\mathbb{E}_g \Big\| \sum_j \Gamma_j^{-1/p} g_j Y_j \otimes Z_j \Big\| \leq 2\sqrt{2} \mathbb{E}_g \Big\| \sum_j \Gamma_j^{-1/p} g_j Y_j \Big\|$$

$$+ 2\sqrt{2} \mathbb{E}_g \Big\| \sum_j \Gamma_j^{-1/p} g'_j Z_j \Big\| .$$

Integrating and using the equivalence of moments of both Gaussian and stable random vectors conclude in this way the proof of Theorem 5.11.

We mention to conclude some comments and open questions about Theorem 5.11. While $\|W\|_{p,\infty}$ always dominates $\sigma_p(U)\sigma_p(V)$, it is not true in general that the inequality of Theorem 5.11 can be reversed. The works [G-Ma-Z] and [M-T] have actually shown a variety of examples with different sizes of $\|W\|_{p,\infty}$. One may introduce weak moments similar to the Gaussian case; a natural lower bound for $\|W\|_{p,\infty}$ would then be, besides $\sigma_p(U)\sigma_p(V)$,

$$\lambda_p(U)\|V\|_{p,\infty} + \lambda_p(V)\|U\|_{p,\infty}$$

where

$$\lambda_p(U) = \sup_{\|f\|\leq 1} \left(\int |f(x)|^p dm_U(x) \right)^{1/p}$$

and similarly for V. However, neither this lower bound nor the upper bound of the theorem seem to provide the exact weight of $\|W\|_{p,\infty}$. The definitive result, if any, should be in between. This question is still under study.

Notes and References

As announced, our exposition of stable distributions and random variables in infinite dimensional spaces is quite restricted. More general expositions based on infinitely divisible distributions, Lévy measures, Lévy-Khintchine representations and related central limit theorems for triangular arrays may be found in the treatises [Ar-G2] and [Li] to which we actually also refer for more accurate references and historical background. See also the work [A-A-G], the paper [M-Z], etc. The survey article [Wer] presents a sample of the topics studied in the rather extensive literature on stable distributions (see also [Sa-T]). More on $L_{p,\infty}$-spaces (and interpolation spaces $L_{p,q}$) may be found e.g. in [S-W].

The theory of stable laws was constructed by P. Lévy [Lé1]. The few facts presented here as an introduction may be found in the classical books on Probability Theory such as [Fe1].

Representation of p-stable variables, $0 < p < 2$, goes back to the work by P. Lévy and was revived recently by R. LePage, M. Woodroofe and J. Zinn [LP-W-Z]; see [LP2] for the history of this representation. For a recent and new representation (in particular used in [Ta18]), see [Ro3]. The proof of Theorem 5.1 is taken from [Pi16] (see also [M-P2]). Theorem 5.2 and the existence of spectral measures is due to P. Lévy [Lé1] (who actually dealt with the Euclidean sphere); our exposition follows [B-DC-K]. Remark 5.4 and uniqueness of the symmetric spectral measure concentrated on the unit sphere follow from the more general results about uniqueness of Lévy measures for Banach space valued random variables, and started with [Ja1] and [Kue1] in

Hilbert space and was then extended to more general spaces by many authors (cf. [Ar-G2], [Li] for the details). (5.9) was noticed in [Pi12].

Proposition 5.6 is due to A. de Acosta [Ac2]; a prior result for $\sum_i \theta_i x_i$ was established by J. Hoffmann-Jørgensen [HJ1] (see also [HJ3] and Chapter 6). A. de Acosta [Ac3] also established the limit (5.11) (by a different method however) while the full conclusion (5.10) was proved by A. Araujo and E. Giné [Ar-G1]. Proposition 5.7 is taken from [G-Ma-Z]. Lemma 5.8 is due to M. B. Marcus and G. Pisier [M-P2] with a simplified proof by J. Zinn (cf. [M-Zi], [Pi16]). The equivalences (5.16) – (5.19) were described by L. Schwartz [Schw1].

Comparison theorems for stable random variables intrigued many people and it was probably known for a long time that Slepian's lemma does not extend to p-stable variables with $0 < p < 2$. The various introductory comments collect informations taken from [E-F], [M-P2], [Li], [Ma3]... Our exposition follows the work by M. B. Marcus and G. Pisier [M-P2]; Theorem 5.10 is theirs (but the case $p = 1$ was only proved in [Ta8]). Theorem 5.11 on tensor product of stable distributions was established by E. Giné, M. B. Marcus and J. Zinn [G-Ma-Z], and further investigated in [M-T].

6. Sums of Independent Random Variables

Sums of independent random variables already appeared in the preceding chapters in some concrete situations (Gaussian and Rademacher averages, representation of stable random variables). On the intuitive basis of central limit theorems which approximate normalized sums of independent random variables by smooth limiting distributions (Gaussian, stable), one would expect that results similar to those presented previously should hold in a sense or in another for sums of independent random variables. The results presented in this chapter go in this direction and the reader will recognize in this general setting the topics covered before: integrability properties, equivalence of moments, concentration, tail behavior, etc. We will mainly describe ideas and techniques which go from simple but powerful observations such as symmetrization (randomization) techniques to more elaborate results like those obtained from the isoperimetric inequality for product measures of Theorem 1.4. Section 6.1 is concerned with symmetrization, Section 6.2 with Hoffmann-Jørgensen's inequalities and the equivalence of moments of sums of independent random variables. In the last and main section, martingale and isoperimetric methods are developed in this context. Many results presented in this chapter will be of basic use in the study of limit theorems later.

Let us emphasize that the infinite dimensional setting is characterized by the *lack* of the orthogonality property $\mathbb{E}|\sum_i X_i|^2 = \sum_i \mathbb{E}|X_i|^2$, where (X_i) is a finite sequence of independent mean zero real valued random variables. This type of identity or equivalence extends to finite dimensional random vectors, and even to Hilbert space valued random variables, but *does not* in general for arbitrary Banach space valued random variables (cf. Chapter 9). With respect to the classical theory which is developed under this orthogonality property, the study of sums of independent Banach space valued random variables undertaken here requires in particular to circumvent this difficulty. Besides the difficult control in probability (which will be discussed further in this book), the various tools introduced in this chapter provide a more than satisfactory extension of the classical theory of sums of independent random variables. Actually, many ideas clarify the real case (for example, the systematic use of symmetrization-randomization) and, as for the isoperimetric approach to exponential inequalities (Section 6.3), go beyond the known results.

Since here we need not be concerned with tightness properties, we present the various results in the setting introduced in Chapter 2 and already used in the previous chapters. That is, let B be a Banach space such that there exists a countable subset D of the unit ball of the dual space such that $\|x\| = \sup_{f \in D} |f(x)|$ for all x in B. We say that a map X from some probability space $(\Omega, \mathcal{A}, \mathbb{P})$ into B is a random variable if $f(X)$ is measurable for each f in D. We recall that this definition covers the case of Radon random variables or equivalently of Borel random variables taking their values in a separable Banach space.

6.1 Symmetrization and Some Inequalities for Sums of Independent Random Variables

One simple but basic idea in the study of sums of independent random variables is the concept of symmetrization. If X is a random variable, one can construct a *symmetric* random variable which is "near" X by looking at $\tilde{X} = X - X'$ where X' denotes an independent copy of X (constructed on some different probability space $(\Omega', \mathcal{A}', \mathbb{P}')$). The distributions of X and $X - X'$ are indeed closely related; for example, for any $t, a > 0$, by independence and identical distributions,

$$(6.1) \qquad \mathbb{P}\{\|X\| \le a\}\, \mathbb{P}\{\|X\| > t + a\} \le \mathbb{P}\{\|X - X'\| > t\}.$$

This inequality is of particular interest when for example a is chosen such that $\mathbb{P}\{\|X\| \le a\} \ge 1/2$ in which case it follows that

$$\mathbb{P}\{\|X\| > t + a\} \le 2\mathbb{P}\{\|X - X'\| > t\}.$$

It also follows in particular that $\mathbb{E}\|X\|^p < \infty$ ($0 < p < \infty$) if and only if $\mathbb{E}\|X - X'\|^p < \infty$.

Actually, (6.1) is somewhat too crude in various applications and the following improvements are noteworthy: for $t, a > 0$,

$$(6.2) \qquad \inf_{f \in D} \mathbb{P}\{|f(X)| \le a\}\, \mathbb{P}\{\|X\| > t + a\} \le \mathbb{P}\{\|X - X'\| > t\}.$$

For a proof, let ω be such that $\|X(\omega)\| > t + a$; then, for some h in D, $|h(X(\omega))| > t + a$. Hence

$$\inf_{f \in D} \mathbb{P}'\{|f(X')| \le a\} \le \mathbb{P}\{\|X(\omega) - X'\| > t\}.$$

Integrating with respect to ω then yields (6.2). Similarly, one can show that for $t, a > 0$,

$$(6.3) \qquad \mathbb{P}\{\|X\| > t + a\} \le \mathbb{P}\{\|X - X'\| > t\} + \sup_{f \in D} \mathbb{P}\{|f(X)| > a\}.$$

When we deal with a sequence $(X_i)_{i \in \mathbb{N}}$ of independent random variables, we construct an associated sequence of independent and symmetric random variables by setting, for each i, $\tilde{X}_i = X_i - X_i'$ where (X_i') is an independent copy of the sequence (X_i). Recall that $(\widetilde{X}_i)$ is then a symmetric sequence in the sense that it has the same distribution as $(\varepsilon_i \widetilde{X}_i)$ where (ε_i) is a Rademacher sequence which is independent of (X_i) and (X_i'). That is, we can *randomize* by independent choices of signs symmetric sequences (X_i). Accordingly and following Chapter 2, we denote by $\mathbb{E}_\varepsilon, \mathbb{P}_\varepsilon$ (resp. $\mathbb{E}_X, \mathbb{P}_X$) partial integration with respect to (ε_i) (resp. (X_i)).

The fact that the symmetric sequence $(\widetilde{X}_i)$ built over (X_i) is useful in the study of (X_i) can be illustrated in different ways. Let us start for example with the Lévy-Itô-Nisio theorem for independent but not necessarily symmetric variables (cf. Theorem 2.4) where symmetrization proves its efficiency. Since weak convergence is involved in this statement we restrict ourselves for simplicity to the case of Radon random variables. However, the equivalence between (i) and (ii) holds in our general setting.

Theorem 6.1. *Let (X_i) be a sequence of independent Borel random variables with values in a separable Banach space B. Set $S_n = \sum_{i=1}^n X_i$, $n \geq 1$. The following are equivalent:*

(i) the sequence (S_n) converges almost surely;

(ii) (S_n) converges in probability;

(iii) (S_n) converges weakly.

Proof. Suppose that (S_n) converges weakly to some random variable S. On some different probability space $(\Omega', \mathcal{A}', \mathbb{P}')$, consider a copy (X_i') of the sequence (X_i) and set $S_n' = \sum_{i=1}^n X_i'$; (S_n') converges weakly to S' which has the same distribution as S. Set $\widetilde{S}_n = S_n - S_n'$, $\widetilde{S} = S - S'$ defined on $\Omega \times \Omega'$. Since $\widetilde{S}_n \to \widetilde{S}$ weakly, by the result for symmetric sequences (Theorem 2.4), $\widetilde{S}_n \to \widetilde{S}$ almost surely. In particular, there exists, by Fubini's theorem, an ω' in Ω' such that

$$S_n - S_n'(\omega') \to S - S'(\omega') \text{ almost surely}.$$

On the other hand $S_n \to S$ weakly. By difference, it follows from these two observations that $(S_n'(\omega'))$ is a relatively compact sequence in B. Moreover, taking characteristic functionals, for every f in B',

$$\exp\Big(if\big(S_n'(\omega')\big)\Big) \to \exp\Big(if\big(S'(\omega')\big)\Big).$$

Hence $f(S_n'(\omega')) \to f(S'(\omega'))$ and thus $S_n'(\omega)$ converges to $S'(\omega')$ in B. Therefore $S_n \to S$ almost surely and the theorem is proved.

While Lévy's inequalities (Proposition 2.3) are one of the main ingredients in the proof of Itô-Nisio's theorem in the symmetrical case, one can actually also prove directly the preceding statement using instead a similar inequality known as Ottaviani's inequality. Its proof follows the pattern of the proof of Lévy's inequalities.

Lemma 6.2. *Let $(X_i)_{i \leq N}$ be independent random variables in B and set $S_k = \sum_{i=1}^{k} X_i$, $k \leq N$. Then, for every $s, t > 0$,*

$$\mathbb{P}\{\max_{k \leq N} \|S_k\| > s + t\} \leq \frac{\mathbb{P}\{\|S_N\| > t\}}{1 - \max_{k \leq N} \mathbb{P}\{\|S_N - S_k\| > s\}} .$$

Proof. Let $\tau = \inf\{k \leq N; \|S_k\| > s + t\}$ ($+\infty$ if no such k exists). Then, as usual, $\{\tau = k\}$ only depends on $X_1, \ldots, X_k$ and $\sum_{k=1}^{N} \mathbb{P}\{\tau = k\} = \mathbb{P}\{\max_{k \leq N} \|S_k\| > s + t\}$. When $\tau = k$ and $\|S_N - S_k\| \leq s$, then $\|S_N\| > t$. Hence, by independence,

$$\mathbb{P}\{\|S_N\| > t\} = \sum_{k=1}^{N} \mathbb{P}\{\tau = k, \|S_N\| > t\}$$

$$\geq \sum_{k=1}^{N} \mathbb{P}\{\tau = k, \|S_N - S_k\| \leq s\}$$

$$\geq \inf_{k \leq N} \mathbb{P}\{\|S_N - S_k\| \leq s\} \sum_{k=1}^{N} \mathbb{P}\{\tau = k\}$$

which is the desired result.

The symmetrization procedure is further illustrated in the next trivial lemma. As always, (ε_i) is a Rademacher sequence which is independent of (X_i). Recall that X is centered if $\mathbb{E}f(X) = 0$ for every f in D.

Lemma 6.3. *Let $F : \mathbb{R}_+ \to \mathbb{R}_+$ be convex. Then, for any finite sequence (X_i) of independent mean zero random variables in B such that $\mathbb{E}F(\|X_i\|) < \infty$ for every i,*

$$\mathbb{E}F\left(\frac{1}{2}\left\|\sum_i \varepsilon_i X_i\right\|\right) \leq \mathbb{E}F\left(\left\|\sum_i X_i\right\|\right) \leq \mathbb{E}F\left(2\left\|\sum_i \varepsilon_i X_i\right\|\right) .$$

Proof. Recall $\widetilde{X}_i = X_i - X_i'$ and let (ε_i) be a Rademacher sequence which is independent from (X_i) and (X_i'). Then, by Fubini, Jensen's inequality and zero mean (cf. (2.5)), and by convexity, we have

$$\mathbb{E}F\left(\left\|\sum_i X_i\right\|\right) \leq \mathbb{E}F\left(\left\|\sum_i \widetilde{X}_i\right\|\right) = \mathbb{E}F\left(\left\|\sum_i \varepsilon_i \widetilde{X}_i\right\|\right) \leq \mathbb{E}F\left(2\left\|\sum_i \varepsilon_i X_i\right\|\right) .$$

Conversely, by the same arguments,

$$\mathbb{E}F\left(\frac{1}{2}\left\|\sum_i \varepsilon_i X_i\right\|\right) \leq \mathbb{E}F\left(\frac{1}{2}\left\|\sum_i \varepsilon_i \tilde{X}_i\right\|\right)$$

$$= \mathbb{E}F\left(\frac{1}{2}\left\|\sum_i \tilde{X}_i\right\|\right) \leq \mathbb{E}F\left(\left\|\sum_i X_i\right\|\right).$$

The lemma is proved.

Note that when the variables X_i are not centered, we have similarly

$$\mathbb{E}F\left(\sup_{f\in D}\left|\sum_i f(X_i) - \mathbb{E}f(X_i)\right|\right) \leq \mathbb{E}F\left(2\left\|\sum_i \varepsilon_i X_i\right\|\right)$$

and also

$$\mathbb{E}F\left(\sup_{f\in D}\left|\sum_i \varepsilon_i\big(f(X_i) - \mathbb{E}f(X_i)\big)\right|\right) \leq \mathbb{E}F\left(2\left\|\sum_i X_i\right\|\right).$$

Thus, symmetrization indicates how results on symmetric random variables can be transferred to general results. In the sequel of this chapter, we therefore mainly concentrate only on symmetrical distributions for which the results are usually clearer and easier to state. We leave it to the interested reader to extend them to the case of general (or mean zero) independent random variables by the techniques just presented.

Before turning to the main object of this chapter, we would like to briefly mention in passing a concentration inequality which is often useful when for example Lévy's inequalities do not readily apply. It is due to M. Kanter [Kan].

Proposition 6.4. *Let (X_i) be a finite sequence of independent symmetric random variables with values in B. Then, for any x in B and any $t > 0$,*

$$\mathbb{P}\left\{\left\|\sum_i X_i - x\right\| \leq t\right\} \leq \frac{3}{2}\left(1 + \sum_i \mathbb{P}\{\|X_i\| > t\}\right)^{-1/2}.$$

Since symmetric sequences of random variables can be randomized by an independent Rademacher sequence, the contraction and comparison properties described for Rademacher averages in Chapter 4 can be extended to this more general setting. The next two lemmas are easy instances of this procedure. The second one will prove extremely useful in the sequel.

Lemma 6.5. *Let (X_i) be a finite symmetric sequence of random variables with values in B. Let further (ξ_i) and (ζ_i) be real random variables such that $\xi_i = \varphi_i(X_i)$ where $\varphi_i : B \to \mathbb{R}$ is symmetric (even), and similarly for ζ_i. Then, if $|\xi_i| \leq |\zeta_i|$ almost surely for every i, for any convex function $F : \mathbb{R}_+ \to \mathbb{R}_+$ (and under some appropriate integrability assumptions),*

$$\mathbb{E}F\left(\left\|\sum_i \xi_i X_i\right\|\right) \leq \mathbb{E}F\left(\left\|\sum_i \zeta_i X_i\right\|\right).$$

We also have, for every $t > 0$,

$$\mathbb{P}\left\{\left\|\sum_i \xi_i X_i\right\| > t\right\} \leq 2\mathbb{P}\left\{\left\|\sum_i \zeta_i X_i\right\| > t\right\}.$$

In particular, these inequalities apply when $\xi_i = I_{\{X_i \in A_i\}} \leq 1 \equiv \zeta_i$ where the sets A_i are symmetric in B (in particular $A_i = \{\|x\| \leq a_i\}$).

Proof. The sequence (X_i) has the same distribution as $(\varepsilon_i X_i)$. By the symmetry assumption on the φ_i's and Fubini's theorem

$$\mathbb{E}F\left(\left\|\sum_i \xi_i X_i\right\|\right) = \mathbb{E}_X \mathbb{E}_\varepsilon F\left(\left\|\sum_i \varepsilon_i \xi_i X_i\right\|\right).$$

By the contraction principle (Theorem 4.4)

$$\mathbb{E}_\varepsilon F\left(\left\|\sum_i \varepsilon_i \xi_i X_i\right\|\right) \leq \mathbb{E}_\varepsilon F\left(\left\|\sum_i \varepsilon_i \zeta_i X_i\right\|\right)$$

from which the first inequality of the lemma follows. The second is established similarly using (4.7).

Lemma 6.6. *Let $F : \mathbb{R}_+ \to \mathbb{R}_+$ be convex and increasing. Let (X_i) be arbitrary random variables in B. Then, if $\mathbb{E}F(\|X_i\|) < \infty$,*

(6.4) $$\mathbb{E}F\left(\frac{1}{2}\sup_{f \in D}\left|\sum_i \varepsilon_i |f(X_i)|\right|\right) \leq \mathbb{E}F\left(\left\|\sum_i \varepsilon_i X_i\right\|\right).$$

When the X_i's are independent and symmetric in $L_2(B)$, we also have

(6.5) $$\mathbb{E}\left(\sup_{f \in D}\left(\sum_i f^2(X_i)\right)\right) \leq \sup_{f \in D}\sum_i \mathbb{E}f^2(X_i) + 8\mathbb{E}\left\|\sum_i X_i\|X_i\|\right\|.$$

Proof. (6.4) is simply Theorem 4.12 applied conditionally. To establish (6.5), we write

$$\mathbb{E}\left(\sup_{f \in D}\left(\sum_i f^2(X_i)\right)\right) \leq \sup_{f \in D}\sum_i \mathbb{E}f^2(X_i) + \mathbb{E}\left(\sup_{f \in D}\left|\sum_i f^2(X_i) - \mathbb{E}f^2(X_i)\right|\right).$$

Lemma 6.3 shows that

$$\mathbb{E}\left(\sup_{f \in D}\left|\sum_i f^2(X_i) - \mathbb{E}f^2(X_i)\right|\right) \leq 2\mathbb{E}\left(\sup_{f \in D}\left|\sum_i \varepsilon_i f^2(X_i)\right|\right)$$

and, by (4.19),

$$\mathbb{E}\left(\sup_{f\in D}\left|\sum_i \varepsilon_i f^2(X_i)\right|\right) \leq 4\mathbb{E}\left\|\sum_i \varepsilon_i X_i \|X_i\|\right\|.$$

The lemma is thus established.

6.2 Integrability of Sums of Independent Random Variables

Integrability of sums of independent vector valued random variables are based on various inequalities. While isoperimetric methods, which are most powerful, will be described in the next section, we present here some more classical and easier ideas. An important result is a set of inequalities due to J. Hoffmann-Jørgensen which is the content of the next statement. Some of its various consequences are presented in the subsequent theorems.

Proposition 6.7. *Let* $(X_i)_{i\leq N}$ *be independent random variables with values in* B. *Set* $S_k = \sum_{i=1}^k X_i$, $k \leq N$. *For every* $s, t > 0$,

$$(6.6) \quad \mathbb{P}\{\max_{k\leq N}\|S_k\| > 3t + s\} \leq \left(\mathbb{P}\{\max_{k\leq N}\|S_k\| > t\}\right)^2 + \mathbb{P}\{\max_{i\leq N}\|X_i\| > s\}.$$

If the variables are symmetric, then, for $s, t > 0$,

$$(6.7) \quad \mathbb{P}\{\|S_N\| > 2t + s\} \leq 4\left(\mathbb{P}\{\|S_N\| > t\}\right)^2 + \mathbb{P}\{\max_{i\leq N}\|X_i\| > s\}.$$

Proof. Let $\tau = \inf\{j \leq N; \|S_j\| > t\}$. By definition, $\{\tau = j\}$ only depends on the random variables $X_1, \dots, X_j$ and $\{\max_{k\leq N}\|S_k\| > t\} = \sum_{j=1}^N \{\tau = j\}$ (disjoint union). On $\{\tau = j\}$, $\|S_k\| \leq t$ if $k < j$ and when $k \geq j$

$$\|S_k\| \leq t + \|X_j\| + \|S_k - S_j\|$$

so that in any case

$$\max_{k\leq N}\|S_k\| \leq t + \max_{i\leq N}\|X_i\| + \max_{j<k\leq N}\|S_k - S_j\|.$$

Hence, by independence,

$$\mathbb{P}\{\tau = j, \max_{k\leq N}\|S_k\| > 3t + s\}$$

$$\leq \mathbb{P}\{\tau = j, \max_{i\leq N}\|X_i\| > s\} + \mathbb{P}\{\tau = j\}\mathbb{P}\{\max_{j<k\leq N}\|S_k - S_j\| > 2t\}.$$

Since $\max_{j<k\leq N}\|S_k - S_j\| \leq 2\max_{k\leq N}\|S_k\|$, a summation over $j = 1, \dots, N$ yields (6.6).

Concerning (6.7), for every $j = 1, \ldots, N$,

$$\|S_N\| \leq \|S_{j-1}\| + \|X_j\| + \|S_N - S_j\|,$$

so that

$$\mathbb{P}\{\tau = j, \|S_N\| > 2t + s\} \leq \mathbb{P}\{\tau = j, \max_{i \leq N} \|X_i\| > s\}$$

$$+ \mathbb{P}\{\tau = j\}\mathbb{P}\{\|S_N - S_j\| > t\}.$$

Using Lévy's inequality (2.6) for symmetric variables and summing over j yields (6.7). The proposition is proved.

The preceding inequalities are mainly used with $s = t$. Their main interest and usefulness stem from the squared probability which make them close in a sense to exponential inequalities (see below).

As a first consequence of the preceding inequalities, the next proposition is still a technical step preceding the integrability statements. It however already expresses, in the context of sums of independent random variables, a property similar to the one presented in the preceding chapters on Gaussian, Rademacher and stable vector valued random variables. Namely, if the sums are controlled in probability (L_0), they are also controlled in L_p, $p > 0$, provided the same holds for the maximum of the individual summands. Applied to Gaussian, Rademacher or stable averages, the next proposition actually gives rise to *new proofs* of the equivalence of moments for these particular sums of independent random variables.

Proposition 6.8. *Let $0 < p < \infty$ and let $(X_i)_{i \leq N}$ be independent random variables in $L_p(B)$. Set $S_k = \sum_{i=1}^{k} X_i$, $k \leq N$. Then, for $t_0 = \inf\{t > 0; \mathbb{P}\{\max_{k \leq N} \|S_k\| > t\} \leq (2 \cdot 4^p)^{-1}\}$,*

$$(6.8) \qquad \mathbb{E} \max_{k \leq N} \|S_k\|^p \leq 2 \cdot 4^p \mathbb{E} \max_{i \leq N} \|X_i\|^p + 2(4t_0)^p.$$

If, moreover, the X_i's are symmetric, and $t_0 = \inf\{t > 0; \mathbb{P}\{\|S_N\| > t\} \leq (8 \cdot 3^p)^{-1}\}$, then

$$(6.9) \qquad \mathbb{E}\|S_N\|^p \leq 2 \cdot 3^p \mathbb{E} \max_{i \leq N} \|X_i\|^p + 2(3t_0)^p.$$

Proof. We only show (6.9), the proof of (6.8) being similar using (6.6). Let $u > t_0$. By integration by parts and (6.7),

$$\mathbb{E}\|S_N\|^p = 3^p \int_0^\infty \mathbb{P}\{\|S_N\| > 3t\}dt^p$$

$$= 3^p \left(\int_0^u + \int_u^\infty \right) \mathbb{P}\{\|S_N\| > 3t\}dt^p$$

$$\leq (3u)^p + 4 \cdot 3^p \int_u^\infty \left(\mathbb{P}\{\|S_N\| > t\} \right)^2 dt^p + 3^p \int_u^\infty \mathbb{P}\{\max_{i \leq N} \|X_i\| > t\}dt^p$$

$$\leq (3u)^p + 4 \cdot 3^p \mathbb{P}\{\|S_N\| > u\} \int_0^\infty \mathbb{P}\{\|S_N\| > t\}dt^p + 3^p \mathbb{E}\max_{i \leq N} \|X_i\|^p$$

$$\leq 2(3u)^p + 2 \cdot 3^p \mathbb{E}\max_{i \leq N} \|X_i\|^p$$

since $4 \cdot 3^p \mathbb{P}\{\|S_N\| > u\} \leq 1/2$ by the choice of u. Since this holds for arbitrary $u > t_0$ the proposition is established.

It is actually possible to obtain a true equivalence of moments for sums of independent symmetric random variables. However, the formulation is somewhat technical. Before we introduce this result, we need a simple lemma on moments of maximum of independent random variables which is of independent interest.

Lemma 6.9. *Let $p > 0$ and let (Z_i) be a finite sequence of independent positive random variables in L_p. Given $\lambda > 0$, let $\delta_0 = \inf\{t > 0; \sum_i \mathbb{P}\{Z_i > t\} \leq \lambda\}$. Then*

$$\lambda(1 + \lambda)^{-1}\delta_0^p + (1 + \lambda)^{-1} \sum_i \int_{\delta_0}^\infty \mathbb{P}\{Z_i > t\}dt^p$$

$$\leq \mathbb{E}\max_i Z_i^p \leq \delta_0^p + \sum_i \int_{\delta_0}^\infty \mathbb{P}\{Z_i > t\}dt^p .$$

Proof. Use integration by parts. The right hand side inequality is trivial (and actually holds for *any* $\delta_0 > 0$). Turning to the left hand side we use Lemma 2.6: the definition of δ_0 indicates that

$$\mathbb{P}\{\max_i Z_i > t\} \geq \begin{cases} (1 + \lambda)^{-1} \sum_i \mathbb{P}\{Z_i > t\} & \text{if } t > \delta_0 \\ \lambda(1 + \lambda)^{-1} & \text{if } t \leq \delta_0 . \end{cases}$$

Lemma 6.9 then clearly follows.

The announced equivalence of moments is as follows.

Proposition 6.10. *Let $0 < p, q < \infty$. Let (X_i) be a finite sequence of independent and symmetric random variables in $L_p(B)$. Then, for some constant $K_{p,q}$ depending on p, q only,*

$$\left\|\sum_i X_i\right\|_p \underset{K_{p,q}}{\sim} \left\|\max_i \|X_i\|\right\|_p + \left\|\sum_i X_i I_{\{\|X_i\| \leq \delta_0\}}\right\|_q$$

where $\delta_0 = \inf\{t > 0; \sum_i \mathbb{P}\{\|X_i\| > t\} \le (8 \cdot 3^p)^{-1}\}$ and where the sign $a \underset{K_{p,q}}{\frown} b$
means that $K_{p,q}^{-1} b \le a \le K_{p,q} b$.

Proof. By the triangle inequality

$$\mathbb{E}\left\|\sum_i X_i\right\|^p \le 2^p \mathbb{E}\left\|\sum_i X_i I_{\{\|X_i\| \le \delta_0\}}\right\|^p + 2^p \mathbb{E}\left\|\sum_i X_i I_{\{\|X_i\| > \delta_0\}}\right\|^p.$$

If we apply (6.9) to the second term of the right hand side of this inequality,
we see that, by the definition of δ_0, we can take $t_0 = 0$ there so that

$$\mathbb{E}\left\|\sum_i X_i I_{\{\|X_i\| > \delta_0\}}\right\|^p \le 2 \cdot 3^p \mathbb{E} \max_i \|X_i\|^p.$$

Turning to the first term and applying again Proposition 6.8, we can take

$$t_0 = \left(8 \cdot 3^p \mathbb{E}\left\|\sum_i X_i I_{\{\|X_i\| \le \delta_0\}}\right\|^q\right)^{1/q}.$$

The first half of the proposition follows. To prove the reverse inequality note
that, by (6.9) again,

$$\mathbb{E}\left\|\sum_i X_i I_{\{\|X_i\| \le \delta_0\}}\right\|^q \le 2 \cdot 3^q \delta_0^q + 2(3t_0)^q$$

where we can choose for t_0,

$$t_0 = \left(8 \cdot 3^q \mathbb{E}\left\|\sum_i X_i\right\|^p\right)^{1/p}.$$

Then, if we draw from Lemma 6.9 the fact that

$$\mathbb{E} \max_i \|X_i\|^p \ge \lambda(1 + \lambda)^{-1} \delta_0^p$$

with $\lambda = (8 \cdot 3^p)^{-1}$, the proof will be complete since we know from Lévy's
inequality (2.7) that

$$\mathbb{E} \max_i \|X_i\|^p \le 2\mathbb{E}\left\|\sum_i X_i\right\|^p.$$

We now summarize in various (integrability) theorems for sums of independent vector valued random variables the preceding powerful inequalities
and arguments.

Theorem 6.11. *Let $(X_i)_{i \in \mathbb{N}}$ be a sequence of independent random variables
with values in B. Set, as usual, $S_n = \sum_{i=1}^n X_i$, $n \ge 1$. Let also $0 < p < \infty$.
Then, if $\sup_n \|S_n\| < \infty$ almost surely, we have an equivalence between:*

(i) $\mathbb{E} \sup_n \|S_n\|^p < \infty$;

(ii) $\mathbb{E} \sup_n \|X_n\|^p < \infty$.

Furthermore, if the sequence (S_n) converges almost surely, (i) and (ii) are also equivalent to

(iii) $\mathbb{E} \left\| \sum_i X_i \right\|^p < \infty$

and in this case (S_n) also converges in L_p.

Proof. That (i) implies (ii) is obvious. Let N be fixed. By Proposition 6.8, t_0 being defined there,

$$\mathbb{E} \max_{n \leq N} \|S_n\|^p \leq 2 \cdot 4^p \mathbb{E} \max_{i \leq N} \|X_i\|^p + 2(4t_0)^p .$$

Since $\mathbb{P}\{\sup_n \|S_n\| < \infty\} = 1$, there is $M > 0$ such that $t_0 \leq M$ independently of N. Letting N tend to infinity shows (ii) $\Rightarrow$ (i). The assertion relative to (iii) follows from Lévy's inequalities for symmetric random variables, and from an easy symmetrization argument based on (6.1) in general.

Corollary 6.12. *Let (a_n) be an increasing sequence of positive numbers tending to infinity. Let (X_i) be independent random variables with values in B and set, as usual, $S_n = \sum_{i=1}^n X_i$, $n \geq 1$. Then, if $\sup_n \|S_n\|/a_n < \infty$ almost surely, for any $0 < p < \infty$, the following are equivalent:*

(i) $\mathbb{E} \sup_n \left(\dfrac{\|S_n\|}{a_n} \right)^p < \infty$;

(ii) $\mathbb{E} \sup_n \left(\dfrac{\|X_n\|}{a_n} \right)^p < \infty$.

Proof. We define a new sequence (Y_i) of independent random variables with values in the Banach space $\ell_\infty(B)$ of all bounded sequences $x = (x_n)$ with the sup-norm $\|x\| = \sup_n \|x_n\|$ by setting

$$Y_i = \left(0, \ldots, 0, \frac{X_i}{a_i}, \frac{X_i}{a_{i+1}}, \frac{X_i}{a_{i+2}}, \ldots \right)$$

where there are $i - 1$ zeroes to start with. Clearly $\|Y_i\| = \|X_i\|/a_i$ for all i, and

$$\sup_n \left\| \sum_{i=1}^n Y_i \right\| = \sup_n \frac{\|S_n\|}{a_n} .$$

Apply then Theorem 6.11 to the sequence (Y_i) in $\ell_\infty(B)$.

Remark 6.13. It is amusing to note that Hoffmann-Jørgensen's inequalities and Theorem 6.11 describe well enough independence to include the Borel-Cantelli lemma! Indeed, if (A_i) is a sequence of independent sets, we get from Theorem 6.11 that if $\sum_i I_{A_i}$ converges almost surely, then $\mathbb{E}(\sum_i I_{A_i}) = \sum_i \mathbb{P}(A_i) < \infty$; this corresponds to the independent part of the Borel-Cantelli lemma.

With the preceding material, let us now consider an almost surely convergent series $S = \sum_i X_i$ of independent symmetric, or only with mean zero, uniformly bounded random variables with values in B and let us try to investigate the integrability properties of $\|S\|$. Assume more precisely that the X_i's are symmetric and that $\|X_i\|_\infty \leq a < \infty$ for every i. For each N, set $S_N = \sum_{i=1}^N X_i$. By (6.7), for every $t > 0$,

$$\mathbb{P}\{\|S_N\| > 2t + a\} \leq \left(2\mathbb{P}\{\|S_N\| > t\}\right)^2.$$

Let t_0 to be specified in a moment and define the sequence $t_n = 2^n(t_0 + a) - a$. The preceding inequality indicates that, for every n,

$$\mathbb{P}\{\|S_N\| > t_n\} \leq \left(2\mathbb{P}\{\|S_N\| > t_{n-1}\}\right)^2.$$

By iteration, we get

$$\mathbb{P}\{\|S_N\| > t_n\} \leq 2^{2^{n+1}-2}\left(\mathbb{P}\{\|S_N\| > t_0\}\right)^{2^n}.$$

If $S = \sum_i X_i$ converges almost surely, there exists t_0 such that, for every N, $\mathbb{P}\{\|S_N\| > t_0\} \leq 1/8$. Summarizing, for every N and n,

$$\mathbb{P}\{\|S_N\| > 2^n(t_0 + a)\} \leq 2^{-2^n}.$$

It easily follows that for some $\lambda > 0$, $\sup_N \mathbb{E}\exp(\lambda\|S_N\|) < \infty$, and thus, by Fatou's lemma, that $\mathbb{E}\exp(\lambda\|S\|) < \infty$. By convergence, this can easily be improved into the same property for *all* $\lambda > 0$. Note actually that $\sup_N \mathbb{E}\exp(\lambda\|S_N\|) < \infty$ as soon as the sequence (S_N) is stochastically bounded.

The preceding iteration procedure may be compared to the proof of (3.5) in the Gaussian case. To complement it, let us present a somewhat neater argument for the same result, which, if only a small variation, completes our ability with the technique. The argument is the following; applied to power functions, it yields an alternate proof of Proposition 6.8 (cf. Remark 6.15 below).

Proposition 6.14. *Let $(X_i)_{i \leq N}$ be independent and symmetric random variables with values in B, $S_k = \sum_{i=1}^k X_i$, $k \leq N$. Assume that $\|X_i\|_\infty \leq a$ for all $i \leq N$. Then, for every $\lambda, t > 0$,*

$$\mathbb{E}\exp(\lambda\|S_N\|) \leq \exp(\lambda t) + 2\exp(\lambda(t+a))\mathbb{P}\{\|S_N\| > t\}\mathbb{E}\exp(\lambda\|S_N\|).$$

Proof. Let as usual $\tau = \inf\{k \leq N; \|S_k\| > t\}$. We can write

$$\mathbb{E}\exp(\lambda\|S_N\|) \leq \exp(\lambda t) + \sum_{k=1}^N \int_{\{\tau=k\}} \exp(\lambda\|S_N\|)\,d\mathbb{P}.$$

On the set $\{\tau = k\}$,

$$\|S_N\| \le \|S_{k-1}\| + \|X_k\| + \|S_N - S_k\| \le t + a + \|S_N - S_k\|$$

so that, by independence,

$$\int_{\{\tau = k\}} \exp(\lambda\|S_N\|)d\mathbb{P} \le \exp(\lambda(t+a))\mathbb{P}\{\tau = k\}\,\mathbb{E}\exp(\lambda\|S_N - S_k\|).$$

By Jensen's inequality and mean zero, $\mathbb{E}\exp(\lambda\|S_N - S_k\|) \le \mathbb{E}\exp(\lambda\|S_N\|)$ (cf. (2.5)), and, summing over k,

$$\sum_{k=1}^{N}\mathbb{P}\{\tau = k\} = \mathbb{P}\{\max_{k\le N}\|S_k\| > t\} \le 2\mathbb{P}\{\|S_N\| > t\}$$

where we have used Lévy's inequality (2.6). The proof is complete. (Note that there is an easy analog when the variables are only centered.)

Remark 6.15. As announced, the proof of Proposition 6.14 applied to power functions yields an alternate proof of the inequalities of Proposition 6.8. It actually yields an inequality in the form of Kolmogorov's converse inequality. That is, under the assumption of the last proposition, for every $t > 0$ and every $1 \le p < \infty$,

$$\mathbb{P}\{\|S_N\| > t\} \ge \frac{1}{2^p}\left[1 - \frac{2^{2p}(t^p + \mathbb{E}\max_{i\le N}\|X_i\|^p)}{\mathbb{E}\|S_N\|^p}\right].$$

Let (X_i) be a sequence of independent symmetric (or only mean zero) random variables, uniformly bounded by a, such that $S = \sum_i X_i$ converges almost surely. Then, as a consequence of Proposition 6.14, we recover the fact that $\mathbb{E}\exp(\lambda\|S\|) < \infty$ for some (actually all) $\lambda > 0$. Indeed, if we choose t in Proposition 6.14 that satisfies $\mathbb{P}\{\|S_N\| > t\} \le (2e)^{-1}$ for every N and let $\lambda = (t+a)^{-1}$, we simply get

$$\sup_N \mathbb{E}(\exp\lambda\|S_N\|) \le 2\exp(\lambda t) < \infty.$$

This exponential integrability result is not quite satisfactory since it is known that for real valued random variables, $\mathbb{E}\exp(\lambda|S|\log^+|S|) < \infty$ for some $\lambda > 0$. This result on the line is one instance of the Poisson behavior of general sums of independent (bounded) random variables as opposed to the normal behavior of more specialized ones, such as Rademacher averages. It originates in the sharp *quadratic* real exponential inequalities (see for example (6.10) below). These results can however be extended to the vector valued case. To this aim, we use isoperimetric methods to obtain sharp exponential estimates for sums of independent random variables, even improving in some places the scalar case.

6.3 Concentration and Tail Behavior

This section is mainly devoted to applications of isoperimetric methods (Theorem 1.4) to the integrability and tail behavior of sums of independent random variables. One of the objectives will be to try to extend to the infinite dimensional setting the classical (quadratic) exponential inequalities as those of Bernstein, Kolmogorov (Lemma 1.6), Prokhorov, Bennett, Hoeffding, etc. (Of course, the lack of orthogonality forces to investigate new arguments, like therefore isoperimetry.) To state one, and for the matter of comparison, let us consider Bennett's inequality. Let (X_i) be a finite sequence of independent mean zero *real valued* random variables such that $\|X_i\|_\infty \leq a$ for every i; then, if $b^2 = \sum_i \mathbb{E} X_i^2$, for all $t > 0$,

$$(6.10) \qquad \mathbb{P}\left\{\sum_i X_i > t\right\} \leq \exp\left[\frac{t}{a} - \left(\frac{t}{a} + \frac{b^2}{a^2}\right)\log\left(1 + \frac{at}{b^2}\right)\right].$$

This inequality is rather typical of the tail behavior of sums of independent random variables. This behavior varies depending on the relative sizes of t and of the ratio b^2/a. Since $\log(1 + x) \geq x - \frac{1}{2}x^2$ when $0 \leq x \leq 1$, if $t \leq b^2/a$, (6.10) implies

$$\mathbb{P}\left\{\sum_i X_i > t\right\} \leq \exp\left(-\frac{t^2}{2b^2} + \frac{at^3}{2b^4}\right),$$

which is furthermore less than $\exp(-t^2/4b^2)$ if $t \leq b^2/2a$ (for example). On the other hand, for every $t > 0$,

$$\mathbb{P}\left\{\sum_i X_i > t\right\} \leq \exp\left[-\frac{t}{a}\left(\log\left(1 + \frac{at}{b^2}\right) - 1\right)\right]$$

which is sharp for large values of t (larger than b^2/a). These two inequalities actually describe the classical *normal* and *Poisson* type behaviors of sums of independent random variables according to the size of t with respect to b^2/a.

Before we turn to the isoperimetric argument in this study, we would like to present some results based on martingales. Although these do not seem to be powerful enough in general for the integrability and tail behavior questions that we have in mind, they are however rather simple and quite useful in many situations. They also present the advantage of being formulated as concentration inequalities, some of which will be useful in this form in Chapter 9.

The key observation in order to use (real valued) martingale inequalities in the study of sums of independent vector valued random variables relies on the following simple but extremely useful observation of V. Yurinskii. Let $(X_i)_{i \leq N}$ be integrable random variables in B. Denote by $\mathcal{A}_i$ the σ-algebra generated by the variables $X_1, \ldots, X_i$, $i \leq N$, and by $\mathcal{A}_0$ the trivial algebra. Write as usual $S_N = \sum_{i=1}^{N} X_i$ and set, for each i,

$$d_i = \mathbb{E}^{\mathcal{A}_i}\|S_N\| - \mathbb{E}^{\mathcal{A}_{i-1}}\|S_N\|.$$

$(d_i)_{i \leq N}$ defines a real valued martingale difference sequence $(\mathbb{E}^{\mathcal{A}_{i-1}} d_i = 0)$ and $\sum_{i=1}^{N} d_i = \|S_N\| - \mathbb{E}\|S_N\|$. Then, we have:

Lemma 6.16. *Assume that the random variables X_i are independent. Then, in the preceding notation, almost surely for every $i \leq N$,*

$$|d_i| \leq \|X_i\| + \mathbb{E}\|X_i\|.$$

Furthermore, if the X_i's are in $L_2(B)$ we also have

$$\mathbb{E}^{\mathcal{A}_{i-1}} d_i^2 \leq \mathbb{E}\|X_i\|^2.$$

Proof. Independence ensures that

$$d_i = \left(\mathbb{E}^{\mathcal{A}_i} - \mathbb{E}^{\mathcal{A}_{i-1}}\right)\left(\|S_N\| - \|S_N - X_i\|\right)$$

and the first inequality of the lemma already follows from the triangle inequality since

$$|d_i| \leq \left(\mathbb{E}^{\mathcal{A}_i} + \mathbb{E}^{\mathcal{A}_{i-1}}\right)\left(\|X_i\|\right) = \|X_i\| + \mathbb{E}\|X_i\|.$$

Since conditional expectation is a projection in L_2 the same argument leads to the second inequality of the lemma.

The philosophy of the preceding observation is that the deviation of the norm of a sum S_N of independent random vectors X_i, $i \leq N$, from its expectation $\mathbb{E}\|S_N\|$ can be written as a *real valued* martingale whose differences d_i are nearly exactly controlled by the norms of the corresponding individual summands X_i of S_N. Therefore, in a sense, up to $\mathbb{E}\|S_N\|$, $\|S_N\|$ is as good as a real valued martingale with differences comparable to $\|X_i\|$. Typical in this regard is the following quadratic inequality, immediate consequence of Lemma 6.16 and of the orthogonality of martingale differences:

(6.11) $$\mathbb{E}\big|\,\|S_N\| - \mathbb{E}\|S_N\|\,\big|^2 \leq \sum_{i=1}^{N} \mathbb{E}\|X_i\|^2.$$

Of course, on the line or even in Hilbert space, if the variables are centered, this inequality (which then becomes an equality) holds true without the centering factor $\mathbb{E}\|S_N\|$.

This remark is a rather important feature of the study of sums of independent Banach space valued random variables (we will also find it in the isoperimetric approach developed later). It shows how, when a control in *expectation* (or only in *probability* by Proposition 6.8) of a sum of independent random variables is given, then one can expect, using some of the classical *scalar* arguments, an *almost sure* control. This was already the line of the integrability theorems discussed in Section 6.2 and those of this section will be similar. In the next two chapters devoted to strong limit theorems for sums of independent random variables, this will lead to various equivalences between almost sure and in probability limiting properties under necessary (and classical) moment assumptions on the individual summands. As in the Gaussian, Rademacher and stable cases, it is then to know how to control in probability (or weakly) sums of independent random variables. On the line or in finite dimensional spaces, this is easily accomplished by orthogonality and moment conditions. This is much more difficult in the infinite dimensional setting and may be considered as a main problem of the theory. It will be studied in some instances in the second part of this work, in Chapters 9, 10 and 14 in particular.

From Lemma 6.16, the various martingale inequalities of Chapter 1, Section 1.3, can be applied to $\|S_N\| - \mathbb{E}\|S_N\|$ yielding concentration properties for norms of sums $S_N = \sum_{i=1}^{N} X_i$ of independent random variables around their expectation in terms of the size of the summands X_i. Let us record some of them at this stage. Lemma 1.6 (together with Lemma 6.16) shows that if $a = \max_{i \leq N} \|X_i\|_\infty$ and $b \leq (\sum_{i=1}^{N} \mathbb{E}\|X_i\|^2)^{1/2}$, then, for every $t > 0$,

$$(6.12) \quad \mathbb{P}\{| \|S_N\| - \mathbb{E}\|S_N\| | > t\} \leq 2 \exp\left[-\frac{t^2}{2b^2}\left(2 - \exp\left(\frac{2at}{b^2}\right)\right)\right].$$

One can also prove a martingale version of (6.10) which, when applied to $\|S_N\| - \mathbb{E}\|S_N\|$, yields

$$(6.13) \quad \begin{aligned} &\mathbb{P}\{| \|S_N\| - \mathbb{E}\|S_N\| | > t\} \\ &\quad \leq 2 \exp\left[\frac{t}{2a} - \left(\frac{t}{2a} + \frac{b^2}{4a^2}\right)\log\left(1 + \frac{2at}{b^2}\right)\right]. \end{aligned}$$

Lemma 1.7 indicates in the same way that if $1 < p < 2$, $q = p/p - 1$ and $a = \sup_{i \geq 1} i^{1/p}\|X_i\|_\infty$ is assumed to be finite, for all $t > 0$,

$$(6.14) \quad \mathbb{P}\{| \|S_N\| - \mathbb{E}\|S_N\| | > t\} \leq 2 \exp(-t^q/C_q a^q)$$

where $C_q > 0$ only depends on q. (6.14) is of particular interest when $X_i = \varepsilon_i x_i$ where (ε_i) is a Rademacher sequence and (x_i) a finite sequence in B. We then have, for all $t > 0$,

$$(6.15) \quad \mathbb{P}\left\{\left| \left\|\sum_i \varepsilon_i x_i\right\| - \mathbb{E}\left\|\sum_i \varepsilon_i x_i\right\| \right| > t\right\} \leq 2 \exp(-t^q/C_q\|(x_i)\|_{p,\infty}^q)$$

where $\|(x_i)\|_{p,\infty} = \|(\|x_i\|)\|_{p,\infty}$. Of course, this inequality may be compared to the concentration property of Theorem 4.7 as well as to the inequalities described in the first section of Chapter 4. The previous two inequalities will be helpful both in this chapter and in Chapter 9 where in particular concentration will be required for the construction of ℓ_p^n-subspaces, $1 < p < 2$, of Banach spaces. Note that we already used (6.11) in the preceding chapter to prove a concentration inequality for stable random variables (Proposition 5.7).

Now, we turn to the main part of this chapter with the applications to sums of independent random variables of the isoperimetric inequality for product measures of Theorem 1.4. This isoperimetric inequality appears as a powerful tool which will be shown to be efficient in many situations. In particular, it will allow us to complete the study of the integrability properties of sums of independent random variables started in the preceding section and to investigate almost sure limit theorems in the next chapters. This isoperimetric approach, which is a priori very different from the classical tools (although similarities can and will be detected), seems to subsume in general the usual arguments.

First, let us briefly recall Theorem 1.4 and (1.13). Let N be an arbitrary but fixed integer and let $\mathbf{X} = (X_i)_{i \leq N}$ be a sample of independent random variables with values in B. (In this setting, we may simply equip B with the σ-algebra generated by the linear functionals $f \in D$.) For A measurable in the product space $B^N = \{x \in B; \; x = (x_i)_{i \leq N}, x_i \in B\}$ and for integers q, k, set

$$H(A, q, k)$$
$$= \left\{ x \in B^N; \exists \, x^1, \ldots, x^q \in A, \; \mathrm{Card}\left\{ i \leq N; x_i \notin \{x_i^1, \ldots, x_i^q\} \right\} \leq k \right\}.$$

Then, if $\mathbb{P}\{\mathbf{X} \in A\} \geq 1/2$ and $k \geq q$,

$$(6.16) \qquad \mathbb{P}_*\left\{ \mathbf{X} \in H(A, q, k) \right\} \geq 1 - \left(\frac{K_0}{q} \right)^k.$$

Recall that, for convenience, the numerical constant K_0 is assumed to be an integer.

On $H(A, q, k)$, the sample $\mathbf{X}$ is controlled by a finite number q of points in A provided k values are neglected. The isoperimetric inequality (6.16) precisely estimates, with an exponential decay in k, the probability that this is satisfied. In applications to sums of independent random variables, the k values for which the isoperimetric inequality does not provide any control may be thought of as the largest elements (in norm) of the sample. This observation is actually the conducting rod to the subsequent developments. We will see how, up to the control of the large values, the isoperimetric inequality provides optimal estimates of the tail behavior of sums of independent random variables. As for vector valued Gaussian variables and Rademacher series, several parameters are used to measure the tail of sums of independent random variables. These involve some quantity in the L_0-topology (median or expectation), information on weak moments, and thus, as is clear from the isoperimetric approach, estimates on large values.

Let us now state and prove the tail estimate on sums of independent random variables which we draw from the isoperimetric inequality (6.16). It may be compared to the inequalities presented at the beginning of the section although we do not explicit this comparison since the vector valued case induces several complications. However, various arguments in both this chapter and the next one provide the necessary methodology and tools towards this goal. Our estimate deals with symmetric variables since Rademacher randomization and conditional use of tail estimates for Rademacher averages are an essential complement to the approach. If $(X_i)_{i \leq N}$ is a finite sequence of random variables, we denote by $(\|X_i\|^*)_{i \leq N}$ the non-increasing rearrangement of $(\|X_i\|)_{i \leq N}$.

Theorem 6.17. *Let* $(X_i)_{i \leq N}$ *be independent and symmetric random variables with values in* B. *Then, for any integers* $k \geq q$ *and real numbers* $s, t > 0$,

(6.17)

$$\mathbb{P}\left\{\left\|\sum_{i=1}^{N} X_i\right\| > 8qM + 2s + t\right\}$$

$$\leq \left(\frac{K_0}{q}\right)^k + \mathbb{P}\left\{\sum_{i=1}^{k}\|X_i\|^* > s\right\} + 2\exp\left(-\frac{t^2}{128qm^2}\right)$$

where

$$M = \mathbb{E}\left\|\sum_{i=1}^{N} u_i\right\|, \quad m = \mathbb{E}\left(\sup_{f \in D}\left(\sum_{i=1}^{N} f^2(u_i)\right)^{1/2}\right)$$

and $u_i = X_i I_{\{\|X_i\| \leq s/k\}}$, $i \leq N$.

Before we turn to the proof of Theorem 6.17, let us make some comments in order to clarify the statement. First note that M and m which are defined with truncated random variables u_i are easily majorized (using the contraction principle for M) by the same expressions with X_i instead of u_i (provided $X_i \in L_2(B)$). (6.17) is often enough for applications in this simpler form. Actually, note also that when we need not be concerned with truncations, for example if we deal with bounded variables, then in (6.17) the parameter $2s$ on the left hand side can be improved to s. This is completely clear from the proof below and is sometimes useful as we will see in Theorem 6.19. The reader recognizes in the first two terms on the right of (6.17) the isoperimetric bound (6.16) and the largest values of the sample. M corresponds to a control in probability of the sum, m to weak moment estimates. Actually m can be used in several ways. These are summarized in the following three estimates. First, by (6.5) and the contraction principle

(6.18)
$$m^2 \leq \sup_{f \in D} \sum_{i=1}^{N} \mathbb{E}f^2(u_i) + 8\frac{Ms}{k}.$$

Then, by (4.3) and symmetry,

(6.19)
$$m^2 \leq 2M^2,$$

while, trivially,

(6.20)
$$m^2 \leq \sum_{i=1}^{N} \mathbb{E}\|u_i\|^2.$$

(6.18) basically corresponds to the sharpest estimate and will prove efficient in limit theorems for example. The two others, especially (6.19), are convenient when no weak moment assumption needs to be taken into account; both include the real valued situation.

Let us now show how Theorem 6.17 is obtained from the isoperimetric inequality.

Proof of Theorem 6.17. We decompose the proof step by step.

First step. This is an elementary observation on truncation and large values. *Recall that if no truncation is needed, this step can be omitted .* If $(X_i)_{i \leq N}$ are (actually arbitrary) random variables and if $s \geq \sum_{i=1}^{k} \|X_i\|^*$, then

$$\left\| \sum_{i=1}^{N} X_i \right\| \leq s + \left\| \sum_{i=1}^{N} u_i \right\|$$

where we recall that $u_i = X_i I_{\{\|X_i\| \leq s/k\}}$, $i \leq N$. Indeed, if J denotes the set of integers $i \leq N$ such that $\|X_i\| > s/k$, then Card $J \leq k$ since, if not, this would contradict $\sum_{i=1}^{k} \|X_i\|^* \leq s$. Then

$$\left\| \sum_{i=1}^{N} X_i \right\| \leq \left\| \sum_{i \in J} X_i \right\| + \left\| \sum_{i \notin J} X_i \right\| \leq \sum_{i=1}^{k} \|X_i\|^* + \left\| \sum_{i=1}^{N} u_i \right\|$$

which gives the result.

Second step. Application of the isoperimetric inequality. By symmetry and independence, recall that the sample $\mathbf{X} = (X_i)_{i \leq N}$ has the same distribution as $(\varepsilon_i X_i)_{i \leq N}$ where (ε_i) is a Rademacher sequence which is independent of $\mathbf{X}$. Suppose that we are given A in B^N such that $\mathbb{P}\{\mathbf{X} \in A\} \geq 1/2$. If then $\mathbf{X} \in H = H(A, q, k)$, there exist, by definition, $j \leq k$ and $x^1, \ldots, x^q \in A$ such that

$$\{1, \ldots, N\} = \{i_1, \ldots, i_j\} \cup I$$

where $I = \bigcup_{\ell=1}^{q} \{i \leq N; X_i = x_i^\ell\}$. Together with the first step we can then write, if $s \geq \sum_{i=1}^{k} \|X_i\|^*$,

$$\left\| \sum_{i=1}^{N} \varepsilon_i X_i \right\| \leq s + \left\| \sum_{i=1}^{N} \varepsilon_i u_i \right\|$$

$$\leq s + \left\| \sum_{\ell=1}^{j} \varepsilon_{i_\ell} u_{i_\ell} \right\| + \left\| \sum_{i \in I} \varepsilon_i u_i \right\|$$

$$\leq 2s + \left\| \sum_{i \in I} \varepsilon_i u_i \right\|.$$

From the isoperimetric inequality (6.16) we then clearly get that, for $k \geq q$, $s, t > 0$,

$$\mathbb{P}\left\{\left\|\sum_{i=1}^{N} X_i\right\| > 2s + t\right\} \leq \left(\frac{K_0}{q}\right)^k + \mathbb{P}\left\{\sum_{i=1}^{k} \|X_i\|^* > s\right\}$$

(6.21)

$$+ \int_{\{\mathbf{X} \in H\}} \mathbb{P}_\varepsilon\left\{\left\|\sum_{i \in I} \varepsilon_i u_i\right\| > t\right\} d\mathbb{P}_X.$$

Here, there is a slight abuse in notation since $\{\mathbf{X} \in H\}$ need not be measurable and we should be somewhat more careful in this application of Fubini's theorem. However, this is irrelevant and for simplicity we skip these details.

Third step. Choice of A and conditional estimates on Rademacher averages. On the basis of (6.21), we are now interested in some conditional estimates of $\mathbb{P}_\varepsilon\{\|\sum_{i \in I} \varepsilon_i u_i\| > t\}$ with some appropriate choice for A. This is the place where randomization appears to be crucial. The tail inequality on Rademacher averages that we use is inequality (4.11) in the following form: if (x_i) is a finite sequence in B and $\sigma = \sup_{f \in D}(\sum_i f^2(x_i))^{1/2}$, for any $t > 0$,

(6.22) $$\mathbb{P}\left\{\left\|\sum_i \varepsilon_i x_i\right\| > 2\mathbb{E}\left\|\sum_i \varepsilon_i x_i\right\| + t\right\} \leq 2\exp(-t^2/8\sigma^2).$$

(With some worse constants, we could also use (4.15).) Of course, this kind of inequality is simpler on the line (cf. the subgaussian inequality (4.1)) and *the interested reader is perhaps invited to consider this simpler case to start with.* Provided with this inequality, let us consider $A = A_1 \cap A_2$ where

$$A_1 = \left\{x = (x_i)_{i \leq N}; \ \mathbb{E}\left\|\sum_{i=1}^{N} \varepsilon_i x_i I_{\{\|x_i\| \leq s/k\}}\right\| \leq 4M\right\},$$

$$A_2 = \left\{x = (x_i)_{i \leq N}; \ \sup_{f \in D}\left(\sum_{i=1}^{N} f^2(x_i)I_{\{\|x_i\| \leq s/k\}}\right)^{1/2} \leq 4m\right\}.$$

The very definitions of M and m clearly show that $\mathbb{P}\{\mathbf{X} \in A\} \geq 1/2$ so that we are in a position to apply (6.21). Now, observe that Rademacher averages are *monotone* in the sense that $\mathbb{E}\|\sum_{i \in J} \varepsilon_i x_i\|$ is an increasing function of $J \subset \mathbb{N}$. We use this property in the following way. By definition of I, for each $i \in I$, we can fix $1 \leq \ell(i) \leq q$ with $X_i = x_i^{\ell(i)}$. Let $I_\ell = \{i; \ell(i) = \ell\}, 1 \leq \ell \leq q$. We have

$$\mathbb{E}_\varepsilon\left\|\sum_{i \in I} \varepsilon_i u_i\right\| = \mathbb{E}_\varepsilon\left\|\sum_{\ell=1}^{q} \sum_{i \in I_\ell} \varepsilon_i u_i\right\| \leq \sum_{\ell=1}^{q} \mathbb{E}_\varepsilon\left\|\sum_{i \in I_\ell} \varepsilon_i x_i^\ell I_{\{\|x_i^\ell\| \leq s/k\}}\right\|.$$

Then, by monotonicity of Rademacher averages, and the definition of A (recall that $x^1, \ldots, x^q$ belong to A), it follows that

$$\mathbb{E}_\varepsilon\left\|\sum_{i \in I} \varepsilon_i u_i\right\| \leq \sum_{\ell=1}^{q} \mathbb{E}_\varepsilon\left\|\sum_{i=1}^{N} \varepsilon_i x_i^\ell I_{\{\|x_i^\ell\| \leq s/k\}}\right\| \leq 4qM.$$

Similarly, but with sums of squares (which are also monotone),

$$\sup_{f \in D} \sum_{i \in I} f^2(u_i) \leq 16qm^2 .$$

Theorem 6.17 now clearly follows from these observations combined with the estimate on Rademacher averages (6.22) and with (6.21).

Remark 6.18. One of the key observations in the third step of the preceding proof is the monotonicity of Rademacher averages. It might be interesting to describe what the isoperimetric inequality directly produces when applied to conditional averages $\mathbb{E}_\varepsilon \| \sum_{i=1}^N \varepsilon_i X_i \|$ which thus satisfy this basic uncondi-tionality property. Let therefore $(X_i)_{i \leq N}$ be independent symmetric variables in $L_1(B)$ and set $M = \mathbb{E} \| \sum_{i=1}^N X_i \|$. Then, for every $k \geq q$ and $s > 0$,

$$(6.23) \quad \mathbb{P}_X \left\{ \mathbb{E}_\varepsilon \left\| \sum_{i=1}^N \varepsilon_i X_i \right\| > 2qM + s \right\} \leq \left(\frac{K_0}{q} \right)^k + \mathbb{P} \left\{ \sum_{i=1}^k \|X_i\|^* > s \right\}.$$

For the proof we let

$$A = \left\{ x = (x_i)_{i \leq N}; \; \mathbb{E} \left\| \sum_{i=1}^N \varepsilon_i x_i \right\| \leq 2M \right\}$$

so that, by symmetry, $\mathbb{P}\{\mathbf{X} \in A\} \geq 1/2$. If $\mathbf{X} \in H(A, q, k)$, there exist $j \leq k$ and $x^1, \ldots, x^q$ in A such that $\{1, \ldots, N\} = \{i_1, \ldots, i_j\} \cup I$ where $I = \bigcup_{\ell=1}^q \{i \leq N; X_i = x_i^\ell\}$. Then, as in the third step,

$$\mathbb{E}_\varepsilon \left\| \sum_{i=1}^N \varepsilon_i X_i \right\| \leq \sum_{i=1}^k \|X_i\|^* + \mathbb{E}_\varepsilon \left\| \sum_{i \in I} \varepsilon_i X_i \right\|$$

$$\leq \sum_{i=1}^k \|X_i\|^* + \sum_{\ell=1}^q \mathbb{E}_\varepsilon \left\| \sum_{i=1}^N \varepsilon_i x_i^\ell \right\|$$

$$\leq \sum_{i=1}^k \|X_i\|^* + 2qM$$

by monotonicity of Rademacher averages and definition of A. (6.23) then simply follows from (6.16). Note that the same applies to sums of independent real *positive* random variables since these share this monotonicity property, and this case is actually one first instructive example for the understanding of the technique.

Next, we turn to several applications of Theorem 6.17. We first solve the integrability question of almost surely convergent series of independent cen-tered bounded random variables. Proposition 6.14 provided an exponential in-tegrability property. It is however known, from (6.10) or Prokhorov's arcsinh inequality e.g. (cf. [Sto]), that in the real case these series are integrable with respect to the function $\exp(x \log^+ x)$. Furthermore, this order of integrability is best possible. Take indeed $(X_i)_{i\geq 1}$ to be a sequence of independent random variables such that, for each $i \geq 1$, $X_i = \pm 1$ with equal probability $(2i^2)^{-1}$ and $X_i = 0$ with probability $1 - i^{-2}$. Then $\sum_i \mathbb{E}X_i^2 < \infty$. However, if $S_N = \sum_{i=1}^{N} X_i$ and $\alpha, \varepsilon > 0$, for every N,

$$\mathbb{E}\exp\big(\alpha|S_N|(\log^+ |S_N|)^{1+\varepsilon}\big) = \sum_{k=0}^{N} \exp\big(\alpha k(\log^+ k)^{1+\varepsilon}\big)\mathbb{P}\{|S_N| = k\}$$

$$\geq \exp\big(\alpha N(\log N)^{1+\varepsilon}\big)\prod_{i=1}^{N} i^{-2}$$

which goes to infinity with N.

The following theorem extends this strong integrability property to the vector valued case.

Theorem 6.19. *Let (X_i) be independent and symmetric random variables with values in B such that $S = \sum_i X_i$ converges almost surely and $\|X_i\|_\infty \leq a$ for every i. Then, for every $\lambda < 1/a$,*

$$\mathbb{E}\exp\big(\lambda\|S\|\log^+ \|S\|\big) < \infty.$$

If the X_i's are merely centered, this holds for all $\lambda < 1/2a$.

Proof. For every N, set $S_N = \sum_{i=1}^{N} X_i$. We show

$$\sup_N \mathbb{E}\exp\big(\lambda\|S_N\|\log^+ \|S_N\|\big) < \infty$$

which is enough by Fatou's lemma. We already know from Theorem 6.11 that $\sup_N \mathbb{E}\|S_N\| = M < \infty$ (this can also be deduced directly from the isoperimetric inequality so that this proof is actually self-contained). We use (6.17) together with (6.19) and the comment after Theorem 6.17 concerning truncation to see that, for all integers $k \geq q$ and all real numbers $s, t > 0$,

$$\mathbb{P}\{\|S_N\| > 8qM + s + t\}$$
$$\leq \left(\frac{K_0}{q}\right)^k + \mathbb{P}\Big\{\sum_{i=1}^{k} \|X_i\|^* > s\Big\} + 2\exp\left(-\frac{t^2}{256qM^2}\right).$$

Since, almost surely, $\sum_{i=1}^{k} \|X_i\|^* \leq ka$,

$$\mathbb{P}\{\|S_N\| > 8qM + ka + t\} \leq \left(\frac{K_0}{q}\right)^k + 2\exp\left(-\frac{t^2}{256qM^2}\right).$$

Let $\varepsilon > 0$. For $u > 0$ large enough, set $t = \varepsilon u$, $k = [(1 - 2\varepsilon)a^{-1}u]$ (integer part) and

$$q = \left[\frac{\varepsilon^2 a}{256 M^2} \cdot \frac{u}{\log u}\right].$$

Then, for $u \geq u_0(M, a, \varepsilon)$ large enough, $k \geq q$ and

$$\mathbb{P}\{\|S_N\| > u\} \leq \mathbb{P}\{\|S_N\| > 8qM + ka + t\}$$
$$\leq \exp\big(-(1 - 3\varepsilon)a^{-1}u \log u\big) + 2\exp(-a^{-1}u \log u).$$

Since this is uniform in N, the conclusion follows. If the random variables are only centered, use for example the symmetrization Lemma 6.3. Note again that $\sup_N \mathbb{E} \exp(\lambda \|S_N\| \log^+ \|S_N\|) < \infty$ as soon as the sequence (S_N) is bounded in probability.

Theorem 6.19 is closely related to the best order of growth as functions of p of the constants in the L_p-inequalities of J. Hoffmann-Jørgensen (Proposition 6.8). This is the content of the next statement.

Theorem 6.20. *There is a universal constant K such that for all $p > 1$ and all finite sequences (X_i) of independent mean zero random variables in $L_p(B)$,*

$$\left\|\sum_i X_i\right\|_p \leq K \frac{p}{\log p}\left(\left\|\sum_i X_i\right\|_1 + \left\|\max_i \|X_i\|\right\|_p\right).$$

Proof. We may and do assume the X_i's, $i \leq N$, to be symmetric. If r is an integer, we set $X_N^{(r)} = \|X_j\|$ whenever $\|X_j\|$ is the r-th maximum of the sample $(\|X_i\|)_{i \leq N}$ (ties being broken by the index and $X_N^{(r)} = 0$ if $r > N$). Then, if $M = \|\sum_{i=1}^N X_i\|_1$, Theorem 6.17 and (6.19) indicate that, for $k \geq q$ and $s, t > 0$,

(6.24)
$$\mathbb{P}\left\{\left\|\sum_{i=1}^N X_i\right\| > 8qM + 2s + t\right\}$$
$$\leq \left(\frac{K_0}{q}\right)^k + \mathbb{P}\left\{\sum_{r=1}^k X_N^{(r)} > s\right\} + 2\exp\left(-\frac{t^2}{256qM^2}\right).$$

To establish Theorem 6.20, we may assume by homogeneity that $M \leq 1$ and $\|X_N^{(1)}\|_p \leq 1$ ($X_N^{(1)} = \max_{i \leq N} \|X_i\|$). Therefore, in particular, $\mathbb{P}\{X_N^{(1)} > u\} \leq u^{-p}$ for all $u > 0$. By induction over r, for every $u > 0$, one easily sees that

$$\mathbb{P}\{X_N^{(r)} > u\} \leq \mathbb{P}\{\max_{i \leq N} \|X_i\| > u\}\mathbb{P}\{X_N^{(r-1)} > u\},$$

from which we deduce, by iteration, that

$$\mathbb{P}\{X_N^{(r)} > u\} \leq \big(\mathbb{P}\{X_N^{(1)} > u\}\big)^r \leq u^{-rp}.$$

(With some worse irrelevant numerical constants, the same result may be obtained, perhaps more simply, from the successive application of Lemmas 2.5 and 2.6.) Let $u \geq 1$ be fixed. We have $\mathbb{P}\{X_N^{(2)} > u^{2/3}\} \leq u^{-4p/3}$. Further, if ℓ is the smallest integer such that $2^\ell \geq u^2$,

$$\mathbb{P}\{X_N^{(\ell)} > 2\} \leq 2^{-\ell p} \leq u^{-2p} \leq u^{-4p/3}.$$

Hence, the probability of the complement of the set $\{X_N^{(1)} \leq u, \ X_N^{(2)} \leq u^{2/3},$ $X_N^{(\ell)} \leq 2\}$ is smaller than $\mathbb{P}\{X_N^{(1)} > u\} + 2u^{-4p/3}$. We now apply (6.24). Let k be the smallest integer $\geq u$. On the set $\{X_N^{(1)} \leq u, X_N^{(2)} \leq u^{2/3}, X_N^{(\ell)} \leq 2\}$,

$$\sum_{r=1}^{k} X_N^{(r)} \leq u + \ell u^{2/3} + 2(k-1) \leq Cu$$

for some constant C. If we now take q in (6.24) to be the smallest integer $\geq \sqrt{u}$, $s = Cu$, $t = u$, it follows from the preceding that

$$\mathbb{P}\left\{\left\|\sum_{i=1}^{N} X_i\right\| > 2(C+10)u\right\}$$

$$\leq \exp\left(-u \log\left(\frac{\sqrt{u}}{K_0}\right)\right) + \mathbb{P}\{X_N^{(1)} > u\} + 5u^{-4p/3} + 2\exp\left(-\frac{u^{3/2}}{256}\right).$$

Standard computations using the integration by parts formula then give the constant $p/\log p$ in the inequality of the theorem. The proof is complete. $\blacksquare$

The preceding inequalities can also be investigated for exponential functions. Recall that for $0 < \alpha < \infty$ we let $\psi_\alpha = \exp(x^\alpha) - 1$ (linear near the origin when $0 < \alpha < 1$ in order for ψ_α to be convex) and denote by $\|\cdot\|_{\psi_\alpha}$ the norm of the Orlicz space L_{ψ_α}.

Theorem 6.21. *There is a constant K_α, depending on α only, such that for all finite sequences (X_i) of independent mean zero random variables in $L_{\psi_\alpha}(B)$, if $0 < \alpha \leq 1$,*

$$(6.25) \qquad \left\|\sum_i X_i\right\|_{\psi_\alpha} \leq K_\alpha\left(\left\|\sum_i X_i\right\|_1 + \left\|\max_i \|X_i\|\right\|_{\psi_\alpha}\right),$$

and, if $1 < \alpha \leq 2$,

$$(6.26) \qquad \left\|\sum_i X_i\right\|_{\psi_\alpha} \leq K_\alpha\left(\left\|\sum_i X_i\right\|_1 + \left(\sum_i \|X_i\|_{\psi_\alpha}^\beta\right)^{1/\beta}\right)$$

where $1/\alpha + 1/\beta = 1$.

Proof. We only give the proof of (6.26). Similar (and even simpler when $0 < \alpha < 1$) arguments are used for (6.25) (cf. [Tal1]). By Lemma 6.3, we reduce ourselves to symmetric random variables $(X_i)_{i \leq N}$. We set $d_i = \|X_i\|_{\psi_\alpha}$ so that $\mathbb{P}\{\|X_i\| > u\} \leq 2\exp(-(u/d_i)^\alpha)$. By homogeneity we can assume that $\|\sum_{i=1}^N X_i\|_1 \leq 1$, $\sum_{i=1}^N d_i^\beta \leq 1$, and there is no loss in generality to assume the sequence $(d_i)_{i \leq N}$ to be decreasing. Hence, $\sum_{2^i \leq N} 2^i d_{2^i}^\beta \leq 2$. We can find a sequence $\gamma_i \geq 2^i d_{2^i}^\beta$ such that $\gamma_i \geq \gamma_{i+1}/\sqrt{2} \geq \gamma_i/2$ for $i \geq 1$ and $\sum_i \gamma_i \leq 10$ (e. g. $\gamma_i = \sum_{j \geq 1} 2^{-|j-i|/2} 2^j d_{2^j}^\beta$). Let $c_i = (2^{-i}\gamma_i)^{1/\beta}$ so that $c_{i+1}^\alpha \leq c_i^\alpha 2^{-\alpha/2\beta}$, $c_i^\alpha \leq c_{i+1}^\alpha 2^{3\alpha/2\beta}$. Then $\sum_i 2^i c_i^\beta \leq 10$ and $d_j \leq c_i$ for $j \geq 2^i$. We observe that $\sum_{\ell \geq 2} 4 c_\ell (4 \log 4^{\ell+1})^{1/\alpha} \leq C_\alpha < \infty$.

It will be enough to show that for some constants a, c depending on α only, if $u \geq c$ then

$$(6.27) \qquad \mathbb{P}\left\{ \sum_{r \leq u^\alpha} X_N^{(r)} > au \right\} \leq \exp(-u^\alpha).$$

Indeed, if we then take, in (6.24), q to be $2K_0$, k to be of the order of u^α, s and t to be of the order of u, we obtain the tail estimate corresponding to (6.26). In order to establish (6.27), since $\mathbb{P}\{X_N^{(1)} > u\} \leq 2\exp(-u^\alpha)$, it is actually enough to find s, a, c such that when $u \geq c$

$$\mathbb{P}\left\{ \sum_{2^s \leq r \leq u^\alpha} X_N^{(r)} > au \right\} \leq \exp(-u^\alpha).$$

Fix u (large enough) and denote by n the largest integer such that $2^n \leq u^\alpha$. Suppose that $\sum_{r \leq u^\alpha} X_N^{(r)} > au$ so that $\sum_{\ell=0}^n 2^\ell X_N^{(2^\ell)} > a$. Let

$$L = \left\{ s \leq \ell \leq n;\, X_N^{(2^\ell)} > 2 c_\ell (4 \log 4^{\ell+1})^{1/\alpha} \right\}.$$

Then $\sum_{\ell \in L} 2^\ell X_N^{(2^\ell)} > au - C_\alpha > au/2$ for u large enough. For $\ell \in L$, we can find a number a_ℓ of the form $2^m (m \in \mathbb{Z})$ such that $X_N^{(2^\ell)} > a_\ell$ and $a_\ell \geq c_\ell (4 \log 4^{\ell+1})^{1/\alpha}$ and $\sum_{\ell \in L} 2^\ell a_\ell > au/4$. There exist disjoint subsets J_ℓ, $\ell \in L$, of $\{1, \ldots, N\}$ such that $\operatorname{Card} J_\ell \geq 2^{\ell-1}$ and $\|X_i\| > a_\ell$ for $i \in J_\ell$. Set $I_\ell = J_\ell \backslash \{1, \ldots, 2^{\ell-2}\}$. We have

$$\mathbb{P}\{\forall \ell \in L, X_N^{(2^\ell)} > a_\ell\} \leq \sum \prod_{\ell \in L} \prod_{i \in I_\ell} \mathbb{P}\{\|X_i\| > a_\ell\}$$

where the summation is taken over all possible choices of $(I_\ell)_{\ell \in L}$. Hence

$$\mathbb{P}\{\forall \ell \in L, X_N^{(2^\ell)} > a_\ell\} \leq \prod_{\ell \in L} \left(\sum_{2^{\ell-2} < i \leq N} \mathbb{P}\{\|X_i\| > a_\ell\} \right)^{2^{\ell-2}}.$$

We know that $\mathbb{P}\{\|X_i\| > a_\ell\} \leq 2 \exp(-(a_\ell/d_i)^\alpha)$, so

$$\sum_{2^{\ell-2} < i \leq N} \mathbb{P}\{\|X_i\| > a_\ell\} \leq \sum_{j \geq \ell-2} 2^{j+1} \exp\left(-(a_\ell/c_j)^\alpha\right).$$

Since $\ell \in L$, we have $(a_\ell/c_\ell)^\alpha \geq (a_\ell/c_\ell)^\alpha/2 + \log 4^{j+2}$ and since $c_{j+1}^\alpha \leq c_j^\alpha 2^{-\alpha/2\beta}$, $c_j^\alpha \leq c_{j+1}^\alpha 2^{3\alpha/2\beta}$ we clearly have

$$(a_\ell/c_j)^\alpha \geq (a_\ell/c_\ell)^\alpha/4 + \log 4^{j+2}$$

for $j \geq \ell - 2$ provided s has been taken large enough. It follows that

$$\sum_{j \geq \ell - 2} 2^{j+1} \exp\left(-(a_\ell/c_j)^\alpha\right) \leq \sum_{j \geq \ell - 2} 2^{-j-1} \exp\left(-\frac{1}{4}\left(\frac{a_\ell}{c_\ell}\right)^\alpha\right)$$

$$\leq \exp\left(-\frac{1}{4}\left(\frac{a_\ell}{c_\ell}\right)^\alpha\right).$$

Hence

$$\mathbb{P}\{\forall \ell \leq n, X_N^{(2^\ell)} > a_\ell\} \leq \exp\left(-\frac{1}{8}\sum_{\ell \in L} 2^\ell\left(\frac{a_\ell}{d_\ell}\right)^\alpha\right).$$

By Hölder's inequality,

$$\frac{au}{4} \leq \sum_{\ell \in L} 2^\ell a_\ell \leq \left(\sum_{\ell \in L} 2^\ell\left(\frac{a_\ell}{d_\ell}\right)^\alpha\right)^{1/\alpha} \left(\sum_{\ell \in L} 2^\ell c_\ell^\beta\right)^{1/\beta}$$

so that $\sum_{\ell \in L} 2^\ell (a_\ell/d_\ell)^\alpha \geq (au/40)^\alpha$, and, if we take for example $a = 640$, we get

$$\mathbb{P}\{\forall \ell \leq n, X_N^{(2^\ell)} > a_\ell\} \leq \exp(-2t^\alpha).$$

The number of possible choices for the sequence $(a_\ell)_{\ell \leq n}$ is less than $\exp(u^\alpha)$ for c large enough. From this, (6.27) follows and the proof of (6.26) is complete.

Remark 6.22. The preceding proof shows that the constant K_α in (6.26) is bounded in any interval $[1 + \varepsilon, 2]$. Applying (6.26) to independent copies of a Gaussian random variable, one can then deduce from this observation (and a finite dimensional approximation) that vector valued Gaussian variables are in $L_{\psi_2}(B)$.

To conclude this chapter, we would like to mention that the previous statements are only a few examples of a seemingly large number of variations that one can try with the isoperimetric approach. As an illustration, let us mention a few more ideas.

The first idea is that in the third step of the proof of Theorem 6.17, possibly other inequalities on Rademacher averages can be used. One may think for example of (6.15). Actually, from the more general inequality (6.14) directly (which was obtained by martingale methods), we already get an interesting inequality in the spirit of Theorem 6.21 which deals with the case $\alpha > 2$. Namely, in the hypotheses of Theorem 6.21, for $2 < \alpha < \infty$ and $1/\alpha + 1/\beta = 1$,

$$(6.28) \qquad \left\|\sum_i X_i\right\|_{\psi_\alpha} \leq K_\alpha\left(\left\|\sum_i X_i\right\|_1 + \left\|(\|X_i\|_\infty)\right\|_{\beta,\infty}\right).$$

Using estimates on large values similar to those used in the proof of Theorem 6.21, it is possible to improve on (6.28) by the isoperimetric method to get

$$(6.29) \qquad \left\|\sum_i X_i\right\|_{\psi_\alpha} \leq K_\alpha\left(\left\|\sum_i X_i\right\|_1 + \left\|(\|X_i\|_{\psi_s})\right\|_{\beta,\infty}\right)$$

for $2 < \alpha < s \leq \infty$ (See [Tal1]).

Another idea concerns the possibility in the three step procedure of the proof of Theorem 6.17 to let s, and at some point t too, be *random*. One random bound for the largest values is given by the inequality

$$\sum_{i=1}^{k} \|X_i\|^* \leq \beta k^{1/\beta} \|(X_i)\|_{\alpha,\infty},$$

$1 < \alpha < \infty$, $\beta = \alpha/\alpha - 1$. With this choice of s, one can show the following inequalities; as usual, (X_i) is a finite sequence of independent symmetric vector valued random variables. If $2 \leq \alpha < \infty$, for some constant $K_\alpha > 0$ depending on α only and every $t > 0$,

$$(6.30) \quad \mathbb{P}\left\{\left\|\sum_i X_i\right\| > t\left(\left\|\sum_i X_i\right\|_1 + \|(X_i)\|_{\alpha,\infty}\right)\right\} \leq K_\alpha \exp(-t^\beta/K_\alpha);$$

if $1 < \alpha < 2$,

$$(6.31) \quad \mathbb{P}\left\{\left\|\sum_i X_i\right\| > K_\alpha\left\|\sum_i X_i\right\|_1 + t\|(X_i)\|_{\alpha,\infty}\right\} \leq K_\alpha \exp(-t^\beta/K_\alpha).$$

To illustrate these inequalities, let us prove the second; it uses (6.15) as opposed to the first inequality which uses the usual quadratic inequality (Theorem 4.7). We start with (6.21), actually without truncated variables since here we need not be concerned with truncation. There, we set s to be random and equal to $\beta k^{1/\beta}\|(X_i)\|_{\alpha,\infty}$. Take further $t = 2qM + k^{1/\beta}\|(X_i)\|_{\alpha,\infty}$, therefore also random, where $M = \|\sum_i X_i\|_1$. It follows that, for $k \geq q$,

$$\mathbb{P}\left\{\left\|\sum_i X_i\right\| > 2qM + (2\beta + 1)k^{1/\beta}\|(X_i)\|_{\alpha,\infty}\right\}$$

$$\leq \left(\frac{K_0}{q}\right)^k + \int_{\{\mathbf{X} \in H\}} \mathbb{P}_\varepsilon\left\{\left\|\sum_{i\in I} \varepsilon_i X_i\right\| > 2qM + k^{1/\beta}\|(X_i)\|_{\alpha,\infty}\right\} d\mathbb{P}_X.$$

Let $A = \{(x_i); \mathbb{E}\|\sum_i \varepsilon_i x_i\| \leq 2M\}$ so that $\mathbb{P}\{\mathbf{X} \in A\} \geq 1/2$. By the definition of I and monotonicity of Rademacher averages,

$$\mathbb{E}_\varepsilon\left\|\sum_{i\in I} \varepsilon_i X_i\right\| \leq 2qM.$$

Now, by the inequality (6.15) conditionally on the X_i's,

$$\mathbb{P}_\varepsilon\left\{\left\|\sum_{i\in I} \varepsilon_i X_i\right\| > 2qM + k^{1/\beta}\|(X_i)\|_{\alpha,\infty}\right\} \leq 2\exp(-k/C_\alpha).$$

Letting for example $q = 2K_0$, we then clearly deduce (6.31). As already mentioned, the proof of (6.30) is similar.

We conclude these applications although many different techniques might now be combined from what we learned to yield some new useful inequalities. We hope that the interested reader has now enough information to establish from the preceding ideas the tools he will need in his own study.

Notes and References

The study of sums of independent Banach space valued random variables and the introduction of symmetrization ideas were initiated by the works of J.-P. Kahane [Ka1] and of J. Hoffmann-Jørgensen [HJ1], [HJ2]. Sections 6.1 and 6.2 basically follow their work (cf. [HJ3]).

Inequality (6.2) has been noticed in [V-C1] and [Ale1]. Theorem 6.1 is due to K. Itô and M. Nisio [I-N] (cf. Chapter 2), extending the classical result of P. Lévy on the line. Lemma 6.2 is usually referred to as Ottaviani's inequality and is part of the folklore. Proposition 6.4 was established by M. Kanter [Kan] as an improvement of various previous concentration inequalities of the same type. It turns out to be rather useful in some cases (cf. Section 10.3). Lemmas 6.5 and 6.6 are easy variations and extensions of the contraction principle and comparison properties (see further [J-M2], [HJ3]).

Proposition 6.7 is due to J. Hoffmann-Jørgensen [HJ2] who used it to establish Theorem 6.11 and Corollary 6.12. Proposition 6.8 is implicit in his proof. Applied to sums $\sum_i \theta_i x_i$ where (θ_i) is a standard p-stable sequence, it yielded historically the first integrability and moment equivalence results for vector valued stable random variables. Lemma 6.9 is classical and is taken in this form from [G-Z1]. In this paper, E. Giné and J. Zinn establish the moment equivalences of sums of independent (symmetric) random variables of Proposition 6.10. Remark 6.13 has been mentioned to us by J. Zinn. The integrability of sums of independent bounded vector valued random variables has been studied by many authors. Iteration of Hoffmann-Jørgensen's inequality has been used in [J-M2], [Pi3] and [Kue5] for example. Proposition 6.14 is due to A. de Acosta [Ac4] while the content of Remark 6.15 is taken from the paper [A-S].

Among the numerous real (quadratic) exponential inequalities, let us mention those by S. Bernstein (cf. [Ho]), A. N. Kolmogorov [Ko1] (cf. [Sto]), Y. Prokhorov [Pro2] (cf. [Sto]), G. Bennett [Ben], etc. We refer to the interesting paper by W. Hoeffding [Ho] for a comparison between these inequalities and for further developments. See also an inequality by D. K. Fuk and S. V. Nagaev [F-N] (cf. [Na2], [Yu2]). The key Lemma 6.16 is due to V. Yurinskii [Yu1] (see also [Yu2] and [Ber]). Inequality (6.11) (and some others) has been put forward in [Ac6]. Inequalities (6.12), (6.13) and (6.14) may be found respectively in [K-Z], [Ac5] and [Pi16] (see also [Pi12]). A. de Acosta [Ac5] used (6.13) to establish Theorem 6.19 in cotype 2 spaces. There is also a version of Prokhorov's arcsinh inequality noticed in [J-S-Z] which may be applied similarly to $\|S_N\| - \mathbb{E}\|S_N\|$.

In the contribution [Led6] (see [L-T2]), on the law of the iterated logarithm, a Gaussian randomization argument is used to decompose the study of the sums of independent random variables into two parts: one for which the Gaussian (or Rademacher) concentration properties can be applied conditionally, and a second part which is enriched by an unconditionality property (monotonicity of Rademacher averages). This kind of decomposition is one important feature of the isoperimetric approach (cf. Remark 6.18) and the discovery of Theorem 1.4 was motivated by this unconditionality property. The isoperimetric approach (Theorem 6.17) and its applications to the integrability of sums of independent vector valued random variables were developed by the second author in [Ta11] and we follow here this work (and [L-T5]). Very recently, the same author [Ta22] established an important extension of Theorem 1.3 to general product measures. This result is much simpler than Theorem 1.4, and, although not directly comparable to it, allows to recover similarly Theorem 6.17 and all the subsequent bounds and integrability theorems. It provides thus, avoiding the difficult Theorem 1.4, a significant simplification in the isoperimetric background of our study of the tails of sums of independent random variables as well as of their applications to strong limit theorems in the next chapters. Theorem 6.19 is in [Ta11] and extends the prior result of [Ac5]. Recently, an alternate proof of Theorem 6.20 has been given by S. Kwapień and J. Szulga [K-S] using very interesting hypercontractive estimates in the spirit of what we discussed in the Gaussian and Rademacher cases (Sections 3.2 and 4.4). On the line, Theorem 6.20 was obtained in [J-S-Z] where a thorough relation to exponential integrability is given. Theorem 6.21 is taken from [Ta11] and improves upon [Kue5], [Ac5]. Inequality (6.29) is also in [Ta11]. Inequalities (6.30) and (6.31) extend to the vector valued case various ideas of [He7], [He8] (see further Lemma 14.4).

7. The Strong Law of Large Numbers

In this chapter and in the next one, we present respectively the strong law of large numbers and the law of the iterated logarithm for sums of independent Banach space valued random variables. In this study, the isoperimetric approach of Section 6.3 demonstrates its efficiency. We only investigate extensions to vector valued random variables of some of the classical limit theorems such as the laws of large numbers of Kolmogorov and Prokhorov.

One main feature of the results we present is the equivalence, under *classical* moment conditions, of the *almost sure* limit theorem and the corresponding property in *probability*. In a sense, this can be seen as yet another instance in which the theory is broken into two parts: starting from a statement in probability (weak statement), prove almost sure properties; then try to understand the weak statement. It is one of the main difficulties of the vector valued setting to control boundedness in probability or tightness of a sum of independent random variables. On the line, this is usually done with orthogonality and moment conditions. In general spaces, one has either to put conditions on the Banach space or to use empirical process methods. Some of these questions will be discussed in the sequel of the work, starting with Chapter 9, especially in the context of the central limit theorem which forms the typical example of a weak statement. As announced, in this chapter and in the next one, almost sure limit properties are investigated under assumptions in probability.

In the first part of this chapter, we study a general statement for almost sure limit theorems for sums of independent random variables. It is directly drawn from the isoperimetric approach of Section 6.3 and already presents some interest for real valued random variables. We introduce it together with some generalities on strong limit theorems such as symmetrization (randomization) and blocking arguments. The second paragraph is devoted to applications to concrete examples such as the independent and identically distributed (iid) Kolmogorov-Marcinkiewicz-Zygmund strong laws of large numbers and the laws of Kolmogorov, Brunk, Prokhorov, etc. for independent random variables.

Apart from one example where Radon random variables will be useful, we can adopt the setting of the last chapter and deal with a Banach space B for which there is a countable subset D of the unit ball of the dual space B' such that $\|x\| = \sup_{f \in D} |f(x)|$ for all $x \in B$. X is a random variable with values in B if $f(X)$ is measurable for every f in D. When $(X_i)_{i \in \mathbb{N}}$ is a sequence of (independent) random variables in B we set, as usual, $S_n = X_1 + \cdots + X_n$, $n \geq 1$.

7.1 A General Statement for Strong Limit Theorems

Let $(X_i)_{i \in \mathbb{N}}$ be a sequence of independent random variables with values in B. Let also (a_n) be a sequence of positive numbers increasing to infinity. We study the *almost sure* behavior of the sequence (S_n/a_n). As described above, such a study in the infinite dimensional setting can be developed reasonably only if one assumes some (*necessary*) boundedness or convergence condition in *probability*. Recall that a sequence (Y_n) is bounded in probability if for every $\varepsilon > 0$ there exists $A > 0$ such that, for every n,

$$\mathbb{P}\{\|Y_n\| > A\} < \varepsilon.$$

This kind of hypothesis will be common to all the limit theorems discussed here. In particular, this condition on the behavior in probability allows to use a simple symmetrization procedure summarized in the next trivial lemma.

Lemma 7.1. *Let $(Y_n), (Y_n')$ be independent sequences of random variables such that the sequence $(Y_n - Y_n')$ is almost surely bounded (resp. convergent to 0) and such that (Y_n) is bounded (resp. convergent to 0) in probability. Then, (Y_n) is almost surely bounded (resp. convergent to 0). More quantitatively, if for some numbers M and A,*

$$\limsup_{n \to \infty} \|Y_n - Y_n'\| \leq M \quad almost\ surely$$

and

$$\limsup_{n \to \infty} \mathbb{P}\{\|Y_n\| > A\} < 1,$$

then

$$\limsup_{n \to \infty} \|Y_n\| \leq 2M + A \quad almost\ surely.$$

Given (X_i), let then (X_i') denote an independent copy of the sequence (X_i) and set, for each i, $\tilde{X}_i = X_i - X_i'$ defining thus independent and symmetric random variables. Lemma 7.1 tells us that under some appropriate assumptions in probability on (S_n/a_n), it is enough to study $(\sum_{i=1}^{n} \tilde{X}_i/a_n)$, reducing ourselves to symmetric random variables. From now on, we therefore only describe the various results and the general theorem we have in mind in the *symmetrical* case. This avoids several unnecessary complications about centerings, but, with some care, it would however be possible to study the general case.

As we learned, properties in probability are often equivalent to properties in L_p which are more convenient. For sums of independent random variables, this is shown by Hoffmann-Jørgensen's inequalities (Proposition 6.8) on which the following useful lemma relies.

Lemma 7.2. *Let (X_i) be independent and symmetric random variables with values in B. If the sequence (S_n/a_n) is bounded (resp. convergent to 0) in* *probability, for any $p > 0$ and any bounded sequence (c_n) of positive numbers, the sequence*

$$\left(\frac{1}{a_n^p} \mathbb{E}\left\|\sum_{i=1}^{n} X_i I_{\{\|X_i\| \leq c_n a_n\}}\right\|^p\right)$$

is bounded (resp. convergent to 0).

Proof. We only show the part of the statement concerning the convergence of the sequence (S_n/a_n). Let (c_n) be bounded by c. By inequality (6.9), for each n,

$$\mathbb{E}\left\|\sum_{i=1}^{n} X_i I_{\{\|X_i\| \leq c_n a_n\}}\right\|^p \leq 2 \cdot 3^p \mathbb{E}\left(\max_{i \leq n} \|X_i\|^p I_{\{\|X_i\| \leq c_n a_n\}}\right) + 2\left(3t_0(n)\right)^p$$

where

$$t_0(n) = \inf\left\{t > 0 : \mathbb{P}\left\{\left\|\sum_{i=1}^{n} X_i I_{\{\|X_i\| \leq c_n a_n\}}\right\| > t\right\} \leq (8 \cdot 3^p)^{-1}\right\}.$$

When (S_n/a_n) converges to 0 in probability, using the contraction principle in the form of Lemma 6.5, for each $\varepsilon > 0$, $t_0(n)$ is seen to be smaller than εa_n for n large enough. Concerning the maximum term, by integration by parts and Lévy's inequality (2.7),

$$\mathbb{E}\left(\max_{i \leq n} \|X_i\|^p I_{\{\|X_i\| \leq c_n a_n\}}\right) \leq \int_0^{c a_n} \mathbb{P}\left\{\max_{i \leq n} \|X_i\| > t\right\} dt^p$$

$$\leq 2a_n^p \int_0^c \mathbb{P}\left\{\|S_n\| > t a_n\right\} dt^p.$$

The conclusion follows by dominated convergence.

A classical and important observation in the study of strong limit theorems for sums S_n of independent random variables is that it can be developed in quite general situations through *blocks of exponential sizes*. More precisely, assume that there exists a subsequence (a_{m_n}) of (a_n) such that for each n

(7.1) $$c a_{m_n} \leq a_{m_{n+1}} \leq C a_{m_n + 1}$$

where $1 < c \leq C < \infty$. This hypothesis is by no means restrictive since it can be shown (cf. [Wi]) that, for any fixed $M > 1$, one can find a strictly increasing sequence (m_n) of integers such that the preceding holds with $c = M, C = M^3$. We thus assume *throughout this section* that (7.1) holds for some subsequence (m_n) and define, for each n, $I(n)$ as the set of integers $\{m_{n-1} + 1, \ldots, m_n\}$. The next lemma describes the reduction to blocks in the study of the almost sure behavior of (S_n/a_n).

Lemma 7.3. *Let (X_i) be independent and symmetric random variables. The sequence (S_n/a_n) is almost surely bounded (resp. convergent to 0) if and only if the same holds for $(\sum_{i \in I(n)} X_i/a_{m_n})$.*

Proof. We only show the convergence statement. By the Borel-Cantelli lemma and the Lévy inequality for symmetric random variables (2.6), the sequence

$$\left(\sup_{k\in I(n)} \frac{1}{a_{m_n}} \left\| \sum_{i=m_{n-1}+1}^{k} X_i \right\| \right)$$

converges almost surely to 0. Hence, for almost every ω and for every $\varepsilon > 0$, there exists ℓ_0 such that for all $\ell \geq \ell_0$,

$$\sup_{k\in I(\ell)} \left\| \sum_{i=m_{\ell-1}+1}^{k} X_i(\omega) \right\| \leq \varepsilon a_{m_\ell} .$$

Let now n and $j \geq \ell_0$ be such that $m_{j-1} < n \leq m_j$. Then

$$\|S_n(\omega)\| \leq \|S_{m_{\ell_0}-1}(\omega)\| + \sum_{\ell=\ell_0}^{j-1} \left\| \sum_{i\in I(\ell)} X_i(\omega) \right\| + \left\| \sum_{i=m_{j-1}+1}^{n} X_i(\omega) \right\|$$

$$\leq \|S_{m_{\ell_0}-1}(\omega)\| + \varepsilon \sum_{\ell=1}^{j} a_{m_\ell}$$

$$\leq \|S_{m_{\ell_0}-1}(\omega)\| + \varepsilon a_n C \sum_{\ell=0}^{\infty} c^{-\ell}$$

where we have used (7.1). Since $c > 1$ and $C < \infty$, the conclusion follows.

As a corollary to Lemmas 7.1 and 7.3 we can state the following equivalence for general independent variables.

Corollary 7.4. *Let (X_i) be independent random variables with values in B. Then $S_n/a_n \to 0$ almost surely if and only if $S_n/a_n \to 0$ in probability and $S_{m_n}/a_{m_n} \to 0$ almost surely, and similarly for boundedness.*

After these preliminaries, we are in a position to describe the general result about the almost sure behavior of (S_n/a_n). Recall that we assume (7.1). By symmetrization and Lemma 7.3, we have to study, thanks to the Borel-Cantelli lemma, convergence of series of the type

$$\sum_n \mathbb{P} \left\{ \left\| \sum_{i\in I(n)} X_i \right\| > \varepsilon a_n \right\}$$

for some, or all, $\varepsilon > 0$. The sufficient conditions which we describe for this to hold are obtained by the isoperimetric inequality for product measures in the form of Theorem 6.17. They are of various types. There is the assumption in probability on the sequence (S_n/a_n). If the sequence (S_n/a_n) is almost surely bounded, this is also the case for $(\max_{i\leq n} \|X_i\|/a_n)$. This necessary condition on the *norm* of the individual summands X_i is unfortunately not powerful enough in general and has to be complemented with some information on the successive maximum of the sample $(\|X_1\|,\ldots,\|X_n\|)$. Once this is given, the last sufficient condition deals with *weak* moment assumptions. These will appear to be almost optimal.

The announced result can now be stated. Recall K_0 is the absolute constant of the isoperimetric inequality (6.16). If r is an integer, we set $X_{I(n)}^{(r)} = \|X_j\|$ whenever $\|X_j\|$ is the r-th maximum of the sample $(\|X_i\|)_{i\in I(n)}$ (breaking ties by priority of index, and setting $X_{I(n)}^{(r)} = 0$ if $r > \operatorname{Card} I(n)$).

Theorem 7.5. *Let (X_i) be a sequence of independent and symmetric random variables with values in B. Assume there exist an integer $q \geq 2K_0$ and a sequence (k_n) of integers such that the following hold:*

$$(7.2) \qquad \sum_n \left(\frac{K_0}{q}\right)^{k_n} < \infty,$$

$$(7.3) \qquad \sum_n \mathbb{P}\left\{\sum_{r=1}^{k_n} X_{I(n)}^{(r)} > \varepsilon a_{m_n}\right\} < \infty$$

for some $\varepsilon > 0$. Set, for each n,

$$M_n = \mathbb{E}\left\|\sum_{i\in I(n)} X_i I_{\{\|X_i\|\leq \varepsilon a_{m_n}/k_n\}}\right\|,$$

$$\sigma_n = \sup_{f\in D}\left(\sum_{i\in I(n)} \mathbb{E}\left(f^2(X_i)I_{\{\|X_i\|\leq \varepsilon a_{m_n}/k_n\}}\right)\right)^{1/2}.$$

Then, if $L = \limsup_{n\to\infty} M_n/a_{m_n} < \infty$, and, for some $\delta > 0$,

$$(7.4) \qquad \sum_n \exp(-\delta^2 a_{m_n}^2/\sigma_n^2) < \infty,$$

we have

$$(7.5) \qquad \sum_n \mathbb{P}\left\{\left\|\sum_{i\in I(n)} X_i\right\| > 10^2 \alpha(\varepsilon,\delta,q,L) a_{m_n}\right\} < \infty$$

where $\alpha(\varepsilon,\delta,q,L) = \varepsilon + qL + (\varepsilon L + \delta^2)^{1/2}(q\log\frac{q}{K_0})^{1/2} \leq \varepsilon + qL + q(\varepsilon L + \delta^2)^{1/2}$. Conversely, if (7.5) holds for some (resp. all) $\alpha > 0$ and if (7.2) and (7.3) are satisfied, then $L < \infty$ (resp. $L = 0$) and (7.4) holds for some (resp. all) $\delta > 0$.

Proof. There is nothing to prove concerning the sufficient part of this theorem which readily follows from the inequality of Theorem 6.17 together with (6.18) applied to the sample $(X_i)_{i\in I(n)}$ with $k = k_n$ ($\geq q$ for n large enough), $s = \varepsilon a_{m_n}$ and

$$t = 10^2 (\varepsilon L + \delta^2)^{1/2}\left(q\log\frac{q}{K_0}\right)^{1/2} a_{m_n}.$$

(Of course, the numerical constant 10^2 is not the best one; it is simply a convenient number.) The necessary part concerning L is contained in Lemma 7.2. The necessity of (7.4) is based on Kolmogorov's exponential minoration inequality (Lemma 8.1 below). Assume, for instance, that for all $\alpha > 0$

$$\sum_n \mathbb{P}\left\{ \left\| \sum_{i \in I(n)} X_i \right\| > \alpha a_{m_n} \right\} < \infty.$$

Set, for simplicity, $X_i^n = X_i I_{\{\|X_i\| \leq \varepsilon a_{m_n}/k_n\}}$, $i \in I(n)$, and choose, for each n, f_n in D such that

$$\sigma_n^2 \leq 2 \sum_{i \in I(n)} \mathbb{E} f_n^2(X_i^n) \quad (\leq 2\sigma_n^2).$$

By the contraction principle (cf. second part of Lemma 6.5), we still have

$$\sum_n \mathbb{P}\left\{ \sum_{i \in I(n)} f_n(X_i^n) > \alpha a_{m_n} \right\} < \infty.$$

Let $\delta > 0$. If $\eta > 0$ is such that $\delta^2 \eta \geq \log(q/K_0)$, by (7.2) it is enough to check (7.4) for the integers n satisfying $a_{m_n}^2 \leq \eta k_n \sigma_n^2$. Let us therefore assume this holds. Recall Lemma 8.1 below and the parameter and constants γ, $\varepsilon(\gamma)$, $K(\gamma)$ therein, γ being arbitrary but fixed for our purposes. Let $\alpha > 0$ be small enough such that $(1 + \gamma)\alpha^2 \leq \delta^2$ and $2\alpha\varepsilon\eta \leq \varepsilon(\gamma)$ so that

$$(\alpha a_{m_n})(\varepsilon a_{m_n}/k_n) \leq \alpha\varepsilon\eta\sigma_n^2 \leq \varepsilon(\gamma) \sum_{i \in I(n)} \mathbb{E} f_n^2(X_i^n).$$

Since $L = 0$, it follows from Lemma 7.2 and orthogonality that $\sum_{i \in I(n)} \mathbb{E} f_n^2(X_i^n)/a_{m_n}^2 \to 0$; thus, for n large enough,

$$\alpha a_{m_n} \geq K(\gamma) \left(\sum_{i \in I(n)} \mathbb{E} f_n^2(X_i^n) \right)^{1/2}.$$

Lemma 8.1 then exactly implies that

$$\mathbb{P}\left\{ \sum_{i \in I(n)} f_n(X_i^n) > \alpha a_{m_n} \right\} \geq \exp\left(-(1 + \gamma)\alpha^2 a_{m_n}^2/\sigma_n^2 \right)$$

which gives the result since $(1 + \gamma)\alpha^2 \leq \delta^2$. This completes the proof of Theorem 7.5.

Theorem 7.5 expresses that, under (7.2) and (7.3), some necessary and sufficient conditions involving the behavior in probability or expectation and weak moments can be given to describe the almost sure behavior of (S_n/a_n). However, conditions (7.2) and (7.3) look rather technical and it would be desirable to find if possible simple, or at least easy to be handled, hypotheses on

(X_i) in order these conditions be fulfilled. There could be many ways to do this. We suggest such a way in terms of the probabilities $\mathbb{P}\{\max_{i \in I(n)} \|X_i\| > t\}$ (or $\sum_{i \in I(n)} \mathbb{P}\{\|X_i\| > t\}$). Of course, no vector valued structure is involved here.

Lemma 7.6. *In the notation of Theorem 7.5, assume that, for some $u > 0$,*

$$(7.6) \qquad \sum_n \mathbb{P}\{\max_{i \in I(n)} \|X_i\| > u a_{m_n}\} < \infty,$$

and that, for some $v > 0$, all n and t, $0 < t \le 1$,

$$(7.7) \qquad \mathbb{P}\{\max_{i \in I(n)} \|X_i\| > t v a_{m_n}\} \le \delta_n \exp\left(\frac{1}{t}\right)$$

where $\sum_n \delta_n^s < \infty$ for some integer s. Then, for each $q > K_0$, there exists a sequence (k_n) of integers such that $\sum_n (K_0/q)^{k_n} < \infty$, which satisfies

$$\sum_n \mathbb{P}\left\{\sum_{r=1}^{k_n} X_{I(n)}^{(r)} > 2s\left(u + v\left(\log \frac{q}{K_0}\right)^{-1}\right) a_{m_n}\right\} < \infty.$$

Proof. The idea is simply that if the largest element of the sample $(X_i)_{i \in I(n)}$ is exactly estimated by (7.6), the $2s$-th largest one is already very small. If $\mathbb{P}\{\max_{i \in I(n)} \|X_i\| > t v a_{m_n}\} \le 1/2$, and since $X_{I(n)}^{(2s)} > t v a_{m_n}$ if and only if $\mathrm{Card}\{i \in I(n);\ \|X_i\| > t v a_{m_n}\} \ge 2s$, we have by Lemmas 2.5 and 2.6

$$\mathbb{P}\{X_{I(n)}^{(2s)} > t v a_{m_n}\} \le \left(\sum_{i \in I(n)} \mathbb{P}\{\|X_i\| > t v a_{m_n}\}\right)^{2s}$$

$$\le \left(2\mathbb{P}\{\max_{i \in I(n)} \|X_i\| > t v a_{m_n}\}\right)^{2s}$$

$$\le \left(2\delta_n \exp\left(\frac{1}{t}\right)\right)^{2s}$$

where we have used hypothesis (7.7) in the last inequality (one can also use the small trick on large values shown in the proof of Theorem 6.20). The choice of $t = t(n) = (\log 1/\sqrt{\delta_n})^{-1}$ bounds the previous probability by $(2\delta_n)^s$ which, by hypothesis, is the current term of a summable series. Define then k_n, for each n, to be the integer part of

$$2s\left(\log \frac{q}{K_0}\right)^{-1} \log \frac{1}{\sqrt{\delta_n}}.$$

It is plain that $\sum_n (K_0/q)^{k_n} < \infty$. Now, we have

$$\sum_{r=1}^{k_n} X_{I(n)}^{(r)} \le 2s X_{I(n)}^{(1)} + k_n X_{I(n)}^{(2s)}$$

from which it follows that, for every n,

$$\mathbb{P}\left\{\sum_{r=1}^{k_n} X_{I(n)}^{(r)} > 2s\left(u + v\left(\log \frac{q}{K_0}\right)^{-1}\right)a_{m_n}\right\}$$
$$\leq \mathbb{P}\{X_{I(n)}^{(1)} > ua_{m_n}\} + \mathbb{P}\{X_{I(n)}^{(2s)} > t(n)va_{m_n}\}.$$

Therefore, the lemma is established.

Remark 7.7. Assume that (7.7) of Lemma 7.6 is strengthened into

$$\mathbb{P}\{\max_{i \in I(n)} \|X_i\| > tva_{m_n}\} \leq \frac{1}{t^p}\delta_n$$

for some $v > 0$, all n and t, $0 < t \leq 1$, and some $p > 0$. Then the preceding proof can easily be improved to yield that Lemma 7.6 holds with a sequence (k_n) satisfying

$$\sum_n \frac{1}{k_n^{2ps}} < \infty$$

(or even $\sum_n k_n^{-p's} < \infty$ for some $p' > p$). This observation is sometimes useful.

When the independent random variabl X_i are identically distributed, and the normalizing sequence (a_n) is regular enough, the first condition in Lemma 7.6 actually implies the second. This is the purpose of the next lemma. The subsequence (m_n) is chosen to be $m_n = 2^n$ for every n. The regularity condition on (a_n) contains the cases of $a_n = n^{1/p}$, $0 < p < \infty$, $a_n = (2nLLn)^{1/2}$, etc, which are the basic examples that we have in mind for the applications.

Lemma 7.8. *Let (a_n) be such that for some $p > 0$ and every $k \leq n$, $a_{2^n}^p \geq 2^{n-k}a_{2^k}^p$. Assume that (X_i) is a sequence of independent and identically distributed (as X) random variables. Then, if for some $u > 0$,*

$$\sum_n 2^n \mathbb{P}\{\|X\| > ua_{2^n}\} < \infty,$$

for all n and $0 < t \leq 1$

$$\mathbb{P}\{\max_{i \in I(n)} \|X_i\| > tua_{2^n}\} \leq 2^n \mathbb{P}\{\|X\| > tua_{2^n}\} \leq \frac{1}{t^{2p}}\delta_n$$

where $\sum_n \delta_n < \infty$.

Proof. For each n set $\gamma_n = 2^n \mathbb{P}\{\|X\| > ua_{2^n}\}$. There exists a sequence (β_n) such that $\beta_n \geq \gamma_n$ and $\beta_n \leq 2\beta_{n+1}$ for every n such that $\sum_n \beta_n < \infty$. Let $0 < t \leq 1$, and $k \geq 1$ be such that $2^{-k} \leq t^p \leq 2^{-k+1}$. If $k < n$,

$$2^n \mathbb{P}\{\|X\| > tua_{2^n}\} \leq 2^k \gamma_{n-k} \leq 2^k \beta_{n-k} \leq 2^{2k}\beta_n \leq 4t^{-2p}\beta_n.$$

If $k \geq n$,

$$2^n \mathbb{P}\{\|X\| > tua_{2^n}\} \leq 2^n \leq 4t^{-2p}2^{-n}.$$

The conclusion follows with $\delta_n = 4\max(\beta_n, 2^{-n})$.

Notice that Lemma 7.8 enters the setting of Remark 7.7.

7.2 Examples of Laws of Large Numbers

This section is devoted to applications to some classical strong laws of large numbers for Banach space valued random variables of the preceding general Theorem 7.5. Issued from sharp isoperimetric methods, this general result is actually already of interest on the line as will be clear from some of the statements we will obtain here. The results which we present follow rather easily from Theorem 7.5.

We do not seek the greatest generality in normalizing sequences and (almost) only deal with the classical strong law of large numbers given by $a_n = n$. That is, we usually say that a sequence (X_i) satisfies the *strong law of large numbers* (in short SLLN) if $S_n/n \to 0$ almost surely. We sometimes speak of the *weak law of large numbers* meaning that $S_n/n \to 0$ in probability. Thus, when $a_n = n$, we may simply take $m_n = 2^n$ as the blocking subsequence. Our applications here deal with the independent and identically distributed (iid) SLLN and the SLLN of Kolmogorov and Prokhorov as typical and classical examples. Further applications can easily be imagined.

The first theorem is the vector valued version of the SLLN of Kolmogorov and Marcinkiewicz-Zygmund for iid random variables. Although this result can be deduced from Theorem 7.5, there is a simpler argument based on Lemma 6.16 and the martingale representation of $\|S_n\| - \mathbb{E}\|S_n\|$. However, to demonstrate the universal character of the isoperimetric approach, we present the two proofs.

Theorem 7.9. *Let $0 < p < 2$. Let (X_i) be a sequence of iid random variables distributed as X with values in B. Then*

$$\frac{S_n}{n^{1/p}} \to 0 \quad \text{almost surely}$$

if and only if

$$\mathbb{E}\|X\|^p < \infty \quad \text{and} \quad \frac{S_n}{n^{1/p}} \to 0 \quad \text{in probability}.$$

Proof. The necessity is obvious; indeed, if $S_n/n^{1/p} \to 0$ almost surely, then $X_n/n^{1/p} \to 0$ almost surely from which it follows by the Borel-Cantelli lemma and identical distributions that

$$\sum_n \mathbb{P}\{\|X\| > n^{1/p}\} < \infty$$

which is equivalent to $\mathbb{E}\|X\|^p < \infty$. Turning to the sufficiency, it is enough, by Lemma 7.1, to prove the conclusion for the symmetric variable $X - X'$ where X' denotes an independent copy of X. Since $X - X'$ satisfies the same conditions as X, we can assume without loss of generality X itself to be symmetric. By Lemma 7.3, or directly by Lévy's inequality (2.6), it suffices to show that for every $\varepsilon > 0$

$$\sum_n \mathbb{P}\left\{\left\|\sum_{i \in I(n)} X_i\right\| > \varepsilon 2^{n/p}\right\} < \infty$$

where $I(n) = \{2^{n-1} + 1, \ldots, 2^n\}$.

First proof. For each n, set $u_i = u_i(n) = X_i I_{\{\|X_i\| \le 2^{n/p}\}}$, $i \in I(n)$. We have:

$$\sum_n \mathbb{P}\{\exists i \in I(n) : u_i \ne X_i\} \le \sum_n 2^n \mathbb{P}\{\|X\| > 2^{n/p}\}$$

which is finite under $\mathbb{E}\|X\|^p < \infty$. It therefore suffices to show that for every $\varepsilon > 0$

$$\sum_n \mathbb{P}\left\{\left\|\sum_{i \in I(n)} u_i\right\| > \varepsilon 2^{n/p}\right\} < \infty.$$

By Lemma 7.2, we know that

$$\lim_{n \to \infty} \frac{1}{2^{n/p}} \mathbb{E}\left\|\sum_{i \in I(n)} u_i\right\| = 0.$$

Hence, it is sufficient to prove that for every $\varepsilon > 0$

$$\sum_n \mathbb{P}\left\{\left\|\sum_{i \in I(n)} u_i\right\| - \mathbb{E}\left\|\sum_{i \in I(n)} u_i\right\| > \varepsilon 2^{n/p}\right\} < \infty.$$

By the quadratic inequality (6.11) and identical distribution, the preceding sum is less than

$$\frac{1}{\varepsilon^2} \sum_n \frac{1}{2^{2n/p}} \sum_{i \in I(n)} \mathbb{E}\|u_i\|^2 \le \frac{1}{\varepsilon^2} \sum_n \frac{1}{2^{n(2/p-1)}} \mathbb{E}\left(\|X\|^2 I_{\{\|X\| \le 2^{n/p}\}}\right)$$

$$\le \frac{1}{\varepsilon^2} \mathbb{E}\left(\|X\|^2 \sum_n \frac{1}{2^{n(2/p-1)}} I_{\{2^n \ge \|X\|^p\}}\right)$$

which is finite under the condition $\mathbb{E}\|X\|^p < \infty$. This concludes the first proof.

Second proof. We apply the general Theorem 7.5 with Lemmas 7.6 and 7.8 for the control of the large values. $\mathbb{E}\|X\|^p < \infty$ is equivalent to saying that for every $\varepsilon > 0$

$$\sum_n 2^n \mathbb{P}\{\|X\| > \varepsilon 2^{n/p}\} < \infty.$$

Let $\varepsilon > 0$ be fixed. By Lemmas 7.6 and 7.8, there is a sequence (k_n) of integers such that $\sum_n 2^{-k_n} < \infty$ and

$$\sum_n \mathbb{P}\left\{\sum_{r=1}^{k_n} X_{I(n)}^{(r)} > 5\varepsilon 2^{n/p}\right\} < \infty.$$

Then, apply Theorem 7.5 with $q = 2K_0$. As before, we can take $L = 0$ by Lemma 7.2. To check condition (7.4) note that

$$\sigma_n^2 \le 2^n \mathbb{E}\left(\|X\|^2 I_{\{\|X\| \le 2^{n/p}\}}\right)$$

(at least for all large enough n's). The same computation as in the first proof shows that $\sum_n 2^{-2n/p}\sigma_n^2 < \infty$ under $\mathbb{E}\|X\|^p < \infty$ so that (7.4) will hold for every $\delta > 0$. The conclusion of Theorem 7.5 then tells us that, for all $\delta > 0$,

$$\sum_n \mathbb{P}\left\{\left\|\sum_{i \in I(n)} X_i\right\| > 10^2(5\varepsilon + 2K_0\delta)\right\} < \infty.$$

The proof is therefore complete.

At this point, we open a digression on the hypothesis $S_n/n^{1/p} \to 0$ in probability in the preceding theorem on which we shall actually come back in Chapter 9 on type and cotype of Banach spaces. Let us assume that we deal here with a Borel random variable X with values in a separable Banach space B. It is known and very easy to show that in finite dimensional spaces, when $\mathbb{E}\|X\|^p < \infty$, $0 < p < 2$ (and $\mathbb{E}X = 0$ for $1 \le p < 2$), then

$$\frac{S_n}{n^{1/p}} \to 0 \quad \text{in probability}.$$

Let us briefly sketch the proof of this fact for real valued random variables. For all $\varepsilon, \delta > 0$

$$\mathbb{P}\{|S_n| > 2\varepsilon n^{1/p}\} \le n\mathbb{P}\{|X| > \delta n^{1/p}\} + \mathbb{P}\left\{\left|\sum_{i=1}^n X_i I_{\{|X_i| \le \delta n^{1/p}\}}\right| > 2\varepsilon n^{1/p}\right\}.$$

If $0 < p < 1$,

$$\left|\sum_{i=1}^n \mathbb{E}\left(X_i I_{\{|X_i| \le \delta n^{1/p}\}}\right)\right| \le n(\delta n^{1/p})^{1-p}\mathbb{E}|X|^p;$$

choose then $\delta = \delta(\varepsilon) > 0$ such that $\delta^{1-p}\mathbb{E}|X|^p = \varepsilon$. If $1 \le p < 2$, by centering,

$$\left|\sum_{i=1}^n \mathbb{E}\left(X_i I_{\{|X_i| \le \delta n^{1/p}\}}\right)\right| \le n\mathbb{E}\left(|X| I_{\{|X| > \delta n^{1/p}\}}\right)$$

which can be made smaller than $\varepsilon n^{1/p}$ for all large enough n's. Hence, in any case, we can center and write that, for n large,

$$\mathbb{P}\left\{\left|\sum_{i=1}^{n}X_i I_{\{|X_i|\leq\delta n^{1/p}\}}\right| > 2\varepsilon n^{1/p}\right\}$$

$$\leq \mathbb{P}\left\{\left|\sum_{i=1}^{n}X_i I_{\{|X_i|\leq\delta n^{1/p}\}} - \mathbb{E}\left(X_i I_{\{|X_i|\leq\delta n^{1/p}\}}\right)\right| > \varepsilon n^{1/p}\right\}$$

$$\leq \frac{1}{\varepsilon^2 n^{2/p}}\sum_{i=1}^{n}\mathbb{E}\left(|X_i|^2 I_{\{|X_i|\leq\delta n^{1/p}\}}\right)$$

by Chebyshev's inequality. By an integration by parts,

$$n^{1-2/p}\mathbb{E}\left(|X|^2 I_{\{|X|\leq\delta n^{1/p}\}}\right) \leq \int_0^\delta nt^p \mathbb{P}\{|X| > tn^{1/p}\}\frac{dt^2}{t^p}$$

so that the conclusion follows by dominated convergence since we know that $\lim_{u\to\infty} u^p \mathbb{P}\{|X| > u\} = 0$ when $\mathbb{E}|X|^p < \infty$.

Note indeed that if X is real symmetric (for example), then $S_n/n^{1/p} \to 0$ in probability as soon as $\lim_{t\to\infty} t^p \mathbb{P}\{|X| > t\} = 0$ (which is actually necessary and sufficient). We shall come back to this and to extensions to the vector valued setting in Chapter 9.

From the preceding observation, a finite dimensional approximation argument shows that in arbitrary separable Banach spaces, when $0 < p \leq 1$, the integrability condition $\mathbb{E}\|X\|^p < \infty$ (and X has mean zero for $p = 1$) also implies that $S_n/n^{1/p} \to 0$ in probability. Indeed, since X is Radon and $\mathbb{E}\|X\|^p < \infty$, we can choose, for every $\varepsilon > 0$, a finite valued random variable Y (with mean zero when $p = 1$ and $\mathbb{E}X = 0$) such that $\mathbb{E}\|X - Y\|^p < \varepsilon$. Letting T_n denote the partial sums associated with independent copies of Y, we have, *by the triangle inequality and since $p \leq 1$*,

$$\mathbb{E}\|S_n - T_n\|^p \leq n\mathbb{E}\|X - Y\|^p \leq \varepsilon n.$$

Y being a finite dimensional random variable, $T_n/n^{1/p} \to 0$ in probability and the claim follows immediately. Theorem 7.9 therefore states in this case (i.e. $0 < p \leq 1$) that

$$\frac{S_n}{n^{1/p}} \to 0 \quad \text{almost surely if and only if} \quad \mathbb{E}\|X\|^p < \infty$$

(and $\mathbb{E}X = 0$ for $p = 1$). In particular, we recover the extension to separable Banach space of the classical iid SLLN of Kolmogorov.

Corollary 7.10. *Let X be a Borel random variable with values in a separable Banach space B. Then*

$$\frac{S_n}{n} \to 0 \quad \text{almost surely}$$

if and only if $\mathbb{E}\|X\| < \infty$ and $\mathbb{E}X = 0$.

The proof of this result presented as a corollary of Theorems 7.5 and 7.9 is of course too complicated and a rather elementary direct proof can be given. For example, choose a finite valued mean zero random variable Y such that $\mathbb{E}\|X - Y\| < \varepsilon$ so that by the triangle inequality and the scalar law of large numbers

$$\limsup_{n\to\infty} \frac{1}{n}\left\|\sum_{i=1}^{n}(X_i - Y_i)\right\| \leq \limsup_{n\to\infty} \frac{1}{n}\sum_{i=1}^{n}\|X_i - Y_i\| < \varepsilon$$

where (Y_i) is a sequence of independent copies of Y; then apply again the finite dimensional SLLN to Y (see e.g. [HJ3]). It is however instructive to deduce it from general methods.

The preceding elementary approximation argument *does not* extend to the case $1 < p < 2$. It would require an inequality such as

$$\mathbb{E}\left\|\sum_i Y_i\right\|^p \leq C \sum_i \mathbb{E}\|Y_i\|^p$$

for every finite sequence (Y_i) of independent centered random variables with values in B, C depending on B and p. Such an inequality does not hold in a general Banach space and actually defines the spaces of *type p* discussed later in Chapter 9. Mimicking the preceding argument we can however already announce that in (separable) spaces of type p, $1 < p < 2$, $S_n/n^{1/p} \to 0$ almost surely if and only if $\mathbb{E}\|X\|^p < \infty$ and $\mathbb{E}X = 0$. We shall see how this property is actually characteristic of type p spaces (see Theorem 9.21 below).

The following example completes this discussion, showing in particular that $S_n/n^{1/p}$, $1 < p < 2$, cannot in general tend to 0 even if X has very strong integrability properties. The example is adapted for further purposes (the random variable is moreover pregaussian in the sense of Section 9.3).

Example 7.11. In c_0, the separable Banach space of all real sequences tending to 0 equipped with the sup-norm, there exists, for all decreasing sequences (α_n) of positive numbers tending to 0, an almost surely bounded and symmetric random variable X such that $(S_n/n\alpha_n)$ does not tend to 0 in probability.

Proof. Let $(\xi_k)_{k\geq 1}$ be independent with distribution

$$\mathbb{P}\{\xi_k = +1\} = \mathbb{P}\{\xi_k = -1\} = \frac{1}{2}(1 - \mathbb{P}\{\xi_k = 0\}) = \frac{1}{\log(k+1)}.$$

Define $\beta_k = \sqrt{\alpha_n}$ whenever $2^{n-1} \leq k < 2^n$ and take X to be the random variable in c_0 with coordinates $(\beta_k\xi_k)_{k\geq 1}$. Then X is clearly symmetric and almost surely bounded. However, $(S_n/n\alpha_n)$ does not tend to 0 in probability. Indeed, denote by $(\xi_{k,i})_i$ independent copies of (ξ_k). Set further, for each $k, n \geq 1$,

$$E_{n,k} = \bigcap_{i=1}^{n}\{\xi_{k,i} = 1\}, \quad A_n = \bigcup_{k\leq 2^n} E_{n,k}.$$

Clearly $\mathbb{P}(E_{n,k}) = (\log(k+1))^{-n}$ and

$$\mathbb{P}(A_n) = 1 - \prod_{k \leq 2^n} \mathbb{P}(E^c_{n,k}) = 1 - \prod_{k \leq 2^n} \left(1 - \frac{1}{(\log(k+1))^n}\right).$$

Therefore, as is easily seen, $\mathbb{P}(A_n) \to 1$. Now, if (e_k) is the canonical basis of c_0,

$$\frac{S_n}{n\alpha_n} = \sum_{k=1}^{\infty} \frac{1}{n\alpha_n} \left(\sum_{i=1}^{n} \xi_{k,i}\right) \beta_k e_k.$$

On A_n,

$$\frac{\|S_n\|}{n\alpha_n} \geq \max_{k \leq 2^n} \frac{1}{n\alpha_n} \beta_k \left|\sum_{i=1}^{n} \xi_{k,i}\right| \geq \frac{\sqrt{\alpha_n}}{\alpha_n} = \frac{1}{\sqrt{\alpha_n}}.$$

Hence, since $\alpha_n \to 0$, for every $\varepsilon > 0$,

$$\liminf_{n\to\infty} \mathbb{P}\left\{\frac{\|S_n\|}{n\alpha_n} > \varepsilon\right\} \geq \liminf_{n\to\infty} \mathbb{P}(A_n) = 1$$

which establishes the claim.

After having extended to the vector valued setting the iid SLLN, we turn to the SLLN for independent but not necessarily identically distributed random variables. Here again, we restrict to classical statements like the SLLN of Kolmogorov. This SLLN states that if (X_i) is a sequence of independent mean zero real valued random variables such that

$$\sum_i \frac{\mathbb{E}X_i^2}{i^2} < \infty,$$

then the SLLN holds, i.e. $S_n/n \to 0$ almost surely. From this result, together with a truncation argument, Kolmogorov deduced his iid theorem. The next theorem points towards a general extension of this result which is already of interest on the line. It is characterized by a careful balance between conditions on the norm of the X_i's and assumptions on their weak moments. The subsequent corollary is perhaps a more practical result. We take up again our framework of non-necessarily Radon random variables.

Theorem 7.12. *Let (X_i) be a sequence of independent random variables with values in B. Assume that*

(7.8) $$\frac{X_i}{i} \to 0 \quad almost \ surely$$

and

(7.9) $$\frac{S_n}{n} \to 0 \quad in \ probability.$$

Assume further that for some $v > 0$, all n and t, $0 < t \leq 1$,

$$(7.10) \qquad \mathbb{P}\{\max_{i \in I(n)} \|X_i\| > tv2^n\} \leq \delta_n \exp\left(\frac{1}{t}\right)$$

where $\sum_n \delta_n^s < \infty$ for some $s > 0$, and that, for each $\delta > 0$,

$$(7.11) \quad \sum_n \exp\left(-\delta 2^{2n} \Big/ \sup_{f \in D} \sum_{i \in I(n)} \mathbb{E}\big(f^2(X_i) I_{\{\|X_i\| \leq 2^n\}}\big)\right) < \infty$$

(where we recall that $I(n) = \{2^{n-1} + 1, \ldots, 2^n\}$). Then the SLLN holds, i.e.

$$\frac{S_n}{n} \to 0 \quad almost \ surely.$$

Proof. We simply apply Theorem 7.5 and Lemma 7.6. If we define $Y_i = X_i I_{\{\|X_i\| \leq i\}}$, by (7.8), almost surely $Y_i = X_i$ for every large enough i so that it clearly suffices to prove the result for the sequence (Y_i) instead of (X_i). We therefore assume that $\|X_i\| \leq i$ almost surely. If (X_i') is an independent copy of the sequence (X_i), the sequence of symmetric variables $(X_i - X_i')$ will satisfy the same kind of hypotheses as (X_i). By (7.9) and Lemma 7.1, we can reduce to the case of a symmetric sequence (X_i). In Theorem 7.5, whatever the choice of (k_n) may be, we can take $L = 0$ thanks to (7.9) (Lemma 7.2). Since $X_i/i \to 0$ almost surely, for every $u > 0$,

$$\sum_n \mathbb{P}\{\max_{i \in I(n)} \|X_i\| > u2^n\} < \infty.$$

Summarizing the conclusions of Lemma 7.6 and Theorem 7.5, for all $u, \delta > 0$ and all $q \geq 2K_0$, and for s assumed to be an integer,

$$\sum_n \mathbb{P}\left\{\Big\|\sum_{i \in I(n)} X_i\Big\| > 10^2\left[2s\left(u + v\left(\log\frac{q}{K_0}\right)^{-1}\right) + q\delta\right]2^n\right\} < \infty.$$

It obviously follows that

$$\sum_n \mathbb{P}\left\{\Big\|\sum_{i \in I(n)} X_i\Big\| > \varepsilon 2^n\right\} < \infty$$

for every $\varepsilon > 0$, hence the conclusion by Lemma 7.3.

Corollary 7.13. *Under the hypotheses of Theorem 7.10, but with (7.10) replaced by*

$$(7.10') \qquad \sum_n \left(\frac{1}{2^{np}} \sum_{i \in I(n)} \mathbb{E}\|X_i\|^p\right)^s < \infty$$

for some $p > 0$ and $s > 0$, the SLLN is satisfied.

Proof. Simply note that, for every n,

$$\sum_{i \in I(n)} \mathbb{P}\{\|X_i\| > tv2^n\} \le \frac{1}{(tv)^p} \cdot \frac{1}{2^{np}} \sum_{i \in I(n)} \mathbb{E}\|X_i\|^p$$

$$\le C(p) \exp\left(\frac{1}{t}\right) \frac{1}{v^p} \cdot \frac{1}{2^{np}} \sum_{i \in I(n)} \mathbb{E}\|X_i\|^p$$

from which (7.10) of Theorem 7.12 follows. Note that the sums $\sum_{i \in I(n)} \mathbb{E}\|X_i\|^p$ in (7.10') can also be replaced, if one wishes it, by expressions of the type

$$\sup_{t>0} t^p \sum_{i \in I(n)} \mathbb{P}\{\|X_i\| > t\}.$$

In Theorem 7.12 (and Corollary 7.13), conditions (7.8) and (7.9) are of course necessary, (7.9) describing the usual assumption in probability on the sequence (S_n). For *real valued* mean zero random variables, this condition (7.9) is *automatically satisfied under* (7.11) and the scalar statement such obtained is sharp. Note also that, under (7.8), it is legitimate (and we used it in the proof of Theorem 7.12) to assume that $\|X_i\|_\infty \le i$ for all i. This is sometimes convenient when various statements have to be compared; for example, (7.10) (via (7.10')) then holds under the stronger condition

$$\sum_i \frac{\mathbb{E}\|X_i\|^p}{i^p} < \infty$$

which in this case is seen to be weaker and weaker as p increases. Then this condition implies (7.8) and (7.11) provided $p \le 2$ and we therefore have the following corollary. (Since no weak moments are involved, it might be obtained simpler from Lemma 6.16, see [K-Z].)

Corollary 7.14. *Let (X_i) be a sequence of independent random variables with values in B. If for some $1 \le p \le 2$*

$$\sum_i \frac{\mathbb{E}\|X_i\|^p}{i^p} < \infty,$$

then the SLLN holds, i.e. $S_n/n \to 0$ almost surely, if and only if the weak law of large numbers holds, i.e. $S_n/n \to 0$ in probability.

Along the same line of ideas, Theorem 7.12 also contains extensions of Brunk's SLLN. Brunk's theorem in the real case states that if (X_i) are independent with mean zero and satisfy

$$\sum_i \frac{\mathbb{E}|X_i|^p}{i^{p/2+1}} < \infty \quad \text{for some} \quad p \ge 2,$$

then the SLLN holds. To include this result, simply note that for $p \geq 2$

$$\frac{1}{2^{2n}} \sum_{i \in I(n)} \mathbb{E} f^2(X_i) \leq \left(\frac{1}{2^{n(p/2+1)}} \sum_{i \in I(n)} \mathbb{E} |f(X_i)|^p \right)^{2/p}$$

for every n and f in D.

One feature of the isoperimetric approach which is at the basis of Theorems 7.5 and 7.12 is a common treatment of the SLLN of Kolmogorov and Prokhorov. As easily as we obtained Theorem 7.12 from the preceding section, we get an extension of Prokhovov's theorem to the vector valued case. We still work under conditions (7.9) and (7.11) but reinforce (7.8) into

$$\|X_i\|_\infty \leq i/LLi \text{ for each } i$$

where $LLt = L(Lt)$ and $Lt = \max(1, \log t)$, $t \geq 0$. This boundedness assumption provides the exact bound on large values and actually fits (7.10) of Theorem 7.12. Indeed, for each n and t, $0 < t \leq 1$,

$$\mathbb{P}\left\{ \max_{i \in I(n)} \|X_i\| > 2t2^n \right\} \leq \delta_n \exp\left(\frac{1}{t} \right)$$

with $\delta_n = \exp(-2LL2^n)$ which is summable. We thus obtain as a last corollary the following version of Prokhorov's SLLN. Note that under the preceding boundedness assumption on the X_i's, condition (7.11) becomes necessary; the proof follows the necessary part in Theorem 7.5 and, therefore, we do not detail it.

Corollary 7.15. *Let (X_i) be a sequence of independent random variables with values in B. Assume that, for every i,*

$$\|X_i\|_\infty \leq i/LLi.$$

Then the SLLN is satisfied if and only if

$$\frac{S_n}{n} \to 0 \text{ in probability}$$

and

$$\sum_n \exp\left(-\delta 2^{2n} \bigg/ \sup_{f \in D} \sum_{i \in I(n)} \mathbb{E} f^2(X_i) \right) < \infty$$

for every $\delta > 0$ (where $I(n) = \{2^{n-1} + 1, \ldots, 2^n\}$).

Notes and References

Various expositions on the strong laws of large numbers (SLLN) for sums of independent real valued random variables may be found in the classical works e.g. [Lé1], [Gn-K], [Re], [Sto], [Pe] etc. In particular, Lemmas 7.1 and 7.3 are clearly presented in [Sto] and the vector valued situation does not make any difference.

This chapter, which is based on the isoperimetric approach of [Ta11] and [Ta9] presented in Section 6.3, follows the paper [L-T5]. In particular, Theorem 7.5 and Lemma 7.6 are taken from there.

The extension of the Marcinkiewicz-Zygmund SLLN (Theorem 7.9) is due independently to A. de Acosta [Ac6] and T. A. Azlarov and N. A. Volodin [A-V], with the first proof. The classical iid SLLN of Kolmogorov in separable Banach spaces (Corollary 7.10) was established by E. Mourier back in the early fifties [Mo] (see also [F-M1], [F-M2]). A simple proof may be found in [HJ3]. The non-separable version of this result, which is not discussed in this text, has given recently rise to many developments related to measure theory; see for example [HJ5], [Ta3], and, in the context of empirical processes, [V-C1], [V-C2], [G-Z2]. Example 7.11 is taken from [C-T1].

Theorem 7.12 comes from [L-T5]. For further developments, see [Al2]. The real valued statement, at least in the form of Corollary 7.13, can be obtained as a consequence of the Fuk-Nagaev inequality (cf. [F-N], [Yu2]). Let us mention that a suitable vector valued version of the SLLN of S. V. Nagaev [Na1], [Na2] still seems to be found; cf. [Al3] and [Ber] in this regard. Corollary 7.14 is due to J. Kuelbs and J. Zinn [K-Z] who extended results of A. Beck [Be] and J. Hoffmann-Jørgensen and G. Pisier [HJ-P] in special classes of spaces (cf. Chapter 9). The work of J. Kuelbs and J. Zinn was important in realizing that under an assumption in probability no condition has to be imposed on the space. In some smooth spaces, see [He5]. Brunk's SLLN appeared in [Br] and was first investigated in Banach spaces in [Wo3]. Extensions of Prokhorov's SLLN [Pro2] were undertaken in [K-Z], [He6], [Al1] (where, in particular, necessity was shown) and the final result was obtained in [L-T4] (see also [L-T5]). Applications of the isoperimetric method to strong limit theorems for trimmed sums of iid random variables are further described in [L-T5].

8. The Law of the Iterated Logarithm

This chapter is devoted to the classical laws of the iterated logarithm of Kolmogorov and Hartman-Wintner-Strassen in the vector valued setting. These extensions both enlighten the scalar statements and describe various new interesting phenomena in the infinite dimensional setting. As in the previous chapter on the strong law of large numbers, the isoperimetric approach proves to be an efficient tool in this study. The main results described here show again how the strong *almost sure* statement of the law of the iterated logarithm reduces to the corresponding (necessary) statement in *probability*, under moment conditions similar to those of the scalar case.

Together with the law of large numbers and the central limit theorem, the law of the iterated logarithm (in short LIL) is a main subject in Probability Theory. Here, we only concentrate on the classical (but typical) forms of the LIL for sums of independent Banach space valued random variables. We first describe, starting from the real valued case, the extension of Kolmogorov's LIL. In Section 8.2, we describe the Hartman-Wintner-Strassen form of the (iid) LIL in Banach space and characterize the random variables which satisfy it. A last survey paragraph is devoted to a discussion of various results and questions about the identification of the limits in the vector valued LIL.

In all this chapter, if $(X_i)_{i \in \mathbb{N}}$ is a sequence of random variables, we set, as usual, $S_n = X_1 + \cdots + X_n$, $n \geq 1$. Recall also that LL denotes the iterated logarithm function, that is, $LLt = L(Lt)$ and $Lt = \max(1, \log t)$, $t \geq 0$.

8.1 Kolmogorov's Law of the Iterated Logarithm

Let $(X_i)_{i \in \mathbb{N}}$ be a sequence of independent real valued mean zero random variables such that $\mathbb{E}X_i^2 < \infty$ for every i. Set, for each n, $s_n = (\sum_{i=1}^n \mathbb{E}X_i^2)^{1/2}$.

Assume that the sequence (s_n) increases to infinity. Assume further that for some sequence (η_i) of positive number tending to 0,

$$\|X_i\|_\infty \leq \eta_i s_i / \left(LLs_i^2\right)^{1/2} \quad \text{for every } i.$$

Then, *Kolmogorov's* LIL *states that*, with probability one,

$$(8.1) \qquad \limsup_{n \to \infty} \frac{S_n}{(2s_n^2 LLs_n^2)^{1/2}} = 1.$$

The proof of the upper bound in (8.1) is based on the exponential inequality of Lemma 1.6 applied to the sums S_n of independent mean zero random variables. The lower bound, which is somewhat more complicated, relies on Kolmogorov's converse exponential inequality described in the following lemma. Its proof (cf. [Sto]) is a precise amplification of the argument leading to (4.2).

Lemma 8.1. *Let (X_i) be a finite sequence of independent mean zero real valued random variables such that $\|X_i\|_\infty \leq a$ for all i. Then, for every $\gamma > 0$, there exist positive numbers $K(\gamma)$ (large enough) and $\varepsilon(\gamma)$ (small enough) depending on γ only, such that for every t satisfying $t \geq K(\gamma)b$ and $ta \leq \varepsilon(\gamma)b^2$ where $b = (\sum_i \mathbb{E}X_i^2)^{1/2}$,*

$$\mathbb{P}\left\{\sum_i X_i > t\right\} \geq \exp\{-(1+\gamma)t^2/2b^2\}.$$

The next theorem presents the extension to Banach space valued random variables of Kolmogorov's LIL. This extension involves a careful balance between conditions on the norms of the random variables and weak moment assumptions. Since tightness properties are unessential in this first section, we describe the result in our setting of a Banach space B for which there exists a countable subset D of the unit ball of the dual space such that $\|x\| = \sup_{f \in D} |f(x)|$ for every x in B. X is a random variable with values in B if $f(X)$ is measurable for every f in D.

Theorem 8.2. *Let B be as before and let $(X_i)_{i \in \mathbb{N}}$ be a sequence of independent random variables with values in B such that $\mathbb{E}f(X_i) = 0$ and $\mathbb{E}f^2(X_i) < \infty$ for each i and each f in D. Set, for each n, $s_n = \sup_{f \in D}(\sum_{i=1}^n \mathbb{E}f^2(X_i))^{1/2}$, assumed to increase to infinity. Assume further that for some sequence (η_i) of positive numbers tending to 0 and every i*

(8.2) $$\|X_i\|_\infty \leq \eta_i s_i / (2LLs_i^2)^{1/2}.$$

Then, if the sequence $(S_n/(2s_n^2 LLs_n^2)^{1/2})$ converges to 0 in probability, with probability one,

(8.3) $$\limsup_{n \to \infty} \frac{\|S_n\|}{(2s_n^2 LLs_n^2)^{1/2}} = 1.$$

This type of statement clearly shows in which direction the extension of the scalar result is looked for. The proof of Theorem 8.2 could seem to be somewhat involved. Let us mention however, and this will be accomplished in a first step, that the proof of the finiteness of the lim sup in (8.3) is rather easy on the basis of the isoperimetric approach of Section 6.3. Then, the fact that it is actually less than 1 requires some technicalities. The lower bound reproduces the scalar case.

Proof. To simplify the notation, let us set, for each n, $u_n = (2LLs_n^2)^{1/2}$. As announced, we first show that

$$(8.4) \qquad \limsup_{n \to \infty} \frac{\|S_n\|}{s_n u_n} \leq M$$

almost surely for some numerical constant M. To this aim, replacing (X_i) by $(X_i - X_i')$ where (X_i') is an independent copy of the sequence (X_i), and since (by centering, cf. (2.5))

$$s_n \leq \sup_{f \in D} \left(\sum_{i=1}^n \mathbb{E} f^2 (X_i - X_i') \right)^{1/2} \leq 2s_n \,,$$

we can assume by Lemma 7.1 that we deal with symmetric variables. For each n, define m_n as the smallest integer m such that $s_m > 2^n$. It is easily seen that

$$\frac{s_{n+1}}{s_n} \sim 1, \quad s_{m_n} \sim 2^n, \quad \frac{s_{m_n+1}}{s_{m_n}} \sim 2 \,.$$

By the Borel-Cantelli lemma, we need show that

$$\sum_n \mathbb{P} \left\{ \max_{m_{n-1} < m \leq m_n} \frac{\|S_m\|}{s_m u_m} > M \right\} < \infty \,.$$

Using the preceding and Lévy's inequality (2.6) (by increasing M), it suffices to prove that

$$\sum_n \mathbb{P} \{ \|S_{m_n}\| > M s_{m_n} u_{m_n} \} < \infty \,.$$

We make use of the isoperimetric inequality of Theorem 6.17 together with (6.18) which we apply to the sample of independent and symmetric random variables $(X_i)_{i \leq m_n}$ for each n. Assuming for simplicity that the sequence (η_i) is bounded by 1, we take there $q = 2K_0$, $k = [u_{m_n}^2] + 1$, $s = t = 20\sqrt{q}\, s_{m_n} u_{m_n}$. For these choices, by (8.2),

$$\sum_{i=1}^k \|X_i\|^* \leq (u_{m_n}^2 + 1) \frac{s_{m_n}}{u_{m_n}} \leq s \,.$$

Since $(S_n/s_n u_n)$ converges to 0 in probability, by Lemma 7.2, $\mathbb{E}\|S_n\|/s_n u_n \to 0$. Hence, at least for n large enough, m^2 in Theorem 6.17 can be bounded, using (6.18), by $2s_{m_n}^2$. Thus, it follows from (6.17) that, for large n's,

$$\mathbb{P} \{ \|S_{m_n}\| > (60\sqrt{2K_0} + 1) s_{m_n} u_{m_n} \} \leq 2^{-u_{m_n}^2} + 2 \exp(-u_{m_n}^2)$$

which gives the result since $u_{m_n}^2 \sim 2LL4^n$.

Note that this proof shows (8.4) already when the sequence $(S_n/s_n u_n)$ is only bounded in probability, which of course is necessary.

We now turn to the more delicate proof that the lim sup is actually equal to 1. We begin by showing it is less than or equal to 1. Since $S_n/s_n u_n \to 0$ in probability, observe first that, by symmetrization, Lemma 7.2 and centering (Lemma 6.3), we have both

$$(8.5) \qquad \lim_{n\to\infty} \frac{1}{s_n u_n} \mathbb{E}\|S_n\| = \lim_{n\to\infty} \frac{1}{s_n u_n} \mathbb{E}\left\|\sum_{i=1}^n \varepsilon_i X_i\right\| = 0$$

where as usual (ε_i) denotes a Rademacher sequence which is independent of (X_i).

For $\rho > 1$, let m_n for each n be the smallest m such that $s_m > \rho^n$. As above for $\rho = 2$ we have

$$s_{m_n} \sim \rho^n, \qquad \frac{s_{m_n+1}}{s_{m_n}} \sim \rho.$$

To establish the claim it will be sufficient to show that for every $\varepsilon > 0$ and $\rho > 1$,

$$(8.6) \qquad \sum_n \mathbb{P}\{\|S_{m_n}\| > (1+\varepsilon)s_{m_n} u_{m_n}\} < \infty.$$

Indeed, in order that $\limsup_{n\to\infty} \|S_n\|/s_n u_n \leq 1$ almost surely, it suffices, by the Borel-Cantelli lemma, that for every $\delta > 0$ there is an increasing sequence (m_n) of integers such that

$$\sum_n \mathbb{P}\left\{ \max_{m_{n-1} < m \leq m_n} \frac{\|S_m\|}{s_m u_m} > (1+2\delta)\right\} < \infty.$$

A simple use of Ottaviani's inequality (Lemma 6.2) together with (8.5) shows that, for all large enough n's,

$$\mathbb{P}\left\{ \max_{m_{n-1} < m \leq m_n} \frac{\|S_m\|}{s_m u_m} > (1+2\delta)\right\}$$
$$\leq \mathbb{P}\left\{ \max_{m_{n-1} < m \leq m_n} \|S_m\| > (1+2\delta)s_{m_{n-1}} u_{m_{n-1}}\right\}$$
$$\leq 2\mathbb{P}\{\|S_{m_n}\| > (1+\delta)s_{m_{n-1}} u_{m_{n-1}}\}.$$

Now, for $\delta > 0$, there exist $\varepsilon > 0$ and $\rho > 1$ such that if m_n is defined from ρ as before, for all large n's,

$$(1+\delta)s_{m_{n-1}} u_{m_{n-1}} \geq (1+\varepsilon)s_{m_n} u_{m_n}.$$

So, it suffices to show (8.6).

The proof is based on a finite dimensional approximation argument via some *entropy* estimate. Let now $\varepsilon > 0$ and $\rho > 1$ be fixed. For f, g in D, and every n, set

$$d_2^n(f,g) = \left(\frac{1}{s_{m_n}^2} \sum_{i=1}^{m_n} \mathbb{E}(f-g)^2(X_i)\right)^{1/2}.$$

Recall that $N(D, d_2^m; \varepsilon)$ denotes the minimal number of elements g in D such that for every f in D there exists such a g with $d_2^m(f, g) < \varepsilon$. For every n, define

$$\alpha_n = \frac{1}{s_{m_n} u_{m_n}} \mathbb{E} \left\| \sum_{i=1}^{m_n} \varepsilon_i X_i \right\|$$

which tends to 0 when n goes to infinity by (8.5).

Lemma 8.3. *For every large enough n,*

$$N(D, d_2^m; \varepsilon) \leq \exp\left(\alpha_n u_{m_n}^2\right).$$

Proof. Suppose this is not the case. Then, infinitely often in n, there exists $D_n \subset D$ such that for any $f \neq g$ in D_n, $d_2^m(f, g) \geq \varepsilon$ and

$$\operatorname{Card} D_n = \left[\exp\left(\alpha_n u_{m_n}^2\right)\right] + 1.$$

By Lemma 1.6 and (8.2), for $h = f - g$, $f \neq g$ in D_n, and n large,

$$\mathbb{P}\left\{ \frac{1}{s_{m_n}^2} \sum_{i=1}^{m_n} h^2(X_i) < \frac{\varepsilon^2}{2} \right\} \leq \mathbb{P}\left\{ \frac{1}{s_{m_n}^2} \sum_{i=1}^{m_n} \left(-h^2(X_i) + \mathbb{E}h^2(X_i)\right) > \frac{\varepsilon^2}{2} \right\}$$

$$\leq \exp\left(-u_{m_n}^2\right).$$

For n large enough,

$$\operatorname{Card} D_n \exp\left(-u_{m_n}^2\right) < 1/4.$$

It follows that, infinitely often in n,

$$\mathbb{P}\left\{ \forall f \neq g \text{ in } D_n, \ \frac{1}{s_{m_n}^2} \sum_{i=1}^{m_n} (f-g)^2(X_i) \geq \frac{\varepsilon^2}{2} \right\} \geq \frac{3}{4}.$$

We would like to apply, conditionally on the X_i's, the Sudakov type minoration of Proposition 4.13. To this aim, first note that by (8.5), with high probability, for example larger than $3/4$,

$$\frac{1}{s_{m_n}} \mathbb{E}_\varepsilon \left\| \sum_{i=1}^{m_n} \varepsilon_i X_i \right\| \leq \frac{\varepsilon^2}{K} \cdot \frac{u_{m_n}}{\max_{i \leq m_n} \eta_i}$$

for every n large enough, and thus, by (8.2), for $i \leq m_n$,

$$\frac{\|X_i\|_\infty}{s_{m_n}} \leq \frac{\max_{j \leq m_n} \eta_j}{u_{m_n}} \leq \varepsilon^2 \left(\frac{K}{s_{m_n}} \mathbb{E}_\varepsilon \left\| \sum_{i=1}^{m_n} \varepsilon_i X_i \right\| \right)^{-1}$$

where $K > 0$ is the numerical constant of Proposition 4.13. This proposition then shows that, with probability bigger than $1/2$,

$$\frac{1}{s_{m_n}} \mathbb{E}_\varepsilon \left\| \sum_{i=1}^{m_n} \varepsilon_i X_i \right\| \geq \frac{1}{K} \cdot \frac{\varepsilon}{\sqrt{2}} \left(\log \operatorname{Card} D_n\right)^{1/2} \geq \frac{\varepsilon \sqrt{\alpha_n} u_{m_n}}{\sqrt{2}\, K}.$$

Therefore, by integrating, we find that, infinitely often in n, $\alpha_n \geq \varepsilon\sqrt{\alpha_n}/2\sqrt{2}K$ which leads to a contradiction since $\alpha_n \to 0$. The proof of Lemma 8.3 is complete.

We can now establish (8.6). According to Lemma 8.3, we denote, for each n (large enough) and f in D, by $g_n(f)$ an element of D such that $d_2^n(f, g_n(f)) < \varepsilon$ in such a way that the set D_n of all $g_n(f)$'s has a cardinality less than $\exp(\alpha_n u_{m_n}^2)$. We write that

$$\left\| \sum_{i=1}^{m_n} X_i \right\| \leq \sup_{g \in D_n} \left| \sum_{i=1}^{m_n} g(X_i) \right| + \sup_{h \in D'_n} \left| \sum_{i=1}^{m_n} h(X_i) \right|$$

where $D'_n = \{f - g_n(f); f \in D\}$. The main observation concerning D'_n is that

$$\sup_{h \in D'_n} \left(\sum_{i=1}^{m_n} \mathbb{E}h^2(X_i) \right)^{1/2} \leq \varepsilon s_{m_n}^2.$$

It is then an easy exercise to see how the proof of the first part can be reproduced (and adapted to the case of a norm of the type $\sup_{h \in D'_n} |h(\cdot)|$, thus depending on n) to yield that for some numerical constant M,

$$\sum_n \mathbb{P}\left\{ \sup_{h \in D'_n} \left| \sum_{i=1}^{m_n} h(X_i) \right| > M\varepsilon s_{m_n} u_{m_n} \right\} < \infty.$$

The proof of (8.6) will be complete if we show that

$$\sum_n \mathbb{P}\left\{ \sup_{g \in D_n} \left| \sum_{i=1}^{m_n} g(X_i) \right| > (1+\varepsilon)s_{m_n} u_{m_n} \right\} < \infty.$$

Now, as in the real valued case, we have by Lemma 1.6 that for all large enough n's (in order to efficiently use (8.2) and $\eta_i \to 0$, first neglect the first terms of the summation),

$$\mathbb{P}\left\{ \sup_{g \in D_n} \left| \sum_{i=1}^{m_n} g(X_i) \right| > (1+\varepsilon)s_{m_n} u_{m_n} \right\} \leq 2\,\mathrm{Card}\,D_n \exp\left(-(1+\varepsilon)LLs_{m_n}^2\right),$$

hence the result since $\mathrm{Card}\,D_n \leq \exp(2\alpha_n LLs_{m_n}^2)$, $\alpha_n \to 0$ and $s_{m_n} \sim \rho^n$ ($\rho > 1$).

In the last part of this proof, we show that

$$\limsup_{n \to \infty} \frac{\|S_n\|}{s_n u_n} \geq 1$$

almost surely, reproducing simply (more or less) the scalar case based on the exponential minoration inequality of Lemma 8.1. Recall that for $\rho > 1$ we let $m_n = \inf\{m; s_m > \rho^n\}$. By the zero-one law (cf. [Sto]), the $\limsup$ we study is almost surely non-random; by the upper bound which we just established, we have that (for example)

$$\mathbb{P}\big\{\|S_{m_n}\| \le 2s_{m_n}u_{m_n} \text{ for every large enough } n\big\} = 1.$$

Suppose it can be proved that for all $\varepsilon > 0$ and all $\rho > 1$ large enough

(8.7)
$$\mathbb{P}\bigg\{\Big\|\sum_{i\in I(n)} X_i\Big\| > (1-\varepsilon)^2 \bigg(2\sup_{f\in D}\sum_{i\in I(n)} \mathbb{E}f^2(X_i) \\ \cdot LL\Big(\sup_{f\in D}\sum_{i\in I(n)} \mathbb{F}^2(X_i)\Big)\bigg)^{1/2} \text{ i.o. in } n\bigg\} = 1$$

where $I(n)$ denotes the set of integers between $m_{n-1}+1$ and m_n. Then, on a set of probability one, i.o. in n,

$$\|S_{m_n}\| \ge \Big\|\sum_{i\in I(n)} X_i\Big\| - \|S_{m_{n-1}}\|$$

$$\ge (1-\varepsilon)^2 \bigg(2\sup_{f\in D}\sum_{i\in I(n)} \mathbb{E}f^2(X_i)LL\Big(\sup_{f\in D}\sum_{i\in I(n)} \mathbb{E}f^2(X_i)\Big)\bigg)^{1/2} \\ - 2s_{m_{n-1}}u_{m_{n-1}}$$

$$\ge (1-\varepsilon)^2 \big[2\big(s_{m_n}^2 - s_{m_{n-1}}^2\big)LL\big(s_{m_n}^2 - s_{m_{n-1}}^2\big)\big]^{1/2} - 2s_{m_{n-1}}u_{m_{n-1}}$$

and, for n large, this lower bound behaves like

$$\bigg[(1-\varepsilon)^2\Big(1 - \frac{1}{\rho^2}\Big)^{1/2} - \frac{2}{\rho}\bigg]s_{m_n}u_{m_n}.$$

For ρ large enough and $\varepsilon > 0$ arbitrarily small, this will therefore show that the lim sup is ≥ 1 almost surely, hence the conclusion. Let us prove (8.7). For each n, let f_n in D be such that

$$\sum_{i\in I(n)} \mathbb{E}f_n^2(X_i) \ge (1-\varepsilon)\sup_{f\in D}\sum_{i\in I(n)} \mathbb{E}f^2(X_i).$$

Thus, the probability in (8.7) is larger than or equal to

$$\mathbb{P}\bigg\{\sum_{i\in I(n)} f_n(X_i) > (1-\varepsilon)\bigg(2\sum_{i\in I(n)} \mathbb{E}f_n^2(X_i) \\ \cdot LL\Big(\sum_{i\in I(n)} \mathbb{E}f_n^2(X_i)\Big)\bigg)^{1/2} \text{ i.o. in } n\bigg\}.$$

Now, we can apply Lemma 8.1 to the independent centered real valued random variables $f_n(X_i)$, $i \in I(n)$. Taking there $\gamma = \varepsilon/1 - \varepsilon$, for all large enough n's,

$$\mathbb{P}\bigg\{\sum_{i\in I(n)} f_n(X_i) > (1-\varepsilon)\bigg(2\sum_{i\in I(n)} \mathbb{E}f_n^2(X_i)LL\Big(\sum_{i\in I(n)} \mathbb{E}f_n^2(X_i)\Big)\bigg)^{1/2}\bigg\} \\ \ge \exp\bigg[-(1-\varepsilon)LL\Big(\sum_{i\in I(n)} \mathbb{E}f_n^2(X_i)\Big)\bigg]$$

where it has been used that

$$s_{m_n} \leq \sup_{f \in D} \left(\sum_{i \in I(n)} \mathbb{E} f^2(X_i) \right)^{1/2} + s_{m_{n-1}}$$

$$\leq \left(\frac{1}{1-\varepsilon} \sum_{i \in I(n)} \mathbb{E} f_n^2(X_i) \right)^{1/2} + s_{m_{n-1}}$$

and that the ratio $s_{m_{n-1}}/s_{m_n}$ is small for large $\rho > 1$. Then, this observation yields the conclusion thanks to the independent half of the Borel-Cantelli lemma. Theorem 8.2 is established.

8.2 Hartman-Wintner-Strassen's Law of the Iterated Logarithm

Having described the fundamental LIL of Kolmogorov for sums of independent random variables and its extension to the vector value case, we now turn to the independent and identically distributed (iid) LIL and the results of Hartman-Wintner and Strassen. Let X be a random variable, and, in the rest of the chapter, $(X_i)_{i \in \mathbb{N}}$ will always denote a sequence of independent *copies* of X. The basic normalization sequence is here, *and for the rest of the chapter,*

$$a_n = (2n LLn)^{1/2}$$

which is seen to correspond to the sequence in (8.1) when $\mathbb{E} X^2 = 1$. The LIL of *Hartman and Wintner states that,* if X is a real valued random variable such that $\mathbb{E} X = 0$ and $\mathbb{E} X^2 = \sigma^2 < \infty$, the sequence (a_n) stabilizes the partial sums S_n in such a way that, with probability one,

$$(8.8) \qquad \limsup_{n \to \infty} \frac{S_n}{a_n} = -\liminf_{n \to \infty} \frac{S_n}{a_n} = \sigma .$$

Conversely, if the sequence (S_n/a_n) is almost surely bounded, then $\mathbb{E} X^2 < \infty$ (and $\mathbb{E} X = 0$ trivially by the SLLN). P. Hartman and W. Wintner deduced their result from Kolmogorov's LIL using a (clever) truncation argument. When $\mathbb{E} X^2 < \infty$, one can find a sequence (η_i) of positive numbers tending to 0 such that

$$(8.9) \qquad \sum_i \frac{1}{LLi} \mathbb{P}\{|X| > \eta_i (i/LLi)^{1/2}\} < \infty .$$

Set then, for each i,

$$Y_i = X_i I_{\{|X_i| \leq \eta_i (i/LLi)^{1/2}\}} - \mathbb{E}\left(X_i I_{\{|X_i| \leq \eta_i (i/LLi)^{1/2}\}} \right) ,$$

and $Z_i = X_i - Y_i$. Since the Y_i's are bounded at a level corresponding to the application of Kolmogorov's LIL, (8.1) already gives that

$$\limsup_{n \to \infty} \frac{1}{a_n} \sum_{i=1}^{n} Y_i = \sigma$$

almost surely. The proof then consists in showing that the contribution of the Z_i's is negligible. To this aim, simply observe that by Cauchy-Schwarz's inequality,

$$\frac{1}{a_n} \left| \sum_{i=1}^{n} X_i I_{\{|X_i| > \eta_i (i/LLi)^{1/2}\}} \right|$$

$$\leq \left(\frac{1}{n} \sum_{i=1}^{n} X_i^2 \right)^{1/2} \left(\frac{1}{2LLn} \sum_{i=1}^{n} I_{\{|X_i| > \eta_i (i/LLi)^{1/2}\}} \right)^{1/2} .$$

The first root on the right of this inequality defines an almost surely bounded (convergent!) sequence by the SLLN and $\mathbb{E}X^2 < \infty$; the second root converges to 0 by Kronecker's lemma (cf. e.g. [Sto]) and (8.9). Since the centerings in Z_i are similarly taken into account, it follows that

$$\lim_{n \to \infty} \frac{1}{a_n} \sum_{i=1}^{n} Z_i = 0 \quad \text{almost surely}$$

and therefore (8.8) holds.

The necessity of $\mathbb{E}X^2 < \infty$ can be obtained from a simple symmetry argument. By symmetrization, we may and do assume X to be symmetric. Let $c > 0$ and define

$$\overline{X} = X I_{\{|X| \leq c\}} - X I_{\{|X| > c\}} .$$

Since X is symmetric, $\overline{X}$ has the same distribution as X. Now, assume that the sequence (S_n/a_n) is almost surely bounded. By the zero-one law, there is a finite number M such that, with probability one, $\limsup_{n \to \infty} |S_n|/a_n = M$. Now $2X I_{\{|X| \leq c\}} = X + \overline{X}$ and since $\overline{X}$ has the same law as X,

$$2 \limsup_{n \to \infty} \frac{1}{a_n} \left| \sum_{i=1}^{n} X_i I_{\{|X_i| \leq c\}} \right| \leq 2M .$$

almost surely. By (8.8), since $\mathbb{E}(X^2 I_{\{|X| \leq c\}}) < \infty$ and X is symmetric, it follows that

$$\mathbb{E}(X^2 I_{\{|X| \leq c\}}) \leq M^2 .$$

Letting c tend to infinity implies that $\mathbb{E}X^2 < \infty$.

While P. Hartman and A. Wintner used the rather deep result of A. N. Kolmogorov, the case of iid random variables should a priori appear as easier. Since then, simpler proofs of (8.8), which even produce more, have been obtained. As an illustration, we would like to intuitively describe in a direct way why the lim sup in (8.8) should be *finite* when $\mathbb{E}X = 0$ and $\mathbb{E}X^2 < \infty$. The idea is based on randomization by Rademacher random variables, a tool extensively used throughout this book. It explains rather easily the common steps and features of LIL's results such as the fundamental use of exponential bounds of Gaussian type (Lemma 1.6 in Kolmogorov's LIL) and the study of (S_n) through blocks of exponential sizes. It suffices for this (modest) purpose to treat the case of a symmetric random variable so that we may assume as usual that (X_i) has the same distribution as $(\varepsilon_i X_i)$ where (ε_i) is a Rademacher sequence which is independent of (X_i). By Lévy's inequality (2.6) for sums of independent symmetric random variables and the Borel-Cantelli lemma, it is enough to find a finite number M such that

$$\sum_n \mathbb{P}\left\{ \left| \sum_{i=1}^{2^n} \varepsilon_i X_i \right| > M a_{2^n} \right\} < \infty.$$

Since $\mathbb{E}X^2 < \infty$, X^2 satisfies the law of large numbers and hence, by the Borel-Cantelli lemma again (independent case),

$$\sum_n \mathbb{P}\left\{ \sum_{i=1}^{2^n} X_i^2 > 2^{n+1} \mathbb{E}X^2 \right\} < \infty.$$

Now, we simply write that, for every n,

$$\mathbb{P}\left\{ \left| \sum_{i=1}^{2^n} \varepsilon_i X_i \right| > M a_{2^n} \right\} \leq \mathbb{P}\left\{ \sum_{i=1}^{2^n} X_i^2 > 2^{n+1} \mathbb{E}X^2 \right\}$$

$$+ \mathbb{P}\left\{ \left| \sum_{i=1}^{2^n} \varepsilon_i X_i \right| > M a_{2^n}, \ \sum_{i=1}^{2^n} X_i^2 \leq 2^{n+1} \mathbb{E}X^2 \right\}.$$

The classical subgaussian estimate (4.1) can be used conditionally on the sequence (X_i) to show that the second probability on the right of the preceding inequality is less than

$$\int_{\left\{ \sum_{i=1}^{n} X_i^2 \leq 2^{n+1} \mathbb{E}X^2 \right\}} 2 \exp\left(-M^2 2^n LL2^n \Big/ \sum_{i=1}^{2^n} X_i^2 \right) d\mathbb{P}$$

$$\leq 2 \exp\left(-M^2 LL2^n / 2\mathbb{E}X^2 \right).$$

If we then choose $M^2 > 2\mathbb{E}X^2$, the claim is established.

This simple approach, which reduces in a sense the iid LIL to the SLLN through Gaussian exponential estimates, can be pushed further in order to show the necessity of $\mathbb{E}X^2 < \infty$ when the sequence (S_n/a_n) is almost surely bounded. One can use the converse subgaussian inequality (4.2). Alternatively, and without going into the details, it is not difficult to see that if (g_i) is an orthogaussian sequence which is independent of (X_i), the two non-random lim sup's

$$\limsup_{n \to \infty} \frac{1}{a_n} \left| \sum_{i=1}^{n} X_i \right| \quad \text{and} \quad \limsup_{n \to \infty} \frac{1}{a_n} \left| \sum_{i=1}^{n} g_i X_i \right|$$

are equivalent. Independence and the stability properties of (g_i) expressed for example by

$$\limsup_{n \to \infty} \frac{|g_n|}{(2 \log n)^{1/2}} = 1 \quad \text{almost surely}$$

can then easily be used to check the necessity of $\mathbb{E}X^2 < \infty$ (cf. [L-T2]).

These rather easy ideas describe some basic facts in the study of the LIL such as exponential estimates of Gaussian type, blocking arguments, the connection of the LIL with the law of large numbers (for squares) and of course the central limit theorem by the introduction of Gaussian randomization. In fact, the LIL can be thought of as some almost sure form of the central limit theorem. The framework of these elementary observations will lead later to the infinite dimensional LIL.

The preceding sketchy proof of Hartman-Wintner's LIL of course only provides qualitative results and not the exact value of the lim sup in (8.8). [Note however, as was pointed out to us by D. Stroock, that if ones takes for granted the exact value of the limsup for *bounded* random variables (as follows from Kolmogorov's result for instance), then the preceding simple approach can yield (8.8) for every mean zero square integrable random variable by simply breaking it into a bounded part and a part with arbitrarily small variance.] Simple proofs of (8.8) have been given in the literature; they include the more precise and so-called *Strassen's form of the* LIL which states that, if and only if $\mathbb{E}X = 0$ and $\mathbb{E}X^2 < \infty$,

$$(8.10) \qquad\qquad \lim_{n \to \infty} d\left(\frac{S_n}{a_n}, \ [-\sigma, \sigma] \right) = 0$$

and

$$(8.11) \qquad\qquad C\left(\frac{S_n}{a_n} \right) = [-\sigma, \sigma]$$

almost surely, where $d(x, A) = \inf\{|x - y|; \ y \in A\}$ is the distance of the point x to the set A and where $C(x_n)$ denotes the set of limit points of the sequence (x_n), i.e. $C(x_n) = \{x \in \mathbb{R}; \ \liminf_{n \to \infty} |x_n - x| = 0\}$. (8.10) and $C(S_n/a_n) \subset [-\sigma, \sigma]$ follow rather easily from the LIL of Hartman-Wintner. The full property (8.11) is more delicate and various arguments can be used to establish it; we will obtain (8.10) and (8.11) in the more general context of Banach space valued random variables below. Strassen's approach used Brownian motion and the Skorohod embedding of a sequence of iid random variables in the Brownian paths.

Our objective in the sequel of this chapter will be to investigate the iid LIL for vector valued random variables. Until the end of the chapter, we deal with Radon variables and even, for more convenience, with separable Banach spaces, although various conclusions still hold in our usual more general setting; these will be indicated in remarks. For Radon random variables, the picture is probably the most complete and satisfactory and we adopt this framework in order not to obscure the main scheme.

Therefore, let B denote a separable Banach space. We start by describing what can be understood by a LIL for independent and identically distributed Banach space valued random variables. Let X be a Borel random variable with values in B, (X_i) a sequence of independent copies of X. As usual, $S_n = X_1 + \cdots + X_n$, $n \geq 1$. According to Hartman-Wintner's LIL, we can say that X satisfies the LIL with respect to the classical normalizing sequence $a_n = (2nLLn)^{1/2}$ if the non-random limit (zero-one law)

$$(8.12) \qquad \Lambda(X) = \limsup_{n \to \infty} \frac{\|S_n\|}{a_n}$$

is finite. (If X is non-degenerate, $\Lambda(X) > 0$ by the scalar case.) We will actually define this property as the *bounded* LIL. Indeed, we might as well say that X satisfies the LIL whenever the sequence (S_n/a_n) is almost surely relatively compact in B since this means the same in finite dimensional spaces. We will say then that X satisfies the *compact* LIL and it will turn out that in infinite dimension bounded and compact LIL are *not* equivalent. Actually, Strassen's formulation (8.10) and (8.11) even suggests a third definition: X satisfies the LIL if there is a compact convex symmetric set K in B such that, almost surely,

$$(8.13) \qquad \lim_{n \to \infty} d\left(\frac{S_n}{a_n}, K\right) = 0$$

and

$$(8.14) \qquad C\left(\frac{S_n}{a_n}\right) = K$$

where $d(x, K) = \inf\{\|x - y\| \; ; \; y \in K\}$ and where $C(S_n/a_n)$ denotes the set of the cluster points of the sequence (S_n/a_n). It is a non-trivial result that the compact LIL and this definition actually coincide. Before describing precisely this result, we would like to study what the limit set K should be. It will turn out to be the unit ball of the so-called reproducing kernel Hilbert space associated with the covariance structure of X. We sketch its construction and properties, some of which go back to the Gaussian setting as described in Chapter 3.

X is therefore a fixed Borel random variable on some probability space $(\Omega, \mathcal{A}, \mathbb{P})$ with values in the separable Banach space B. Recall that the separability allows to assume $\mathcal{A}$ to be countably generated and thus $L_2(\Omega, \mathcal{A}, \mathbb{P})$ to be separable. Suppose that for all f in B', $\mathbb{E}f(X) = 0$ and $\mathbb{E}f^2(X) < \infty$. Let us observe, as a remark, that these hypotheses are natural in the context of the LIL since if for example X satisfies the bounded LIL, then, for each f in B', $(f(S_n/a_n))$ is almost surely bounded and therefore $\mathbb{E}f(X) = 0$ and $\mathbb{E}f^2(X) < \infty$ by the scalar case. Under these hypotheses,

$$(8.15) \qquad \sigma(X) = \sup_{\|f\| \leq 1} \left(\mathbb{E}f^2(X)\right)^{1/2} < \infty.$$

Indeed, if we consider the operator $A = A_X$ which is defined as $A : B' \to L_2 = L_2(\Omega, \mathcal{A}, \mathbb{P})$, $Af = f(X)$, then $\|A\| = \sigma(X)$ and A is bounded by an easy closed graph argument. Let $A^* = A_X^*$ denote the adjoint of A. First note that since X defines a Radon random variable, A^* actually maps L_2 into $B \subset B''$ (cf. Section 2.1). Indeed, there exists a sequence (K_n) of compact sets in B such that $\mathbb{P}\{X \notin K_n\} \to 0$. If ξ is in L_2, $A^*(\xi I_{\{X \in K_n\}})$ belongs to B since it can be identified with the expectation (in the strong sense) $\mathbb{E}(\xi X I_{\{X \in K_n\}})$. Now $\mathbb{E}(\xi X I_{\{X \in K_n\}})$ converges to $\mathbb{E}(\xi X)$ (weak integral) in B'' since

$$\sup_{\|f\| \leq 1} f\big(\mathbb{E}(\xi X I_{\{X \notin K_n\}})\big) \leq \sigma(X)\big(\mathbb{E}(\xi^2 I_{\{X \notin K_n\}})\big)^{1/2} \to 0.$$

Hence $A^*\xi = \mathbb{E}(\xi X)$ belongs to B.

On the image $A^*(L_2) \subset B$ of L_2 by A^*, consider the scalar product $\langle \cdot, \cdot \rangle_X$ transferred from L_2: if $\xi, \zeta \in L_2$,

$$\langle A^*\xi, A^*\zeta \rangle_X = \langle \xi, \zeta \rangle_{L_2} = \int \xi\zeta \, d\mathbb{P}.$$

Denote by $H = H_X$ the (separable) Hilbert space $A^*(L_2)$ equipped with $\langle \cdot, \cdot \rangle_X$. H is called the *reproducing kernel Hilbert space* associated with the covariance structure of X. The word "reproducing" stems from the fact that H reproduces the covariance of X in the sense that for f, g in B', if $x = A^*(g(X)) \in H$, $f(x) = \mathbb{E}f(X)g(X)$. In particular, if X and Y are random variables with the same covariance structure, i.e. $\mathbb{E}f(X)g(X) = \mathbb{E}f(Y)g(Y)$ for all f, g in B', this reproducing property implies that $H_X = H_Y$. Note that since $A(B')^\perp = \operatorname{Ker} A^*$, we also have that H is the completion, with respect to the scalar product $\langle \cdot, \cdot \rangle_X$, of the image of B' by the composition $S = A^*A : B \to B'$. Observe further that for any x in H,

$$(8.16) \qquad \|x\| \leq \sigma(X)\langle x, x \rangle_X.$$

Denote by $K = K_X$ the closed unit ball of H, i.e. $K = \{x \in B;\ x = \mathbb{E}(\xi X),\ \|\xi\|_2 \leq 1\}$, which thus defines a bounded convex symmetric set in B. By the Hahn-Banach theorem, we also have that $K = \{x \in B;\ f(x) \leq \|f(X)\|_2 \text{ for all } f \text{ in } B'\}$, and, by separability, this can be achieved by taking only a (well-chosen) *sequence* (f_k) in B'. As the image of the unit ball of L_2 by A^*, K is weakly compact and therefore also closed for the topology of the norm on B. K is separable in B by (8.16). Moreover, it is easily verified that for any f in B',

$$\|f(X)\|_2 = \sup_{x \in K} f(x), \quad \sigma(X) = \sup_{x \in K} \|x\|.$$

Although K is weakly compact, it is not always compact. The next easy lemma describes for further references equivalent instances for K to be compact.

Lemma 8.4. *The following are equivalent:*

(i) K *is compact;*

(ii) A *(resp. A^*) is compact;*

(iii) $S = A^*A$ *is compact;*

(iv) *the covariance function $T(f,g) = \mathbb{E}f(X)g(X)$ is weakly sequentially continuous;*

(v) *the family of real valued random variables $\{f^2(X);\ f \in B',\ \|f\| \le 1\}$ is uniformly integrable.*

Proof. (i) and (ii) are clearly equivalent and imply (iii). To see that (iv) holds under (iii), it suffices to show that $\|f_n(X)\|_2 \to 0$ when $f_n \to 0$ weakly in B'. By the uniform boundedness principle, we may assume that $\|f_n\| \le 1$ for every n. The compactness of S ensures that we can extract from the sequence (x_n) defined by $x_n = \mathbb{E}(Xf_n(X))$ a subsequence, still indexed by n, which converges to some x. Then,

$$\mathbb{E}f_n^2(X) = f_n(x_n) \le \|x_n - x\| + |f_n(x)| \to 0\,.$$

Assume (v) is not satisfied. Then, there exist $\varepsilon > 0$ and a sequence (c_n) of positive numbers increasing to infinity such that for every n

$$\sup_{\|f\|\le 1} \int_{\{\|X\|>c_n\}} f^2(X)\,d\mathbb{P} \ge \sup_{\|f\|\le 1} \int_{\{|f(X)|>c_n\}} f^2(X)\,d\mathbb{P} > \varepsilon\,.$$

Hence, for every n, one can find f_n, $\|f_n\| \le 1$, such that

$$\int_{\{\|X\|>c_n\}} f_n^2(X)\,d\mathbb{P} > \varepsilon\,.$$

Extract then from the sequence (f_n) in the unit ball of B' a weakly convergent subsequence, still denoted as (f_n), and convergent to some f. By (iv), $f_n(X) \to f(X)$ in L_2 and this clearly yields a contradiction since

$$\lim_{n\to\infty} \int_{\{\|X\|>c_n\}} f^2(X)d\mathbb{P} = 0\,.$$

Finally, (v) easily implies (ii); indeed, if (f_n) is a sequence in the unit ball of B', for some subsequence and some f, $f_n \to f$ weakly, so $f_n(X) \to f(X)$ almost surely and hence in L_2 by uniform integrability; A is therefore compact. The proof of Lemma 8.4 is complete.

Note that when $\mathbb{E}\|X\|^2 < \infty$ (in the case of Gaussian random vectors for instance), K is compact.

As simple examples, consider the case where $B = \mathbb{R}^N$ and the covariance matrix of X is the identity matrix; K is then simply the Euclidean unit ball of $\mathbb{R}^N$. When X follows the Wiener distribution on $C[0,1]$, H can be identified with the so-called Cameron-Martin Hilbert space of the absolutely continuous elements x in $C[0,1]$ such that $x(0) = 0$ and $\int_0^1 x'(t)^2 dt < \infty$, and K is known in this case as Strassen's limit set ([C-M], [St1]).

Having described the natural limit set in (8.13) and (8.14), we now present the theorem, due to J. Kuelbs, connecting the definition of the compact LIL with (8.13) and (8.14).

Theorem 8.5. *Let X be a Borel random variable with values in a separable Banach space B. If the sequence (S_n/a_n) is almost surely relatively compact in B, then, with probability one,*

$$\lim_{n \to \infty} d\left(\frac{S_n}{a_n}, K\right) = 0 \quad and \quad C\left(\frac{S_n}{a_n}\right) = K$$

where $K = K_X$ is the unit ball of the reproducing kernel Hilbert space associated with the covariance structure of X and K is compact. Conversely, if the preceding holds for some compact set K, then X satisfies the compact LIL and $K = K_X$.

According to this theorem, when we speak of the compact LIL we mean one of the equivalent properties of this statement.

Proof. As we have seen, when X satisfies the bounded LIL, which is always the case under one of the properties of Theorem 8.5, $K = K_X$ is well defined. Let us first show that, with probability one, $C(S_n/a_n) \subset K$. As was observed in the definition of K, there is a sequence (f_k) in B' such that a point x belongs to K as soon as

(8.17) $$f_k(x) \le \|f_k(X)\|_2$$

for every k. Denote by Ω_0 the set of full probability (by the scalar LIL) of the ω's such that for every k

$$\limsup_{n \to \infty} \frac{|f_k(S_n(\omega))|}{a_n} \le \|f_k(X)\|_2 .$$

So if $x \in C(S_n(\omega)/a_n)$ and $\omega \in \Omega_0$, x clearly satisfies (8.17) and therefore belongs to K. This first property easily implies

(8.18) $$\lim_{n \to \infty} d\left(\frac{S_n}{a_n}, K\right) = 0$$

with probability one. Indeed, if this is not the case, one can find, by the relative compactness of the sequence (S_n/a_n), a subsequence of (S_n/a_n) converging to some point in the complement of K and this is impossible as we have just seen.

We are thus left with the proof that $C(S_n/a_n) = K$. To this aim, it suffices, by density, to show that any x in K belongs almost surely to $C(S_n/a_n)$.

First, let us assume that B is finite dimensional. Since the covariance matrix of X is symmetric positive definite, it may be diagonalized in some orthonormal basis. We are therefore reduced to the case where B is Hilbertian and K is its unit ball. Let then x be in B with $|x| = 1$ and let $\varepsilon > 0$. By (8.18), for n large enough, $|S_n/a_n|^2 \leq 1 + \varepsilon$. By Hartman-Wintner's LIL (8.8), along a subsequence,

$$\left\langle \frac{S_n}{a_n}, x \right\rangle \geq \|\langle x, X \rangle\|_2 - \varepsilon = 1 - \varepsilon.$$

Hence, along this subsequence and for n large,

$$\left| \frac{S_n}{a_n} - x \right|^2 = \left| \frac{S_n}{a_n} \right|^2 + |x|^2 - 2 \left\langle \frac{S_n}{a_n}, x \right\rangle$$
$$\leq 1 + \varepsilon + 1 - 2 + 2\varepsilon = 3\varepsilon$$

and therefore $x \in C(S_n/a_n)$ almost surely. To reach the interior points of K, we increase the dimension and consider the random variable in $B \times \mathbb{R}$ given by $Y = (X, \varepsilon)$ where ε is a Rademacher variable independent of X. Let x be in K with $|x| = \theta < 1$; then $y = (x, (1 - \theta^2)^{1/2})$ belongs to the unit sphere of $B \times \mathbb{R}$ and thus, by the preceding step, to the cluster set associated to Y. By projection, $x \in C(S_n/a_n)$.

Now, we complete the proof of Theorem 8.5 and use this finite dimensional result to get the full conclusion concerning the cluster set $C(S_n/a_n)$. When X satisfies the LIL, it also satisfies the strong law of large numbers and therefore $\mathbb{E}\|X\| < \infty$. There exists an increasing sequence $(\mathcal{A}_N)$ of finite σ-algebras of $\mathcal{A}$ such that $X^N = \mathbb{E}^{\mathcal{A}_N} X$ converges almost surely and in $L_1(B)$ to X. Note that if (S_n/a_n) is almost surely relatively compact, the property (iv) of Lemma 8.4 is fulfilled; indeed, if $f_k \to 0$ weakly, by compactness,

$$\lim_{k \to \infty} \sup_n \left| f_k \left(\frac{S_n}{a_n} \right) \right| = 0$$

with probability one. By (8.8), it follows that $\|f_k(X)\|_2 \to 0$ which gives (iv). By Lemma 8.4, K is therefore compact, or equivalently, the family $\{f^2(X); f \in B', \|f\| \leq 1\}$ is uniformly integrable. Therefore, there exists (Lemma of La Vallé-Poussin, cf. e.g. [Me]) a positive convex function ψ on $\mathbb{R}_+$ with $\lim_{t \to \infty} \psi(t)/t = \infty$ such that $\sup_{\|f\| \leq 1} \mathbb{E}\psi(f^2(X)) < \infty$. By Jensen's inequality, it follows that the family $\{f^2(X - X^N); \|f\| \leq 1, N \in \mathbb{N}\}$ is also uniformly integrable. This can be used to show that $\sigma(X - X^N) \to 0$ when $N \to \infty$ (where $\sigma(\cdot)$ is defined in (8.15)). Let now x in $K : x = \mathbb{E}(\xi X)$, $\|\xi\|_2 \leq 1$. If $x^N = \mathbb{E}(\xi X^N)$, $\|x - x^N\| \leq \sigma(X - X^N) \to 0$. Furthermore, by (8.18), $\Lambda(X) \leq \sup_{x \in K} \|x\| = \sigma(X)$, and since the X^N's are finite dimensional they also satisfy the compact LIL and therefore $\Lambda(X - X^N) \leq \sigma(X - X^N)$

for every N. Now, we simply write, by the triangle inequality, that for every N,

$$\liminf_{n\to\infty}\left\|\frac{S_n}{a_n} - x\right\| \le \liminf_{n\to\infty}\left\|\frac{S_n^N}{a_n} - x^N\right\| + \Lambda(X - X^N) + \|x - x^N\|$$

$$\le \liminf_{n\to\infty}\left\|\frac{S_n^N}{a_n} - x^N\right\| + 2\sigma(X - X^N)$$

where S_n^N are the partial sums associated to X^N. By the previous step in finite dimension, for each N,

$$\liminf_{n\to\infty}\left\|\frac{S_n^N}{a_n} - x^N\right\| = 0$$

almost surely. Letting N tend to infinity then shows that $x \in C(S_n/a_n)$ which completes the proof of Theorem 8.5.

The preceding discussion and Theorem 8.5 present the definitions of Hartman-Wintner-Strassen's LIL for Banach space valued random variables. We now would like to turn to the crucial question of knowing *when* a random variable X with values in a Banach space B satisfies the bounded or compact LIL in terms of minimal conditions, depending if possible on the distribution of X only.

If B is the line or, more generally, a finite dimensional space, X satisfies the bounded or compact LIL if and only if $\mathbb{E}X = 0$ and $\mathbb{E}\|X\|^2 < \infty$. However, already in Hilbert space, while these conditions are sufficient, the integrability condition $\mathbb{E}\|X\|^2 < \infty$ is no longer necessary. It happens conversely that in some spaces, bounded mean zero random variables do not satisfy the LIL. Moreover, examples disprove the equivalence between bounded and compact LIL in the infinite dimensional setting. All these examples will actually become clear with the final characterization. They however pointed out historically the difficulty in finding what this characterization should be. In particular, the issue is based on a careful examination of the necessary conditions for a Banach space valued random variable to satisfy the LIL which we now would like to describe.

Assume first that X satisfies the bounded LIL in B. Then clearly, for each f in B', $f(X)$ satisfies the scalar LIL and thus $\mathbb{E}f(X) = 0$ and $\mathbb{E}f^2(X) < \infty$. These *weak* integrability conditions are complemented by a necessary integrability property on the *norm*; indeed, it is necessary that the sequence (X_n/a_n) is bounded almost surely and thus, by independence, the Borel-Cantelli lemma and identical distribution, for some finite M,

$$\sum_n \mathbb{P}\{\|X\| > Ma_n\} < \infty.$$

As is easily seen, this turns out to be equivalent to the integrability condition

$$\mathbb{E}\big(\|X\|^2/LL\|X\|\big) < \infty.$$

These are the best moment conditions which can be deduced from the bounded LIL. However, as we mentioned it, there are almost surely bounded (mean zero) random variables which do not satisfy the LIL. This unfortunate fact forces, in order to expect some characterization, to complete the preceding integrability conditions, which depend only on the distribution of X, by some condition involving the laws of the partial sums S_n instead of X only. As we will see, this can be avoided in some spaces, but it is necessary to proceed along these lines in general as actually could have been expected from the previous chapters. The third necessary condition is then simply (and trivially) that the sequence (S_n/a_n) should be bounded in probability.

In finite dimension, the weak L_2 integrability of course implies the strong L_2 integrability, and therefore $\mathbb{E}(\|X\|^2/LL\|X\|) < \infty$, as well as the stochastic boundedness of (S_n/a_n) (as is easily seen, for example, from the central limit theorem). It is remarkable that this easily obtained set of necessary conditions is also sufficient for X to satisfy the bounded LIL. Before stating this characterization, let us complete the discussion on necessary conditions by the case of the compact LIL. We keep that $\mathbb{E}(\|X\|^2/LL\|X\|) < \infty$. By Theorem 8.5, $K = K_X$ should be compact, or equivalently $\{f^2(X);\ f \in B',\ \|f\| \leq 1\}$ should be uniformly integrable (this can also be proved directly as is clear from the last part of the proof of Theorem 8.5). Finally, under the compact LIL, it is necessary that the sequence (S_n/a_n) is not only bounded in probability, but converges to 0; indeed, the sequence of the laws of S_n/a_n is necessarily tight with 0 as only possible limit point since $\mathbb{E}f(X) = 0$, $\mathbb{E}f^2(X) < \infty$ for all f in B'. Let us note that the stochastic boundedness (and a fortiori convergence to 0) of the sequence (S_n/a_n) also contains the fact that X is centered. (To see this, use the analog of Lemma 10.1 for the normalization a_n.)

Now, we can present the characterization of random variables satisfying the bounded or compact LIL. As is typical in Probability in Banach spaces, it reduces in a sense, under necessary and natural moment conditions, the *almost sure* behavior of the sequence (S_n/a_n) to its behavior in *probability*.

Theorem 8.6. *Let X be a Borel random variable with values in a separable Banach space B. In order that X satisfy the bounded LIL, it is necessary and sufficient that the following conditions are fulfilled:*

(i) $\mathbb{E}(\|X\|^2/LL\|X\|) < \infty$;

(ii) for each f in B', $\mathbb{E}f(X) = 0$ and $\mathbb{E}f^2(X) < \infty$;

(iii) the sequence (S_n/a_n) is bounded in probability.

In order that X satisfy the compact LIL, it is necessary and sufficient that (i) holds and that (ii) and (iii) are replaced by

(ii') $\mathbb{E}X = 0$ and $\{f^2(X);\ f \in B',\ \|f\| \leq 1\}$ is uniformly integrable;

(iii') $S_n/a_n \to 0$ in probability.

Proof. Necessity has been discussed above. The proof of the sufficiency for the bounded LIL is the main point of the whole theorem. We will show indeed that for some numerical constant M, for all symmetric random variables X satisfying (i), we have

$$(8.19) \qquad \Lambda(X) = \limsup_{n \to \infty} \frac{\|S_n\|}{a_n} \le M\big(\sigma(X) + L(X)\big)$$

where $\sigma(X) = \sup_{\|f\| \le 1} (\mathbb{E}f^2(X))^{1/2}$ and

$$L(X) = \limsup_{n \to \infty} \frac{1}{a_n} \mathbb{E} \Big\| \sum_{i=1}^{n} X_i I_{\{\|X_i\| \le a_n\}} \Big\|.$$

Since $\sigma(X) < \infty$ under (ii) ((8.15)) and since (iii) implies that $L(X) < \infty$ by Lemma 7.2, this inequality (8.19) contains the bounded LIL, at least for symmetric random variables but actually also in general by symmetrization and Lemma 7.1. From (8.19) also follows the compact version. Indeed, by Lemma 7.2, $L(X)$ can be chosen to be 0 by (iii′) and, by symmetrization and Lemma 7.1, the inequality also holds in this form (with $L(X) = 0$) for non-necessarily symmetric random variables satisfying (i) and (iii′). This estimate applied to quotient norms by finite dimensional subspaces then yields the conclusion. More precisely, if F denotes a finite dimensional subspace of B and $T = T_F$ the quotient map $B \to B/F$, we get from (8.19) that $\Lambda(T(X)) \le M\sigma(T(X))$. Under (ii′), $\sigma(T(X))$ can be made arbitrarily small with large enough F (show for example, as in the proof of Theorem 8.5, that $\sigma(X - X^N) \to 0$ where $X^N = \mathbb{E}^{\mathcal{A}_N} X$). The sequence (S_n/a_n) then appears as being arbitrarily close to some bounded set in the finite dimensional subspace F, and is therefore relatively compact (Lemma 2.2). Hence X satisfies the compact LIL under (i), (ii′) and (iii′).

Thus, we are left with the proof of (8.19). To this aim, we use the isoperimetric approach as developed for example in the preceding chapter on the SLLN. We intend more precisely to apply Theorem 7.5. In order to verify the first set of conditions there, we employ Lemmas 7.6 and 7.8. The integrability condition $\mathbb{E}(\|X\|^2/LL\|X\|) < \infty$ is equivalent to saying that for every $\varepsilon > 0$

$$\sum_n 2^n \mathbb{P}\{\|X\| > \varepsilon a_{2^n}\} < \infty.$$

Let $\varepsilon > 0$ be fixed. Setting together the conclusions of Lemmas 7.6 and 7.8 we see that (taking $q = 2K_0$) there exists a sequence of integers (k_n) such that $\sum_n 2^{-k_n} < \infty$ for which

$$\sum_n \mathbb{P}\Big\{ \sum_{r=1}^{k_n} \|X_{2^{n-1}}^{(r)}\| > 5\varepsilon a_{2^n} \Big\} < \infty$$

where $X_{2^{n-1}}^{(r)}$ is the r-th largest element of the sample $(\|X_i\|)_{i \le 2^{n-1}}$. L in Theorem 7.5 is less than $L(X)$ (contraction principle) and, for each n, $\sigma_n^2 \le$

$2^n \sigma(X)^2$ so that (7.4) is satisfied with, for example, $\delta = \sigma(X)$. The conclusion of Theorem 7.5, with $q = 2K_0$, is then

$$\sum_n \mathbb{P}\{\|S_{2^{n-1}}\| > 10^2 [\varepsilon + 2K_0(\sigma(X) + L(X) + (5\varepsilon L(X))^{1/2})] a_{2^n}\} < \infty.$$

Since $\varepsilon > 0$ is arbitrary and $a_{2^n} \sim \sqrt{2} a_{2^{n-1}}$, inequality (8.19) follows from the Borel-Cantelli lemma and the maximal inequality of Lévy (2.6). Theorem 8.6 is thus established.

While Theorem 8.6 provides a rather complete characterization of random variables satisfying the LIL, hypotheses on the distribution of the partial sums rather than only on the variable have to be used. It is worthwhile to point out at this stage that in a special class of Banach spaces, it is possible to get rid of these assumptions and to state the characterization in terms of the moment conditions (i) and (ii) (or (ii')) only. Anticipating on the next chapter as we did with the law of large numbers, say that a Banach space B is of *type* 2 if there is a constant C such that for any finite sequence (Y_i) of independent mean zero random variables with values in B, we have

$$\mathbb{E}\left\|\sum_i Y_i\right\|^2 \leq C \sum_i \mathbb{E}\|Y_i\|^2.$$

Hilbert spaces are clearly of type 2; further examples are discussed in Chapter 9. In type 2 spaces the integrability condition $\mathbb{E}(\|X\|^2/LL\|X\|) < \infty$ implies, with $\mathbb{E}X = 0$, that $S_n/a_n \to 0$ in probability, hence the nicer form of Theorem 8.6 in this case. Let us prove this implication.

Lemma 8.7. *Let X be a mean zero random variable with values in a type 2 Banach space B such that $\mathbb{E}(\|X\|^2/LL\|X\|) < \infty$ Then $S_n/a_n \to 0$ in probability.*

Proof. We show that if X is symmetric and $\mathbb{E}(\|X\|^2/LL\|X\|) < \infty$, then $\mathbb{E}\|S_n\|/a_n \to 0$ which, by Lemma 6.3, implies the lemma. For each n,

$$\mathbb{E}\|S_n\| \leq \mathbb{E}\left\|\sum_{i=1}^n X_i I_{\{\|X_i\| \leq a_n\}}\right\| + n\mathbb{E}(\|X\| I_{\{\|X\| > a_n\}}).$$

A simple integration by parts shows that, under $\mathbb{E}(\|X\|^2/LL\|X\|) < \infty$,

$$\lim_{n \to \infty} \frac{n}{a_n} \mathbb{E}(\|X\| I_{\{\|X\| > a_n\}}) = 0.$$

By the type 2 inequality (and symmetry),

$$\frac{1}{a_n} \mathbb{E}\left\|\sum_{i=1}^n X_i I_{\{\|X_i\| \leq a_n\}}\right\| \leq \left(\frac{C}{2LLn} \mathbb{E}(\|X\|^2 I_{\{\|X\| \leq a_n\}})\right)^{1/2}.$$

For each $t > 0$, the square of the right hand side of this inequality is seen to be smaller than

$$\frac{Ct^2}{2LLn} + \frac{C}{2LLn}\mathbb{E}\left(\|X\|^2 I_{\{t<\|X\|\leq a_n\}}\right) \leq \frac{Ct^2}{2LLn} + C\mathbb{E}\left(\frac{\|X\|^2}{LL\|X\|}I_{\{\|X\|>t\}}\right).$$

Letting n, and then t, go to infinity concludes the proof.

As announced, Lemma 8.7 implies the next corollary.

Corollary 8.8. *Let X be a Borel random variable with values in a separable type 2 Banach space B. Then X satisfies the bounded (resp. compact) LIL if and only if $\mathbb{E}(\|X\|^2/LL\|X\|) < \infty$ and $\mathbb{E}f(X) = 0$, $\mathbb{E}f^2(X) < \infty$ for all f in B' (resp. $\{f^2(X);\ f \in B',\ \|f\| \leq 1\}$ is uniformly integrable).*

Remark 8.9. Theorem 8.6 has been presented in the context of Radon random variables. However, its proof clearly indicates some possible extensions to some more general settings. This is in particular completely obvious for the bounded version which, as in the case of Kolmogorov's LIL, does not require any approximation argument. With some precautions, extensions of the compact LIL can also be imagined. We leave this to the interested reader.

8.3 On the Identification of the Limits

In this last paragraph, we would like to describe various results and examples concerning the limits of the sequence (S_n/a_n) in the bounded form of the LIL for Banach space valued random variables. We learned from Theorem 8.5 that when the sequence (S_n/a_n) is almost surely relatively compact in B, then, with probability one,

$$(8.20) \qquad \lim_{n\to\infty} d\left(\frac{S_n}{a_n}, K\right) = 0$$

and

$$(8.21) \qquad C\left(\frac{S_n}{a_n}\right) = K$$

where $K = K_X$ is the unit ball of the reproducing kernel Hilbert space associated with the covariance structure of X and K is compact in this case. In particular also,

$$(8.22) \qquad \Lambda(X) = \limsup_{n\to\infty} \frac{\|S_n\|}{a_n} = \sigma(X) = \sup_{\|f\|\leq 1}\left(\mathbb{E}f^2(X)\right)^{1/2}$$

(recall $\sigma(X) = \sup_{x\in K}\|x\|$). One might now be interested in knowning if these properties still hold, or what they become, when for example X only satisfies the bounded LIL and not the compact one, or even, for (8.21), when X simply satisfies $\mathbb{E}f(X) = 0$ and $\mathbb{E}f^2(X) < \infty$ for all f in B' (in order for K to be well defined). To put the question in a clearer perspective, let us mention the example (cf. [Kue6]) of a bounded random variable satisfying the bounded LIL but for which the cluster set $C(S_n/a_n)$ is empty. Further examples of pathological situations have been observed in the literature. Here, we would like to briefly describe some positive results as well as some open problems.

We start with the remarkable results of K. Alexander [Ale3] on the cluster set. Let X be a Borel random variable with values in a separable Banach space B such that $\mathbb{E}f(X) = 0$ and $\mathbb{E}f^2(X) < \infty$ for every f in B'. As we have seen in the first part of the proof of Theorem 8.5, almost surely $C(S_n/a_n) \subset K$ where $K = K_X$. From an easy zero-one law, it can be shown that the cluster set $C(S_n/a_n)$ is almost surely non-random. It can be K, and can also be the empty set as alluded to above. As a main result, it is shown in [Ale3] that $C(S_n/a_n)$ can actually only be empty or αK for some α in $[0, 1]$, and examples are given in [Ale4] showing that every value of α can indeed occur. Moreover, a series condition involving the laws of the partial sums S_n determines the value of α. More precisely, we have the following theorem which we state without proof refering to [Ale3].

Theorem 8.10. *Let X be a Borel random variable with values in a separable Banach space B such that $\mathbb{E}f(X) = 0$ and $\mathbb{E}f^2(X) < \infty$ for every f in B'. Let*

$$\alpha^2 = \sup\left\{\beta \geq 0; \ \sum_n n^{-\beta}\mathbb{P}\{\|S_m/a_m\| < \varepsilon \ \text{for some} \ 2^{n-1} < m \leq 2^n\}\right.$$

$$\left. = \infty \ \text{for all} \ \varepsilon > 0\right\}$$

whenever this set is not empty. Then $C(S_n/a_n) = \alpha K$, or $\emptyset$ when this set is empty. In particular, $\alpha = 1$ when $S_n/a_n \to 0$ in probability.

These results settle the nature of the cluster set $C(S_n/a_n)$. Similar questions can of course be asked concerning (8.20) and (8.22). Although the results are less complete here, one positive fact is available. We have of course to assume here that X satisfies the bounded LIL, that is, by Theorem 8.6, that $\mathbb{E}(\|X\|^2/LL\|X\|) < \infty$, $\sigma(X) < \infty$ and (S_n/a_n) is bounded in probability. It turns out that when this last condition is strengthened into $S_n/a_n \to 0$ in probability, one can prove (8.20) and (8.22) with $K = K_X$, whether K is compact *or not*. This is the object of the following theorem, the proof of which amplifies some of the techniques of the proof of Theorem 8.2 and which provides rather a complete description of the limits in this case. As we will see, the general situation may be quite different.

Theorem 8.11. *Let X be a Borel random variable with values in a separable Banach space B. Assume that $\mathbb{E}X = 0$, $\mathbb{E}(\|X\|^2/LL\|X\|) < \infty$, $\sigma(X) = \sup_{\|f\| \leq 1}(\mathbb{E}f^2(X))^{1/2} < \infty$ and that $S_n/a_n \to 0$ in probability. Then*

(8.23) $$\Lambda(X) = \limsup_{n \to \infty} \frac{\|S_n\|}{a_n} = \sigma(X) \quad \textit{almost surely}.$$

Moreover,

(8.24) $$\lim_{n \to \infty} d\left(\frac{S_n}{a_n}, K\right) = 0 \quad \textit{and} \quad C\left(\frac{S_n}{a_n}\right) = K$$

with probability one where $K = K_X$ is the unit ball of the reproducing kernel Hilbert space associated with the covariance structure of X.

Proof. It is enough to prove (8.23); indeed, replacing the norm of B by the gauge of $K + \varepsilon B_1$ where B_1 is the unit ball of B, it is easily seen that $d(S_n/a_n, K) \to 0$. Identification of the cluster set follows from Theorem 8.10 since $S_n/a_n \to 0$ in probability. To establish (8.23), by homogeneity and the scalar LIL, we need only show that $\Lambda(X) \leq 1$ when $\sigma(X) = 1$. As in the proof of Theorem 8.2 (see (8.6)), by the Borel-Cantelli lemma and Ottaviani's inequality (Lemma 6.2), it suffices to prove that for all $\varepsilon > 0$ and $\rho > 1$

$$\sum_n \mathbb{P}\left\{ \left\| \sum_{i=1}^{m_n} X_i \right\| > (1+\varepsilon) a_{m_n} \right\} < \infty$$

where $m_n = [\rho^n]$, $n \geq 1$ (integer part).

Let $0 < \varepsilon \leq 1$ and $\rho > 1$ be fixed. For every integer n and every f, h in the unit ball U of B', set

$$d_2^n(f, h) = \left(\mathbb{E}(f - h)^2(X) I_{\{\|X\| \leq a_{m_n}\}} \right)^{1/2}.$$

Let further $N(U, d_2^n; \varepsilon)$ be the minimal number of elements h in U such that for every f in U there exists such an h with $d_2^n(f, h) < \varepsilon$. As in the proof of Theorem 8.2, we first need an estimate of the size of these entropy numbers when n varies. However, with respect to Lemma 8.3, it does not seem possible to use the Sudakov minoration for Rademacher processes since the truncations do not appear to fit the right levels. Instead, we will rather use the Gaussian minoration via an improved randomization property which is made possible since we are working with identically distributed random variables. Let respectively (ε_i) and (g_i) be Rademacher and standard Gaussian sequences independent of (X_i). Under the assumptions of the theorem, we have

(8.25) $$\lim_{n \to \infty} \frac{1}{a_n} \mathbb{E} \left\| \sum_{i=1}^n \varepsilon_i X_i \right\| = \lim_{n \to \infty} \frac{1}{a_n} \mathbb{E} \left\| \sum_{i=1}^n g_i X_i \right\| = 0.$$

If X is symmetric, the limit on the left of (8.25) is seen to be 0 using Lemma 7.2 (since $S_n/a_n \to 0$ in probability) and the elementary fact that $\lim_{n \to \infty} n a_n^{-1} \mathbb{E}(\|X\| I_{\{\|X\| > a_n\}}) = 0$ under $\mathbb{E}(\|X\|^2 / LL\|X\|) < \infty$. By symmetrization, the left of (8.25) also holds in general since X is centered (Lemma 6.3). One can also use for this result Corollary 10.2 for a_n. Concerning Gaussian randomization, we refer to Proposition 10.4 below and the comments thereafter. Using the latter property and the Gaussian Sudakov minoration, the proof of Lemma 8.3 is trivially modified to this setting to yield the existence of a sequence (α_n) of positive numbers tending to 0 such that for all large enough n's

$$(8.26) \qquad N(U, d_2^n; \varepsilon) \leq \exp(\alpha_n LLm_n).$$

According to this result, we denote, for each n and f in U, by $h_n(f)$ an element of U such that $d_2^n(f, h_n(f)) < \varepsilon$ in such a way that the set U_n of all $h_n(f)$'s has a cardinality less than $\exp(\alpha_n LLm_n)$. We can write

$$\left\| \sum_{i=1}^{m_n} X_i \right\| \leq \sup_{f \in U_n} \left| \sum_{i=1}^{m_n} f(X_i) \right| + \sup_{h \in V_n} \left| \sum_{i=1}^{m_n} h(X_i) \right|$$

where $V_n = \{f - h_n(f); f \in U\} \subset 2U$. The main observation concerning V_n is that $\mathbb{E}(h^2(X) I_{\{\|X\| \leq a_{m_n}\}}) \leq \varepsilon^2$ for all h in V_n and all n. Although the proof of Theorem 8.6 and (8.19) through Theorem 7.5 is described in the setting of a single true norm of a Banach space, it is clear that it also applies to more general norms which might depend on n on the block of size m_n. In this way, it is just a mere exercise to verify that for some numerical constant C,

$$\sum_n \mathbb{P}\left\{ \sup_{h \in V_n} \left| \sum_{i=1}^{m_n} h(X_i) \right| > C\varepsilon a_{m_n} \right\} < \infty.$$

We are thus left to show that, for some $C > 0$,

$$(8.27) \qquad \sum_n \mathbb{P}\left\{ \sup_{f \in U_n} \left| \sum_{i=1}^{m_n} f(X_i) \right| > (1 + C\varepsilon) a_{m_n} \right\} < \infty.$$

Let $\delta = \delta(\varepsilon) > 0$ to be specified in a moment and set, for each n, $c_n = \delta m_n / a_{m_n}$. Define, for every n, $i \leq m_n$, and f in U,

$$Y_i(f, n) = \max\left(-c_n, \min(f(X_i), c_n)\right) - \mathbb{E}\left(\max(-c_n, \min(f(X_i), c_n))\right).$$

Note that $|Y_i(f, n)| \leq 2c_n$ and $\mathbb{E}(Y_i(f, n)^2) \leq 1$. By Lemma 1.6 applied to the sum of the independent mean zero random variables $Y_i(f, n)$, $i \leq m_n$, it follows that

$$\mathbb{P}\left\{ \sup_{f \in U_n} \left| \sum_{i=1}^{m_n} Y_i(f, n) \right| > (1 + \varepsilon) a_{m_n} \right\} \leq 2 \text{Card } U_n \exp\left(-(1 + \varepsilon) LLm_n\right)$$

provided $\delta = \delta(\varepsilon) > 0$ is small enough so that $2 - \exp(2(1 + \varepsilon)\delta) \geq (1 + \varepsilon)^{-1}$. By (8.26), it already follows that

$$(8.28) \qquad \sum_n \mathbb{P}\left\{ \sup_{f \in U_n} \left| \sum_{i=1}^{m_n} Y_i(f, n) \right| > (1 + \varepsilon) a_{m_n} \right\} < \infty.$$

Now consider $Z_i(f, n) = f(X_i) - Y_i(f, n)$, $i \leq m_n$, $f \in U$, $n \geq 1$. Note that, by the centering of the $f(X_i)$'s,

$$\mathbb{E}|Z_i(f,n)| \le 2\mathbb{E}\big(\|X\| I_{\{\|X\|>c_n\}}\big).$$

The integrability condition $\mathbb{E}(\|X\|^2/LL\|X\|) < \infty$ is equivalent to saying that $\sum_n \beta_n < \infty$ where

$$\beta_n = \frac{m_n}{(LLm_n)^2}\, \mathbb{P}\{\|X\| > c_n\}$$

(elementary verification). There exists a sequence (γ_n) such that $\gamma_n \ge \beta_n$, $\sum_n \gamma_n < \infty$, which satisfies the regularity property $\gamma_{n+1} \le \rho^{1/3}\gamma_n$ for every n (recall that $\rho > 1$). It is then easily seen that for every n,

$$\mathbb{E}\big(\|X\| I_{\{\|X\|>c_n\}}\big) \le \sum_{\ell \ge n} c_{\ell+1}\mathbb{P}\{\|X\| > c_\ell\}$$

$$\le \sum_{\ell \ge n} \frac{c_{\ell+1}(LLm_\ell)^2}{m_\ell}\,\gamma_\ell$$

$$\le C_1(\rho,\delta)\gamma_n \sum_{\ell \ge n} \frac{(LL\rho^\ell)^{3/2}}{\rho^{\ell/2}}\,\rho^{(\ell-n)/3}$$

$$\le C_2(\rho,\delta)\gamma_n \frac{(LLm_n)^{3/2}}{m_n^{1/2}}$$

for some constants $C_1(\rho,\delta)$, $C_2(\rho,\delta) > 0$. Consider the set of integers $L = \{n;\ 2C_2(\rho,\delta)\gamma_n LLm_n \le \varepsilon\}$. The preceding estimate indicates that for all $n \in L$, $f \in U$ and $i \le m_n$,

$$\mathbb{E}|Z_i(f,n)| \le \varepsilon a_{m_n}/m_n.$$

We use this property to show that if $n \in L$ is large enough,

(8.29) $$\mathbb{E}\left(\sup_{f\in U} \sum_{i=1}^{m_n} |Z_i(f,n)|\right) \le 2\varepsilon a_{m_n}.$$

Indeed, from Theorem 4.12, (8.25) implies

$$\lim_{n\to\infty} \frac{1}{a_{m_n}}\mathbb{E}\left(\sup_{f\in U}\left|\sum_{i=1}^{m_n} \varepsilon_i |Z_i(f,n)|\right|\right) = 0;$$

hence, by Lemma 6.3,

$$\lim_{n\to\infty} \frac{1}{a_{m_n}}\mathbb{E}\left(\sup_{f\in U}\left|\sum_{i=1}^{m_n} |Z_i(f,n)| - \mathbb{E}|Z_i(f,n)|\right|\right) = 0$$

from which the announced property (8.29) follows.

The main interest in the introduction of the absolute values in (8.29) is that it allows to use the isoperimetric inequality (6.16) in a very simple manner. It provides us indeed with the crucial monotonicity property (cf. Remark 6.18 about positive random variables). More precisely, let $n \in L$ and set

$$A = \left\{ \omega \in \Omega;\ \sup_{f \subset U} \sum_{i=1}^{m_n} |Z_i(f,n)(\omega)| \leq 4\varepsilon a_{m_n} \right\}.$$

Then $\mathbb{P}(A) \geq 1/2$ by (8.29) (at least for n large). Now, if, for $\omega \in \Omega$, there exist $\omega^1, \ldots, \omega^q$ in A such that $X_i(\omega) \in \{X_i(\omega^1), \ldots, X_i(\omega^q)\}$ except perhaps for at most k values of $i \leq m_n$, then

$$\sup_{f \in U} \sum_{i=1}^{m_n} |Z_i(f,n)(\omega)| \leq \sum_{i=1}^{k} \|Z_i(\omega)\|^* + \sum_{\ell=1}^{q} \sup_{f \in U} \sum_{i=1}^{m_n} |Z_i(f,n)(\omega^\ell)|$$

$$\leq \sum_{i=1}^{k} \|Z_i(\omega)\|^* + 4q\varepsilon a_{m_n}$$

where $(\|Z_i\|^*)$ denotes the non-increasing rearrangement of the family $(\|X_i\| + \mathbb{E}\|X_i\|)_{i \leq m_n}$. Hence, the isoperimetric inequality (6.16) ensures that for $k \geq q$

$$\mathbb{P}\left\{ \sup_{f \in U} \sum_{i=1}^{m_n} |Z_i(f,n)| > (1 + 4q)\varepsilon a_{m_n} \right\} \leq \left(\frac{K_0}{q} \right)^k + \mathbb{P}\left\{ \sum_{i=1}^{k} \|Z_i\|^* > \varepsilon a_{m_n} \right\}.$$

If we now choose $q = 2K_0$ and $k = k_n$ as in the proof of Theorem 8.6 using the integrability condition $\mathbb{E}(\|X\|^2/LL\|X\|) < \infty$, we get

$$(8.30) \qquad \sum_{n \in L} \mathbb{P}\left\{ \sup_{f \in U} \sum_{i=1}^{m_n} |Z_i(f,n)| > (1 + 8K_0)\varepsilon a_{m_n} \right\} < \infty.$$

Combining (8.28) and (8.30), we see that in order to establish (8.27) and to conclude the proof of the theorem, we have to show that for some numerical constant $C > 0$,

$$(8.31) \qquad \sum_{n \notin L} \mathbb{P}\left\{ \sup_{f \in U_n} \left| \sum_{i=1}^{m_n} f(X_i) \right| > C\varepsilon a_{m_n} \right\} < \infty.$$

We follow very much the pattern of the case $n \in L$. Let now $c_n' = m_n/4\varepsilon a_{m_n}$, and define $Y_i'(f,n)$, $Z_i'(f,n)$ as $Y_i(f,n)$, $Z_i(f,n)$ before but with c_n' instead of c_n. Now, we observe, since $\sigma(X) \leq 1$, that

$$\mathbb{E}|Z_i'(f,n)| \leq 8\varepsilon a_{m_n}/m_n,\ i \leq m_n.$$

Exactly as what we did with $Z_i(f,n)$, we can get from the isoperimetric approach that

$$(8.32) \qquad \sum_n \mathbb{P}\left\{ \sup_{f \in U} \sum_{i=1}^{m_n} |Z_i'(f,n)| > C\varepsilon a_{m_n} \right\} < \infty.$$

Concerning $Y_i'(f,n)$, the exponential inequality of Lemma 1.6 shows that

$$\mathbb{P}\left\{\sup_{f\in U_n}\left|\sum_{i=1}^{m_n}Y_i'(f,n)\right|>\varepsilon a_{m_n}\right\}\leq 2\operatorname{Card}U_n\exp(-\varepsilon^2(2-\sqrt{e})LLm_n)$$

$$\leq 2\exp\left(-\left(\frac{\varepsilon^2}{4}-\alpha_n\right)LLm_n\right)$$

where we have used (8.26). Now, if $n\notin L$, $LLm_n>\varepsilon(2C_2(\rho,\delta)\gamma_n)^{-1}$ where $\sum_n\gamma_n<\infty$. Since $\alpha_n\to 0$, we clearly get that

$$\sum_{n\notin L}\mathbb{P}\left\{\sup_{f\in U_n}\left|\sum_{i=1}^{m_n}Y_i'(f,n)\right|>\varepsilon a_{m_n}\right\}<\infty$$

from which, together with (8.32), (8.31) follows. This completes the proof of Theorem 8.11.

Theorem 8.11 settles the question of the identification of the limits when $S_n/a_n\to 0$ in probability. However, very little is known at the present time about the limit (8.20), or just only (8.22), in the case of the bounded LIL, that is, in the setting of Theorem 8.11, when (S_n/a_n) is only bounded in probability (cf. Theorem 8.6). The limsup $\Lambda(X)$ need not be equal to $\sigma(X)$ and has to take into account the stochastic boundedness of (S_n/a_n), for example through $\Gamma(X)=\limsup_{n\to\infty}\mathbb{E}\|S_n/a_n\|$ (cf. [A-K-L]). One might wonder which function of $\sigma(X)$ and $\Gamma(X)$ (or some other quantity equivalent to $\Gamma(X)$) $\Lambda(X)$ could be. We believe that this study could lead to some rather intricate situations as suggested by the following example with which we close this chapter. Here, we construct an example showing that the condition $S_n/a_n\to 0$ in probability in Theorem 8.11 is *not* necessary for the limit $\lim_{n\to\infty}d(S_n/a_n,K)$ to be almost surely 0. Let us note that K. Alexander [Ale4] showed that when $\lim_{n\to\infty}d(S_n/a_n,K)=0$, then necessarily $C(S_n/a_n)=K$.

Example 8.12. There exists a random variable X satisfying the bounded LIL in the Banach space c_0 such that

$$\lim_{n\to\infty}d\left(\frac{S_n}{a_n},K\right)=0$$

with probability one, with $K=K_X$ the unit ball of the reproducing kernel Hilbert space associated with the covariance structure of X, but for which the sequence (S_n/a_n) does not converge in probability to 0.

The construction of this example is based on the following preliminary study which appears to be almost canonical and could possibly be useful for related constructions. It will be convenient for this study, as well as for the example itself, to use the language and notation of empirical processes (cf. Chapter 14).

Let I be a subinterval of $[0,1]$ of length b divided into p equal subintervals. Consider the class $\mathcal{G}$ of functions on $[0,1]$ defined by $\mathcal{G}=\{I_A;A$ is the union of h subintervals (of I)}. It is implicit that p and h are large enough for all the computations that we will develop. With some abuse of notation, we denote by (X_i) a sequence of independent variables which are uniformly distributed on $[0, 1]$. We study here, for every n, the quantity

$$\mathbb{E}\left\|\sum_{i=1}^{n} \varepsilon_i f(X_i)\right\|_{\mathcal{G}}$$

where $\|\sum_{i=1}^{n} \varepsilon_i f(X_i)\|_{\mathcal{G}} = \sup_{f \in \mathcal{G}} |\sum_{i=1}^{n} \varepsilon_i f(X_i)|$ and, as usual, (ε_i) is a Rademacher sequence which is independent of (X_i). First, note that, obviously, $\|\sum_{i=1}^{n} \varepsilon_i f(X_i)\|_{\mathcal{G}} \leq \mathrm{Card}\{i \leq n;\ X_i \in I\}$ so that

$$(8.33) \qquad \mathbb{E}\left\|\sum_{i=1}^{n} \varepsilon_i f(X_i)\right\|_{\mathcal{G}} \leq bn.$$

Now, let us try to improve this general estimate for relative values of n. To this aim, we use Bennett's inequality (6.10) from which we easily deduce that, for all f in $\mathcal{G}$ and all $t > 0$,

$$\mathbb{P}\left\{\left|\sum_{i=1}^{n} \varepsilon_i f(X_i)\right| > t\right\} \leq 2 \exp\left[-\frac{t}{C_1}\theta\left(\frac{t}{n\sigma^2}\right)\right]$$

where $\sigma^2 = hb/p$, $\theta(u) = u$ when $0 < u \leq e$, $\theta(u) = e\log u$ when $u \geq e$ and where C_1 is some (large) numerical constant. Consider now t_0 such that $t_0\theta(t_0/n\sigma^2)/C_1 \geq h\log p$ (≥ 1). Since $\mathrm{Card}\,\mathcal{G} \leq p^h = \exp(h\log p)$ and θ is increasing, for all $t \geq t_0$,

$$\mathbb{P}\left\{\left\|\sum_{i=1}^{n} \varepsilon_i f(X_i)\right\|_{\mathcal{G}} > t\right\} \leq 2 \exp\left[h\log p - \frac{t}{C_1}\theta\left(\frac{t_0}{n\sigma^2}\right)\right].$$

Now, by an integration by parts and the definition of t_0, it follows that $\mathbb{E}\|\sum_{i=1}^{n} \varepsilon_i f(X_i)\|_{\mathcal{G}} \leq 3t_0$. It is easily verified that when $n \geq C_1 p\log p/e^2 b$, one may take t_0 to be $\sqrt{C_1}\sqrt{n}h(b\log p/p)^{1/2}$ whereas when $n \leq C_1 p\log p/e^2 b$ we can take

$$t_0 = \frac{C_1 eh\log p}{\log\left(\frac{C_1 ep\log p}{bn}\right)}.$$

Thus, we have obtained, combining with (8.33), that:

$$(8.34) \qquad \frac{1}{\sqrt{n}}\mathbb{E}\left\|\sum_{i=1}^{n} \varepsilon_i f(X_i)\right\|_{\mathcal{G}} \leq b\sqrt{n} \quad \text{if } n \leq \frac{h}{b},$$

$$(8.35) \qquad \frac{1}{\sqrt{n}}\mathbb{E}\left\|\sum_{i=1}^{n} \varepsilon_i f(X_i)\right\|_{\mathcal{G}} \leq \frac{3C_1 eh\log p}{\sqrt{n}\log\left(\frac{C_1 ep\log p}{bn}\right)} \quad \text{if } n \leq \frac{C_1}{e^2} \cdot \frac{p\log p}{b},$$

$$(8.36) \qquad \frac{1}{\sqrt{n}} \mathbb{E}\left\| \sum_{i=1}^{n} \varepsilon_i f(X_i)\right\|_{\mathcal{G}} \le 3\sqrt{C_1} h\left(\frac{b\log p}{p}\right)^{1/2} \quad \text{if } n \ge \frac{C_1}{e^2} \cdot \frac{p\log p}{b}.$$

Since the right hand side of (8.35) is decreasing in n (for the values of n that we consider), we obtain a first consequence of this investigation. Below, C denotes some numerical constant possibly varying from line to line.

Corollary 8.13. *If $m \ge h/b$,*

$$\sup_{n \ge m} \frac{1}{a_n} \mathbb{E}\left\| \sum_{i=1}^{n} \varepsilon_i f(X_i)\right\|_{\mathcal{G}}$$

$$\le C \max\left(\frac{h\log p}{\sqrt{LL\frac{h}{b}}\sqrt{m}\log\left(\frac{C_1 e p\log p}{m}\right)}, \frac{h}{\sqrt{LL\frac{h}{b}}}\left(\frac{b\log p}{p}\right)^{1/2}\right).$$

Corollary 8.14. *If $m \ge h/b$ and, in addition, $C_1 p \ge m^2$,*

$$\sup_{n \ge m} \frac{1}{a_n} \mathbb{E}\left\| \sum_{i=1}^{n} \varepsilon_i f(X_i)\right\|_{\mathcal{G}} \le C \max\left(\frac{h}{\sqrt{LL\frac{h}{b}}\sqrt{m}}, \frac{h}{\sqrt{LL\frac{h}{b}}}\left(\frac{b\log p}{p}\right)^{1/2}\right).$$

To obtain a control which is uniform in n, consider the bound (8.34) for $r = h/b$ and the previous corollary.

Corollary 8.15. *If $C_1 p \ge h^2$, then*

$$\sup_{n} \frac{1}{a_n} \mathbb{E}\left\| \sum_{i=1}^{n} \varepsilon_i f(X_i)\right\|_{\mathcal{G}} \le C \frac{\sqrt{hb}}{\sqrt{LL\frac{h}{b}}}.$$

With this preliminary study and the preceding statements, we now start the construction of Example 8.12. Consider increasing sequences of integers $(n(q)), (p(q)), (s(q))$ to be specified later on. Let $I_q, J_q, q \in \mathbb{N}$, be disjoint intervals in $[0, 1]$ where, for each q, I_q has length $b(q) = LLn(q)/n(q)$ and J_q, $b(q)/s(q)$. We divide I_q in $p(q)$ equal subintervals and denote by $\mathcal{I}_q$ the family of these subintervals. We set

$$a(q) = a_{n(q)} = \left(2n(q)LLn(q)\right)^{1/2} \quad \text{and} \quad c(q) = \frac{1}{10} \cdot \frac{a(q)}{LLn(q)}.$$

We set further, always for every q,

$$\mathcal{G}_q = \left\{ c(q) I_A ; A \text{ is union of } [n(q)b(q)] \text{ intervals of } \mathcal{I}_q \right\}$$

where $[\,\cdot\,]$ is the integer part function. We note that for every f in $\mathcal{G}_q$, $\|f\|_2 = d(q)$ where

$$d(q)^2 = c(q)^2 [n(q)b(q)] \frac{b(q)}{p(q)}$$

which is equivalent, for q large, to $2LLn(q)/10^2 p(q)$. Let T_q be the affine map from J_q into I_q. For a function f with support in I_q and constant on the intervals of J_q, set

$$U_q(f) = f + \left(\frac{1 - d(q)^2}{d(q)^2}\right)^{1/2} \sqrt{s(q)} f \circ T_q$$

so that $\|U_q(f)\|_2 = \|f\|_2/d(q)$. We further set $\mathcal{F}_q = \{U_q(f); \ f \in \mathcal{G}_q\}$, $\mathcal{G}'_q = \{U_q(f) - f; \ f \in \mathcal{G}_q\}$. In particular $\|f\|_2 = 1$ if $f \in \mathcal{F}_q$. Let $\mathcal{F} = \bigcup_q \mathcal{F}_q$. We will show that one can choose appropriately the sequences $(n(q)), (p(q)), (s(q))$ such that

(8.37)
$$\frac{1}{a_{n(q)}} \left\| \sum_{i=1}^{n(q)} \varepsilon_i f(X_i) \right\|_{\mathcal{F}} \nrightarrow 0 \quad \text{in probability}$$

and, for every $u > 0$,

(8.38)
$$\limsup_{n \to \infty} \frac{1}{a_n} \left\| \sum_{i=1}^{n} \varepsilon_i f(X_i) \right\|_{\mathcal{F}_u} \leq 1 \quad \text{almost surely}$$

where $\mathcal{F}_u = \{f \in u\mathrm{Conv}\, \mathcal{F}; \ \|f\|_2 \leq 1\}$. (The notation $\|\cdot\|_{\mathcal{F}}$ has the same meaning as in the preliminary study.)

Before we turn to the somewhat technical details of the construction, let us show why (8.37) and (8.38) give rise to Example 8.12. Observe that $\mathcal{F}$ is countable. Let X be the map from $[0,1] \times \{-1,+1\}$ in $c_0(\mathcal{F})$ defined by $X(x,\varepsilon) = (\varepsilon f(x))_{f \in \mathcal{F}}$. Since the intervals I_q, J_q are disjoint, X actually takes its values in the space of finite sequences. That $S_n/a_n \nrightarrow 0$ in probability follows from (8.37). To establish that $\lim_{n \to \infty} d(S_n/a_n, K) = 0$ almost surely where $K = K_X$, it suffices to show (as in Theorem 8.11) that, for every $\varepsilon > 0$, $\limsup_{n \to \infty} \|\|S_n/a_n\|\| \leq 1$ with probability one where $\|\|\cdot\|\|$ is the gauge of $K + \varepsilon B_1$ and B_1 is the unit ball of $c_0(\mathcal{F})$. But the unit ball of the dual norm is $V = \{g \in \ell_1(\mathcal{F}); \ \|g\| \leq 1/\varepsilon, \ \mathbb{E}g(X)^2 \leq 1\}$ so that it suffices to have

$$\limsup_{n \to \infty} \sup_{g \in V} \frac{1}{a_n} \left| \sum_{i=1}^{n} \varepsilon_i g(X_i) \right| \leq 1 \quad \text{almost surely}.$$

Let V_0 be the elements of V with finite support. Then (8.38) exactly means that

$$\limsup_{n \to \infty} \sup_{g \in V_0} \frac{1}{a_n} \left| \sum_{i=1}^{n} \varepsilon_i g(X_i) \right| \leq 1 \quad \text{almost surely}.$$

Since V_0 is easily seen to be dense in norm in V, the conclusion follows.

Now, let us turn to the construction and the proof of (8.37) and (8.38). We will actually construct (by induction) the sequences $(p(q))$ and $(s(q))$ from sequences $(r(q))$ and $(m(q))$ such that $r(q-1) < n(q) < m(q) < r(q)$ (where each strict inequality actually means much larger). The case $q = 1$, which is similar to the general case, is left to the reader. Therefore, let us assume that $r(q-1)$ has been constructed. We take $n(q)$ large enough such that

$$(8.39) \qquad LLn(q) \geq 2^{4q}, \quad LLc(q) \geq 2^q, \quad b(q) \leq 2^q p(q-1)$$

(this is possible since $c(q) = a(q)/10LLn(q)$ and $b(q) = LLn(q)/n(q)$) and such that

$$(8.40) \qquad \frac{r(q-1)}{a_{r(q-1)}} \cdot \frac{a(q)}{n(q)} \leq 2^{-q}.$$

Then we take $p(q)$ sufficiently large such that

$$(8.41) \qquad p(q) \geq n(q)^4.$$

Set $m(q) = [(2^{-q}p(q))^{1/2}]$ and choose then $s(q)$ large enough so that

$$(8.42) \qquad s(q) \geq 2^q m(q)b(q), \quad s(q)d(q)^2 \geq 1$$

and (what is actually a stronger condition)

$$(8.43) \qquad LL\sqrt{s(q)} \geq p(q).$$

We are left with the choice of $r(q)$. To this aim, set $\mathcal{H}_q = \{g \in 2^q \text{ Conv } (\bigcup_{\ell \leq q} \mathcal{F}_\ell); \|g\|_2 \leq 1\}$. $\mathcal{H}_q$ is a convex set of finite dimension. There exists a finite set $\mathcal{H}'_q$ such that $\mathcal{H}_q \subset \text{Conv } \mathcal{H}'_q$ and $\|g\|_2 \leq (1-2^{-q-1})^{-1/2}$ for $g \in \mathcal{H}'_q$. The exponential inequality of Kolmogorov (Lemma 1.6), applied to each g in $\mathcal{H}'_q$, easily implies that one can find $r(q)$ such that, for every $n \geq r(q)$ and every t with $a_n/2 \leq t \leq 2a_n$,

$$(8.44) \qquad \mathbb{P}\left\{\left\|\sum_{i=1}^{n} \varepsilon_i f(X_i)\right\|_{\mathcal{H}_q} > t\right\} \leq 2\exp\left[-\frac{t^2}{2n}(1-2^{-q})\right]$$

and

$$(8.45) \qquad \mathbb{E}\left\|\sum_{i=1}^{n} \varepsilon_i f(X_i)\right\|_{\mathcal{H}_q} \leq C\sqrt{n}$$

where C is a numerical constant. This completes the construction (by induction) of the various sequences of integers we will work with. Now, we can turn to the proofs of (8.37) and (8.38).

Let T_q^1 be the event

$$T_q^1 = \{\forall i \leq m(q), \text{ each interval of } \mathcal{I}_q \text{ contains at most one } X_i\}.$$

Clearly, if $T_q^{1,c}$ is the complement of T_q^1,

$$\mathbb{P}(T_q^{1,c}) \leq \frac{1}{2}m(q)(m(q)-1)p(q)\left(\frac{b(q)}{p(q)}\right)^2$$

and thus, by definition of $m(q)$ and the fact that $b(q) \leq 1$, $\mathbb{P}(T_q^{1,c}) \leq 2^{-q}$. Similarly, let

$$T_q' = \{\forall i \leq m(q), \ X_i \notin J_q\}.$$

Then $\mathbb{P}(T_q'^{,c}) \leq m(q)b(q)/s(q) \leq 2^{-q}$ by (8.42). We can then already show (8.37). From these estimates indeed, for every q large enough, on a set of probability bigger than $1/3$ (for example), there are at least $[n(q)b(q)]$ points X_i, $i \leq n(q)$, in I_q which are in different intervals of $\mathcal{I}_q$ and not in J_q. Therefore, there exists a union A of $[n(q)b(q)]$ intervals of $\mathcal{I}_q$ such that $\left|\sum_{i=1}^{n(q)} \varepsilon_i I_A(X_i)\right| \geq \frac{1}{2}[n(q)b(q)]$. Since $c(q)n(q)b(q) = a(q)/10$, it follows that, for all large enough q's,

$$\mathbb{P}\left\{\left\|\sum_{i=1}^{n(q)} \varepsilon_i f(X_i)\right\|_{\mathcal{F}_q} \geq \frac{a(q)}{40}\right\} \geq \frac{1}{3}$$

from which (8.37) clearly follows.

We turn to (8.38) which is of course the most difficult part. Let us fix $u > 0$ and recall that $\mathcal{F}_u = \{f \in u\mathrm{Conv}\mathcal{F}; \ \|f\|_2 \leq 1\}$. As in the proof of Theorem 8.11, it suffices to show that for all $\xi > 1$, $\rho > 1$,

$$(8.46) \qquad \limsup_{n \to \infty} \frac{1}{a_{\rho(n)}} \left\|\sum_{i=1}^{\rho(n)} \varepsilon_i f(X_i)\right\|_{\mathcal{F}_u} \leq \xi \quad \text{almost surely}$$

where $\rho(n) = [\rho^n]$. We set $\mathbb{N}_1 = \bigcup_q[r(q-1), m(q)]$ and $\mathbb{N}_2 = \bigcup_q[m(q), r(q)]$ (as subsets of integers) and to study separately $\limsup_{\rho(n)\in\mathbb{N}_1}$ and $\limsup_{\rho(n)\in\mathbb{N}_2}$. Let us first consider the limsup when $\rho(n) \in \mathbb{N}_2$. We shall use the proof of Theorem 8.11 from which we know that we will get a limsup less than or equal to 1 under the conditions $\mathbb{E}(g^2/LLg) < \infty$ where $g = \|f\|_{\mathcal{F}} = \sup_{f\in\mathcal{F}} |f|$ and

$$(8.47) \qquad \lim_{n \in \mathbb{N}_2} \frac{1}{a_n} \mathbb{E}\left\|\sum_{i=1}^{n} \varepsilon_i f(X_i)\right\|_{\mathcal{F}} = 0.$$

To check the integrability condition, let $g_q = \sup_{f\in\mathcal{F}_q} |f|$. The g_q's have disjoint supports. Moreover,

$$g_q = c(q)I_{I_q} + c(q)\left(\frac{1 - d(q)^2}{d(q)^2}\right)^{1/2} \sqrt{s(q)}I_{J_q}$$

$$\leq c(q)I_{I_q} + \frac{c(q)}{d(q)}\sqrt{s(q)}I_{J_q}.$$

It follows that, for all large q's,

$$\mathbb{E}(g_q^2/LLg_q^2) \leq \frac{c(q)^2 b(q)}{LLc(q)} + \frac{c(q)^2 b(q)}{d(q)^2 LL\sqrt{s(q)}}$$

$$\leq \frac{1}{LLc(q)} + \frac{p(q)}{LLn(q)LL\sqrt{s(q)}}$$

which gives rise to a summable series by (8.39) and (8.43). Turning to (8.47), let $n \in [m(q), r(q)]$. Note that $\mathcal{F} \subset \mathcal{H}_q \cup \mathcal{F}_q \cup \bigcup_{\ell > q} \mathcal{F}_\ell$ so that

$$\frac{1}{a_n} \mathbb{E} \left\| \sum_{i=1}^{n} \varepsilon_i f(X_i) \right\|_{\mathcal{F}} \leq \frac{1}{a_n} \mathbb{E} \left\| \sum_{i=1}^{n} \varepsilon_i f(X_i) \right\|_{\mathcal{H}_q} + \frac{1}{a_n} \mathbb{E} \left\| \sum_{i=1}^{n} \varepsilon_i f(X_i) \right\|_{\mathcal{F}_q}$$

$$+ \frac{1}{a_n} \mathbb{E} \left\| \sum_{i=1}^{n} \varepsilon_i f(X_i) \right\|_{\bigcup_{\ell > q} \mathcal{F}_\ell}$$

$$= (I) + (II) + (III).$$

By (8.45), the limit of (I) is zero. Concerning (III), for $\ell > q$ we have (cf. (8.33))

$$\mathbb{E} \left\| \sum_{i=1}^{n} \varepsilon_i f(X_i) \right\|_{\mathcal{G}_\ell} \leq nc(\ell)b(\ell) \leq n\frac{a(\ell)}{n(\ell)}$$

so that, since $n \leq r(q)$,

$$\frac{1}{a_n} \mathbb{E} \left\| \sum_{i=1}^{n} \varepsilon_i f(X_i) \right\|_{\mathcal{G}_\ell} \leq \frac{n}{a_n} \cdot \frac{a(\ell)}{n(\ell)} \leq \frac{r(q)}{a_{r(q)}} \cdot \frac{a(\ell)}{n(\ell)} \leq 2^{-\ell}$$

by (8.40). On the other hand, by (8.42),

$$\mathbb{E} \left\| \sum_{i=1}^{n} \varepsilon_i f(X_i) \right\|_{\mathcal{G}'_\ell} \leq n\frac{c(\ell)}{d(\ell)} \cdot \frac{b(\ell)}{\sqrt{s(\ell)}} \leq nc(\ell)b(\ell) \leq n\frac{a(\ell)}{n(\ell)},$$

and thus, as before, we get $\mathbb{E} \| \sum_{i=1}^{n} \varepsilon_i f(X_i) \|_{\mathcal{G}'_\ell} \leq a_n 2^{-\ell}$. It follows that $\mathbb{E} \| \sum_{i=1}^{n} \varepsilon_i f(X_i) \|_{\mathcal{F}_\ell} \leq a_n 2^{-\ell+1}$ and therefore that $(III) \leq 2^{-q+2}$. Hence, the limit of (III) is also zero. We are left with (II). As for (III), we write that

$$(III) \leq \frac{1}{a_n} \mathbb{E} \left\| \sum_{i=1}^{n} \varepsilon_i f(X_i) \right\|_{\mathcal{G}_q} + \frac{1}{a_n} \mathbb{E} \left\| \sum_{i=1}^{n} \varepsilon_i f(X_i) \right\|_{\mathcal{G}'_q} = (IV) + (V).$$

We evaluate (IV) and (V) by the preliminary study. For (IV), we let there $b = b(q)$, $h = [n(q)b(q)]$, $p = p(q)$ and $m = m(q)$. For q large enough, the definition of $m(q)$ ($m(q) = [(2^{-q}p(q))^{1/2}]$) shows that we are in a position to apply Corollary 8.14. Since $b(q) \leq 1$ and $c(q)n(q)b(q) = a(q)/10$, it follows that, for some numerical constant C and all large enough q's,

$$(IV) \leq C \max \left(\left(\frac{n(q)}{m(q)} \right)^{1/2}, \left(\frac{n(q)}{p(q)} \log p(q) \right)^{1/2} \right).$$

By the choice of $p(q) \geq n(q)^4$, we have $m(q) \geq 2^{-q-1}n(q)^2$, from which, together with (8.39), we deduce that (IV) tends to 0. Using Corollary 8.15 with $b = b(q)/s(q)$, $h = [n(q)b(q)]$ and $p = p(q)$, the control of (V) is similar by (8.43). We have therefore established in this way that (8.47) holds and thus (8.46) holds along $\rho(n) \in \mathbb{N}_2$.

In the last part of this proof, we establish (8.46) when $\rho(n) \in \mathbb{N}_1$. For each q, consider $T_q = \bigcap_{i=1}^{5} T_q^i$ where

$T_q^1 = \{\forall i \leq m(q), \text{ each interval of } \mathcal{I}_q \text{ contains at most one } X_i\}$,

$T_q^2 = \{\forall i \leq m(q), X_i \notin J_q \cup \bigcup_{\ell>q}(I_\ell \cup J_\ell)\}$,

$T_q^3 = \{\text{for } 2^{-2q}n(q) \leq n \leq m(q), \text{ Card}\{i \leq n; X_i \in I_q\} \leq 2e^2nb(q)\}$,

$T_q^4 = \{\text{for } 2^{-q}n(q)/LLn(q) \leq n \leq 2^{-2q}n(q),$
$$\text{Card}\{i \leq n; X_i \in I_q\} \leq e^2 2^{-q+1} a_n/c(q)\},$$

$T_q^5 = \{\forall i \leq 2^{-q}n(q)/LLn(q), X_i \notin I_q\}$.

We would like to show that $\sum_q \mathbb{P}(T_q^c) < \infty$. It suffices to prove that $\sum_q \mathbb{P}(T_q^{i,c}) < \infty$, $i = 1, \ldots, 5$, where $T_q^{i,c}$ is the complement of T_q^i. We have already seen that $\mathbb{P}(T_q^{1,c}) \leq 2^{-q}$. For $i = 2$, note that, when $\ell \geq q$,

$$\mathbb{P}\{\exists i \leq m(q), X_i \in J_\ell\} \leq m(q)\frac{b(\ell)}{s(\ell)} \leq m(\ell)\frac{b(\ell)}{s(\ell)} \leq 2^{-\ell}$$

by (8.42). When $\ell > q$, by (8.39),

$$\mathbb{P}\{\exists i \leq m(q), X_i \in I_\ell\} \leq m(q)b(\ell) \leq p(q)b(\ell) \leq p(\ell-1)b(\ell) \leq 2^{-\ell}.$$

Hence $\mathbb{P}(T_q^{2,c}) \leq 3 \cdot 2^{-q}$. Concerning T_q^5, one need simply note that $|I_q| = b(q) = LLn(q)/n(q)$. For $i = 3, 4$, we use the binomial bound (1.16) which implies in particular

(8.48) $\mathbb{P}\{B(n, \tau) \geq tn\} \leq \exp\left[-tn\left(\frac{\tau}{t} - 1 - \log\frac{\tau}{t}\right)\right] \leq \exp(-tn)$

when $t \geq e^2\tau$ (for example), where $B(n, \tau)$ is the number of successes in a run of n Bernoulli trials with probability of success τ and $0 < t < 1$. We have

$$T_q^3 \subset \bigcap_{\ell \geq 0}\{\text{Card}\{i \leq 2^{-2q+\ell}n(q); X_i \in I_q\} \leq e^2 2^{-2q+\ell}b(q)\}.$$

Then, taking in (8.48) $\tau = b(q)$ and $t = e^2\tau$ (q large), we have

$$\mathbb{P}(T_q^{3,c}) \leq \sum_{\ell \geq 0} \exp\left(-e^2 2^{-2q+\ell}n(q)b(q)\right).$$

Since $n(q)b(q) = LLn(q) \geq 2^{4q}$ by (8.39), it follows that $\sum_q \mathbb{P}(T_q^3) < \infty$. Concerning T_q^4, set $v(\ell) = [2^{-q+\ell}n(q)/LLn(q)]$. Then

$$T_q^4 \subset \bigcap\{\text{Card}\{i \leq v(\ell); X_i \in I_q\} \leq e^2 2^{-q}a_{v(\ell)}/c(q)\}$$

where the intersection is over every $\ell \geq 0$ such that $v(\ell) \leq 2^{-2q}n(q)$, i.e. $2^\ell \leq 2^{-q}LLn(q)$. Take then in (8.48), $t = e^2 2^{-q}a_{v(\ell)}/v(\ell)c(q)$, $\tau = b(q)$ and $n = v(\ell)$. Since $2^\ell \leq 2^{-q}LLn(q)$, one verifies that, at least for q large enough (by (8.39)), $t \geq e^2\tau$. Hence

$$\mathbb{P}(T_q^{4,c}) \le \sum_{\ell \ge 0} \exp\left(-e^2 2^{-q} a_{v(\ell)}/c(q)\right).$$

Since $a_{v(\ell)}/c(q) \ge 2^{-q/2+\ell/2}(LLn(q))^{1/2}$, one concludes as for $i = 3$ that $\sum_q \mathbb{P}(T_q^{4,c}) < \infty$. Thus, we have established that $\sum_q \mathbb{P}(T_q^c) < \infty$.

For every q (large enough), set now

$$P_q = \sum_{r(q-1) \le \rho(n) \le m(q)} \mathbb{P}\left\{ \left\| \sum_{i=1}^{\rho(n)} \varepsilon_i f(X_i) \right\|_{\mathcal{F}_u} > \xi a_{\rho(n)} ; \ T_q \right\}.$$

The proof will be completed once we will have shown that $\sum_q P_q < \infty$. Take q large enough so that $2^q > u$. On T_q, none of the X_i's, $i \le \rho(n) \le m(q)$, is in J_ℓ for $\ell \ge q$ or I_ℓ for $\ell > q$ (definition of T_q^2). On the other hand, the restriction of a function of $\mathcal{F}_u$ to $I_q \cap J_q$ belongs to $\mathcal{F}_{q,u} = \{f \in u\text{Conv}\mathcal{F}_q ; \ \|f\|_2 \le 1\}$ and its restriction to $\bigcup_{\ell < q} I_\ell \cup J_\ell$ is in $\mathcal{H}_{q-1}$. Thus, we have

$$(8.49) \quad \left\| \sum_{i=1}^{\rho(n)} \varepsilon_i f(X_i) \right\|_{\mathcal{F}_u} \le \left\| \sum_{i=1}^{\rho(n)} \varepsilon_i f(X_i) \right\|_{\mathcal{H}_{q-1}} + \left\| \sum_{i=1}^{\rho(n)} \varepsilon_i f(X_i) \right\|_{\mathcal{F}_{q,u}}.$$

Lemma 8.16. *For every $\varepsilon > 0$, there exists $q_0 = q_0(\varepsilon)$ such that for all $q \ge q_0$ and $r(q-1) \le \rho(n) \le m(q)$, one can find numbers $\xi(n,q)$ with the following properties:*

$$(8.50) \quad on \ T_q , \quad \left\| \sum_{i=1}^{\rho(n)} \varepsilon_i f(X_i) \right\|_{\mathcal{F}_{u,q}} \le \xi(n,q) ;$$

$$(8.51) \quad there \ exists \ M = M(\varepsilon, q_0)$$
$$such \ that \ \text{Card}\{n; \ \xi(n,q) > \varepsilon a_{\rho(n)}\} \le M ;$$

$$(8.52) \quad \xi(n,q) \le \tfrac{2}{3} a_{\rho(n)} .$$

Before we prove this lemma, let us show how to conclude with it. By (8.49) and (8.50)

$$\mathbb{P}\left\{ \left\| \sum_{i=1}^{\rho(n)} \varepsilon_i f(X_i) \right\|_{\mathcal{F}_u} > \xi a_{\rho(n)} ; \ T_q \right\}$$

$$\le \mathbb{P}\left\{ \left\| \sum_{i=1}^{\rho(n)} \varepsilon_i f(X_i) \right\|_{\mathcal{H}_{q-1}} > \xi a_{\rho(n)} - \xi(n,q) \right\}.$$

Using $\rho(n) \ge r(q-1)$, we may now use (8.44) and estimate the latter probability by

$$2 \exp\left[-\frac{1}{2\rho(n)} \left(\xi a_{\rho(n)} - \xi(n,q)\right)^2 (1 - 2^{-q+1})\right].$$

In (8.51), take $\varepsilon > 0$ to be such that $\xi - \varepsilon > 1$. If $\xi(n,q) \leq \varepsilon a_{\rho(n)}$, $\xi a_{\rho(n)} - \xi(n,q) \geq (\xi - \varepsilon) a_{\rho(n)}$ and the preceding gives rise to a summable series. For each q ($\geq q_0$), the number of $\xi(n,q) > \varepsilon a_{\rho(n)}$ is controlled by M and thus, using (8.52), there is a contribution of at most

$$M \exp\left(-\frac{1}{18} LLn(q - 1)\right)$$

which defines the current term of a summable series by (8.39).

Proof of Lemma 8.16. The crucial point is the following. If $g \in \mathcal{F}_{q,u}$, let f be the restriction of g to I_q. Then $g = U_q f$. Therefore, $1 \geq \|U_q f\|_2 = \|f\|_2 / d(q)$ so that $\|f\|_2 \leq d(q)$. (The first reader who has comprehensively reached this point of the construction is invited, first to bravely pursue his effort until the end, and then to contact the authors (the second one!) for a reward as a token of his perseverance.) Let n be fixed, $n \leq m(q)$, and assume that we are on T_q. Let $N = \mathrm{Card}\{i \leq n; X_i \in I_q\}$ so that, by Cauchy-Schwarz, $\sum_{i=1}^{n} |f(X_i)| \leq (N \sum_{i=1}^{n} f(X_i)^2)^{1/2}$. Since on T_q (T_q^1) the X_i's are in distinct intervals of $\mathcal{I}_q$, we have

$$\sum_{i=1}^{n} f(X_i)^2 \leq \sum_{I \in \mathcal{I}_q} (\text{value of } f \text{ on } I)^2 = \|f\|_2^2 \frac{p(q)}{b(q)}$$

$$\leq d(q)^2 \frac{p(q)}{b(q)} \leq \frac{2}{100} \cdot \frac{LLn(q)}{b(q)}.$$

When $n \geq 2^{-2q} n(q)$, $N \leq 2e^2 nb(q) \leq 20 nb(q)$ (by T_q^3) so that

$$\sum_{i=1}^{n} |f(X_i)| \leq \left(\frac{2}{5} nLLn(q)\right)^{1/2}$$

and therefore in this case

$$\frac{1}{a_n} \left\|\sum_{i=1}^{n} \varepsilon_i f(X_i)\right\|_{\mathcal{F}_{u,q}} \leq \left(\frac{1}{5} \cdot \frac{LLn(q)}{LLn}\right)^{1/2}.$$

Therefore, when $n \geq 2^{-2q} n(q)$ and $n(q)$ is so large that $LL(2^{-2q} n(q)) \geq \frac{1}{2} LLn(q)$, we have

$$(8.53) \qquad \frac{1}{a_n} \left\|\sum_{i=1}^{n} \varepsilon_i f(X_i)\right\|_{\mathcal{F}_{u,q}} \leq \frac{2}{3}.$$

On the other hand, obviously (cf. (8.33)),

$$\left\|\sum_{i=1}^{n} \varepsilon_i f(X_i)\right\|_{\mathcal{F}_{u,q}} \leq u \left\|\sum_{i=1}^{n} \varepsilon_i f(X_i)\right\|_{\mathcal{F}} \leq uc(q) n(q) b(q) \leq \frac{u}{10} a(q)$$

and thus, when $n \geq n(q)$,

$$(8.54) \qquad \frac{1}{a_n}\left\|\sum_{i=1}^{n}\varepsilon_i f(X_i)\right\|_{\mathcal{F}_{u,q}} \leq \frac{u}{10}\left(\frac{n(q)}{n}\right)^{1/2}.$$

Finally, we can write again

$$\left\|\sum_{i=1}^{n}\varepsilon_i f(X_i)\right\|_{\mathcal{F}_{u,q}} \leq u\left\|\sum_{i=1}^{n}\varepsilon_i f(X_i)\right\|_{\mathcal{F}} \leq uc(q)\text{Card}\{i \leq n;\ X_i \in I_q\}.$$

When $n \geq 2^{-2q}n(q)$ (by T_q^3),

$$\left\|\sum_{i=1}^{n}\varepsilon_i f(X_i)\right\|_{\mathcal{F}_{u,q}} \leq 20unc(q)b(q) = 2un\frac{a(q)}{n(q)}$$

and therefore

$$(8.55) \qquad \frac{1}{a_n}\left\|\sum_{i=1}^{n}\varepsilon_i f(X_i)\right\|_{\mathcal{F}_{u,q}} \leq 2u\left(\frac{n}{n(q)}\right)^{1/2}$$

(where it is assumed as before that q is large enough so that $LL(2^{-q}n(q)) \geq \frac{1}{2}LLn(q)$). When $n \leq 2^{-2q}n(q)$, by T_q^3, T_q^4, Card $\{i \leq N;\ X_i \in I_q\} \leq 20 \cdot 2^{-q}a_n/c(q)$ and thus

$$(8.56) \qquad \frac{1}{a_n}\left\|\sum_{i=1}^{n}\varepsilon_i f(X_i)\right\|_{\mathcal{F}_{u,q}} \leq 20u2^{-q}.$$

Recall that u is fixed. Then, we can simply take

$$\xi(n,q) = \begin{cases} 20u2^{-q}a_{\rho(n)} & \text{if } \rho(n) \leq 2^{-2q}n(q), \\ \min\left(\frac{2}{3}a_{\rho(n)}, 2u\left(\frac{\rho(n)}{n(q)}\right)^{1/2}a_{\rho(n)}\right) & \text{if } 2^{-2q}n(q) \leq \rho(n) \leq n(q), \\ \min\left(\frac{2}{3}a_{\rho(n)},\ \frac{u}{10}\left(\frac{n(q)}{\rho(n)}\right)^{1/2}a_{\rho(n)}\right) & \text{if } \rho(n) \geq n(q). \end{cases}$$

By (8.53) – (8.56), the numbers $\xi(n,q)$ satisfy all the required properties. This completes the proof of Lemma 8.16 and therefore of Example 8.12.

Notes and References

Before reaching the rather complete description that we give in this chapter,
the study of the law of the iterated logarithm (LIL) for Banach space valued
random variables went through several stages of partial results and understanding. We do not try here to give a detailed account of the contributions
of all the authors but rather concentrate only on the main steps of the history
of these results. The exposition of this chapter is basically taken from the papers [L-T2] and [L-T5]. Let us mention that the LIL is an important topic in
Probability Theory and one particular aspect only is developed in this work.
For a survey of various LIL topics, we refer to [Bin].

Kolmogorov's LIL appeared in 1929 [Ko] and is of extraordinary accuracy
for the time. A. N. Kolmogorov used sharp both upper and lower exponential
inequalities (carefully described in [Sto]) presented here as Lemma 1.6 and
Lemma 8.1. The extension to the vector valued setting in the form of Theorem
8.2 first appeared in [L-T4] with a limsup only finite. The best result presented
here is new. A first form of the vector valued extension is due to J. Kuelbs
[Kue4].

The independent and identically distributed scalar LIL is due to P. Hartman and A. Wintner [H-W] (with the proof sketched in the text) who extended previous early results in particular by Hardy and Littlewood and by
Khintchine. In particular, A. Khintchine showed (8.8) for a Rademacher sequence and was the first to use blocks of exponential sizes. The necessity of
the moment conditions was established by V. Strassen [St2]; the simple argument presented here is taken from [Fe2]. The simple qualitative proof by
randomization seems to be part of the folklore. For the converse using Gaussian randomization, we refer to [L-T2]. Strassen's paper [St1] is a milestone
in the study of the LIL and strongly influenced the infinite dimensional developments. Various simple proofs of the Hartman-Wintner-Strassen scalar LIL
are now available, cf. e.g. [Ac7] and the references therein.

The setting up of the framework of the study of the LIL for Banach space
valued random variables was undertaken in the early seventies by J. Kuelbs
(cf. e.g. [Kue2]). Theorem 8.5 belongs to him [Kue3] (however assuming a
finite strong second moment, a hypothesis removed in [Pi3]). The definition
of the reproducing kernel Hilbert space of a weak-L_2 random variable and
the Lemma 8.4 on the compactness of its unit ball combine observations of
[Kue3], [Pi3] and [G-K-Z]. The progresses leading to the final characterization of Theorem 8.6 were numerous. To mention some of them, R. LePage
[LP1] first showed the result for Gaussian random variables and G. Pisier
[Pi3] established the LIL for square integrable random variables satisfying the
central limit theorem, a condition weakened later into the boundedness or
convergence to 0 in probability of the sequence (S_n/a_n) by J. Kuelbs [Kue4].
The first true characterization of the LIL in some infinite dimensional space
is due to V. Goodman, J. Kuelbs and J. Zinn [G-K-Z] in Hilbert space after
a preliminary investigation by G. Pisier and J. Zinn [P-Z]. Then, succesively,
several authors extended the conclusion to various classes of smooth normed

spaces [A-K], [Led2], [Led5]. The final characterization was obtained in [L-T2] using a Gaussian randomization technique already suggested in [Pi2] and put forward in the unpublished manuscript [Led6] where the case of type 2 spaces (Corollary 8.8) was settled. Lemma 8.7 is taken from [G-K-Z]. The short proof given here based on the isoperimetric approach of Section 6.3 is taken from [L-T5]. Its consequence to the relation between the LIL and the CLT is discussed in Chapter 10, with results of [Pi3], [G-K-Z], [He2] in particular.

The remarkable results on the cluster set $C(S_n/a_n)$ presented here as Theorem 8.10 are due to K. Alexander [Ale3], [Ale4]. Among other results, he further showed that, when $\mathbb{E}\|X\|^2 < \infty$, $C(S_n/a_n)$ is almost surely empty if it does not contain 0 or, equivalently, if (and only if) $\liminf_{n\to\infty} \mathbb{E}\|S_n\|/a_n > 0$. Prior observations appeared in [Kue6], [G-K-Z] and in [A-K] where it was first shown that $C(S_n/a_n) = K$ when $S_n/a_n \to 0$ in probability (a result to which the first author of this monograph contributed). Theorem 8.11 is taken from [L-T5]. Some observations on possible values of $\Lambda(X)$ in case of the bounded LIL may be found in [A-K-L]. Example 8.12, in the spirit of [Ale4], is new and due to the second author.

Part II

Tightness of Vector Valued Random Variables and Regularity of Random Processes

9. Type and Cotype of Banach Spaces

The notion of type of a Banach space already appeared in the last chapters on the law of large numbers and the law of the iterated logarithm. We observed there that, in quite general situations, almost sure properties can be reduced to properties in probability or in L_p, $0 \leq p < \infty$. Starting with this chapter, we will now study the possibility of a control in probability, or in the weak topology, of probability distributions of sums of independent random variables. On the line or in finite dimensional spaces, such a control is usually easily verified through moment conditions by the orthogonality property

$$\mathbb{E}\left|\sum_i X_i\right|^2 = \sum_i \mathbb{E}|X_i|^2$$

where (X_i) is a finite sequence of independent mean zero real valued random variables (in L_2). This property extends to Hilbert space with norm instead of absolute values but *does not* extend to general Banach spaces.

This observation already indicates the difficulties in showing tightness or boundedness in probability of sums of independent vector valued random variables. This will be in particular illustrated in the next chapter on the central limit theorem, which is indeed one typical example of this tightness question.

However, in some classes of Banach spaces, this general question has reasonable answers. This classification of Banach spaces is based on the concept of type and cotype. These notions have their origin in the preceding orthogonality property which they extend in various ways. They are closely related to the geometric property of knowing whether a given Banach space contains or not subspaces isomorphic to ℓ_p^n. Starting with Dvoretzky's theorem on spherical sections of convex bodies, we describe these relations in the first paragraph of this chapter as a geometric background. A short exposition of some general properties of type and cotype is given in Section 9.2. In the last section, we come back to our starting question and investigate some results on sums of independent random variables in spaces with some type or cotype. In particular, we complete the results on the law of large numbers as announced in Chapter 7. Pregaussian random variables and stable random variables in spaces of stable type are also discussed. We also briefly discuss spaces which do not contain c_0.

9.1 ℓ_p^n-Subspaces of Banach Spaces

Given $1 \leq p \leq \infty$, recall that ℓ_p^n denotes $\mathbb{R}^n$ equipped with the norm $\left(\sum_{i=1}^n |\alpha_i|^p\right)^{1/p}$ ($\max_{i \leq n} |\alpha_i|$ if $p = \infty$), $\alpha = (\alpha_1, \ldots, \alpha_n) \in \mathbb{R}^n$. When $1 \leq p \leq \infty$, n is an integer and $\varepsilon > 0$, a Banach space B is said to *contain a subspace which is $(1 + \varepsilon)$-isomorphic to ℓ_p^n* if there exist $x_1, \ldots, x_n$ in B such that for all $\alpha = (\alpha_1, \ldots, \alpha_n)$ in $\mathbb{R}^n$

$$\left(\sum_{i=1}^n |\alpha_i|^p\right)^{1/p} \leq \left\|\sum_{i=1}^n \alpha_i x_i\right\| \leq (1 + \varepsilon) \left(\sum_{i=1}^n |\alpha_i|^p\right)^{1/p}$$

($\max_{i \leq n} |\alpha_i|$ if $p = \infty$). B contains ℓ_p^n's *uniformly* if it contains subspaces $(1 + \varepsilon)$-isomorphic to ℓ_p^n for all n and $\varepsilon > 0$.

The purpose of this paragraph is to present some results which relate the set of p's for which a Banach space contains ℓ_p^n's to some probabilistic inequalities satisfied by the norm of B. The fundamental result at the basis of this theory is the following theorem of A. Dvoretzky [Dv].

Theorem 9.1. *Any infinite dimensional Banach space B contains ℓ_2^n's uniformly.*

It should be mentioned that the various subspaces isomorphic to ℓ_2^n do not form a net and therefore cannot in general be patched together to form an infinite dimensional Hilbertian subspace of B. We shall give two different proofs of Theorem 9.1, both of them based on Gaussian variables and their properties described in Chapter 3. The first one uses isoperimetric and concentration inequalities (Section 3.1), the second comparison theorems (Section 3.3). They actually yield a stronger finite dimensional version of Theorem 9.1 which may be interpreted geometrically as a result on almost spherical sections of convex bodies (cf. [F-L-M], [Mi-S], [Pi18]). This finite dimensional statement is the following.

Theorem 9.2. *For each $\varepsilon > 0$, there exists $\eta(\varepsilon) > 0$ such that every Banach space B of dimension N contains a subspace $(1 + \varepsilon)$-isomorphic to ℓ_2^n where $n = [\eta(\varepsilon) \log N]$.*

Theorem 9.1 clearly follows from this result. In the two proofs of Theorem 9.2 we will give, we use a crucial intermediate result known as the Dvoretzky-Rogers lemma. It will be convenient to immediately interpret this result in terms of Gaussian variables. Recall from Chapter 3 that if X is a Gaussian Radon random variable with values in a Banach space B, we set $\sigma(X) = \sup_{\|f\| \leq 1} (\mathbb{E} f^2(X))^{1/2}$, and X has strong moments of all orders (Corollary 3.2). We may then consider the "dimension" (or "concentration dimension") of a Gaussian variable X as the ratio

$$d(X) = \frac{\mathbb{E}\|X\|^2}{\sigma(X)^2}.$$

Note that $d(X)$ depends on both X and the norm of B. Since all moments of X are equivalent and equivalent to the median $M(X)$ of X, the replacement of $\mathbb{E}\|X\|^2$ by $(\mathbb{E}\|X\|)^2$ or $M(X)^2$ in $d(X)$ gives rise to a dimension equivalent, up to numerical constants, to $d(X)$. We freely use this observation below. We already mentioned in Chapter 3 that the strong moment of a Gaussian vector valued variable is usually much bigger than the weak moments $\sigma(X)$. Recall for example that if X follows the canonical distribution in $\mathbb{R}^N$ and if $B = \ell_2^N$, then $\sigma(X) = 1$ and $\mathbb{E}\|X\|^2 = N$ so that $d(X) = N$. Note that if $B = \ell_\infty^N$, then $\sigma(X)$ is still 1 but $\mathbb{E}\|X\|^2$ is of the order of $\log N$ (cf. (3.14)). One of the conclusions of Dvoretzky-Rogers' lemma is that the case of ℓ_∞^N is extremal. Let us state without proof the result (cf. [D-R], [F-L-M], [Mi-S], [Pi16], [TJ]).

Lemma 9.3. *Let B be a Banach space of dimension N and let $\overline{N} = [N/2]$ (integer part). There exist points $(x_i)_{i \leq \overline{N}}$ in B such that $\|x_i\| \leq 1/2$ for all $i \leq \overline{N}$ and satisfying*

$$(9.1) \qquad \left\| \sum_{i=1}^{\overline{N}} \alpha_i x_i \right\| \leq \left(\sum_{i=1}^{N} |\alpha_i|^2 \right)^{1/2}$$

for all $\alpha = (\alpha_1, \ldots, \alpha_{\overline{N}})$ in $\mathbb{R}^{\overline{N}}$. In particular, there exists a Gaussian random vector X with values in B whose dimension $d(X)$ is larger than $c \log N$ where $c > 0$ is some numerical constant.

It is easily seen how the second assertion of the lemma follows from the first one. Let indeed $X = \sum_{i=1}^{\overline{N}} g_i x_i$ where (g_i) is orthogaussian. By (9.1),

$$\sigma(X) = \sup_{\|f\| \leq 1} \left(\mathbb{E} f^2(X) \right)^{1/2} = \sup_{\|f\| \leq 1} \left(\sum_{i=1}^{\overline{N}} f^2(x_i) \right)^{1/2} = \sup_{|\alpha| \leq 1} \left\| \sum_{i=1}^{\overline{N}} \alpha_i x_i \right\| \leq 1.$$

On the other hand, since $\|x_i\| \leq 1/2$ for all $i \leq \overline{N}$, by Lévy's inequality (2.7) and (3.14),

$$\mathbb{E}\|X\|^2 \geq \frac{1}{8} \mathbb{E} \max_{i \leq \overline{N}} |g_i|^2 \geq c \log \overline{N}.$$

With this tool, we can now attack the proofs of Theorem 9.2.

First proof of Theorem 9.2. It is based on the concentration properties of the norm of a Gaussian random vector around its median or mean and stability properties of Gaussian distributions. We use Lemma 3.1 but the simpler inequality (3.2) can be used completely similarly. As a first step, we need two technical lemmas to discretize the problem of finding ℓ_2^n-subspaces. We state them in a somewhat more general setting for further purposes. A δ-net ($\delta > 0$) of a set A in $(B, \|\cdot\|)$ is a set S such that for every x in A one can find y in S with $\|x - y\| \leq \delta$.

Lemma 9.4. *For each $\varepsilon > 0$ there is $\delta = \delta(\varepsilon) > 0$ with the following property. Let n be an integer and let $\|\cdot\|$ be a norm on $\mathbb{R}^n$. Let further S be a δ-net of the unit sphere of $(\mathbb{R}^n, \|\cdot\|)$. Then, if for some $x_1, \ldots, x_n$ in a Banach space B, we have $1 - \delta \le \|\sum_{i=1}^n \alpha_i x_i\| \le 1 + \delta$ for all α in S, then*

$$(1 + \varepsilon)^{-1/2} \|\alpha\| \le \left\|\sum_{i=1}^n \alpha_i x_i\right\| \le (1 + \varepsilon)^{1/2} \|\alpha\|$$

for all α in $\mathbb{R}^n$.

Proof. By homogeneity, we can and do assume that $\|\alpha\| = 1$. By definition of S, there exists α^0 in S such that $\|\alpha - \alpha^0\| \le \delta$, hence $\alpha = \alpha^0 + \lambda_1 \beta$ with $\|\beta\| = 1$ and $|\lambda_1| \le \delta$. Iterating the procedure, we get that $\alpha = \sum_{j=0}^\infty \lambda_j \alpha^j$ with $\alpha^j \in S$, $|\lambda_j| \le \delta^j$ for all $j \ge 0$. Hence

$$\left\|\sum_{i=1}^n \alpha_i x_i\right\| \le \sum_{j=0}^\infty \delta^j \left\|\sum_{i=1}^n \alpha_i^j x_i\right\| \le \frac{1 + \delta}{1 - \delta}$$

($\delta < 1$). In the same way, $\|\sum_{i=1}^n \alpha_i x_i\| \ge (1 - 3\delta)/(1 - \delta)$. It therefore suffices to choose appropriately δ in function of $\varepsilon > 0$ only to get the conclusion.

The size of a δ-net of spheres in finite dimension is easily estimated in terms of $\delta > 0$ and the dimension, as is shown in the next lemma which follows from a simple volume argument.

Lemma 9.5. *Let $\|\cdot\|$ be any norm on $\mathbb{R}^n$. There is a δ-net S of the unit sphere of $(\mathbb{R}^n, \|\cdot\|)$ of cardinality less than $(1 + 2/\delta)^n \le \exp(2n/\delta)$.*

Proof. Let U denote the unit ball of $(\mathbb{R}^n, \|\cdot\|)$ and let $(x_i)_{i \le m}$ be maximal in the unit sphere of U under the relations $\|x_i - x_j\| > \delta$ for $i \ne j$. Then the balls $x_i + (\delta/2)U$ are disjoint and are all contained in $(1 + 2/\delta)U$. By comparing the volumes, we get $m(\delta/2)^n \le (1 + 2/\delta)^n$ from which the lemma follows.

We are in a position to perform the first proof of Theorem 9.2. The main argument is the concentration property of Gausian vectors. Let B be of dimension N. By Lemma 9.3 there exists a Gaussian variable X with values in B whose dimension is larger than $c \log N$. Let $M = M(X)$ denote the median of X. Since (Corollary 3.2) M is equivalent to $\|X\|_2$, we also have that $M \ge c(\log N)^{1/2} \sigma(X)$ where $c > 0$ is numerical (possibly different from the preceding). Let n to be specified and consider independent copies $X_1, \ldots, X_n$ of X. With positive probability, the sample $(X_1, \ldots, X_n)$ will be shown to span a subspace which is almost isometric to ℓ_2^n. More precisely, the basic rotational invariance property of Gaussian distributions indicates that if $\sum_{i=1}^n \alpha_i^2 = 1$, then $\sum_{i=1}^n \alpha_i X_i$ has the same distribution as X. In particular, by Lemma 3.1, for all $t > 0$ and $\alpha_1, \ldots, \alpha_n$ with $\sum_{i=1}^n \alpha_i^2 = 1$,

$$\mathbb{P}\left\{\left|\left\|\sum_{i=1}^{n}\alpha_i X_i\right\| - M\right| > t\right\} \le \exp\left(-\frac{t^2}{2\sigma(X)^2}\right).$$

Now, let $\varepsilon > 0$ be fixed and choose $\delta = \delta(\varepsilon) > 0$ according to Lemma 9.4. Let furthermore S be a δ-net of the unit sphere of ℓ_2^n which can be chosen of cardinality less than $\exp(2n/\delta)$ (Lemma 9.5). Let $t = \delta M$; the preceding inequality implies

$$\mathbb{P}\left\{\exists\,\alpha \in S : \left|\left\|\sum_{i=1}^{n}\alpha_i\frac{X_i}{M}\right\| - 1\right| > \delta\right\} \le (\text{Card } S)\exp\left(-\frac{\delta^2 M^2}{2\sigma(X)^2}\right)$$

$$\le \exp\left(\frac{2n}{\delta} - \frac{\delta^2 c^2}{2}\log N\right)$$

since $M \ge c(\log N)^{1/2}\sigma(X)$. If we assume N large enough (otherwise there is nothing to prove), we can choose $n = [\eta \log N]$ for $\eta = \eta(\delta) = \eta(\varepsilon)$ small enough such that the preceding probability is strictly less than 1. It follows that there exists an ω such that for all α in S

$$1 - \delta \le \left\|\sum_{i=1}^{n}\alpha_i\frac{X_i(\omega)}{M}\right\| \le 1 + \delta.$$

Hence, for $x_i = X_i(\omega)/M$, $i \le n$, we are in the hypotheses of Lemma 9.4 so that the conclusion readily follows.

The second proof is shorter and neater but relies on the somewhat more involved tools of Theorem 3.16 and Corollary 3.21. Indeed, while isoperimetric arguments were used in the first proof, this was only through concentration inequalities which we know, following e.g. (3.2), can actually be established rather easily. The discretization lemmas are not necessary in this second approach.

Second proof of Theorem 9.2. If X is a Gaussian random variable with values in B, denote by $X_1, \ldots, X_n$ independent copies of X. We apply Corollary 3.21. Set $\varphi = \inf_{|\alpha|=1}\|\sum_{i=1}^{n}\alpha_i X_i\|$, $\Phi = \sup_{|\alpha|=1}\|\sum_{i=1}^{n}\alpha_i X_i\|$. Clearly, $0 \le \varphi \le \Phi$ and, by Corollary 3.21,

$$\mathbb{E}\|X\| - \sqrt{n}\sigma(X) \le \mathbb{E}\varphi \le \mathbb{E}\Phi \le \mathbb{E}\|X\| + \sqrt{n}\sigma(X).$$

It follows that for some ω, if $\mathbb{E}\|X\| > \sqrt{n}\sigma(X)$,

$$\frac{\Phi(\omega)}{\varphi(\omega)} \le \frac{\mathbb{E}\|X\| + \sqrt{n}\sigma(X)}{\mathbb{E}\|X\| - \sqrt{n}\sigma(X)}.$$

Choose now X in B according to Lemma 9.3 so that $\mathbb{E}\|X\| > c(\log N)^{1/2}\sigma(X)$ for some numerical $c > 0$. Then, if $\varepsilon > 0$ and $n = [\eta(\varepsilon)\log N]$ where $\eta(\varepsilon) > 0$ can be chosen depending on $\varepsilon > 0$ only,

$$\frac{\mathbb{E}\|X\| + \sqrt{n}\sigma(X)}{\mathbb{E}\|X\| - \sqrt{n}\sigma(X)} \leq \frac{c(\log N)^{1/2} + \sqrt{n}}{c(\log N)^{1/2} - \sqrt{n}} \leq 1 + \varepsilon,$$

so that $\Phi(\omega) \leq (1+\varepsilon)\varphi(\omega)$. But now, by definition of φ and Φ, for every α in $\mathbb{R}^n$ with $|\alpha| = 1$,

$$\varphi(\omega) \leq \left\|\sum_{i=1}^{n} \alpha_i X_i(\omega)\right\| \leq \Phi(\omega).$$

Hence, setting $x_i = X_i(\omega)/\varphi(\omega)$, $i \leq n$, for every $|\alpha| = 1$,

$$1 \leq \left\|\sum_{i=1}^{n} \alpha_i x_i\right\| \leq 1 + \varepsilon$$

which means, by homogeneity, that B contains a subspace $(1+\varepsilon)$-isomorphic to ℓ_2^n. The proof is complete.

Let us note that this second proof provides a dependence of $\eta(\varepsilon)$ in function of ε in Theorem 9.2 of the order of $c\varepsilon^2$ where c is numerical. This is best possible. Recently, G. Schechtman [Sch5] has shown how the more classical isoperimetric approach of the first proof can be modified so as to also yield this best possible dependence.

According thus to Theorem 9.1, every infinite dimensional Banach space contains ℓ_2^n's uniformly. Clearly, this does not extend to ℓ_p^n's for $p \neq 2$ as can be seen from the example of Hilbert space. Related to this question, note that if, for $0 < p \leq 2$, (θ_i) denotes a sequence of independent standard p-stable random variables defined on some probability space $(\Omega, \mathcal{A}, \mathbb{P})$, then, when $p = 2$, (θ_i) (which is then simply an orthogaussian sequence) spans ℓ_2 in $L_q = L_q(\Omega, \mathcal{A}, \mathbb{P})$ for every q, whereas for $p < 2$, (θ_i) spans ℓ_p in L_q only for $q < p$.

It is remarkable that, at least in the case $1 \leq p < 2$, the set of p's for which a Banach space B contains ℓ_p^n's uniformly can be characterized by a probabilistic inequality satisfied by the norm of B. This is what we would like to describe in the rest of this section. The case $p > 2$ will be shortly presented once the notion of cotype will be introduced in the next paragraph.

Let (θ_i) denote as usual a sequence of independent standard p-stable random variables. For $1 \leq p \leq 2$, a Banach space B is said to be of *stable type p* if there is a constant C such that for every finite sequence (x_i) in B

$$(9.2) \qquad \left\|\sum_i \theta_i x_i\right\|_{p,\infty} \leq C \left(\sum_i \|x_i\|^p\right)^{1/p}.$$

The integrability properties and moment equivalences of stable random vectors (Chapter 5) tell us that an equivalent definition of the stable type property is obtained when $\|\cdot\|_{p,\infty}$ is replaced by any $\|\cdot\|_r$, $r < p$. Further, in terms of infinite sequences and using a closed graph argument, B is of stable type p if and only if $\sum_i \theta_i x_i$ converges almost surely whenever $\sum_i \|x_i\|^p < \infty$. In other words, the existence of the spectral measure of the stable variable

$\sum_i \theta_i x_i$ determines its convergence. (As we know, cf. (5.13), this is automatic when $0 < p < 1$ and this is why the range of the stable type is $1 \leq p \leq 2$.) A Banach space B of stable type p is also of stable type p' for every $p' < p$. This is contained in Proposition 9.12 below but we may briefly anticipate one argument in this regard here. We may assume that $p > 1$. Then, as a consequence of the contraction principle (Lemma 4.5), for some $C, r > 0$ and all finite sequences (x_i) in B,

$$(9.3) \qquad \left\| \sum_i \varepsilon_i x_i \right\|_r \leq C \left(\sum_i \|x_i\|^p \right)^{1/p}.$$

In particular, since $p' < p$,

$$\mathbb{E} \left\| \sum_i \varepsilon_i x_i \right\|^r \leq C' \|(x_i)\|_{p',\infty}^r$$

where $C' = C'(r, p, p')$. Let now (θ_i') be a standard p'-stable sequence. Since (θ_i) has the same distribution as $(\varepsilon_i \theta_i')$, this inequality applied conditionally yields

$$\mathbb{E} \left\| \sum_i \theta_i' x_i \right\|^r \leq C' \mathbb{E} \|(\theta_i' x_i)\|_{p',\infty}^r$$

(choose $r < p'$). Using now Lemma 5.8 and the basic fact that $\|\theta_1'\|_{p',\infty} < \infty$, B is seen to be of stable type p'.

As a consequence of the preceding, there is some interest to consider the number

$$(9.4) \qquad p(B) = \sup\{p;\ B \text{ is of stable type } p\}.$$

Examples of spaces of stable type p will be given in the next section once the general theory of type and cotype has been developed. Let us however mention an important example at this stage. Let $1 \leq q < 2$ and let $L_q = L_q(S, \Sigma, \mu)$ on some measure space (S, Σ, μ). Then L_q is of stable type p for all $p < q$. This can be seen for example from the preceding; indeed, a simple use of Fubini's theorem together with Khintchine's inequalities shows that (9.3) holds with $r = p = q$. It then follows that L_q is of stable type p. Unless finite dimensional, L_q is however *not* of stable type q. Indeed, according to (5.19), the canonical basis of ℓ_q cannot satisfy the q-stable type inequality (9.2). Since the stable type property clearly only depends on the collection of the finite dimensional subspaces of a given Banach space, we have similarly that a Banach space containing ℓ_p^n's uniformly ($1 \leq p < 2$) is *not* of stable type p. The following theorem expresses the striking fact that the converse also holds.

Theorem 9.6. *Let $1 \leq p < 2$. A Banach space B contains ℓ_p^n's uniformly if and only if B is not of stable type p.*

Before we turn to the proof, let us first state some important and useful consequences of Theorem 9.6. The first one expresses that the set of p's for which a Banach space is of stable type p is open. The second answers the question addressed in connection with Dvoretzky's theorem for the p's such that $1 \leq p \leq 2$. Recall $p(B)$ of (9.4).

Corollary 9.7. *Let $1 \leq p < 2$. If a Banach space is of stable type p, then it is also of stable type p_1 for some $p_1 > p$.*

Proof. It is rather elementary to check that the set of p's in $[1, 2]$ for which a Banach space contains ℓ_p^n's uniformly is a closed subset of $[1, 2]$. Therefore its complement is open and Theorem 9.6 allows to conclude.

Corollary 9.8. *The set of p's of $[1, 2]$ for which an infinite dimensional Banach space B contains ℓ_p^n's uniformly is equal to $[p(B), 2]$.*

Proof. If $p(B) = 2$, B contains ℓ_2^n's uniformly by Theorem 9.1 and if $p < 2$, B is of stable type p and Theorem 9.6 can be applied. By definition of $p(B)$ and Corollary 9.7, if $p(B) \leq p < 2$, B is not of stable type p and therefore contains ℓ_p^n's uniformly whereas for $p < p(B)$, B is of stable type p and Theorem 9.6 applies again.

We next turn the proof of Theorem 9.6 and, as for Theorem 9.1, we will deduce this result from some stronger finite dimensional version. If B is a Banach space and $1 < p < 2$, denote by $ST_p(B)$ the smallest constant C for which (9.2) *with the L_1-norm on the left* (yielding, as we know, an equivalent condition) holds.

Theorem 9.9. *Let $1 < p < 2$ and $q = p/p - 1$ be the conjugate of p. For each $\varepsilon > 0$ there exists $\eta_p(\varepsilon) > 0$ such that every Banach space B of stable type p contains a subspace which is $(1 + \varepsilon)$-isomorphic to ℓ_p^n where $n = [\eta_p(\varepsilon) ST_p(B)^q]$.*

This statement clearly contains Theorem 9.6. Indeed, if $1 < p < 2$ and B is not of stable type p, then $ST_p(B) = \infty$ and one can find finite dimensional subspaces of B with corresponding stable type constants which are arbitrarily large and hence, by Theorem 9.9, B contains ℓ_p^n's uniformly. Since the set of p's of $[1, 2]$ for which B contains ℓ_p^n's uniformly is closed, we reach the case $p = 1$ as well. That Banach spaces containing ℓ_p^n's uniformly are not of stable type p has been discussed prior to Theorem 9.6.

Proof of Theorem 9.9. It will follow the pattern of the first proof of Theorem 9.2 and will rely, by the series representation of stable random vectors, on the concentration inequality (6.14) for sums of independent random variables. Let $S = \frac{1}{2} ST_p(B)$; by definition of $ST_p(B)$, there exist some non-zero $x_1, \ldots, x_N$ in B such that

$$\sum_{i=1}^{N} \|x_i\|^p = 1 \quad \text{and} \quad \mathbb{E}\left\|\sum_{i=1}^{N} \theta_i x_i\right\| > S$$

(recall that (θ_i) is a standard p-stable sequence). (In a sense, this is the step corresponding to Lemma 9.3 in Theorem 9.2 but, whereas Lemma 9.3 holds true in any Banach space, the stable type constant enters the question here.) Let Y with values on the unit sphere of B be defined as $Y = \pm x_i/\|x_i\|$ with probability $\|x_i\|^p/2$, $i \leq N$. Let (Y_j) be independent copies of Y. Then, as a consequence of Corollary 5.5, $X = c_p^{-1} \sum_{i=1}^{N} \theta_i x_i$ has the same distribution as $\sum_{j=1}^{\infty} \Gamma_j^{-1/p} Y_j$. Now, consider

$$Z = \sum_{j=1}^{\infty} j^{-1/p} Y_j$$

and let (X_i) (resp. (Z_i)) denote independent copies of X (resp. Z). Let furthermore $\alpha = (\alpha_1, \ldots, \alpha_n)$ in $\mathbb{R}^n$ be such that $\sum_{i=1}^{n} |\alpha_i|^p = 1$. We first claim that inequality (6.14) implies that for every $t > 0$,

(9.5) $$\mathbb{P}\left\{\left|\left\|\sum_{i=1}^{n} \alpha_i Z_i\right\| - \mathbb{E}\left\|\sum_{i=1}^{n} \alpha_i Z_i\right\|\right| > t\right\} \leq 2\exp(-t^q/C_q).$$

Indeed, if we have independent copies $(Y_{j,i})_i$ of the sequence (Y_j),

$$\sum_{i=1}^{n} \alpha_i Z_i = \sum_{j=1}^{\infty} \sum_{i=1}^{n} \alpha_i j^{-1/p} Y_{j,i}$$

which, since the $Y_{j,i}$ are iid and symmetric, has the same distribution as

$$\sum_{k=1}^{\infty} \beta_k Y_k$$

where $(\beta_k)_{k \geq 1}$ is the non-increasing rearrangement of the doubly-indexed collection $\{|\alpha_i| j^{-1/p}; j \geq 1, 1 \leq i \leq n\}$. It is easily seen, since $\sum_{i=1}^{n} |\alpha_i|^p = 1$, that $\beta_k \leq k^{-1/p}$, so that (6.14) applied to the preceding sum of independent random variables indeed yields (9.5).

From (9.5), the idea of the proof is exactly the same as that of the proof of Theorem 9.2 and the subspace isomorphic to ℓ_p^n will be found at random. However, while the X_i's are stable random variables and therefore, by the fundamental stability property, for $\sum_{i=1}^{n} |\alpha_i|^p = 1$,

(9.6) $$\mathbb{E}\left\|\sum_{i=1}^{n} \alpha_i X_i\right\| = \mathbb{E}\|X\| > c_p^{-1} S,$$

this is no longer exactly the case for the Z_i's. This would however be needed in view of (9.5). But Z is actually close enough to X so that this stability property can almost be saved for the Z_i's. Indeed, we know from (5.8) that

$$\mathbb{E}\left\|\sum_{j=1}^{\infty} Y_j\left(\Gamma_j^{-1/p} - j^{-1/p}\right)\right\| \le \sum_{j=1}^{\infty} \mathbb{E}|\Gamma_j^{-1/p} - j^{-1/p}| = D_p$$

where D_p is a finite constant depending on p only. Hence, by the triangle inequality and Hölder's inequality, for $\sum_{i=1}^{n} |\alpha_i|^p = 1$,

$$\left| \mathbb{E}\left\|\sum_{i=1}^{n} \alpha_i X_i\right\| - \mathbb{E}\left\|\sum_{i=1}^{n} \alpha_i Z_i\right\| \right| \le D_p \sum_{i=1}^{n} |\alpha_i| \le D_p n^{1/q}.$$

Let now $\delta > 0$ and impose, as a first condition on n, that

$$(9.7) \qquad\qquad D_p n^{1/q} \le \delta S/2c_p.$$

By (9.6), setting $M = \mathbb{E}\|X\|$, we see that

$$\left| \mathbb{E}\left\|\sum_{i=1}^{n} \alpha_i Z_i\right\| - M \right| \le \frac{\delta M}{2}.$$

Hence (9.5) for $t = \delta M/2$ yields

$$\mathbb{P}\left\{ \left| \left\|\sum_{i=1}^{n} \alpha_i Z_i\right\| - M \right| > \delta M \right\} \le 2\exp(-\delta^q M^q/2^q C_q)$$

$$\le 2\exp(-\delta^q S^q/2^q c_p^q C_q).$$

The proof is almost complete. Let R be a δ-net of the unit sphere of ℓ_p^n which can be chosen, according to Lemma 9.5, with a cardinality less than $\exp(2n/\delta)$. Then

$$\mathbb{P}\left\{ \forall \alpha \in R; \left| \left\|\sum_{i=1}^{n} \alpha_i Z_i\right\| - M \right| \le \delta M \right\} \ge 1 - 2\exp\left(\frac{2n}{\delta} - \frac{\delta^q S^q}{2^q c_p^q C_q} \right).$$

Given $\varepsilon > 0$, choose $\delta = \delta(\varepsilon) > 0$ small enough according to Lemma 9.4. Take then $\eta = \eta(\delta) = \eta(\varepsilon) > 0$ such that if $n = [\eta S T_p(B)^q]$ ($S T_p(B)$ being assumed large enough otherwise there is nothing to prove), (9.7) holds and the preceding probability is strictly positive. Then, it follows that one can find an ω such that for every α in $\mathbb{R}^n$

$$(1+\varepsilon)^{-1/2}\left(\sum_{i=1}^{n} |\alpha_i|^p \right)^{1/p} \le \left\| \sum_{i=1}^{n} \alpha_i \frac{Z_i(\omega)}{M} \right\| \le (1+\varepsilon)^{1/2}\left(\sum_{i=1}^{n} |\alpha_i|^p \right)^{1/p}$$

which gives the conclusion. Theorem 9.9 is thus established.

9.2 Type and Cotype

In the preceding section, we discovered how the probabilistic conditions of stable type are related to some geometric properties of Banach spaces. We start in this paragraph a systematic study of related probabilistic conditions named type and cotype (or Rademacher type and cotype).

As usual, (ε_i) denotes a Rademacher sequence. Let $1 \leq p < \infty$. A Banach space B is said to be of *type p* (or *Rademacher type p*) if there is a constant C such that for all finite sequences (x_i) in B,

$$(9.8) \qquad \left\| \sum_i \varepsilon_i x_i \right\|_p \leq C \left(\sum_i \|x_i\|^p \right)^{1/p}.$$

By the triangle inequality, every Banach space is of type 1. On the other hand, Khintchine's inequalities indicate that the definition makes sense for $p \leq 2$ only. Note moreover that the Khintchine-Kahane inequalities in the form of the equivalence of moments of Rademacher series (Theorem 4.7) show that, replacing the p-th moment of $\sum_i \varepsilon_i x_i$ by any other moment, leads to an equivalent definition. Furthermore, by a closed graph argument, B is of type p if and only if $\sum_i \varepsilon_i x_i$ converges almost surely when $\sum_i \|x_i\|^p < \infty$.

Let $1 \leq q \leq \infty$. A Banach space B is said to be of *(Rademacher) cotype q* if there is a constant C such that for all finite sequences (x_i) in B

$$(9.9) \qquad \left(\sum_i \|x_i\|^q \right)^{1/q} \leq C \left\| \sum_i \varepsilon_i x_i \right\|_q$$

$$\left(\sup_i \|x_i\| \leq C \left\| \sum_i \varepsilon_i x_i \right\|_1 \quad \text{when} \quad q = \infty \right).$$

By Lévy's inequalities (Proposition 2.7, (2.7)), or actually by some easy direct argument based on the triangle inequality, every Banach space is of infinite cotype whereas, by Khintchine's inequalities, the definition of cotype q actually reduces to $q \geq 2$. The same comments as for the type apply: any moment of the Rademacher average in (9.9) leads to an equivalent definition and B is of cotype q if and only if the almost sure convergence of the series $\sum_i \varepsilon_i x_i$ implies $\sum_i \|x_i\|^q < \infty$.

It is clear from the preceding comments that a Banach space of type p (resp. cotype q) is also of type p' for every $p' \leq p$ (resp. of cotype q' for every $q' \geq q$). Thus, the "best" possible spaces in terms of the type and cotype conditions are the spaces of type 2 and cotype 2. Hilbert spaces have this property since there, by *orthogonality*,

$$\mathbb{E} \left\| \sum_i \varepsilon_i x_i \right\|^2 = \sum_i \|x_i\|^2.$$

It is a basic result of the theory due to S. Kwapień [Kw1] that the converse (up to isomorphism) also holds.

Theorem 9.10. *A Banach space is of type 2 and cotype 2 if only if it is isomorphic to a Hilbert space.*

If a Banach space is of type p and cotype q so are all its subspaces. Actually, the type and cotype properties are seen to depend only on the collection of the finite dimensional subspaces. It is not difficult to verify that quotients of a Banach space of type p are also of type p, with the same constant. This is no longer the case however for the cotype as is clear for example from the fact that every Banach space can be identified to a quotient of L_1 and that (see below) L_1 is of best possible cotype, that is cotype 2.

To mention some examples, let (S, Σ, μ) be a measure space and let $L_p = L_p(S, \Sigma, \mu)$, $1 \leq p \leq \infty$. Fubini's theorem and Khintchine's inequalities can easily be used to determine the type and cotype of the L_p-spaces. Assume that $1 \leq p < \infty$. Then, L_p is of type p when $p \leq 2$ and of type 2 for $p \geq 2$. Let us briefly check these assertions. Let (x_i) be a finite sequence in L_p. Using Lemma 4.1, we can write

$$\mathbb{E}\left\|\sum_i \varepsilon_i x_i\right\|^p = \int_S \mathbb{E}\left|\sum_i \varepsilon_i x_i(s)\right|^p d\mu(s) \leq B_p^p \int_S \left(\sum_i |x_i(s)|^2\right)^{p/2} d\mu(s).$$

If $p \leq 2$,

$$\int_S \left(\sum_i |x_i(s)|^2\right)^{p/2} d\mu(s) \leq \int_S \sum_i |x_i(s)|^p d\mu(s) = \sum_i \|x_i\|^p,$$

whereas, when $p \geq 2$, by the triangle inequality,

$$\left(\int_S \left(\sum_i |x_i(s)|^2\right)^{p/2} d\mu(s)\right)^{2/p} \leq \sum_i \left(\int_S |x_i(s)|^p d\mu(s)\right)^{2/p} = \sum_i \|x_i\|^2.$$

By considering the canonical basis of ℓ_p one can easily show that the preceding result cannot be improved, i.e., if L_p is infinite dimensional, it is of no type $p' > p$. It can be shown similarly that L_p is of cotype p for $p \geq 2$ and cotype 2 for $p \leq 2$ (and nothing better). Note that L_1 is of cotype 2 as we mentioned it previously. We are left with the case $p = \infty$. It is obvious on the canonical basis that ℓ_1 is of no type $p > 1$ and c_0 (or ℓ_∞) of no cotype $q < \infty$. Since L_∞ isometrically contains any separable Banach space, in particular ℓ_1 and c_0, L_∞ is of type 1 and cotype ∞ and nothing more, and so is c_0. Similarly, $C(S)$ the space of all continuous functions on a compact metric space S equipped with the sup-norm has no non-trivial type or cotype.

Using the equivalence of moments of vector valued Rademacher averages (Theorem 4.7) instead of Khintchine's inequalities, the preceding examples can easily be generalized. Let B be a Banach space of type p and cotype q. Then, for $1 \leq r \leq \infty$, $L_r(S, \Sigma, \mu; B)$ is of type $\min(r, p)$ and of cotype $\max(r, q)$.

Let us mention further that the type and cotype properties appear as dual notions. Indeed, if a Banach space B is of type p, its dual space B' is of cotype $q = p/p - 1$. To check this, let (x_i') be a finite sequence in B'. For each $\varepsilon > 0$ and each i, let x_i in B, $\|x_i\| = 1$, such that $x_i'(x_i) = \langle x_i', x_i \rangle \geq (1 - \varepsilon)\|x_i'\|$ where $\langle \cdot, \cdot \rangle$ is duality. We then have:

$$(1 - \varepsilon) \sum_i \|x_i'\|^q \leq \sum_i \langle x_i', x_i \rangle \|x_i'\|^{q-1}$$

$$= \mathbb{E}\left(\sum_{i,j} \varepsilon_i \varepsilon_j \langle x_j', x_i \rangle \|x_i'\|^{q-1} \right)$$

$$= \mathbb{E}\left(\left\langle \sum_i \varepsilon_i x_i \|x_i'\|^{q-1}, \sum_j \varepsilon_j x_j' \right\rangle \right).$$

Hence, by Hölder's inequality (assuming $p > 1$) and the type p property of B,

$$(1 - \varepsilon) \sum_i \|x_i'\|^q \leq \left\| \sum_i \varepsilon_i x_i \|x_i'\|^{q-1} \right\|_p \left\| \sum_i \varepsilon_i x_i' \right\|_q$$

$$\leq C \left(\sum_i \|x_i'\|^{p(q-1)} \right)^{1/p} \left\| \sum_i \varepsilon_i x_i' \right\|_q.$$

It follows that B' is of cotype q (with constant C). The converse assertion is not true in general since ℓ_1 is of cotype 2 but ℓ_∞ is of no type $p > 1$. A deep result of G. Pisier [Pi11] implies that the cotype is dual to the type if the Banach space does not contain ℓ_1^n's uniformly (i.e., by Theorem 9.6, if it is of some non-trivial type).

After these preliminaries and examples on the notions of type and cotype, we now examine several questions concerning the replacement in the definitions (9.8) and (9.9) of the Rademacher sequence by some other sequences of random variables. We start with an elementary general result. For reasons that will become clearer in the next section, we only deal with Radon random variables.

Proposition 9.11. *Let B be a Banach space of type p and cotype q with type constant C_1 and cotype constant C_2. Then, for every finite sequence (X_i) of independent mean zero Radon random variables in $L_p(B)$ (resp. $L_q(B)$),*

$$\mathbb{E}\left\| \sum_i X_i \right\|^p \leq (2C_1)^p \sum_i \mathbb{E}\|X_i\|^p$$

and

$$\mathbb{E}\left\| \sum_i X_i \right\|^q \geq (2C_2)^{-q} \sum_i \mathbb{E}\|X_i\|^q$$

Proof. We show the assertion relative to the type. By Lemma 6.3,

$$\mathbb{E}\left\| \sum_i X_i \right\|^p \leq 2^p \mathbb{E}\left\| \sum_i \varepsilon_i X_i \right\|^p$$

where (ε_i) is a Rademacher sequence which is independent of (X_i). Applying (9.8) conditionally on the X_i's then immediately yields the result.

Now, we investigate the case where the Rademacher sequence (ε_i) is replaced by a p-stable standard sequence (θ_i) $(1 \leq p \leq 2)$. This will in particular clarify the close relationship between the Rademacher type (9.8) and the stable type discussed in the preceding section. Let $1 \leq p \leq 2$. Recall that a Banach space B is said to be of stable type p if there is a constant C such that for all finite sequences (x_i) in B,

$$\left\|\sum_i \theta_i x_i\right\|_{p,\infty} \leq C\left(\sum_i \|x_i\|^p\right)^{1/p}$$

where (θ_i) is a sequence of independent standard p-stable variables.

Proposition 9.12. *Let $1 \leq p < 2$ and let B be a Banach space. Then, we have:*

(i) If B is of stable type p, it is of type p.

(ii) If B is of type p, it is of stable type p' for every $p' < p$.

(iii) B is of stable type p if and only if there is a constant C such that for all finite sequences (x_i) in B,

$$\mathbb{E}\left\|\sum_i \varepsilon_i x_i\right\| \leq C\|(x_i)\|_{p,\infty} .$$

Proof. Both (i) and (ii) follow from the more difficult claim (iii) but can however be given simple proofs. Indeed, concerning (i), we may assume that $p > 1$ so that $\mathbb{E}|\theta_i| < \infty$ and (i) follows from Lemma 4.5. For (ii), let $1 < p' < p$ and let (θ_i') denote a standard p'-stable sequence. Recall that since $p > p'$,

$$\left(\sum_i |\alpha_i|^p\right)^{1/p} \leq \left(\frac{p}{p-p'}\right)^{1/p} \|(\alpha_i)\|_{p',\infty} .$$

Applying conditionally the type p inequality and the preceding, for (x_i) a finite sequence in B and for some constant C not necessarily the same at each line,

$$\mathbb{E}\left\|\sum_i \theta_i' x_i\right\| = \mathbb{E}\left\|\sum_i \varepsilon_i \theta_i' x_i\right\| \leq C\mathbb{E}\left(\left(\sum_i |\theta_i'|^p \|x_i\|^p\right)^{1/p}\right)$$

$$\leq C\mathbb{E}(\|(|\theta_i'|\|x_i\|)\|_{p',\infty}) .$$

We can conclude by Lemma 5.8 since $p' > 1$ and since $\|\theta_i'\|_{p',\infty} < \infty$.

The "if" part of (iii) reproduces the proof of (ii) that we just gave (with $p' = p$). Conversely, if B is of stable type p, by Corollary 9.7, then B is also of stable type p_1 for some $p_1 > p$, hence of type p_1 by (i). By the comparison between the ℓ_{p_1} and $\ell_{p,\infty}$-norms, the proof is complete.

Note, as a consequence of this proposition, that the number $p(B)$ introduced in (9.4) is also given by

$$p(B) = \sup\{p; \ B \text{ is of type } p\}.$$

Moreover, we know that ℓ_1 is of no type $p > 1$; by Proposition 9.12 and Theorem 9.6, we actually have that a Banach space B is of some type $p > 1$, or equivalently of stable type 1 or of stable type p for some $p > 1$, *if and only if B does not contain ℓ_1^n's uniformly.*

As a consequence of Proposition 9.12, note the following version of Proposition 9.11 for the stable type.

Proposition 9.13. *Let $1 \leq p < 2$. A Banach space B is of stable type p if and only if there is a constant C such that for every finite sequence (X_i) of independent symmetric Radon random variable in $L_{p,\infty}(B)$,*

$$\left\|\sum_i X_i\right\|_{p,\infty}^p \leq C \sup_{t>0} t^p \sum_i \mathbb{P}\{\|X_i\| > t\},$$

or, equivalently, if and only if,

$$\left\|\sum_i X_i\right\|_{p,\infty}^p \leq C \sum_i \|X_i\|_{p,\infty}^p.$$

Proof. The "if" part follows by simply letting $X_i = \theta_i x_i$ in the second inequality. We establish the first inequality (which clearly implies the second) in spaces of stable type p. Assume by homogeneity that we have $\sup_{t>0} t^p \sum_i \mathbb{P}\{\|X_i\| > t\} \leq 1$. Let $t > 0$. Since $t^p \mathbb{P}\{\max_i \|X_i\| > t\} \leq 1$,

$$t^p \mathbb{P}\left\{\left\|\sum_i X_i\right\| > t\right\} \leq 1 + t^p \mathbb{P}\left\{\left\|\sum_i X_i I_{\{\|X_i\| \leq t\}}\right\| > t\right\}.$$

Since B is of stable type p, it is also of type p' for some $p' > p$. Hence by Proposition 9.11, for some C,

$$t^p \mathbb{P}\left\{\left\|\sum_i X_i\right\| > t\right\} \leq 1 + t^{p-p'} \mathbb{E}\left\|\sum_i X_i I_{\{\|X_i\| \leq t\}}\right\|^{p'}$$

$$\leq 1 + C t^{p-p'} \sum_i \mathbb{E}\left(\|X_i\|^{p'} I_{\{\|X_i\| \leq t\}}\right).$$

Now

$$\sum_i \mathbb{E}\left(\|X_i\|^{p'} I_{\{\|X_i\| \leq t\}}\right) \leq \int_0^t \sum_i \mathbb{P}\{\|X_i\| > s\} ds^{p'}$$

$$\leq \int_0^t \frac{ds^{p'}}{s^p} = \frac{p'}{p'-p} t^{p'-p}$$

and the conclusion follows.

Next, we turn to the case of a 2-stable standard sequence, that is an or-thogaussian sequence (g_i), which will lead, by the same way, to some questions analogous to the preceding ones in the context of the cotype. We first complete the simple case of the type for orthogaussian sequences. Indeed, if B is of type p, by Proposition 9.11,

$$\mathbb{E}\left\|\sum_i g_i x_i\right\|^p \leq C \sum_i \|x_i\|^p$$

for some constant C. Conversely, since Gaussian averages always dominate Rademacher averages ((4.8)), such an inequality implies that B must be of type p. In particular, stable type 2 and Rademacher type 2 are equivalent notions.

We have seen in a discussion after Lemma 4.5 that, conversely, Rademacher averages do not always dominate the corresponding Gaussian averages, in particular in ℓ_∞. That is, if in a Banach space B, for some constant C and all finite sequences (x_i),

$$(9.10) \qquad \sum_i \|x_i\|^q \leq C \mathbb{E}\left\|\sum_i g_i x_i\right\|^q ,$$

then B is not obviously of cotype q. This is however true and we now would like to describe some of the deep steps leading to this conclusion, mainly without proofs. The next proposition already covers various applications. Recall that for a (real valued) random variable ξ, we set

$$\|\xi\|_{q,1} = \int_0^\infty \left(\mathbb{P}\{|\xi| > t\}\right)^{1/q} dt .$$

Note that if $s > q$, $\|\xi\|_{q,1} \leq q/(s-q)\|\xi\|_s$.

Proposition 9.14. *Let $1 \leq r < \infty$ and let (ξ_i) be a sequence of independent symmetric real valued random variables distributed as ξ. If B is a Banach space of finite cotype q_0 and if $q = \max(r, q_0)$, then, there is a constant C such that for all finite sequences (x_i) in B,*

$$\left\|\sum_i \xi_i x_i\right\|_r \leq C\|\xi\|_{q,1}\left\|\sum_i \varepsilon_i x_i\right\|_r .$$

Proof. On some (rich enough) probability space, let A be measurable and such that $\mathbb{P}(A) > 0$. Set $\varphi = I_A$; consider independent copies (φ_i) of φ and assume first that $\xi_i = \varepsilon_i \varphi_i$ where (ε_i) is an independent Rademacher sequence. We show that in this case, for some constant C,

$$(9.11) \qquad \left\|\sum_i \xi_i x_i\right\|_r \leq C(\mathbb{P}(A))^{1/q}\left\|\sum_i \varepsilon_i x_i\right\|_r .$$

By an easy approximation and the contraction principle, we may and do assume that $\mathbb{P}(A) = 1/N$ for some integer N. Let then $\{A^1, \ldots, A^N\}$ be a partition of the probability space into sets of probability $1/N$ with $A^1 = A$. Let further $(\varphi_i^j)_i$ be independent copies of I_{A^j} for all $j \leq N$. Using that $L_r(B)$ is of cotype q, with constant C say, we see that

$$\left(\sum_{j=1}^{N}\left\|\sum_i \varepsilon_i \varphi_i^j x_i\right\|_r^q\right)^{1/q} \leq C\left\|\sum_{j=1}^{N} \varepsilon_j'\left(\sum_i \varepsilon_i \varphi_i^j x_i\right)\right\|_r$$

where (ε_j') is another Rademacher sequence which is independent of (φ_i^j) and (ε_i). The left hand side of this inequality is just $N^{1/q}\|\sum_i \xi_i x_i\|_r$. Since $|\sum_{j=1}^{N} \varepsilon_j' \varphi_i^j| = 1$ for every i, by symmetry the right hand side is $C\|\sum_i \varepsilon_i x_i\|_r$. Thus inequality (9.11) holds.

We can conclude the proof of the proposition. Note that

$$|\xi_i| = \int_0^\infty I_{\{|\xi_i|>t\}} dt.$$

For every $t > 0$, by (9.11),

$$\left\|\sum_i \varepsilon_i I_{\{|\xi_i|>t\}} x_i\right\|_r \leq C(\mathbb{P}\{|\xi| > t\})^{1/q}\left\|\sum_i \varepsilon_i x_i\right\|_r.$$

Therefore, by the triangle inequality,

$$\left\|\sum_i \varepsilon_i |\xi_i| x_i\right\|_r \leq C\left(\int_0^\infty (\mathbb{P}\{|\xi| > t\})^{1/q} dt\right)\left\|\sum_i \varepsilon_i x_i\right\|_r$$

from which the conclusion follows since $(\varepsilon_i|\xi_i|)$ has the same distribution as (ξ_i).

Before we turn back to the discussion leading to Proposition 9.14, let us note that this result has a dual version for the type. Namely

Proposition 9.15. *Let $1 \leq r < \infty$ and let (ξ_i) be a sequence of independent symmetric real valued random variables distributed as ξ. If B is a Banach space of type p_0 and if $p = \min(r, p_0)$, there is a constant C such that for all finite sequences (x_i) in B,*

$$\|\xi\|_{p,\infty}\left\|\sum_i \varepsilon_i x_i\right\|_r \leq C\left\|\sum_i \xi_i x_i\right\|_r.$$

This result thus appears as an improvement, in spaces of some type, of the usual contraction principle in which we get $\|\xi\|_1$ on the left (see Lemma 4.5). The proof is entirely similar to the proof of Proposition 9.14; in the last step, simply use the contraction principle to see that, for every $t > 0$,

$$\left\|\sum_i \varepsilon_i I_{\{|\xi_i|>t\}} x_i\right\|_r \leq \frac{1}{t}\left\|\sum_i \varepsilon_i |\xi_i| x_i\right\|_r.$$

It should be noticed that Propositions 9.14 and 9.15 are optimal in the sense that they characterize cotype q_0 and type p_0 whenever $r = q_0$, resp. $r = p_0$. Indeed, if $x_1, \ldots, x_N$ are points in a Banach space and if A is such that $\mathbb{P}(A) = 1/N$, let (φ_i) be independent copies of I_A and $\xi_i = \varepsilon_i \varphi_i$. Then clearly

$$\mathbb{E}\left\|\sum_{i=1}^N \xi_i x_i\right\|^r = \mathbb{E}\left\|\sum_{i=1}^N \varepsilon_i \varphi_i x_i\right\|^r = \sum_{i=1}^N \int_{\{\varphi_j=1\,;\ \forall i\neq j\,,\ \varphi_i=0\}} \left\|\sum_{i=1}^N \varepsilon_i \varphi_i x_i\right\|^r d\mathbb{P}$$

$$= \sum_{j=1}^N \|x_j\|^r \left[\frac{1}{N}\left(1 - \frac{1}{N}\right)^{N-1}\right]$$

which is of the order of $\frac{1}{N}\sum_{i=1}^N \|x_i\|^r$. This easily shows the above claim.

Turning back to the question behind Proposition 9.14, we therefore know that if B has some finite cotype, inequality (9.10) will imply that B is of cotype q. Inequality (9.10) actually easily implies that B cannot contain ℓ_∞^n's uniformly (simply because it cannot hold for the canonical basis of ℓ_∞). A deep result of B. Maurey and G. Pisier [Mau-Pi] shows that this last property characterizes spaces having a non-trivial cotype. The theorem, which is the counterpart for the cotype of the results detailed previously for the type, can be stated as follows. It completes the ℓ_p^n-subspaces question for $p > 2$ although the set of $p > 2$ for which a given Banach space contains ℓ_p^n's uniformly seems to be rather arbitrary. A more probabilistic proof of this theorem is still to be found. We refer to [Mau-Pi], [Mi-Sh], [Mi-S].

Theorem 9.16. *A Banach space B is of cotype q for some $q < \infty$ if and only if B does not contain ℓ_∞^n's uniformly. More precisely, if*

$$q(B) = \inf\{q;\, B \text{ is of cotype } q\}$$

and if B is infinite dimensional, then B contains ℓ_q^n's uniformly for $q = q(B)$.

Summarizing in particular some of the (dual) conclusions of Corollary 9.8 and Theorem 9.16, we retain that an (infinite dimensional) Banach space has some non-trivial type (resp. cotype) if and only if it does not contain ℓ_1^n's (resp. ℓ_∞^n's) uniformly. Furthermore, combining Theorem 9.16 with Proposition 9.14, if a Banach space B does not contain ℓ_∞^n's uniformly, there is a constant C such that for all finite sequences (x_i) in B,

$$(9.12) \qquad \mathbb{E}\left\|\sum_i g_i x_i\right\| \le C\mathbb{E}\left\|\sum_i \varepsilon_i x_i\right\|.$$

This is, in those spaces, an improvement over the, in general, best possible inequality (4.9). Conversely thus, if (9.12) holds in a Banach space B, B does not contain ℓ_∞^n's uniformly. By Proposition 9.14, this characterization easily extends to some more general sequences of independent random variables.

To conclude this section, we would like to briefly indicate the (easy) extension of the notions of type and cotype to operators between Banach spaces. A bounded linear operator $u : E \to F$ between two Banach spaces E and F is said to be of (Rademacher) type p, $1 \leq p \leq 2$, if there is a constant C such that for all finite sequences (x_i) in E,

$$\left\| \sum_i \varepsilon_i u(x_i) \right\|_p \leq C \left(\sum_i \|x_i\|^p \right)^{1/p} .$$

Similarly, u is said to be of cotype q if

$$\left(\sum_i \|x_i\|^q \right)^{1/q} \leq C \left\| \sum_i \varepsilon_i u(x_i) \right\|_q .$$

Some of the easy properties of type and cotype clearly extend without modifications to operators. In particular, this is trivially the case for Proposition 9.11, a fact that we freely use below. One can also consider operators of stable type but, on the basis for example of Proposition 9.12, one may consider (possibly) different definitions. We can say that $u : E \to F$ is of stable type p, $1 \leq p < 2$, if for some constant C and all finite sequences (x_i) in E,

$$(9.13) \qquad \left\| \sum_i \theta_i u(x_i) \right\|_{p,\infty} \leq C \left(\sum_i \|x_i\|^p \right)^{1/p}$$

where (θ_i) is a standard p-stable sequence. We can also say that it is p-stable if

$$(9.14) \qquad \left\| \sum_i \varepsilon_i u(x_i) \right\|_1 \leq C \|(x_i)\|_{p,\infty} .$$

(9.13) and (9.14) are thus equivalent for the identity operator of a given Banach space but, from the lack of a geometric characterization analogous to Theorem 9.6, these two definitions are actually different in general. We refer to [P-R1] for a discussion of this difference as well as of related definitions.

9.3 Some Probabilistic Statements in Presence of Type and Cotype

In this last section, we try to answer some of the questions we started with. Namely, we will establish tightness and convergence in probability of various sums of independent random variables taking their values in Banach spaces having some type or cotype. As we know, this question is motivated by the strong limit theorems which were reduced in the preceding chapters to weak statements as well as by the central limit theorem investigated in the next chapter. We thus revisit now the strong laws of Kolmogorov and of Marcinkiewicz-Zygmund. Type 2 and cotype 2 will also be examined in their relations to pregaussian random variables, as well as spectral measures of stable distributions in spaces of stable type. Finally, but however not directly in relation with type and cotype, we present some results on the almost sure boundedness and convergence of sums of independent random variables in spaces which do not contain isomorphic copies of c_0.

As announced, since we will be dealing with tightness and weak convergence properties, we only consider in this chapter *Radon* random variables. Equivalently, we may assume that we are working with a *separable* Banach space.

We start with Kolmogorov's SLLN for independent random variables. We saw in Corollary 7.14 that if (X_i) is a sequence of independent random variables with values in a Banach space B such that for some $1 \leq p \leq 2$,

$$(9.15) \qquad \sum_i \frac{\mathbb{E}\|X_i\|^p}{i^p} < \infty,$$

then the SLLN holds (i.e. $S_n/n \to 0$ almost surely) if and only if the weak law of large numbers holds (i.e. $S_n/n \to 0$ in probability). In type p spaces, and actually only in those spaces, the series condition (9.15) implies the weak law $S_n/n \to 0$ in probability (provided the variables are centered). This is the conclusion of the next theorem.

Theorem 9.17. *Let $1 \leq p \leq 2$. A Banach space B is of type p if and only if for every sequence (X_i) of independent mean zero (or only symmetric) Radon random variables with values in B, the condition $\sum_i \mathbb{E}\|X_i\|^p / i^p < \infty$ implies the SLLN.*

Proof. Assume first that B is of type p. As we have seen, by Corollary 7.14 we need only show that $S_n/n \to 0$ in probability when $\sum_i \mathbb{E}\|X_i\|^p / i^p < \infty$ and when the X_i's have mean zero. But this is rather trivial under the type p condition. Indeed, by Proposition 9.11, for some constant C and every n,

$$\mathbb{E}\left\|\frac{S_n}{n}\right\|^p \leq C \frac{1}{n^p} \sum_{i=1}^n \mathbb{E}\|X_i\|^p.$$

The result follows from the classical Kronecker lemma (cf. [Sto]). To prove the converse, we assume the SLLN property for random variables of the form $X_i = \varepsilon_i x_i$ where $x_i \in B$ and (ε_i) is a Rademacher sequence. Then, if $\sum_i \|x_i\|^p / i^p < \infty$, we know that $\sum_{i=1}^n \varepsilon_i x_i / n \to 0$ almost surely, and also in $L_1(B)$ (or $L_r(B)$ for any $r < \infty$) by Theorem 4.7 together with Lemma 4.2. Hence, by the closed graph theorem, for some constant C,

$$\sup_n \frac{1}{n} \mathbb{E}\left\|\sum_{i=1}^n \varepsilon_i x_i\right\| \leq C \left(\sum_i \frac{\|x_i\|^p}{i^p}\right)^{1/p}$$

for every sequence (x_i) in B. Given $y_1, \ldots, y_m$ in B, apply this inequality to the sequence (x_i) defined by $x_i = 0$ if $i \leq m$ or $i > 2m$, $x_{m+1} = y_1$, $x_{m+2} = y_2, \ldots, x_{2m} = y_m$. We get

$$\mathbb{E}\left\|\sum_{i=1}^{m}\varepsilon_i y_i\right\| \le 2C\left(\sum_{i=1}^{m}\|y_i\|^p\right)^{1/p}$$

and B is therefore of type p. Theorem 9.17 is established.

As a consequence of the preceding and of Corollary 9.8 we can state

Corollary 9.18. *A Banach space B is of type p for some $p > 1$ if and only if every sequence (X_i) of independent symmetric uniformly bounded Radon random variables with values in B satisfies the SLLN.*

Proof. The necessity follows from Theorem 9.17. Conversely, it suffices to prove that if B is of no type $p > 1$ there exists a bounded sequence (x_i) in B such that $(\sum_{i=1}^{n}\varepsilon_i x_i/n)$ does not converge almost surely or in $L_1(B)$. By Corollary 9.8, together with Proposition 9.12, if B has no non-trivial type, then B contains ℓ_1^n's uniformly. Hence, for every n, there exist $y_1^n, \ldots, y_{2^n}^n$ in B such that $\|y_i^n\| \le 1$, $i = 1, \ldots, 2^n$, and

$$\mathbb{E}\left\|\sum_{i=1}^{2^n}\varepsilon_i y_i^n\right\| \ge 2^{n-1}.$$

Then define (x_i) by letting $x_i = y_j^n$, $j = i + 1 - 2^n$, $2^n \le i < 2^{n+1}$. It follows from Jensen's inequality that if $2^n \le m < 2^{n+1}$, then

$$\frac{1}{m}\mathbb{E}\left\|\sum_{i=1}^{m}\varepsilon_i x_i\right\| \ge \frac{1}{2^{n+1}}\mathbb{E}\left\|\sum_{i=1}^{2^n}\varepsilon_i y_i^n\right\| \ge \frac{1}{4}.$$

This proves Corollary 9.18.

Further results in Chapter 7 can be interpreted similarly under type conditions. We do not detail everything but we would like to describe the independent and identically distributed (iid) case following the discussion next to Theorem 7.9. To this aim, it is convenient to record that Proposition 9.11 is still an equivalent description of the type and cotype properties when the random variables X_i have the *same* distribution. This is, for the type, the content of the following statement.

Proposition 9.19. *Let $1 \le p \le 2$ and let B be a Banach space. Assume that there is a constant C such that for every finite sequence $(X_1, \ldots, X_N)$ of iid symmetric Radon random variables in $L_p(B)$,*

$$\mathbb{E}\left\|\sum_{i=1}^{N}X_i\right\| \le CN^{1/p}\left(\mathbb{E}\|X_1\|^p\right)^{1/p}.$$

Then B is of type p.

Proof. Let $(\varphi_j)_{j \leq N}$ be real valued symmetric random variables with disjoint supports such that for every $j = 1, \ldots, N$,

$$\mathbb{P}\{\varphi_j = 1\} = \mathbb{P}\{\varphi_j = -1\} = \frac{1}{2}(1 - \mathbb{P}\{\varphi_j = 0\}) = \frac{1}{2N}.$$

Let $(x_j)_{j \leq N}$ be points in B. Then $X = \sum_{j=1}^{N} \varphi_j x_j$ is such that $\mathbb{E}\|X\|^p = \sum_{j=1}^{N} \|x_j\|^p / N$ so that it is enough to show that if $X_1, \ldots, X_N$ are independent copies of X,

$$\mathbb{E}\left\|\sum_{i=1}^{N} X_i\right\| \geq c\mathbb{E}\left\|\sum_{j=1}^{N} \varepsilon_j x_j\right\|$$

for some $c > 0$. To this aim, denote by (φ_j^i), $i \leq N$, independent copies of (φ_j), assumed to be independent from a Rademacher sequence (ε_i). By symmetry

$$\mathbb{E}\left\|\sum_{i=1}^{N} X_i\right\| = \mathbb{E}\left\|\sum_{j=1}^{N} \varepsilon_j \left(\sum_{i=1}^{N} \varphi_j^i\right) x_j\right\|$$

and therefore, by Lemma 4.5 for symmetric sequences,

$$\mathbb{E}\left\|\sum_{i=1}^{N} X_i\right\| \geq \left(\mathbb{E}\left|\sum_{i=1}^{N} \varphi_1^i\right|\right) \mathbb{E}\left\|\sum_{j=1}^{N} \varepsilon_j x_j\right\|.$$

Now, by symmetry and Khintchine's inequalities $((4.3))$,

$$\mathbb{E}\left|\sum_{i=1}^{N} \varphi_1^i\right| = \mathbb{E}\left|\sum_{i=1}^{n} \varepsilon_i \varphi_1^i\right| \geq \frac{1}{\sqrt{2}}\mathbb{E}\left(\left(\sum_{i=1}^{N} |\varphi_1^i|^2\right)^{1/2}\right)$$

$$= \frac{1}{\sqrt{2}}\mathbb{E}\left(\left(\sum_{i=1}^{N} |\varphi_1^i|\right)^{1/2}\right).$$

Since

$$\mathbb{P}\left\{\sum_{i=1}^{N} |\varphi_1^i| = 0\right\} = \left(1 - \frac{1}{N}\right)^N \leq \frac{1}{e},$$

we get

$$\mathbb{E}\left|\sum_{i=1}^{N} \varphi_1^i\right| \geq \frac{1}{\sqrt{2}}\left(1 - \frac{1}{e}\right) \geq \frac{1}{3},$$

hence the announced claim with $c = 1/3$. The proof of Proposition 9.19 is complete.

There is an analogous statement for the cotype but the proof involves the deeper Theorem 9.16. The main idea of the proof is the so-called "Poissonization" technique.

Proposition 9.20. *Let $2 \leq q < \infty$ and let B be a Banach space. Assume that there is a constant C such that for every finite sequence $(X_1, \ldots, X_N)$ of iid symmetric Radon random variables in $L_q(B)$,*

$$N\mathbb{E}\|X_1\|^q \leq C\mathbb{E}\left\|\sum_{i=1}^N X_i\right\|^q.$$

Then B is of cotype q.

Proof. Let $x_1, \ldots, x_N$ be points in B and let X with the distribution $(2N)^{-1}\sum_{i=1}^N(\delta_{x_i} + \delta_{-x_i})$. Then

$$N\mathbb{E}\|X\|^q = \sum_{i=1}^N \|x_i\|^q.$$

Take $(X_1, \ldots, X_N)$ to be independent copies of X and let us consider $\sum_{i=1}^N X_i$. Let $(N_i)_{i \leq N}$ be independent Poisson random variables with parameter 1, independent of the X_i's. Let further $(X_{i,j})_{1 \leq i,j \leq N}$ be independent copies of X and set $X_{i,0} = 0$ for each i. Then, as is easy to check on characteristic functionals,

$$\sum_{i=1}^N \sum_{j=0}^{N_i} X_{i,j} \quad \text{has the same distribution as} \quad \sum_{i=1}^N \tilde{N}_i x_i$$

where $\tilde{N}_i = N_i(1/2) - N_i'(1/2)$ and $N_i(1/2), N_i'(1/2)$, $i \leq N$, are independent Poisson random variables with parameter $1/2$. Now, by Jensen's inequality conditionally on (N_i) (cf. (2.5)),

$$\mathbb{E}\left\|\sum_{i=1}^N \sum_{j=0}^{N_i} X_{i,j}\right\|^q \geq \mathbb{E}\left\|\sum_{i=1}^N \sum_{j=0}^{N_i \wedge 1} X_{i,j}\right\|^q.$$

Furthermore, $\mathbb{P}\{N_i \wedge 1 = 0\} = 1 - \mathbb{P}\{N_i \wedge 1 = 1\} = e^{-1}$. Hence, since $X_{i,0} = 0$,

$$\mathbb{E}\left\|\sum_{i=1}^N \sum_{j=0}^{N_i \wedge 1} X_{i,j}\right\|^q = \mathbb{E}\left\|\sum_{i=1}^N \delta_i X_i\right\|^q$$

where δ_i, $i \leq N$, are independent random variables with $\mathbb{P}\{\delta_i = 0\} = 1 - \mathbb{P}\{\delta_i = 1\} = e^{-1}$ and are independent of the sequence (X_i). Again by Jensen's inequality (conditionally on the X_i's),

$$\mathbb{E}\left\|\sum_{i=1}^N \delta_i X_i\right\|^q \geq e^{-q}\mathbb{E}\left\|\sum_{i=1}^N X_i\right\|^q.$$

Summarizing, we have obtained

$$\sum_{i=1}^N \|x_i\|^q \leq C\mathbb{E}\left\|\sum_{i=1}^N X_i\right\|^q \leq Ce^q \mathbb{E}\left\|\sum_{i=1}^N \tilde{N}_i x_i\right\|.$$

Now, this inequality clearly cannot hold for all finite sequences (x_i) in a Banach space B which contains ℓ_∞^n's uniformly since it does not hold for the canonical basis of ℓ_∞. Therefore, by Theorem 9.16, B is of finite cotype and we are in a position to apply Proposition 9.14 since $\mathbb{E}\tilde{N}_i^p < \infty$ for all p's. The proof is complete.

After this digression, we now come back to the main application of Proposition 9.19. Namely, we investigate the relationship between the type condition and the iid SLLN of Marcinkiewicz-Zygmund. If X is a Radon random variable with values in a Banach space B, (X_i) denotes below a sequence of independent copies of X and, as usual, $S_n = X_1 + \cdots + X_n$, $n \geq 1$. Let $1 \leq p < 2$. In Theorem 7.9, we have seen that $S_n/n^{1/p} \to 0$ almost surely if and only if $\mathbb{E}\|X\|^p < \infty$ and $S_n/n^{1/p} \to 0$ in probability. Moreover, as already discussed thereafter, in type p spaces, $S_n/n^{1/p} \to 0$ in probability when $\mathbb{E}\|X\|^p < \infty$ and $\mathbb{E}X = 0$. Now, we show that this result is characteristic of the type p property.

Theorem 9.21. *Let $1 \leq p < 2$ and let B be a Banach space. The following are equivalent:*

(i) B is of type p;

(ii) for every Radon random variable X with values in B, $S_n/n^{1/p} \to 0$ almost surely if and only if $\mathbb{E}\|X\|^p < \infty$ and $\mathbb{E}X = 0$.

Note of course that since every Banach space is of type 1 we recover Corollary 7.10.

Proof. Let us briefly recall the argument leading to (i) $\Rightarrow$ (ii) already discussed after Theorem 7.9. Under the type assumption and the moment conditions $\mathbb{E}\|X\|^p < \infty$ and $\mathbb{E}X = 0$, the sequence $(S_n/n^{1/p})$ was shown to be arbitrarily close to a finite dimensional sequence, and thus to be tight. Since for every linear functional f, $f(S_n/n^{1/p}) \to 0$ in probability, it follows that $S_n/n^{1/p} \to 0$ in probability and we can conclude the proof of (i) $\Rightarrow$ (ii) by Theorem 7.9. Conversely, let us first show that when $\mathbb{E}\|X\|^p < \infty$, $\mathbb{E}X = 0$ and $S_n/n^{1/p} \to 0$ almost surely (or only in probability), then the sequence $(S_n/n^{1/p})$ is bounded in $L_1(B)$. Since X is centered, by a symmetrization argument it is enough to treat the case of a symmetric variable X. Then, by Lemma 7.2, we already know that

$$\sup_n \frac{1}{n^{1/p}} \mathbb{E}\left\|\sum_{i=1}^n X_i I_{\{\|X_i\| \leq n^{1/p}\}}\right\| < \infty.$$

Moreover, it is easily seen by an integration by parts that, under $\mathbb{E}\|X\|^p < \infty$,

$$\frac{1}{n^{1/p}} \mathbb{E}\left\|\sum_{i=1}^n X_i I_{\{\|X_i\| > n^{1/p}\}}\right\| \leq n^{1-1/p} \mathbb{E}\left(\|X\| I_{\{\|X\| > n^{1/p}\}}\right)$$

is uniformly bounded in n. The claim follows. (One can also invoke for this result the version of Corollary 10.2 below for $n^{1/p}$.) Therefore, by the closed graph theorem, there exists a constant C such that for all centered random variables X with values in B

$$\sup_n \frac{1}{n^{1/p}} \mathbb{E}\|S_n\| \leq C \left(\mathbb{E}\|X\|^p\right)^{1/p}.$$

We conclude that B is of type p by Proposition 9.19.

The preceding theorem has an analog for the stable type. Let us briefly state this result and sketch its proof.

Theorem 9.22. *Let $1 \leq p < 2$ and let B be a Banach space. The following are equivalent:*

(i) B is of stable type p;

(ii) for every symmetric Radon random variable X with values in B, $S_n/n^{1/p} \to 0$ in probability if and only if $\lim_{t\to\infty} t^p \mathbb{P}\{\|X\| > t\} = 0$.

Proof. We have noticed next to Theorem 7.9 that (ii) holds true in finite dimensional spaces. The implication (i) $\Rightarrow$ (ii) is then simply based on Proposition 9.13 and a finite dimensional argument as in Theorem 9.21. That $\lim_{t\to\infty} t^p \mathbb{P}\{\|X\| > t\} = 0$ when $S_n/n^{1/p} \to 0$ in probability is a simple consequence of Lévy's inequality (2.7) and Lemma 2.6; indeed, for each $\varepsilon > 0$ and all large enough n's,

$$\frac{1}{4} \geq \mathbb{P}\{\|S_n\| > \varepsilon n^{1/p}\} \geq \frac{1}{2} \mathbb{P}\{\max_{i \leq n} \|X_i\| > \varepsilon n^{1/p}\} \geq \frac{1}{4} n \mathbb{P}\{\|X\| > \varepsilon n^{1/p}\}.$$

The implication (ii) $\Rightarrow$ (i) is obtained as in the last theorem via (iii) of Proposition 9.12 and some care in the closed graph argument.

Some more applications of the type condition in the case of the law of the iterated logarithm were described in Chapter 8 (Corollary 8.8) and we need not recall them here. Now we would like to investigate some consequences in relation with the next chapter on the central limit theorem. They deal with pregaussian variables.

A Radon random variable X with values in a Banach space B such that for every f in B', $\mathbb{E}f(X) = 0$ and $\mathbb{E}f^2(X) < \infty$, is said to be *pregaussian* if there exists a Gaussian Radon variable G in B with the same covariance structure as X, i.e. for all f, g in B', $\mathbb{E}f(X)g(X) = \mathbb{E}f(G)g(G)$ (or just $\mathbb{E}f^2(X) = \mathbb{E}f^2(G)$). Since the distribution of a Gaussian variable is entirely determined by the covariance structure, we denote with some abuse in notation by $G(X)$ a Gaussian random variable with the same covariance structure as the pregaussian variable X. The concept of pregaussian variables and their integrability properties are closely related to type 2 and cotype 2. The following easy lemma is useful in this study.

Lemma 9.23. *Let X be a pregaussian random variable with values in a Banach space B and with associated Gaussian variable $G(X)$. Let Y be a Radon random variable in B such that for every f in B', $\mathbb{E}f^2(Y) \leq \mathbb{E}f^2(X)$ (and $\mathbb{E}f(Y) = 0$). Then Y is pregaussian and, for every $p > 0$, $\mathbb{E}\|G(Y)\|^p \leq 2\mathbb{E}\|G(X)\|^p$.*

Proof. Since Y is Radon, we may assume that it is constructed on some probability space $(\Omega, \mathcal{A}, \mathbb{P})$ with $\mathcal{A}$ countably generated. In particular, $L_2(\Omega, \mathcal{A}, \mathbb{P})$ is separable and we can find a countable orthonormal basis $(h_i)_{i \geq 1}$ of $L_2(\Omega, \mathcal{A}, \mathbb{P})$. Since Y is Radon (cf. Section 2.1), $\mathbb{E}(h_i Y)$ defines an element of B for every i. For every N, let now $G_N = \sum_{i=1}^{N} g_i \mathbb{E}(h_i Y)$ where (g_i) is an orthogaussian sequence. By Bessel's inequality, for every N and f in B',

$$\mathbb{E}f^2(G_N) \leq \mathbb{E}f^2(Y) \leq \mathbb{E}f^2(X) = \mathbb{E}f^2(G(X)) \,.$$

Using (3.11), the Gaussian sequence (G_N) is seen to be tight in B. Since $\mathbb{E}f^2(G_N) \to \mathbb{E}f^2(Y)$, the limit is unique and (G_N) converges almost surely (Itô-Nisio) and in $L_2(B)$ to a Gaussian Radon random variable $G(Y)$ in B with the same covariance as Y. Since the inequality $\mathbb{P}\{\|G(Y)\| > t\} \leq 2\mathbb{P}\{\|G(X)\| > t\}$, $t > 0$, also follows from (3.11), the proof is complete.

Note that the constant 2 in Lemma 9.23 (and its proof) is actually not needed as follows from the deeper result described after (3.11). As a consequence of Lemma 9.23, note also that the sum of two pregaussian variables is also pregaussian. Furthermore, if X is pregaussian and if A is a Borel set such that $XI_{\{X \in A\}}$ has still mean zero, then $XI_{\{X \in A\}}$ is also pregaussian.

To introduce to the following, let us briefly characterize pregaussian variables in the sequence spaces ℓ_p, $1 \leq p < \infty$. Let $X = (X_k)_{k \geq 1}$ be weakly centered and square integrable with values in ℓ_p. Let $G = (G_k)_{k \geq 1}$ be a Gaussian sequence of real valued variables with covariance structure determined by $\mathbb{E}G_k G_\ell = \mathbb{E}X_k X_\ell$ for all k, ℓ. G is the natural candidate for $G(X)$. Note that $\mathbb{E}|G_k|^p = c_p(\mathbb{E}|G_k|^2)^{p/2} = c_p(\mathbb{E}|X_k|^2)^{p/2}$, where $c_p = \mathbb{E}|g|^p$ and g is standard normal. It follows that if

$$(9.16) \qquad \sum_{k=1}^{\infty} (\mathbb{E}|X_k|^2)^{p/2} < \infty \,,$$

then, by a simple approximation, G is seen to define a Gaussian random variable with values in ℓ_p with the same covariance structure as X and such that

$$\mathbb{E}\|G\|^p = \sum_{k=1}^{\infty} \mathbb{E}|G_k|^p = c_p \sum_{k=1}^{\infty} (\mathbb{E}|X_k|^2)^{p/2} < \infty \,.$$

Therefore, $X = (X_k)_{k \geq 1}$ in ℓ_p, $1 \leq p < \infty$, is pregaussian if and only if (9.16) holds. More generally, it can be shown that X with values in $L_p =$

$L_p(S, \Sigma, \mu)$, $1 \leq p < \infty$, where μ is σ-finite, is pregaussian if and only if (it has weak mean zero and)

$$\int_S \left(\mathbb{E}|X(s)|^2\right)^{p/2} d\mu(s) < \infty.$$

The next two propositions are the main observations on pregaussian random variables and their relations to type 2 and cotype 2.

Proposition 9.24. *Let B be a Banach space. Then B is of type 2 if and only if every mean zero Radon random variable X with values in B such that $\mathbb{E}\|X\|^2 < \infty$ is pregaussian and we have the inequality*

$$\mathbb{E}\|G(X)\|^2 \leq C\mathbb{E}\|X\|^2$$

for some constant C depending only on the type 2 constant of B. The equivalence is still true when $\mathbb{E}\|X\|^2 < \infty$ is replaced by X bounded.

Proposition 9.25. *Let B be a Banach space. Then B is of cotype 2 if and only if each pregaussian (Radon) random variable X with values in B satisfies $\mathbb{E}\|X\|^2 < \infty$ and we have the inequality*

$$\mathbb{E}\|X\|^2 \leq C\mathbb{E}\|G(X)\|^2$$

for some constant C depending only on the cotype 2 constant of B. The equivalence is still true when $\mathbb{E}\|X\|^2 < \infty$ is replaced by $\mathbb{E}\|X\|^p < \infty$ for some $0 < p \leq 2$.

Proof of Proposition 9.24. We know from Proposition 9.11 that if B is of type 2, for some $C > 0$ and all finite sequences (x_i) in B,

$$\mathbb{E}\left\|\sum_i g_i x_i\right\|^2 \leq C \sum_i \|x_i\|^2.$$

Now, let X with $\mathbb{E}X = 0$ and $\mathbb{E}\|X\|^2 < \infty$. There exists an increasing sequence $(\mathcal{A}_N)$ of finite σ-algebras such that if $X^N = \mathbb{E}^{\mathcal{A}_N} X$, $X^N \to X$ almost surely and in $L_2(B)$. Since $\mathcal{A}_N$ is finite, X^N can be written as a finite sum $\sum_i x_i I_{A_i}$ with the A_i's disjoint. Then

$$G(X^N) = \sum_i g_i \left(\mathbb{P}(A_i)\right)^{1/2} x_i \quad \text{and} \quad \mathbb{E}\|X^N\|^2 = \sum_i \|x_i\|^2 \mathbb{P}(A_i)$$

so that $\mathbb{E}\|G(X^N)\|^2 \leq C\mathbb{E}\|X^N\|^2 \leq C\mathbb{E}\|X\|^2$ which thus holds for every N. Since the type 2 inequality also holds for quotient norms (with the same constant), the sequence $(G(X^N))$ of Gaussian random variables is seen to be tight. Since further $\mathbb{E}f^2(G(X^N)) = \mathbb{E}f^2(X^N) \to \mathbb{E}f^2(X)$ for every f in B', $(G(X^N))$ necessarily converges weakly to some Gaussian variable $G(X)$ with the same covariance structure as X. By Skorokhod's theorem, cf. Section 2.1, one obtains that $\mathbb{E}\|G(X)\|^2 \leq C\mathbb{E}\|X\|^2$. This establishes the first part of Proposition 9.24.

Conversely, it is sufficient to show that if every centered random variable X such that $\|X\|_\infty \leq 1$ is pregaussian, then B is type 2. Let (x_i) be a sequence in B such that $\sum_i \|x_i\|^2 = 1$. Consider X with distribution $X = \pm x_i/\|x_i\|$ with probability $\|x_i\|^2/2$. By the hypothesis, X is pregaussian and $G(X)$ must be $\sum_i g_i x_i$ which therefore defines a convergent series. Then, so is $\sum_i \varepsilon_i x_i$ and the proof is complete.

Proof of Proposition 9.25. Assume first that B is of cotype 2. Then for some constant C and all finite sequences (x_i) in B

$$(9.17) \qquad \sum_i \|x_i\|^2 \leq C\mathbb{E}\left\|\sum_i g_i x_i\right\|^2 .$$

(This is a priori *weaker* than the cotype 2 inequality with Rademacher random variables – see (9.10) and Remark 9.26 below). Given X pregaussian with values in B, let $Y = \varepsilon X I_{\{\|X\| \leq t\}}$ where $t > 0$ and ε is a Rademacher random variable which is independent of X. By Lemma 9.23, Y is again pregaussian and $\mathbb{E}\|G(Y)\|^2 \leq 2\mathbb{E}\|G(X)\|^2$. Arguing then exactly as in the first part of the proof of Proposition 9.24 by finite dimensional approximations, one obtains

$$\mathbb{E}\|Y\|^2 \leq C\mathbb{E}\|G(Y)\|^2 \leq 2C\mathbb{E}\|G(X)\|^2 .$$

Since $t > 0$ is arbitrary, it follows that $\mathbb{E}\|X\|^2 < \infty$ and the inequality of the proposition holds.

Turning to the converse, let us first show that, given $0 < p \leq 2$, if every pregaussian variable X in B satisfies $\mathbb{E}\|X\|^p < \infty$, then (9.17) holds. We actually show that if $\sum_i g_i x_i$ converges almost surely, then $\sum_i \|x_i\|^2 < \infty$. We assume that $p < 2$, which is the most difficult case, and set $r = 2/2 - p$. Let α_i be positive numbers such that $\sum_i \alpha_i^r = 1$ and define X by:

$$X = \pm\alpha_i^{(1-r)/p} x_i \quad \text{with probability} \quad \alpha_i^r/2 .$$

It is easily seen that $G(X)$ is precisely $\sum_i g_i x_i$ and therefore, by the hypothesis,

$$\mathbb{E}\|X\|^p = \sum_i \alpha_i \|x_i\|^p < \infty .$$

Since this holds of each such sequence (α_i), by duality it must be that $\sum_i \|x_i\|^2 < \infty$. This is the announced claim.

To conclude the proof, let us show how (9.17) implies that B is of cotype 2. First recall that we proved above that (9.17) implies

$$(9.18) \qquad \mathbb{E}\|X\|^2 \leq 2C\mathbb{E}\|G(X)\|^2$$

for every pregaussian variable X in B. Let (x_i) be a finite sequence in B. For every $t > 0$,

$$\left\|\sum_i g_i x_i\right\|_2 \leq \left\|\sum_i g_i I_{\{|g_i|\leq t\}} x_i\right\|_2 + \left\|\sum_i g_i I_{\{|g_i|>t\}} x_i\right\|_2$$

$$\leq t \left\|\sum_i \varepsilon_i x_i\right\|_2 + \left\|\sum_i g_i I_{\{|g_i|>t\}} x_i\right\|_2$$

where we have used the contraction principle. If we now apply (9.18) to $X = \sum_i g_i I_{\{|g_i|>t\}} x_i$, we see that

$$\mathbb{E}\left\|\sum_i g_i I_{\{|g_i|>t\}} x_i\right\|^2 \leq 2C \mathbb{E}(|g|^2 I_{\{|g|>t\}}) \mathbb{E}\left\|\sum_i g_i x_i\right\|^2 .$$

Now choose $t > 0$ small enough in order that $\mathbb{E}(|g|^2 I_{\{|g|>t\}})$ is less than $(8C)^{-1}$. Together with (9.17), we then obtain

$$\sum_i \|x_i\|^2 \leq C\mathbb{E}\left\|\sum_i g_i x_i\right\|^2 \leq 4t^2 C \mathbb{E}\left\|\sum_i \varepsilon_i x_i\right\|^2 .$$

Proposition 9.25 is established.

Remark 9.26. The preceding proof indicates in particular that a Banach space B is of cotype 2 *if and only if* for all finite sequences (x_i) in B

$$\sum_i \|x_i\|^2 \leq C\mathbb{E}\left\|\sum_i g_i x_i\right\|^2 .$$

This can also be obtained by the conjunction of Proposition 9.14 and Theorem 9.16. The preceding direct proof in this case is however simpler since it does not use Theorem 9.16.

After pregaussian random variables, we discuss some results on spectral measures of stable distributions in spaces of stable type. If B is a Banach space of stable type p, $1 \leq p < 2$, and if (x_i) is a sequence in B such that $\sum_i \|x_i\|^p < \infty$, then the series $\sum_i \theta_i x_i$ converges almost surely and defines a p-stable Radon random variable in B. In other words, if m is the finite discrete measure

$$m = \sum_i \frac{\|x_i\|^p}{2} \left(\delta_{-x_i/\|x_i\|} + \delta_{+x_i/\|x_i\|}\right),$$

m is the spectral measure of a p-stable random vector X in B. Moreover

$$\|X\|_{p,\infty} \leq C|m|^{1/p} = C\left(\int \|x\|^p dm(x)\right)^{1/p}$$

for some constant C depending only on the stable type p property of B. Recall from Chapter 5 that m symmetrically distributed on the unit sphere of B is unique. Recall further (cf. Corollary 5.5) that if X is a p-stable Radon random variable with values in a Banach space B, there exists a spectral measure m of X such that $\int \|x\|^p dm(x) < \infty$. The parameter $\sigma_p(X)$ of X is defined as $\sigma_p(X) = (\int \|x\|^p dm(x))^{1/p}$ (which is unique among all possible spectral measures) and we always have (cf. (5.13))

$$K_p^{-1}\sigma_p(X) \le \|X\|_{p,\infty} \,.$$

This inequality is two-sided when $0 < p < 1$ and we have just seen that when $1 \le p < 2$ and $\sum_i \|x_i\|^p < \infty$ in a Banach space of stable type p, $X = \sum_i \theta_i x_i$ converges almost surely and $\|X\|_{p,\infty} \le C\sigma_p(X)$. This property actually extends to general measures on stable type p Banach spaces.

Theorem 9.27. *Let $1 \le p < 2$. A Banach space B is of stable type p if and only if every positive finite Radon measure m on B such that $\int \|x\|^p dm(x) < \infty$ is the spectral measure of a p-stable Radon random vector X in B. Furthermore, if this is the case, there exists a constant C such that*

$$\|X\|_{p,\infty} \le C\left(\int \|x\|^p dm(x)\right)^{1/p}.$$

Proof. The choice of a discrete measure m as above proves the "if" part of the statement. Suppose now that B is of stable type p. Let m_1 denote the image of the measure $\|x\|^p dm(x)$ by the map $x \to x/\|x\|$. Let further Y_j be independent random variables distributed as $m_1/|m_1|$. The natural candidate for X is given by the series representation

$$c_p|m_1|^{1/p} \sum_{j=1}^{\infty} \Gamma_j^{-1/p} \varepsilon_j Y_j$$

(cf. Corollary 5.5). In order to show that this series converges, note the following: since B is of stable type p, by Corollary 9.7 together with Proposition 9.12 (i), B is of Rademacher type p_1 for some $p_1 > p$. Therefore

$$\mathbb{E}\left\|\sum_{j=1}^{\infty} j^{-1/p}\varepsilon_j Y_j\right\| \le C\left(\sum_{j=1}^{\infty} j^{-p_1/p}\right)^{1/p_1} < \infty$$

from which the required convergence easily follows. Thus X defines a p-stable random variable with spectral measure m_1 and therefore also m. The inequality of the theorem follows from the same argument (and from some of the elementary material in Chapter 5).

As yet another application, let us briefly mention an alternate approach to p-stable random variables in Banach spaces of stable type p. This approach goes through Wiener integrals and follows a classical construction of S. Kakutani. Let (S, Σ, m) be a measure space. Define a p-stable *random measure* M based on (S, Σ, m) in the following way: $(M(A))_{A \in \Sigma}$ is a collection of real valued random variables such that for every A, $M(A)$ is p-stable with parameter $m(A)^{1/p}$ and whenever (A_i) are disjoint, the sequence $(M(A_i))$ is independent. For a step function φ of the form $\varphi = \sum_i \alpha_i I_{A_i}, \alpha_i \in \mathbb{R}, A_i \in \Sigma$ disjoint, the stochastic integral $\int \varphi dM$ is well-defined as

$$\int \varphi dM = \sum_i \alpha_i M(A_i).$$

It is a p-stable random variable with parameter $\|\varphi\|_p = (\int |\varphi|^p dm)^{1/p}$. Therefore, by a density argument, it is easy to define the stochastic integral $\int \varphi dM$ for any φ in $L_p(S, \Sigma, m)$.

The question now raises of the possibility of this construction for functions φ taking their values in a Banach space B. As suspected, the class of Banach spaces B of stable type p is the class in which the preceding stochastic integral $\int \varphi dM$ can be defined when $\int \|\varphi\|^p dm < \infty$. In the case φ is the identity map on B, $\int \varphi dM$ is a p-stable random variable in B with spectral measure m so that we recover the conclusion of Theorem 9.27.

Proposition 9.28. *Let $1 \leq p < 2$. A Banach space B is of stable type p if and only if for any measure space (S, Σ, m) and any p-stable random measure M based on (S, Σ, m) the stochastic integral $\int \varphi dM$ with $\int \|\varphi\|^p dm < \infty$ defines a p-stable Radon random variable with values in B and*

$$\left\| \int \varphi dM \right\|_{p,\infty} \leq C \left(\int \|\varphi\|^p dm \right)^{1/p}.$$

Proof. Sufficiency is embedded in Theorem 9.27 with the choice for φ of the identity on (B, m). Conversely, if φ is a step function $\sum_i x_i I_{A_i}$ with x_i in B and A_i mutually disjoint,

$$\int \varphi dM = \sum_i x_i M(A_i)$$

which is equal in distribution to $\sum_i m(A_i)^{1/p} \theta_i x_i$. Now, if B is of stable type p,

$$\left\| \int \varphi dM \right\|_{p,\infty} \leq C \left(\sum_i m(A_i) \|x_i\|^p \right)^{1/p} = C \left(\int \|\varphi\|^p dm \right)^{1/p}.$$

Hence the map $\varphi \rightarrow \int \varphi dM$ can be extended to a bounded operator from $L_p(m; B)$ into $L_{p,\infty}(B)$, and a fortiori into $L_0(B)$. This concludes the proof of Proposition 9.28.

We conclude this chapter with a result on the almost sure boundedness and convergence of sums of independent random variables in spaces which do not contain subspaces isomorphic to the space c_0 of all scalar sequences convergent to 0. This result is not directly related to type and cotype since these are local properties in the sense that they only depend on the collection of finite dimensional subspaces of the given space whereas the property discussed here involves *infinite* dimensional subspaces. It is however natural and convenient for further purposes to record this result at this stage.

Let (X_i) be a sequence of independent *real valued* symmetric random variables such that the sequence (S_n) of the partial sums is almost surely bounded, i.e. $\mathbb{P}\{\sup_n |S_n| < \infty\} = 1$. Then, it is well-known that (S_n) converges almost surely. This is actually contained in one of the steps of the proof of Theorem 2.4 but let us emphasize the argument here. Let (ε_i) be a Rademacher sequence which is independent of (X_i) and recall the partial integration notation $\mathbb{P}_\varepsilon, \mathbb{P}_X, \mathbb{E}_\varepsilon, \mathbb{E}_X$. By symmetry and the assumption, for every $\delta > 0$, there exists a finite number M such that for every n

$$\mathbb{P}\left\{\left|\sum_{i=1}^n \varepsilon_i X_i\right| \le M\right\} \ge 1 - \delta^2 .$$

Let n be fixed for a moment. By Fubini's theorem, if

$$A = \left\{\omega; \ \mathbb{P}_\varepsilon\left\{\left|\sum_{i=1}^n \varepsilon_i X_i(\omega)\right| \le M\right\} \ge 1 - \delta\right\},$$

then $\mathbb{P}_X(A) \ge 1 - \delta$. Now, if $\delta \le 1/8$, by Lemma 4.2 and (4.3), for every ω in A,

$$\sum_{i=1}^n X_i(\omega)^2 = \mathbb{E}_\varepsilon\left|\sum_{i=1}^n \varepsilon_i X_i(\omega)\right|^2 \le 2\sqrt{2}M .$$

Hence, for all n, $\mathbb{P}\{\sum_{i=1}^n X_i(\omega)^2 \le 2\sqrt{2}M\} \ge 1 - \delta$. It follows that $\sum_i X_i^2 < \infty$ almost surely. Thus, by Fubini's theorem, $\sum_i \varepsilon_i X_i$ converges almost surely and the claim follows.

It can easily be shown that this argument extends, for example, to Hilbert space valued random variables. Actually, this property is satisfied for random variables taking their values in, and only in, Banach spaces which do not contain subspaces isomorphic to c_0. This is the content of the next theorem. Recall that a sequence (Y_n) of random variables is almost surely bounded if $\mathbb{P}\{\sup_n \|Y_n\| < \infty\} = 1$.

Theorem 9.29. *Let B be a Banach space. The following are equivalent:*

(i) B does not contain subspaces isomorphic to c_0;

(ii) for every sequence (X_i) of independent symmetric Radon random variables in B, the almost sure boundedness of the sequence (S_n) of the partial sums implies its convergence;

(iii) for every sequence (x_i) in B, if $(\sum_{i=1}^n \varepsilon_i x_i)$ is almost surely bounded, $\sum_i \varepsilon_i x_i$ converges almost surely;

(iv) for every sequence (x_i) in B, if $(\sum_{i=1}^n \varepsilon_i x_i)$ is almost surely bounded, $x_i \to 0$.

Proof. The implications (ii) $\Rightarrow$ (iii) $\Rightarrow$ (iv) are obvious. Let us show that (iv) $\Rightarrow$ (ii). Let (X_i) in B with $\sup_n \|S_n\| < \infty$ with probability one and let (ε_i) be a Rademacher sequence independent of (X_i). By symmetry and Fubini's theorem, for almost every ω on the probability space supporting the X_i's, $\sup_n \| \sum_{i=1}^n \varepsilon_i X_i(\omega) \| < \infty$ almost surely. Hence, by (iv), $X_i(\omega) \to 0$. Similarly, if we take blocks for any strictly increasing sequence (n_k) of integers, $\sum_{n_k < i \le n_{k+1}} X_i \to 0$ almost surely when $k \to \infty$. Hence $S_{n_{k+1}} - S_{n_k} \to 0$ in probability and by Lévy's inequality (2.6), (S_n) is a Cauchy sequence in probability, and thus converges. Therefore (ii) holds by Theorem 2.4.

The main point in this proof is the equivalence between (i) and (iv). That (iv) $\Rightarrow$ (i) is clear by the choice of $x_i = e_i$, the canonical basis of c_0. Let us then show the converse implication and let us proceed by contradiction. Let (x_i) be a sequence in B such that $\inf_i \|x_i\| > 0$ and $\mathbb{P}\{\sup_n \| \sum_{i=1}^n \varepsilon_i x_i \| < \infty\} = 1$. We may assume that the probability space $(\Omega, \mathcal{A}, \mathbb{P})$ is the canonical product space $\{-1, +1\}^{\mathbb{N}}$ equipped with its natural σ-algebra and product measure. It is easy to see that for every $C \in \mathcal{A}$,

$$\lim_{i \to \infty} \mathbb{P}\big(C \cap \{\varepsilon_i = -1\}\big) = \lim_{i \to \infty} \mathbb{P}\big(C \cap \{\varepsilon_i = +1\}\big) = \tfrac{1}{2}\mathbb{P}(C).$$

Let us pick $M < \infty$ so that $\mathbb{P}(A) > 1/2$ where

$$A = \left\{ \sup_n \left\| \sum_{i=1}^n \varepsilon_i x_i \right\| \le M \right\}.$$

By the previous observation, we can define inductively an increasing sequence of integers (n_i) such that for every sequence of signs (a_i) and every k,

$$(9.19) \qquad \mathbb{P}\big(A \cap \{\varepsilon_{n_1} = a_1\} \cap \cdots \cap \{\varepsilon_{n_k} = a_k\}\big) > 2^{-k-1}.$$

Put $\varepsilon_i' = \varepsilon_i$ if i is one of the n_j's, $\varepsilon_i' = -\varepsilon_i$ if not. The sequences (ε_i) and (ε_i') are equidistributed. Therefore, if

$$A' = \left\{ \sup_n \left\| \sum_{i=1}^n \varepsilon_i' x_i \right\| \le M \right\},$$

(9.19) also holds for A' with respect to (ε_i'). Since $\varepsilon_{n_i} = \varepsilon_{n_i}'$ and $\mathbb{P}\{\varepsilon_{n_i} = a_i, \, i \le k\} = 2^{-k}$, it follows by intersection that there is an ω in $A \cap A'$ such that $\varepsilon_{n_i} = a_i$ for all $i = 1, \ldots, k$. Thus

$$\left\| \sum_{i=1}^k a_i x_{n_i} \right\| = \left\| \frac{1}{2}\left(\sum_{j=1}^{n_k} \varepsilon_j(\omega) x_j + \sum_{j=1}^{n_k} \varepsilon_j'(\omega) x_j \right) \right\| \le M.$$

Since the integer k and the signs $a_1, a_2, \ldots, a_k$ have been fixed arbitrarily, this inequality implies that the series $\sum_i x_{n_i}$ is weakly unconditionally convergent, that is $\sum_i |f(x_{n_i})| < \infty$ for every f in B', while $\inf_i \|x_{n_i}\| > 0$. The conclusion is then obtained from the following classical result on basic sequences in Banach spaces.

Lemma 9.30. *Let (y_i) be a sequence in a Banach space B such that for every f in B', $\sum_i |f(y_i)| < \infty$, and such that $\inf_i \|y_i\| > 0$. Then, there exists a* subsequence (y_{i_k}) of (y_i) which is equivalent to the canonical basis of c_0 in the sense that, for some constant $C > 0$ and all finite sequences (α_k) of real numbers,

$$C^{-1} \max_k |\alpha_k| \le \left\| \sum_k \alpha_k y_{i_k} \right\| \le C \max_k |\alpha_k|.$$

Proof. As a consequence of the hypotheses, we know in particular that $\inf_i \|y_i\| > 0$ while $y_i \to 0$ weakly. It is a well-known and important result (cf. e.g. [Li-T1], p.5) that one can then extract a subsequence (y_{n_i}) which is basic in the sense that every element in the span of (y_{n_i}) can be written uniquely as $y = \sum_i \alpha_i y_{n_i}$ for some sequence of scalars (α_i). Then, necessarily $\alpha_i \to 0$ since $\inf_i \|y_{n_i}\| > 0$, and, by the closed graph theorem we already have the lower inequality in the statement of the lemma. Since $\sum_i |f(y_i)| < \infty$ for all f in B', by another application of the closed graph theorem, for some C and all f in B', $\sum_i |f(y_i)| \le C\|f\|$. The conclusion is then obvious: for all finite sequences (α_k) of scalars,

$$\left\| \sum_k \alpha_k y_{i_k} \right\| = \sup_{\|f\| \le 1} \left| \sum_k \alpha_k f(y_{i_k}) \right|$$

$$\le \max_k |\alpha_k| \sup_{\|f\| \le 1} \sum_k |f(y_{i_k})| \le C \max_k |\alpha_k|.$$

This proves the lemma which thus concludes the proof of Theorem 9.29.

Remark 9.31. As a consequence of Theorem 9.29, and more precisely of its proof, if (X_i) is a sequence of independent symmetric Radon random variables with values in a Banach space B such that $\sup_n \|S_n\| < \infty$ almost surely but such that (S_n) does *not* converge, there exist an ω and a subsequence $(i_k) = (i_k(\omega))$ such that $(X_{i_k}(\omega))$ is equivalent to the canonical basis of c_0. Indeed, the various assertions of the theorem are obviously equivalent to saying that if $\sup_n \|S_n\| < \infty$, then $X_i \to 0$ almost surely. By Fubini's theorem, there exists an ω of the space supporting the X_i's such that $\sup_n \|\sum_{i=1}^n \varepsilon_i X_i(\omega)\| < \infty$ but $\inf_i \|X_i(\omega)\| > 0$. The remark then follows from the proof of the implication (i) $\Rightarrow$ (iv).

Notes and References

This chapter reproduces, although not up to the original, parts of the excellent notes [Pi16] by G. Pisier where the interested reader can find more Banach space theory oriented results and in particular quantitative finite dimensional results. A complete exposition of type and cotype and their relations to the local theory of Banach spaces is the book [Mi-S] by V. D. Milman and G. Schechtman. A more recent "volumic" description of the local theory is the book [Pi18] by G. Pisier. We refer to these works for accurate references. For a more operator theoretical point of view, see [Pie], [Pi15], [TJ2].

The Lecture Notes [Schw3] by L. Schwartz surveys much of the connections between Probability and Geometry in Banach spaces until 1980. See also the exposition [Wo2] by W. A. Woyczyński.

Dvoretzky's theorem was established in 1961 [Dv]. The new proof by V. D. Milman [Mil] using isoperimetric methods and amplified later on in the paper [F-L-M] considerably influenced the developments of the local theory of Banach spaces. A detailed account of applications of isoperimetric inequalities and concentration of measure phenomena to Geometry of Banach spaces may be found in [Mi-S]. The "concentration dimension" of a Gaussian variable was introduced by G. Pisier [Pi16] (see also [Pi18]) in a Gaussian version of Dvoretzky's theorem and the first proof of Theorem 9.2 is taken from [Pi16]. The second proof is due to Y. Gordon [Gor1], [Gor2]. The Dvoretzky-Rogers lemma appeared in [D-R]; various simple proofs are given in the modern literature, e.g. [F-L-M], [Mi-S], [Pi16], [TJ2]. The fundamental Theorems 9.6 and 9.16 are due to B. Maurey and G. Pisier [Mau-Pi], with an important contribution by J.-L. Krivine [Kr]. The proof of Theorem 9.6 through stable distributions and their representation is due to G. Pisier [Pi12] and was motivated by the results of W. B. Johnson and G. Schechtman on embedding ℓ_p^m into ℓ_1^n [J-S1]. Embeddings via stable variables had already been used in [B-DC-K].

The notions of type and cotype of Banach spaces were explicitly introduced by B. Maurey in the Maurey-Schwartz Seminar 1972/73 (see also [Mau1]) and independently by J. Hoffmann-Jørgensen [HJ1] (cf. [Pi15]). The basic Theorem 9.10 is due to S. Kwapień [Kw1]. Proposition 9.12 (iii) was known since the paper [M-P2] in which Lemma 5.8 is established. Proposition 9.13 comes from [Ro1] (improved in this form in [Led4] and [Pi16]). Comparison of averages of symmetric random variables in Banach spaces not containing ℓ_∞^n's is described in [Mau-Pi]. The proof of Proposition 9.14 is however due to S. Kwapień [Pi16]. We learned about its optimality as well as of its dual version (Proposition 9.15) from G. Schechtman and J. Zinn (personal communication). Operators of stable type and their possible different definitions (in the context of Probability in Banach spaces) are examined in [P-R1].

The relation of the strong law of large numbers (SLLN) with geometric convexity conditions goes back to the origin of Probability in Banach spaces. In 1962, A. Beck [Be] showed that the SLLN of Corollary 9.18 holds if and only if B is B-convex; a Banach space B is called B-convex if for some $\varepsilon > 0$ and some integer n, for all sequences $(x_i)_{i \leq n}$ in the unit ball of B, one can find a choice of signs $(\varepsilon_i)_{i \leq n}$, $\varepsilon_i = \pm 1$, with $\| \sum_{i=1}^n \varepsilon_i x_i \| \leq (1 - \varepsilon)n$. This property was identified to B not containing ℓ_1^n's in [Gi] and then completely elucidated with the concept of type by G. Pisier [Pi1]. Theorem 9.17 is due to J. Hoffmann-Jørgensen and G. Pisier [HJ1], [Pi1], [HJ-P]. Note the prior contribution [Wo1] in smooth normed spaces. Proposition 9.19 was observed in [Pi3] while Proposition 9.20 is taken from the paper [A-G-M-Z]. More on "Poissonization" may be found there as well as in, e.g., [A-A-G] and [Ar-G2]. A. de Acosta established Theorem 9.21 in [Ac6], partly motivated by the results of M. B. Marcus and W. A. Woyczyński [M-W] on weak laws of large numbers and stable type (Theorem 9.22, but [M-W] goes beyond this statement). More SLLN's are discussed in [Wo2]. See also [Wo3].

Lemma 9.23 is part of the folklore on pregaussian covariances. (9.16) goes back to [Va1]. Propositions 9.24 and 9.25 have been noticed by many authors, e.g. [Pi3], [Ja2], [C-T2], [A-G] (attributed to X. Fernique), [Ar-G2] (via the central limit theorem), etc. Theorem 9.27 has been deduced by several authors from various more general statements on Lévy measures and their integrability properties (cf. [Ar-G2], [Li]). The proof with the representation is borrowed from [Pi16] as also the approach through stochastic integrals (see also [M-P3], [Ro2]). Note that while L_p-spaces, $1 \leq p < 2$, are not of stable type p, one can still describe spectral measures of p-stable random variables in L_p. The proof again goes through the representation together with arguments similar to those used in (5.19) (that this study actually extends). One can show for example in this way that if m is a (say) probability measure on the unit sphere of ℓ_p and if $Y = (Y_k)$ has law m, then m is the spectral measure of a p-stable random variable in ℓ_p, $1 \leq p < 2$, if and only if

$$\sum_k \mathbb{E}\left(|Y_k|^p\left(1 + \log^+ \frac{|Y_k|}{\|Y_k\|_p}\right)\right) < \infty.$$

This has been known for some time by S. Kwapień and G. Pisier and is presented in [C-R-W] and [G-Z1].

That S_n converges almost surely when $\sup_n |S_n| < \infty$ for real valued random variables is classically deduced from Kolmogorov's converse inequality (and the three series theorem). Theorem 9.29 on Banach spaces not containing c_0 is due to J. Hoffmann-Jørgensen [H-J2] and S. Kwapień [Kw2] (for the main implication (i) ⇒ (iv)). Lemma 9.30 goes back to [B-P] (cf. [Li-T1]).

10. The Central Limit Theorem

The study of strong limit theorems for sums of independent random variables such as the strong law of large numbers or the law of the iterated logarithm in the preceding chapters showed that in Banach spaces these can only be reasonably understood when the corresponding weak property, that is tightness or convergence in probability, is satisfied. It was shown indeed that under some natural moment conditions, the strong statements actually reduce to the corresponding weak ones. On the line, or in finite dimensional spaces, the moment conditions usually automatically ensure the weak limiting property. As we pointed out, this is no longer the case in general Banach spaces.

There is some point, therefore, to attempt to investigate one typical tightness question in Banach space. One such example is provided by the central limit theorem (in short CLT). The CLT is of course one of the main topics in Probability Theory. Here also, its study will indicate the typical problems and difficulties to achieve tightness in Banach spaces. We only investigate here the very classical CLT for sums of independent and identically distributed random variables with normalization $\sqrt{n}$. This framework is actually rich enough already to analyze the main questions. In the first section of this chapter, we present some general facts on the CLT. In the second one, we make use of the type and cotype conditions to extend to certain classes of Banach spaces the classical characterization of the CLT. In the last paragraph, we describe a small ball criterion, which might be of independent interest, as well as an almost sure randomized CLT of some possible interest in Statistics. Let us mention that this study of the classical CLT will be further developed in the empirical process framework in Chapter 14.

In the whole chapter, we deal with Radon random variables, and actually for more convenience, with Borel random variables with values in a *separable* Banach space. Some results actually extend, with only minor modifications, to our usual more general setting, such as the results of Section 10.3. We leave this to the interested reader. Thus, let B denote a separable Banach space. If X is a Borel random variable with values in B, we denote by (X_i) a sequence of independent copies of X, and set, as usual, $S_n = X_1 + \cdots + X_n$, $n \geq 1$.

10.1 Some General Facts About the Central Limit Theorem

We start with the fundamental definition of the central limit property. Let X be a (Borel) random variable with values in a separable Banach space B. X is said to satisfy the *central limit theorem* (CLT) in B if the sequence $(S_n/\sqrt{n})$ converges weakly in B.

Once this definition has been given, one of the main questions is of course to decide when a random variable X satisfies the CLT, and if possible in terms only of the distribution of X. It is well-known that on the line a random variable X satisfies the CLT if and only if $\mathbb{E}X = 0$ and $\mathbb{E}X^2 < \infty$, and if X satisfies the CLT, the sequence $(S_n/\sqrt{n})$ converges weakly to a normal distribution with mean zero and variance $\mathbb{E}X^2$. The sufficient part can be established by various methods, for example Lévy's method of characteristic functions or Lindeberg's truncation approach.

We would like to outline here the necessity of $\mathbb{E}X^2 < \infty$ which is particularly clear using the methods developed so far in this book (and somewhat less clear by the usual methods). Let us show more precisely that if the sequence $(S_n/\sqrt{n})$ is stochastically bounded (this is of course necessary for X to satisfy the CLT), then $\mathbb{E}X = 0$ and $\mathbb{E}X^2 < \infty$. Once $\mathbb{E}X^2 < \infty$ has been shown, the centering will be obvious from the strong law of large numbers. Furthermore, replacing X by $X - X'$ where X' is an independent copy of X, we can assume without loss of generality that X is symmetric. For every $n \geq 1$ and $i = 1, \ldots, n$, let

$$u_i = u_i(n) = \frac{X_i}{\sqrt{n}} I_{\{|X_i| \leq \sqrt{n}\}} \, .$$

By the contraction principle (Lemma 6.5), for any $t > 0$,

$$\mathbb{P}\left\{ \left| \sum_{i=1}^{n} u_i \right| > t \right\} \leq 2\mathbb{P}\left\{ |S_n/\sqrt{n}| > t \right\} .$$

By the hypothesis, choose $t = t_0$ independent of n such that the right hand side of this inequality is less than $1/72$. By Proposition 6.8, we get

$$\mathbb{E}\left| \sum_{i=1}^{n} u_i \right|^2 \leq 18(1 + t_0^2)$$

uniformly in n. Hence, by orthogonality and identical distribution,

(10.1) $$\mathbb{E}\big(|X|^2 I_{\{|X| \leq \sqrt{n}\}}\big) \leq 18(1 + t_0^2)$$

and the result follows letting n tend to infinity.

The sufficiency of the conditions $\mathbb{E}X = 0$ and $\mathbb{E}\|X\|^2 < \infty$ for a random variable X to satisfy the CLT clearly extends to the case where X takes values in a finite dimensional space. It is not too difficult to see that this extends also to Hilbert space. Concerning necessity, we note that the preceding argument allows to conclude that $\mathbb{E}X = 0$ and $\mathbb{E}\|X\|^2 < \infty$ for any random variable X satisfying the CLT in a Banach space of *cotype* 2. Indeed, the orthogonality property leading to (10.1) is just the cotype 2 inequality. There are however spaces in which the CLT does not necessarily imply that $\mathbb{E}\|X\|^2 < \infty$. Rather than to give an example at this stage, we refer to the forthcoming Proposition 10.8 in which we will actually realize that $\mathbb{E}\|X\|^2 < \infty$ is necessary for X to satisfy the CLT (in and) *only* in cotype 2 spaces.

However, if X satisfies the CLT in B, for any linear functional f in B', the scalar random variable $f(X)$ satisfies the CLT with limiting Gaussian law with variance $\mathbb{E}f^2(X) < \infty$. Hence, the sequence $(S_n/\sqrt{n})$ actually converges weakly to a Gaussian random variable $G = G(X)$ with the same covariance structure as X. In other words and in the terminology of Chapter 9, a random variable X satisfying the CLT is necessarily *pregaussian*. By Proposition 9.25, we can then recover in particular that a random variable X with values in a Banach space B of cotype 2 satisfying the CLT is such that $\mathbb{E}\|X\|^2 < \infty$.

We mentioned above that in general a random variable satisfying the CLT does not necessarily have a strong second moment. What can then be said on the integrability properties of the norm of X when X satisfy the CLT in an arbitrary Banach space? The next lemma describes the best possible result in this direction. It shows that if the strong second moment is not always available, nevertheless a close property holds. The gap however induces conceptually a rather deep difference.

Lemma 10.1. *Let X be a random variable in B satisfying the CLT. Then X has mean zero and*
$$\lim_{t\to\infty} t^2 \mathbb{P}\{\|X\| > t\} = 0.$$
In particular, $\mathbb{E}\|X\|^p < \infty$ for every $0 < p < 2$.

Proof. The mean zero property follows from the second result together with the law of large numbers. Replacing X by $X - X'$ where X' is an independent copy of X and with some trivial desymmetrization argument, we need only consider the case of a symmetric random variable X. Let $0 < \varepsilon \leq 1$. For any $t > 0$, if $G = G(X)$ denotes the limiting Gaussian distribution of the sequence $(S_n/\sqrt{n})$,
$$\limsup_{n\to\infty} \mathbb{P}\left\{\frac{\|S_n\|}{\sqrt{n}} > t\right\} \leq \mathbb{P}\{\|G\| \geq t\}.$$
Since G is Gaussian, $\mathbb{E}\|G\|^2 < \infty$ and we can therefore find $t_0 = t_0(\varepsilon)$ (≥ 3) large enough so that $\mathbb{P}\{\|G\| \geq t_0\} \leq \varepsilon t_0^{-2}$. (We are thus actually only using that $\lim_{t\to\infty} t^2 \mathbb{P}\{\|G\| > t\} = 0$, the property which we would like to establish for X.) Hence, there exists $n_0 = n_0(\varepsilon)$ such that for all $n \geq n_0$
$$\mathbb{P}\left\{\frac{\|S_n\|}{\sqrt{n}} > t_0\right\} \leq 2\varepsilon t_0^{-2}.$$
By Lévy's inequality (2.7) for symmetric random variables,

$$\mathbb{P}\{\max_{i \leq n} \|X_i\| > t_0\sqrt{n}\} \leq 4\varepsilon t_0^{-2} \quad (\leq 1/2).$$

By Lemma 2.6,

$$n\mathbb{P}\{\|X\| > t_0\sqrt{n}\} \leq 8\varepsilon t_0^{-2}$$

which therefore holds for all $n \geq n_0$. Let now t be such that $t_0\sqrt{n} \leq t < t_0\sqrt{n+1}$ for some $n \geq n_0$. Then

$$t^2\mathbb{P}\{\|X\| > t\} \leq t_0^2(n+1)\mathbb{P}\{\|X\| > t_0\sqrt{n}\} \leq 16\varepsilon ,$$

that is,

$$\limsup_{t \to \infty} t^2\mathbb{P}\{\|X\| > t\} \leq 16\varepsilon .$$

Since $\varepsilon > 0$ is arbitrarily small, this proves the lemma.

It is useful to observe that the preceding argument can also be applied to $S_k/\sqrt{k}$ instead of X for *each fixed* k with bounds *uniform in* k. Indeed, if $Y_1, \ldots, Y_m$ denote independent copies of $S_k/\sqrt{k}$, then $\sum_{i=1}^m Y_i/\sqrt{m}$ has the same distribution as $S_{mk}/\sqrt{mk}$ and the argument remains the same. In this way, we can state the following corollary to the proof of Lemma 10.1.

Corollary 10.2. *If X satisfies the CLT, then*

$$\lim_{t \to \infty} t^2 \sup_n \mathbb{P}\left\{\frac{\|S_n\|}{\sqrt{n}} > t\right\} = 0 .$$

In particular, for any $0 < p < 2$,

$$\sup_n \mathbb{E}\left\|\frac{S_n}{\sqrt{n}}\right\|^p < \infty .$$

It will be convenient for the sequel to retain a simple quantitative version of this result that immediately follows from the proof of Lemma 10.1 and the preceding observation; namely, for $\varepsilon > 0$,

$$(10.2) \quad \text{if} \quad \sup_n \mathbb{P}\left\{\frac{\|S_n\|}{\sqrt{n}} > \varepsilon\right\} \leq \frac{1}{8} , \quad \text{then} \quad \sup_n \mathbb{E}\left\|\frac{S_n}{\sqrt{n}}\right\| \leq 20\varepsilon .$$

Let us mention further that the preceding argument leading to Corollary 10.2 extends to more general normalizing sequences (a_n) like e.g. $a_n = n^{1/p}$, $0 < p < 2$, or $a_n = (2nLLn)^{1/2}$ since the only property really used is that $a_{mk} \leq Ca_m a_k$ for some constant C. Finally, as an alternate approach to the proof of Corollary 10.2, one can use Hoffmann-Jørgensen's inequalities. Combine to this aim Proposition 6.8, Lemma 7.2 and Lemma 10.1.

In the following, we adopt the notation

$$CLT(X) = \sup_n \mathbb{E}\left\|\frac{S_n}{\sqrt{n}}\right\| .$$

By Corollary 10.2, $CLT(X) < \infty$ when X satisfies the CLT. It will be seen below that $CLT(\cdot)$ defines a norm on the linear space of all random variables satisfying the CLT.

At this point, we would like to open a parenthesis and mention a few words about what can be called the bounded form of the CLT. We can say indeed that a random variable X with values in B satisfies the *bounded* CLT if the sequence $(S_n/\sqrt{n})$ is bounded in probability, that is, if for each $\varepsilon > 0$ there is a positive finite number M such that

$$\sup_n \mathbb{P}\left\{\frac{\|S_n\|}{\sqrt{n}} > M\right\} < \varepsilon.$$

The proofs of Lemma 10.1 and Corollary 10.2 of course carry over to conclude that if X satisfies the bounded CLT, then $\mathbb{E}X = 0$ and

$$\sup_{t>0} \sup_n t^2 \mathbb{P}\left\{\frac{\|S_n\|}{\sqrt{n}} > t\right\} < \infty.$$

In particular, we also have that $CLT(X) < \infty$ as already indicated by (10.2). As we have seen to start with, on the line, the bounded CLT also implies that $\mathbb{E}X^2 < \infty$, and similarly $\mathbb{E}\|X\|^2 < \infty$ in cotype 2 spaces. In the scalar case therefore, the bounded and true CLT are equivalent, and, as will follow from subsequent results, this equivalence actually extends to Hilbert spaces and even cotype 2 spaces. However, it is not difficult to see that this equivalence already fails in L_p-spaces when $p > 2$. Rather than to detail an example at this stage, we refer to Theorem 10.10 below where a characterization of the CLT and bounded CLT in L_p-spaces will clearly indicate the difference between these two properties. It is an open problem to characterize those Banach spaces in which the bounded and the true CLT are equivalent.

The bounded CLT does not in general imply that X is pregaussian. Since however $\mathbb{E}f(X) = 0$ and $\mathbb{E}f^2(X) < \infty$ for every f in B', there exists by the finite dimensional CLT a Gaussian random process $G = (G_f)_{f \in B_1'}$ indexed by the unit ball B_1' of B' with the same covariance structure as X. Moreover, G is almost surely bounded since by the finite dimensional CLT and the convergence of the moments (using Skorokhod's theorem),

$$\mathbb{E} \sup_{f \in B_1'} |G_f| \leq CLT(X) < \infty.$$

In particular, by Sudakov's minoration (Theorem 3.18), the family $\{f(X); f \in B_1'\}$ is relatively compact in L_2. However, this Gaussian process G need not define in general a Radon probability distribution on B. Let us consider indeed the following easy but meaningful example. Denote by $(e_k)_{k \geq 1}$ the canonical basis of c_0 and consider the random variable X with values in c_0 defined by

(10.3)
$$X = \left(\frac{\varepsilon_k e_k}{\left(2\log(k+1)\right)^{1/2}}\right)_{k \geq 1}$$

where (ε_k) is a Rademacher sequence. Let us show that X satisfies the bounded CLT. By independence and identical distribution, for every $M > 0$,

$$\mathbb{P}\left\{\frac{\|S_n\|}{\sqrt{n}} > M\right\} = 1 - \prod_{k}\left(1 - \mathbb{P}\left\{\left|\sum_{i=1}^{n}\frac{\varepsilon_i}{\sqrt{n}}\right| > M\left(2\log(k+1)\right)^{1/2}\right\}\right).$$

From the subgaussian inequality (4.1),

$$\mathbb{P}\left\{\left|\sum_{i=1}^{n}\frac{\varepsilon_i}{\sqrt{n}}\right| > M\left(2\log(k+1)\right)^{1/2}\right\} \leq 2\exp\left(-M^2\log(k+1)\right).$$

Therefore, if M is large enough independently of n,

$$\mathbb{P}\left\{\frac{\|S_n\|}{\sqrt{n}} > M\right\} \leq 4\sum_{k}\exp\left(-M^2\log(k+1)\right)$$

from which it clearly follows that X satisfies the bounded CLT. However, X does not satisfy the CLT in c_0 since X is not pregaussian. Indeed, the natural Gaussian structure with the same covariance as X should be given by

$$G = \left(\frac{g_k e_k}{\left(2\log(k+1)\right)^{1/2}}\right)_{k \geq 1}$$

where (g_k) is an orthogaussian sequence. But we know that

$$\limsup_{k \to \infty}\frac{|g_k|}{\left(2\log(k+1)\right)^{1/2}} = 1 \quad \text{almost surely}.$$

Therefore G is a bounded Gaussian random process but does not define a Gaussian random variable with values in c_0.

The choice of c_0 in this example is not casual as is shown by the next result.

Proposition 10.3. *Let B be a separable Banach space. In order that every random variable X with values in B that satisfies the bounded CLT is pregaussian, it is necessary and sufficient that B does not contain an isomorphic copy of c_0.*

Proof. Example (10.3) provides necessity. Assume B does not contain c_0 and let X be defined on $(\Omega, \mathcal{A}, \mathbb{P})$ satisfying the bounded CLT in B. Since $\mathbb{E}\|X\| < \infty$, there exists a sequence $(\mathcal{A}_N)$ of finite sub-σ-algebras of $\mathcal{A}$ such that if $X^N = \mathbb{E}^{\mathcal{A}_N}X$, the sequence (X^N) converges almost surely and in $L_1(B)$ to X. By Jensen's inequality, $CLT(X^N) \leq CLT(X)$ for every N. Set $Y^N = X^N - X^{N-1}$, $N \geq 1$ ($X^0 = 0$). Since for each N, Y^N only takes finitely many values, Y^N is pregaussian. Denote by (G_N) independent Gaussian random variables in B such that G_N has the same covariance structure as Y^N. As is easily seen, for every f in B' and every N,

$$\mathbb{E}f^2\left(\sum_{i=1}^{N} G_i\right) = \mathbb{E}f^2(X^N).$$

Hence, by the finite dimensional CLT (and convergence of moments), for every N,

$$\mathbb{E}\left\|\sum_{i=1}^{N} G_i\right\| \leq CLT(X^N) \leq CLT(X).$$

Thus the sequence $(\sum_{i=1}^{N} G_i)$ is almost surely bounded. Since B does not contain c_0, it converges almost surely (Theorem 9.29) to a Gaussian random variable G which satisfies $\mathbb{E}f^2(G) = \mathbb{E}f^2(X)$ for every f in B'. Hence X is pregaussian.

After this short digression on the bounded CLT we come back to the general study of the CLT. We first recall from Chapter 2 some of the criteria which might be used in order to establish weak convergence of the sequence $(S_n/\sqrt{n})$. From the finite dimensional CLT and Theorem 2.1, a Borel random variable X with values in a separable Banach space B satisfies the CLT if and only if for each $\varepsilon > 0$ one can find a compact set K in B such that

$$\mathbb{P}\left\{\frac{S_n}{\sqrt{n}} \in K\right\} \geq 1 - \varepsilon \quad \text{for every } n$$

(or only every n large enough). Alternatively, and in terms of finite dimensional approximation, a random variable X with values in B such that $\mathbb{E}f(X) = 0$ and $\mathbb{E}f^2(X) < \infty$ for every f in B' (i.e. $f(X)$ satisfies the scalar CLT for every f) satisfies the CLT if and only if for every $\varepsilon > 0$ there is a finite dimensional subspace F of B such that

$$\mathbb{P}\left\{\left\|T\left(\frac{S_n}{\sqrt{n}}\right)\right\| > \varepsilon\right\} < \varepsilon \quad \text{for every } n$$

(or only n large enough) where $T = T_F$ denotes the quotient map $B \to B/F$ (cf. (2.4)). By (10.2), equivalently,

$$(10.4) \qquad\qquad CLT\big(T(X)\big) < \varepsilon.$$

Note that such a property is satisfied as soon as there exists, for every $\varepsilon > 0$, a step mean zero random variable Y such that $CLT(X - Y) < \varepsilon$. (The converse actually also holds cf. [Pi3].) Note further from these considerations that X satisfies the CLT as soon as for each $\varepsilon > 0$ there is a random variable Y satisfying the CLT such that $CLT(X - Y) < \varepsilon$; in particular, the linear space of all random variables satisfying the CLT equipped with the norm $CLT(\cdot)$ defines a Banach space.

Before we turn to the next section, let us continue with these easy observations and mention some comments about symmetrization. By centering and Jensen's inequality on (10.4) for example, clearly, X satisfies the CLT if and only if $X - X'$ does where X' is an independent copy of X. When trying to establish that a random variable X satisfies the CLT, it will thus basically be enough to deal with a symmetric X. In the same spirit, we also see from Lemma 6.3 that X satisfies the CLT if and only if εX does where ε denotes a Rademacher random variable independent of X. This property is actually one example of a general randomization argument which might be worthwhile to detail at this point. If $0 < p, q < \infty$, denote by $L_{p,q}$ the Lorentz space of all real valued random variables ξ such that

$$\|\xi\|_{p,q} = \left(q \int_0^\infty \left(t^p \mathbb{P}\{|\xi| > t\} \right)^{q/p} \frac{dt}{t} \right)^{1/q} < \infty.$$

$L_{p,p}$ is just L_p by the usual integration by parts formula and $L_{p,q_1} \subset L_{p,q_2}$ if $q_1 \leq q_2$.

Proposition 10.4. *Let X be a Borel random variable with values in B such that $CLT(X) < \infty$ (in particular $\mathbb{E}X = 0$). Let further ξ be a non-zero real valued random variable in $L_{2,1}$ independent of X. Then*

$$\tfrac{1}{2}\mathbb{E}|\xi| \, CLT(X) \leq CLT(\xi X) \leq 2\|\xi\|_{2,1}CLT(X).$$

In particular, ξX satisfies the CLT (and $\mathbb{E}X = 0$) if and only if X does.

Proof. The second assertion follows from (10.4) and the inequalities applied to $T(X)$ for quotient maps T. We first prove the right hand side inequality. Assume to begin with that ξ is the indicator function I_A of some set A. Then clearly, by independence and identical distribution, for every n,

$$\mathbb{E}\left\|\sum_{i=1}^n \xi_i X_i\right\| = \mathbb{E}\left\|\sum_{i=1}^{S_n(\xi)} X_i\right\| \leq \mathbb{E}\left((S_n(\xi))^{1/2}\right) CLT(X)$$

$$\leq \sqrt{n}\sqrt{\mathbb{P}(A)} \, CLT(X)$$

where ξ_i are independent copies of ξ and $S_n(\xi) = \xi_1 + \cdots + \xi_n$. Hence, in this case,

$$CLT(\xi X) \leq \sqrt{\mathbb{P}(A)} \, CLT(X).$$

Now, some classical extremal properties of indicators in the spaces $L_{p,1}$ yield the conclusion. Supposing first $\xi \geq 0$, let, for each $\varepsilon > 0$,

$$\xi^{(\varepsilon)} = \sum_{k=1}^\infty \varepsilon k I_{\{\varepsilon(k-1) < \xi \leq \varepsilon k\}} = \sum_{k=1}^\infty \varepsilon I_{\{\xi > \varepsilon(k-1)\}}.$$

By the triangle inequality and the preceding,

$$CLT(\xi^{(\varepsilon)} X) \leq \sum_{k=1}^\infty \varepsilon \left(\mathbb{P}\{\xi > \varepsilon(k-1)\}\right)^{1/2} CLT(X)$$

$$\leq \|\xi\|_{2,1}CLT(X).$$

Letting ε tend to 0 yields the result in this case. The general case follows by writing $\xi = \xi^+ - \xi^-$.

To establish the reverse inequality, note first that by centering and Lemma 6.3,

$$CLT(\varepsilon\xi X) \leq 2CLT(\xi X),$$

where ε is a Rademacher variable which is independent of X and ξ. Use then the contraction principle conditionally on X and ξ in the form of Lemma 4.5. The conclusion follows.

Proposition 10.4 in particular applies when ξ is a standard normal variable. Therefore, the normalized sums $S_n/\sqrt{n}$ can be regarded as conditionally Gaussian and several Gaussian tools and techniques can be used. This will be one of the arguments in the empirical process approach to the CLT developed in Chapter 14. Note further that this Gaussian randomization is similar to the one put forward in the series representation of stable random variables. This relation to stable random variables is not fortuitous and the difficulty in proving tightness in the CLT resembles in some sense the difficulty in showing the existence of a p-stable random vector with a given spectral measure. Let us mention also that while for general sums of independent random variables, Gaussian randomization is heavier than Rademacher randomization (cf. (4.9)), the crucial point in Proposition 10.4 is that we are dealing with independent *identically distributed* random variables. The condition ξ in $L_{2,1}$ has been shown to be best possible in general [L-T1], although L_2 is (necessary and) sufficient in various classes of spaces; this L_2-multiplication property is perhaps related to some geometric properties of the underlying Banach space. Note finally that the argument of the proof of Proposition 10.4 is not really limited to the normalization $\sqrt{n}$ of the CLT and that similar statements can be obtained in case for example of the iid laws of large numbers with normalization $n^{1/p}$, $0 < p < 2$, and in case of the law of the iterated logarithm (see [Led5]).

10.2 Some Central Limit Theorems in Certain Banach Spaces

In this paragraph, we try to find conditions on the distribution only of a Borel random variable X with values in a separable Banach space B in order that it satisfies the CLT. As we know, this is a difficult question in general spaces and at the present time it has a clear cut answer only for special classes of Banach spaces. In this section, we present a sample of these results. We start with some examples and negative facts in order to set up the framework of the study.

Let us first mention that a Gaussian random variable clearly satisfies the CLT. A random variable X with values in a finite dimensional Banach space B satisfies the CLT if and only if $\mathbb{E}X = 0$ and $\mathbb{E}\|X\|^2 < \infty$. As will be shown below, this equivalence extends to infinite dimensional Hilbert spaces, but actually only to them! In general, very bad situations can occur and strong assumptions on the distribution of a random variable X have no reason to ensure that X satisfies the CLT. For example, the random variable in c_0 defined by (10.3) is symmetric and *almost surely bounded* but does not satisfy the CLT. It fails the CLT since it is not pregaussian, but if we go back to Example 7.11, we have a bounded symmetric pregaussian (elementary verification) variable in c_0 which does not satisfy the bounded CLT, hence the CLT. (In [Ma-Pl], there is even an example of a bounded symmetric pregaussian random variable in c_0 satisfying the bounded CLT but failing the CLT.) Even the fact that these examples are constructed in c_0, a space with "bad" geometric properties (of no non-trivial type or cotype) is not restrictive. There exist indeed spaces of type $2 - \varepsilon$ and cotype $2 + \varepsilon$ for every $\varepsilon > 0$ in which one can find bounded pregaussian random variables failing the CLT [Led3].

In spite of these negative examples, some positive results can be obtained. In particular, and as indicated by the last mentioned example, spaces of type 2 and/or cotype 2 play a special rôle. This is also made clear by Propositions 9.24 and 9.25 connecting type 2 and cotype 2 with pregaussian random variables and some inequalities involving those. The first theorem extends to type 2 spaces the sufficient conditions on the line for the CLT.

Theorem 10.5. *Let X be a mean zero random variable such that $\mathbb{E}\|X\|^2 < \infty$ with values in a separable Banach space B of type 2. Then X satisfies the CLT. Conversely, if in a (separable) Banach space B, every random variable X such that $\mathbb{E}X = 0$ and $\mathbb{E}\|X\|^2 < \infty$ satisfies the CLT, then B must be of type 2.*

Proof. The definition of type 2 in the form of Proposition 9.11 immediately implies that for X such that $\mathbb{E}X = 0$ and $\mathbb{E}\|X\|^2 < \infty$,

$$(10.5) \qquad CLT(X) \leq C\left(\mathbb{E}\|X\|^2\right)^{1/2}.$$

If, given $\varepsilon > 0$, we choose a mean zero random variable Y with finite range such that $\mathbb{E}\|X - Y\|^2 < \varepsilon^2/C^2$, the preceding inequality applied to $X - Y$ yields $CLT(X - Y) < \varepsilon$ and thus X satisfies the CLT ((10.4)). Conversely, if, in B, each mean zero Borel random variable X with $\mathbb{E}\|X\|^2 < \infty$ satisfies the CLT, such a random variable is necessarily pregaussian. The second part of the theorem therefore simply follows from the corresponding one in Proposition 9.24. Alternatively, one can invoke a closed graph argument to obtain (10.5) and apply then Proposition 9.19.

We would like to mention at this stage that Theorem 10.5 easily extends to operators of type 2. We state , for later reference, the following corollary of the proof of Theorem 10.5.

Corollary 10.6. *Let $u : E \to F$ be an operator of type 2 between two separable Banach spaces E and F. Let X be a Borel random variable with values in E*

such that $\mathbb{E}X = 0$ *and* $\mathbb{E}\|X\|^2 < \infty$. *Then the random variable* $u(X)$ *satisfies the CLT in* F.

The next statement is the dual result of Theorem 10.5 for cotype 2 spaces.

Theorem 10.7. *Let* X *be a pregaussian random variable with values in a separable cotype 2 Banach space* B. *Then* X *satisfies the CLT. Conversely, if in a (separable) Banach space* B, *any pregaussian random variable satisfies the CLT, then* B *must be of cotype 2.*

Proof. By Proposition 9.25, in a cotype 2 space,

$$\mathbb{E}\|X\|^2 \leq C\mathbb{E}\|G(X)\|^2 < \infty$$

for any pregaussian random variable X with associated Gaussian variable $G(X)$. Now, for each n, $S_n/\sqrt{n}$ is pregaussian and associated to $G(X)$ too. Hence,

$$CLT(X) \leq \left(C\mathbb{E}\|G(X)\|^2\right)^{1/2}.$$

Let now $X^N = \mathbb{E}^{\mathcal{A}_N} X$ where $(\mathcal{A}_N)$ is a sequence of finite σ-algebras generating the σ-algebra of X. Then (X^N) converges almost surely and in $L_2(B)$ to X. For each N, $X - X^N$ is still pregaussian and since

$$\mathbb{E}f^2(X - X^N) \leq 2\mathbb{E}f^2(X) = 2\mathbb{E}f^2(G(X))$$

for every f in B', it follows from (3.11) that the sequence $(G(X - X^N))$ is tight (since $G(X)$ is). It can only converge to 0 since, for every f, $\mathbb{E}f^2(X - X^N) \to 0$. By the Gaussian integrability properties (Corollary 3.2), for each $\varepsilon > 0$ one can find an N such that $\mathbb{E}\|G(X - X^N)\|^2 < \varepsilon^2/C$. Thus $CLT(X - X^N) < \varepsilon$ and X satisfies the CLT by the tightness criterion described in Section 10.1. Conversely, recall that a random variable X satisfying the CLT is such that $\mathbb{E}\|X\|^p < \infty$, $p < 2$ (Lemma 10.1). We need then simply recall one assertion of Proposition 9.25. Theorem 10.7 is thus established. ∎

In cotype 2 spaces, random variables satisfying the CLT have a strong second moment (because they are pregaussian, or cf. Section 10.1). This actually only happens in cotype 2 spaces as is shown by the next statement. This result complements Theorem 10.7.

Proposition 10.8. *If in a (separable) Banach space* B *every Borel random variable* X *satisfying the CLT is in* $L_2(B)$, *then* B *is of cotype 2.*

Proof. Assume that B is not of cotype 2. Since (cf. Remark 9.26) the cotype 2 definitions with either Gaussian or Rademacher averages are equivalent, there exists a sequence (x_j) in B such that $\sum_j g_j x_j$ converges almost surely but for which $\sum_j \|x_j\|^2 = \infty$. Consider, on some suitable probability space,

$$X = \sum_{j=1}^{\infty} 2^{j/2} I_{A_j} g_j x_j$$

where A_j are disjoints sets with $\mathbb{P}(A_j) = 2^{-j}$ and independent from the orthogaussian sequence (g_j). Then $\mathbb{E}\|X\|^2 = \sum_j \|x_j\|^2 = \infty$. However, let us show that X satisfies the CLT which therefore leads to a contradiction and proves the proposition. Let (g_j^i) be independent copies of (g_j) and (A_j^i) independent copies of (A_j), all of them assumed to be independent. For every n, let $N(n)$ be the smallest integer such that $2^{N(n)} \geq 2^5 n$. Let

$$\Omega_0 = \bigcup_{j=N(n)+1}^{\infty} \bigcup_{i=1}^{n} A_j^i .$$

Ω_0 only depends on the A_j^i's and $\mathbb{P}(\Omega_0) \leq 2^{-5}$. Moreover, on the complement of Ω_0,

$$S_n = \sum_{j=1}^{N(n)} 2^{j/2} \left(\sum_{i=1}^{n} I_{A_j^i} g_j^i \right) x_j .$$

We now use the law of large numbers to show that, conditionally on (A_j^i), the Gaussian variables $\sum_{i=1}^{n} I_{A_j^i} g_j^i$ have a variance close to $n2^{-j}$. For every $t > 0$,

$$\mathbb{P}\left\{ \bigcup_{j=1}^{N(n)} \left(\frac{1}{n} \sum_{i=1}^{n} I_{A_j^i} > (t+1)2^{-j} \right) \right\} \leq \sum_{j=1}^{N(n)} \mathbb{P}\left\{ \sum_{i=1}^{n} (I_{A_j^i} - \mathbb{P}(A_j^i)) > tn2^{-j} \right\}$$

$$\leq \frac{1}{t^2 n} \sum_{j=1}^{N(n)} 2^j \leq \frac{2^7}{t^2} .$$

Let us take $t = 2^6$. Hence, for every n, there is a set of probability bigger than $1 - 2^{-4}$ such that conditionally on the A_j^i's, $S_n/\sqrt{n}$ has the same distribution as $\sum_{j=1}^{N(n)} \eta_j x_j$ where the η_j's are independent normal random variables with variances less than $2^6 + 1 \leq 2^8$. We can then write, for every n and $s > 0$,

$$\mathbb{P}\left\{ \frac{\|S_n\|}{\sqrt{n}} > s \right\} \leq 2^{-4} + \mathbb{P}\left\{ \left\| \sum_{j=1}^{N(n)} \eta_j x_j \right\| > s \right\}$$

$$\leq 2^{-4} + \frac{2^4}{s} \mathbb{E}\left\| \sum_{j=1}^{\infty} g_j x_j \right\|$$

where we have used the contraction principle in the last step. If we now choose $s = 2^8 \mathbb{E}\| \sum_{j=1}^{\infty} g_j x_j \|$ (for example), we deduce from (10.2) that

$$CLT(X) \leq 20 \cdot 2^8 \mathbb{E}\left\| \sum_{j=1}^{\infty} g_j x_j \right\| .$$

It is now easy to conclude that X satisfies the CLT. The same inequality when a finite number of x_j are 0 indeed allows to build a finite dimensional approximation of X in the $CLT(\cdot)$-norm. X therefore satisfies the CLT and the proof is thereby completed.

In the spirit of the preceding proof and as a concrete example, let us note in passing that a convergent Rademacher series $X = \sum_i \varepsilon_i x_i$ satisfies the CLT if and only if the corresponding Gaussian series $\sum_i g_i x_i$ converges. The necessity is obvious since $\sum_i g_i x_i$ is a Gaussian variable with the same covariance as X. Concerning the sufficiency, if (ε_{ij}) and (g_{ij}) are respectively doubly-indexed Rademacher and orthogaussian sequences, and starting with a finite sequence (x_i), by (4.8), for every n,

$$\mathbb{E}\left\|\frac{S_n}{\sqrt{n}}\right\| = \mathbb{E}\left\|\sum_i \left(\sum_{j=1}^{n} \frac{\varepsilon_{ij}}{\sqrt{n}}\right) x_i\right\|$$

$$\leq \left(\frac{\pi}{2}\right)^{1/2} \mathbb{E}\left\|\sum_i \left(\sum_{j=1}^{n} \frac{g_{ij}}{\sqrt{n}}\right) x_i\right\|$$

$$= \left(\frac{\pi}{2}\right)^{1/2} \mathbb{E}\left\|\sum_i g_i x_i\right\|$$

where the last step follows from the Gaussian rotational invariance. By approximation, X is then easily seen to satisfy the CLT when $\sum_i g_i x_i$ converges almost surely.

If we recall (Theorem 9.10) that a Banach space of type 2 and cotype 2 is isomorphic to a Hilbert space, the conjunction of Theorem 10.5 and Proposition 10.8 yields an isomorphic characterization of Hilbert space by the CLT.

Corollary 10.9. *A separable Banach space B is isomorphic to a Hilbert space if and only if for every Borel random variable X with values in B the conditions $\mathbb{E}X = 0$ and $\mathbb{E}\|X\|^2 < \infty$ are necessary and sufficient for X to satisfy the CLT.*

The preceding results are perhaps the most satisfactory ones on the CLT in Banach spaces although they actually concern rather small classes of spaces. While the case of cotype 2 spaces may be considered as completely understood, this is not exactly true for type 2 spaces. Indeed, while Theorem 10.4 indicates that the conditions $\mathbb{E}X = 0$ and $\mathbb{E}\|X\|^2 < \infty$ are sufficient for X to satisfy the CLT in a type 2 space, the integrability condition $\mathbb{E}\|X\|^2 < \infty$ need not conversely be necessary (it is only so in cotype 2 spaces). As we have seen, in general, when X satisfies the CLT, one only knows that (Lemma 10.1),

$$\lim_{t \to \infty} t^2 \mathbb{P}\{\|X\| > t\} = 0.$$

It is therefore of some interest to try to understand the spaces in which the best possible necessary conditions, i.e. X is pregaussian and $\lim_{t \to 0} t^2 \mathbb{P}\{\|X\|$

$> t\} = 0$, are also sufficient for a random variable X to satisfy the CLT. One convenient way to investigate this class of spaces is an inequality, similar in some sense to the type and cotype inequalities, but which combines moment assumptions with the pregaussian character. More precisely, let us say that a separable Banach space B satisfies the inequality Ros(p), $1 \leq p < \infty$, if there is a constant C such that for any finite sequence (X_i) of independent pregausssian random variables with values in B with associated Gaussian variables $(G(X_i))$ (which may be assumed to be independent)

$$(10.6) \qquad \mathbb{E}\Big\|\sum_i X_i\Big\|^p \leq C\Big(\sum_i \mathbb{E}\|X_i\|^p + \mathbb{E}\Big\|\sum_i G(X_i)\Big\|^p\Big).$$

This inequality is the vector valued version of an inequality discovered by H. P. Rosenthal (hence the appellation) on the line. That (10.6) holds on the line for any p is easily deduced from, for example, Proposition 6.8 and the observation that

$$\mathbb{E}\Big|\sum_i G(X_i)\Big|^p = C_p\Big(\sum_i \mathbb{E}G(X_i)^2\Big)^{p/2}$$

$$= C_p\Big(\sum_i \mathbb{E}X_i^2\Big)^{p/2} = C_p\Big(\mathbb{E}\Big|\sum_i X_i\Big|^2\Big)^{p/2}.$$

The same argument together with Proposition 9.25 shows that cotype 2 spaces also satisfy Ros(p) for every p, $1 \leq p < \infty$. It can be shown actually that the spaces of cotype 2 are the only ones with this property (cf. [Led4]).

The main interest of the inequality Ros(p) in connection with the CLT lies in the following observation.

Theorem 10.10. *Let B be a separable Banach space satisfying* Ros(p) *for some $p > 2$. Then, a Borel random variable X with values in B satisfies the CLT if and only if it is pregaussian and* $\lim_{t \to \infty} t^2 \mathbb{P}\{\|X\| > t\} = 0$.

Proof. The necessity has been discussed in Section 10.1 and holds, as we know, in any space. Turning to sufficiency, our aim is to show that for any symmetric pregaussian variable X with values in B

$$(10.7) \qquad CLT(X) \leq C(\|X\|_{2,\infty} + \mathbb{E}\|G(X)\|)$$

for some constant C. This property easily implies the conclusion. Indeed, given X symmetric, pregaussian and such that $\lim_{t \to \infty} t^2 \mathbb{P}\{\|X\| > t\} = 0$ and given $\varepsilon > 0$, let us first choose t large enough in order that, if $Y = XI_{\{\|X\| \leq t\}}$ (which is still pregaussian by Lemma 9.23),

$$\|X - Y\|_{2,\infty} + \mathbb{E}\|G(X - Y)\| < \frac{\varepsilon}{2C}.$$

To this aim, use that $\lim_{t \to \infty} t^2 \mathbb{P}\{\|X\| > t\} = 0$ and $\lim_{t \to \infty} \mathbb{E}f^2(XI_{\{\|X\| > t\}}) = 0$ for every f in B'. Consider then $(\mathcal{A}_N)$ a sequence of finite σ-algebras generating the σ-algebra of X and set $Y^N = \mathbb{E}^{\mathcal{A}_N} Y$. Then $Y^N \to Y$ almost surely and in $L_2(B)$, and, as usual, $G(Y - Y^N) \to 0$ in $L_2(B)$. Applying (10.7) to $X - Y$ and $Y - Y^N$ for N large enough yields $CLT(X - Y^N) < \varepsilon$ and therefore X satisfies the CLT from this finite dimensional approximation. If X is not symmetric, replace it for example by εX where ε is a Rademacher random variable which is independent of X.

Therefore, it suffices to establish (10.7). Assume by homogeneity that $\|X\|_{2,\infty} \le 1$. For each n, we have that,

$$\frac{1}{\sqrt{n}} \mathbb{E}\left\| \sum_{i=1}^{n} X_i \right\| \le \mathbb{E}\left\| \sum_{i=1}^{n} u_i \right\| + \sqrt{n}\,\mathbb{E}\left(\|X\| I_{\{\|X\| > \sqrt{n}\}} \right)$$

where $u_i = u_i(n) = n^{-1/2} X_i I_{\{\|X_i\| \le \sqrt{n}\}}$, $i = 1, \dots, n$. Since $\|X\|_{2,\infty} \le 1$, by integration by parts, it is easily seen that

$$\sup_n \sqrt{n}\,\mathbb{E}\left(\|X\| I_{\{\|X\| > \sqrt{n}\}} \right) \le 2.$$

Applying the inequality Ros(p), $p > 2$, to the u_i's, we get

$$\mathbb{E}\left\| \sum_{i=1}^{n} u_i \right\|^p \le C\left(n \mathbb{E}\|u_1\|^p + 2\mathbb{E}\|G(X)\|^p \right)$$

since the u_i's are pregaussian and $\mathbb{E}\|G(u_1)\|^p \le 2\mathbb{E}\|G(X)\|^p$ (Lemma 9.23). Now, since $p > 2$ and $\|X\|_{2,\infty} \le 1$,

$$n\mathbb{E}\|u_1\|^p \le n^{1-p/2} \int_0^{\sqrt{n}} \mathbb{P}\{\|X\| > t\}\, dt^p$$

$$\le n^{1-p/2} \int_0^{\sqrt{n}} \frac{dt^p}{t^2} = \frac{p}{p-2}.$$

(10.7) thus follows (recall Gaussian random vectors have all their moments equivalent). Theorem 10.10 is established.

Type 2 spaces of course satisfy Ros(2) but the important property in Theorem 10.10 is Ros(p) for $p > 2$. We already noticed that cotype 2 spaces verify Ros(p) for all p; in particular L_p-spaces with $1 \le p \le 2$. When $2 < p < \infty$, L_p satisfies Ros(p) for the corresponding p. This follows from the scalar inequality together with Fubini's theorem. Indeed, if (X_i) are independent pregaussian random variables in $L_p = L_p(S, \Sigma, \mu)$ where (S, Σ, μ) is σ-finite,

$$\mathbb{E}\left\| \sum_i X_i \right\|^p = \int_S \mathbb{E}\left| \sum_i X_i(s) \right|^p d\mu(s)$$

$$\le \int_S C\left(\sum_i \mathbb{E}|X_i(s)|^p + \mathbb{E}\left| \sum_i G(X_i)(s) \right|^p \right) d\mu(s)$$

$$= C\left(\sum_i \mathbb{E}\|X_i\|^p + \mathbb{E}\left\| \sum_i G(X_i) \right\|^p \right).$$

When $2 < p < \infty$, L_p is of type 2 and enters the setting of Theorem 10.5 but since it satisfies $\text{Ros}(p)$ and $p > 2$ we can also apply the more precise Theorem 10.10. Together with the characterization (9.16) of pregaussian structures, the CLT is therefore completely understood in the L_p-spaces, $1 \le p < \infty$.

One might wonder from the preceding about some more examples of spaces satisfying $\text{Ros}(p)$ for some $p > 2$, especially among the class of type 2 spaces. It can be shown that Banach lattices which are r-convex and s-concave for some $r > 2$ and $s < \infty$ (cf. [Li-T2]) belong to this class. However, already $\ell_2(\ell_r)$, $r > 2$, which is of type 2, verifies $\text{Ros}(p)$ for *no* $p > 2$. Actually, this is true of $\ell_2(B)$ as soon as B is a Banach space which is *not* of cotype 2. To see this, note that if B is not of cotype 2, for every $\varepsilon > 0$, one can find a pregaussian random variable Y in B such that

$$\|Y\| = 1 \quad \text{almost surely and} \quad \mathbb{E}\|G(Y)\|^2 < \varepsilon.$$

Consider then independent copies Y_i of Y and set $X_i = Y_i e_i$ in $\ell_2(B)$ where (e_i) is the canonical basis. Assume then that $\ell_2(B)$ satisfies $\text{Ros}(p)$ for some $p > 2$ and apply this inequality to the sample $(X_i)_{i \le N}$. Since Gaussian moments are all equivalent, we should have, for some constant C and all N,

$$\mathbb{E}\left\|\sum_{i=1}^N X_i\right\|^p \le C\left(\sum_{i=1}^N \mathbb{E}\|X_i\|^p + \left(\mathbb{E}\left\|\sum_{i=1}^N G(X_i)\right\|^2\right)^{p/2}\right).$$

That is, since $G(X_i) = G(Y_i)e_i$,

$$N^{p/2} \le C\left(N + (N\mathbb{E}\|G(Y)\|^2)^{p/2}\right)$$
$$\le C\left(N + (\varepsilon N)^{p/2}\right).$$

Hence, if ε is small enough and N tends to infinity, this leads to a contradiction.

Thus finding spaces satisfying $\text{Ros}(p)$ for $p > 2$ seems a difficult task, and so the CLT under the best possible necessary conditions. One interesting problem in this context would be to know whether Theorem 10.10 has some converse; that is, if in a Banach space B, the conditions X pregaussian and $\lim_{t\to\infty} t^2\mathbb{P}\{\|X\| > t\} = 0$ are sufficient for X to satisfy the CLT, does B satisfy $\text{Ros}(p)$ for some $p > 2$? This could be analogous to theorems on the laws of large numbers and the type of a Banach space (cf. Corollary 9.18).

As a remark we would like to briefly come back at this stage to the bounded CLT and show that the bounded CLT and true CLT already differ in L_p for $p > 2$. It is clear from the proof of Theorem 10.10 (cf. (10.7)) that a pregaussian variable in a $\text{Ros}(p)$-space, $p > 2$, such that $\|X\|_{2,\infty} < \infty$ satisfies the bounded CLT. To prove our claim, it is therefore simply enough to construct a pregaussian random variable X in, for example, ℓ_p, $2 < p < \infty$, such that

$$(10.8) \qquad 0 < \limsup_{t\to\infty} t^2\mathbb{P}\{\|X\| > t\} < \infty.$$

To this aim, let N be an integer valued random variable such that $\mathbb{P}\{N = i\} = ci^{-1-2/p}$, $i \ge 1$. Let (ε_i) be a Rademacher sequence which is independent of N and consider X in ℓ_p given by

$$X = \sum_{i=1}^{\infty} \varepsilon_i I_{\{N^2 \le i < N^2 + N\}} e_i$$

where (e_i) is the canonical basis of ℓ_p. Then $\|X\| = N^{1/p}$ and (10.8) clearly holds. We are left to show that X is pregaussian and we use (9.16). That is, it suffices to show that

$$\sum_{i=1}^{\infty} (\mathbb{P}\{N \le i < N^2 + N\})^{p/2} < \infty;$$

but this is clear since by definition of N, $\mathbb{P}\{N \le i < N^2 + N\}$ is of the order of $i^{-1/2-1/p}$ when $i \to \infty$, and $p > 2$.

In conclusion to this section, we present some remarks on the relation between the CLT and the LIL (for simplicity we understand here by LIL only the compact law of the iterated logarithm) in Banach spaces. On the line and in finite dimensional spaces, CLT and LIL are of course equivalent, that is, a random variable X satisfies the CLT if and only if it satisfies the LIL, since they are both characterized by the moment conditions $\mathbb{E}X = 0$ and $\mathbb{E}\|X\|^2 < \infty$. However, the conjunction of Corollary 10.9 and Corollary 8.8 indicates that this equivalence already fails in infinite dimensional Hilbert space where one can find a random variable satisfying the LIL but failing the CLT. This observation together with Dvoretzky's theorem (Theorem 9.1) actually shows that the implication LIL $\Rightarrow$ CLT only holds in finite dimensional spaces. Indeed, if B is an infinite dimensional Banach space in which every Borel random variable satisfying the LIL also satisfies the CLT, by a closed graph argument, for some constant C and every X in B,

$$CLT(X) \le C\Lambda(X)$$

where we recall that $\Lambda(X) = \limsup_{n\to\infty} \|S_n\|/a_n$ (non-random). By Theorem 9.1, the same inequality would hold for all step random variables with values in a Hilbert space, and hence, by approximation, for all random variables. But this is impossible as we have seen. Hence we can state

Theorem 10.11. *Let B be a separable Banach space in which every Borel random variable satisfying the LIL also satisfies the CLT. Then B is finite dimensional.*

Concerning the implication CLT $\Rightarrow$ LIL, a general statement is available. Indeed, if X satisfies the CLT, then trivially $S_n/a_n \to 0$ in probability, and since X is pregaussian the unit ball of the reproducing kernel Hilbert space associated to X is compact. The characterization of the LIL in Banach spaces (Theorem 8.6) then yields the following theorem.

Theorem 10.12. *Let X be a Borel random variable with values in a separable Banach space B satisfying the CLT. Then X satisfies the LIL if and only if $\mathbb{E}(\|X\|^2/LL\|X\|) < \infty$.*

The moment condition $\mathbb{E}(\|X\|^2/LL\|X\|) < \infty$ is of course necessary in this statement since it is not comparable to the tail behavior $\lim_{t\to\infty} t^2\mathbb{P}\{\|X\| > t\} = 0$ necessary for the CLT. Despite this general satisfactory result, the question of the implication CLT $\Rightarrow$ LIL is not solved for all that. Theorem 10.12 indicates that the spaces in which random variables satisfying the CLT also satisfy the LIL are exactly those in which the CLT implies the integrability property $\mathbb{E}(\|X\|^2/LL\|X\|) < \infty$. This is of course the case for cotype 2 spaces but the characterization of the CLT in L_p-spaces shows that L_p with $p > 2$ does not satisfy this property. An argument similar to the one used for Theorem 10.11, but this time with Theorem 9.16 instead of Dvoretzky's theorem, then shows that the spaces satisfying CLT $\Rightarrow$ LIL are necessarily of cotype $2 + \varepsilon$ for every $\varepsilon > 0$. But a final characterization is still to be obtained.

10.3 A Small Ball Criterion
for the Central Limit Theorem

In this last paragraph, we develop a criterion for the CLT which, while certainly somewhat difficult to verify in practice, involves in its elaboration several interesting arguments and ideas developed throughout this book. The result therefore presents some interest from a theoretical point of view. The idea of its proof can be used further for an almost sure randomized version of the CLT.

Recall that we deal in all this chapter with a separable Banach space B. We noticed, prior to Theorem 3.3, that for a Gaussian Radon random variable G, with values in B each ball centered at the origin has a positive mass for the distribution of G. Therefore, it follows that if X is a Borel random variable satisfying the CLT in B, for every $\varepsilon > 0$,

$$(10.9) \qquad \liminf_{n\to\infty} \mathbb{P}\left\{ \frac{\|S_n\|}{\sqrt{n}} < \varepsilon \right\} > 0.$$

It turns out that, conversely, if a Gaussian cylindrical measure charges each ball centered at the origin, then it is Radon. Surprisingly, this converse extends to the CLT. Namely, if (10.9) holds for every $\varepsilon > 0$, and if the necessary tail condition $\lim_{t\to\infty} t^2\mathbb{P}\{\|X\| > t\} = 0$ holds, then X satisfies the CLT. This is theoretical small ball criterion for the CLT. It can thus be stated as follows.

Theorem 10.13. *Let X be a Borel random variable with values in a separable Banach space B. Then X satisfies the CLT if and only if the following two properties are satisfied:*

(i) $\lim_{t\to\infty} t^2\mathbb{P}\{\|X\| > t\} = 0$;

(ii) for each $\varepsilon > 0$, $\alpha(\varepsilon) = \liminf_{n\to\infty} \mathbb{P}\{\|S_n/\sqrt{n}\| < \varepsilon\} > 0$.

Before we turn to the proof of this result, we would like to mention a few facts, one of which will be of help in the proof. As we have seen, (i) and (ii) are necessary for X to satisfy the CLT and are optimal. Indeed, the tail condition (i) cannot be suppressed in general, i.e. (ii) does not necessarily imply (i) (cf. [L-T3]). However, and we would like to detail this point, (ii), and for *one* $\varepsilon > 0$ only, already implies the bounded CLT, that is that the sequence $(S_n/\sqrt{n})$ is bounded in probability (and therefore (ii) implies $\sup_{t>0} t^2 \mathbb{P}\{\|X\| > t\} < \infty$). This claim is based on the inequality of Proposition 6.4. Replacing X by $X - X'$ where X' is an independent copy of X, it is enough to deal with the symmetrical case. Let $Y_1, \ldots, Y_m$ be independent copies of $S_n/\sqrt{n}$. Since $\sum_{i=1}^m Y_i/\sqrt{m}$ has the same distribution as $S_{mn}/\sqrt{mn}$, by Proposition 6.4, for every large enough n,

$$\frac{\alpha(\varepsilon)}{2} \leq \mathbb{P}\left\{\left\|\sum_{i=1}^m Y_i\right\| < \varepsilon\sqrt{m}\right\} \leq \frac{3}{2}\left(1 + \sum_{i=1}^m \mathbb{P}\{\|Y_i\| > \varepsilon\sqrt{m}\}\right)^{-1/2}$$

and thus, for every large enough n and every m,

$$m\mathbb{P}\left\{\frac{\|S_n\|}{\sqrt{n}} > \varepsilon\sqrt{m}\right\} \leq \frac{9}{\alpha(\varepsilon)^2}.$$

Therefore, as announced, $(S_n/\sqrt{n})$ is bounded in probability. In particular, $CLT(X) < \infty$ and while X is not necessarily pregaussian, at least there is a bounded Gaussian process with the same covariance structure as X and the family $\{f(X); f \in B', \|f\| \leq 1\}$ is totally bounded in L_2. All that was noticed in Section 10.1 when we discussed the bounded CLT.

Let us note that the preceding argument based on Proposition 6.4 shows similarly that a random variable X satisfies the CLT if there is a compact set K in B such that

$$\liminf_{n\to\infty} \mathbb{P}\left\{\frac{S_n}{\sqrt{n}} \in K\right\} > 0.$$

This improved version of the usual tightness criterion may be useful to understand the intermediate level of Theorem 10.13.

Proof of Theorem 10.13. Replacing as usual X by $X - X'$ we may and do assume that X is symmetric. The sequence (X_i) of independent copies of X has therefore the same distribution as $(\varepsilon_i X_i)$ where (ε_i) is a Rademacher sequence which is independent of (X_i). Recall that we denote by $\mathbb{P}_\varepsilon$, $\mathbb{E}_\varepsilon$ (resp. $\mathbb{P}_X$, $\mathbb{E}_X$) conditional probability and expectation with respect to the sequence (X_i) (resp. (ε_i)). We show, and this is enough, that there is a numerical constant C such that, given $\delta > 0$, one can find a finite dimensional subspace F of B with quotient map $T = T_F : B \to B/F$ such that

(10.10) $$\limsup_{n\to\infty} \mathbb{P}_X\left\{\mathbb{E}_\varepsilon\left\|\sum_{i=1}^n \varepsilon_i \frac{T(X_i)}{\sqrt{n}}\right\| > C\delta\right\} \leq \delta.$$

The proof is based on the isoperimetric inequality for product measures of Theorem 1.4 which was one of the main tools in the study of the strong limit theorems and which proves to be also of some interest in weak statements such as the CLT. Here, we use the full statement of Theorem 1.4 and not only (1.13). The main step in this proof will be to show that if

$$A = \left\{ x = (x_i)_{i \leq n} \in B^n; \; \mathbb{E}_\varepsilon \left\| \sum_{i=1}^n \varepsilon_i \frac{T(x_i)}{\sqrt{n}} \right\| \leq 2\delta \right\},$$

for each $\delta > 0$, there exists a finite dimensional subspace F of B such that if $T = T_F$,

(10.11) $$\mathbb{P}\{(X_i)_{i \leq n} \in A\} \geq \theta(\delta) > 0$$

for all large enough n's, where $\theta(\delta) > 0$ depends on $\delta > 0$ only. Let us show how to conclude when (10.11) is satisfied. For integers q, k, recall $H(A, q, k)$. If $(X_i)_{i \leq n} \in H(A, k, q)$, there exist $j \leq k$ and $x^1, \ldots, x^q$ in A such that

$$\{1, \ldots, n\} = \{i_1, \ldots, i_j\} \cup I$$

where $I = \bigcup_{\ell=1}^q \{i \leq n; \; X_i = x_i^\ell\}$. By monotonicity of Rademacher averages (cf. Remark 6.18),

$$\mathbb{E}_\varepsilon \left\| \sum_{i=1}^n \varepsilon_i \frac{T(X_i)}{\sqrt{n}} \right\| \leq k \max_{i \leq n} \frac{\|X_i\|}{\sqrt{n}} + \mathbb{E}_\varepsilon \left\| \sum_{i \in I} \varepsilon_i \frac{T(X_i)}{\sqrt{n}} \right\|$$

$$\leq k \max_{i \leq n} \frac{\|X_i\|}{\sqrt{n}} + \sum_{\ell=1}^q \mathbb{E}_\varepsilon \left\| \sum_{i=1}^n \varepsilon_i \frac{T(x_i^\ell)}{\sqrt{n}} \right\|$$

$$\leq k \max_{i \leq n} \frac{\|X_i\|}{\sqrt{n}} + 2q\delta.$$

Hence

$$\mathbb{P}_X \left\{ \mathbb{E}_\varepsilon \left\| \sum_{i=1}^n \varepsilon_i \frac{T(X_i)}{\sqrt{n}} \right\| > (2q + 1)\delta \right\}$$

$$\leq \mathbb{P}^* \{(X_i)_{i \leq n} \notin H(A, q, k)\} + \mathbb{P} \left\{ k \max_{i \leq n} \frac{\|X_i\|}{\sqrt{n}} > \delta \right\}.$$

Now, under (10.11), Theorem 1.4 tells us that for some numerical constant K, which we might choose to be an integer for convenience,

$$\mathbb{P}^* \{(X_i)_{i \leq n} \notin H(A, q, k)\} \leq \left[K \left(\frac{\log(1/\theta(\delta))}{k} + \frac{1}{q} \right) \right]^k.$$

Let us choose q to be $2K$, and then take $k = k(\delta)$ large enough depending on $\delta > 0$ only such that the preceding probability is less than $\delta/2$. Since, by (i),

$$\lim_{n \to \infty} \mathbb{P}\{\max_{i \leq n} \|X_i\| > \delta\sqrt{n}/k(\delta)\} = 0,$$

(10.10) will be satisfied and the CLT for X will hold.

Therefore, we have to establish (10.11). We write, for every n,

$$\mathbb{P}_X\left\{\mathbb{E}_\varepsilon\left\|\sum_{i=1}^n \varepsilon_i \frac{T(X_i)}{\sqrt{n}}\right\| > 2\delta\right\}$$

$$\leq \mathbb{P}\left\{\left\|\sum_{i=1}^n \varepsilon_i \frac{X_i}{\sqrt{n}}\right\| < \delta\right\} + \mathbb{P}\left\{\mathbb{E}_\varepsilon\left\|\sum_{i=1}^n \varepsilon_i \frac{T(X_i)}{\sqrt{n}}\right\| - \left\|\sum_{i=1}^n \varepsilon_i \frac{T(X_i)}{\sqrt{n}}\right\| > \delta\right\}$$

since $\|T\| \leq 1$. By (ii), (10.11) will hold as soon as

$$\limsup_{n\to\infty} \mathbb{P}\left\{\mathbb{E}_\varepsilon\left\|\sum_{i=1}^n \varepsilon_i \frac{T(X_i)}{\sqrt{n}}\right\| - \left\|\sum_{i=1}^n \varepsilon_i \frac{T(X_i)}{\sqrt{n}}\right\| > \delta\right\} < \alpha(\delta)$$

for some appropriate choice of T. Setting, for every n,

$$u_i = u_i(n) = \frac{X_i}{\sqrt{n}} I_{\{\|X_i\| \leq c(\delta)\sqrt{n}\}}, \quad i = 1, \dots, n$$

where $c(\delta) > 0$ is to be specified, it is actually enough by (i) to check that

$$(10.12) \quad \limsup_{n\to\infty} \mathbb{P}\left\{\mathbb{E}_\varepsilon\left\|\sum_{i=1}^n \varepsilon_i T(u_i)\right\| - \left\|\sum_{i=1}^n \varepsilon_i T(u_i)\right\| > \delta\right\} < \alpha(\delta).$$

To this aim, we use the concentration properties of Rademacher averages (Theorem 4.7). We have to switch to expectations instead of medians, but this is easy. Conditionally on (X_i), denote by M a median of $\|\sum_{i=1}^n \varepsilon_i T(u_i)\|$ and let $\sigma = \sup_{\|f\|\leq 1}\left(\sum_{i=1}^n f^2(T(u_i))\right)^{1/2}$ where the supremum runs here over the unit ball of the dual space of B/F for an F to be chosen. By Theorem 4.7, for every $t > 0$,

$$\mathbb{P}_\varepsilon\left\{\left|\left\|\sum_{i=1}^n \varepsilon_i T(u_i)\right\| - M\right| > t\right\} \leq 4\exp\left(-\frac{t^2}{8\sigma^2}\right) \leq 32\frac{\sigma^2}{t^2}.$$

In particular, by an integration by parts,

$$\left|\mathbb{E}_\varepsilon\left\|\sum_{i=1}^n \varepsilon_i T(u_i)\right\| - M\right| \leq 12\sigma.$$

Hence, if $\sigma \leq \delta/24$,

$$\mathbb{P}_\varepsilon\left\{\mathbb{E}_\varepsilon\left\|\sum_{i-1}^n \varepsilon_i T(u_i)\right\| - \left\|\sum_{i=1}^n \varepsilon_i T(u_i)\right\| > \delta\right\} \leq 128\frac{\sigma^2}{\delta^2}.$$

Thus, integrating with respect to $\mathbb{P}_X$,

$$\mathbb{P}\left\{\mathbb{E}_\varepsilon\left\|\sum_{i=1}^n \varepsilon_i T(u_i)\right\| - \left\|\sum_{i=1}^n \varepsilon_i T(u_i)\right\| > \delta\right\}$$

$$(10.13)$$

$$\leq \mathbb{P}\left\{\sigma > \frac{\delta}{24}\right\} + \frac{128}{\delta^2}\mathbb{E}\sigma^2 \leq \frac{10^3}{\delta^2}\mathbb{E}\sigma^2.$$

To prove (10.12), we have thus simply to show that $\mathbb{E}\sigma^2$ can be made arbitrarily small independently of n for a well chosen large enough subspace F of B. To this aim, recall from the discussion prior to this proof that, under (ii), $CLT(X) < \infty$ and that this implies that $\{f(X); f \in B', \|f\| \le 1\}$ is relatively compact in L_2. We use Lemma 6.6 (6.5) (and the contraction principle) to see that, for every n,

$$\mathbb{E}\sigma^2 = \mathbb{E} \sup_{\|f\| \le 1} \sum_{i=1}^{n} f^2(T(u_i))$$
$$\le \sup_{\|f\| \le 1} \mathbb{E} f^2(T(X)) + 8c(\delta)CLT(X).$$

Choose then $c(\delta) > 0$ to be less than $\delta^2 \alpha(\delta)/16 \cdot 10^3 CLT(X)$, and choose also $T : B \to B/F$ associated to some large enough finite dimensional subspace F of B such that $\sup_{\|f\| \le 1} \mathbb{E} f^2(T(X)) \le \delta^2 \alpha(\delta)/2 \cdot 10^3$. According to (10.13), we see that (10.12) is satisfied. This was the property that we had to prove. Theorem 10.13 is therefore established.

We conclude this chapter with an almost sure randomized version of the CLT. The interest in such a result lies in its proof itself, which is similar in nature to the preceding proof, and in possible statistical applications. It might be worthwhile to recall Proposition 10.4 before the statement. As there also, if X is a Banach space valued random variable and ξ a real valued random variable independent of X, and if (X_i) (resp. (ξ_i)) are independent copies of X (resp. ξ), the sequences (X_i) and (ξ_i) are understood to be independent (constructed on different probability spaces).

Theorem 10.14. *Let X be a mean zero Borel random variable with values in a separable Banach space B and let ξ be a real valued random variable in $L_{2,1}$ independent of X such that $\mathbb{E}\xi = 0$ and $\mathbb{E}\xi^2 = 1$. The following are equivalent:*

(i) $\mathbb{E}\|X\|^2 < \infty$ and X satisfies the CLT;

(ii) for almost every ω of the probability space supporting the X_i's, the sequence $\left(\sum_{i=1}^{n} \xi_i X_i(\omega)/\sqrt{n}\right)$ converges in distribution.

In either case, the limit of $\left(\sum_{i=1}^{n} \xi_i X_i(\omega)/\sqrt{n}\right)$ does not depend on ω and is distributed as $G(X)$, the Gaussian distribution with the same covariance structure as X.

Proof. It is plain that under (ii) the product ξX satisfies the CLT. Hence, since X has mean zero and $\xi \not\equiv 0$, X also satisfies the CLT by Proposition 10.4. To show that $\mathbb{E}\|X\|^2 < \infty$, replacing ξ by $\xi - \xi'$ where ξ' is an independent copy of ξ, we may assume ξ to be symmetric. For almost every ω, $\left(\sum_{i=1}^{n} \xi_i X_i(\omega)/\sqrt{n}\right)$ is bounded in probability. Hence, by Lévy's inequality (2.7), the same holds for the sequence $(|\xi_n|\|X_n(\omega)\|/\sqrt{n})$. Since ξ is non-zero, it follows that for almost every ω

$$\sup_n \frac{\|X_n(\omega)\|}{\sqrt{n}} < \infty.$$

Therefore, by independence and the Borel-Cantelli lemma, $\mathbb{E}\|X\|^2 < \infty$. This proves the implication (ii) $\Rightarrow$ (i).

The main tool in the proof of the converse implication (i) $\Rightarrow$ (ii) is the following lemma. This lemma may be considered as some vector valued extension of a strong law of large numbers for squares.

Lemma 10.15. *In the setting of Theorem 10.14, if $\mathbb{E}\|X\|^2 < \infty$, for some numerical constant K, almost surely,*

$$\limsup_{n\to\infty} \frac{1}{\sqrt{n}} \mathbb{E}_\xi \left\| \sum_{i=1}^n \xi_i X_i \right\| \leq K \limsup_{n\to\infty} \frac{1}{\sqrt{n}} \mathbb{E} \left\| \sum_{i=1}^n \xi_i X_i \right\|$$

where, as usual, $\mathbb{E}_\xi$ denotes partial integration with respect to the sequence (ξ_i).

Proof. Set $M = \limsup_{n\to\infty} \mathbb{E}\| \sum_{i=1}^n \xi_i X_i / \sqrt{n}\|$, assumed to be finite. By Lemma 6.3, since $\mathbb{E}\xi = 0$, we may and do assume that ξ is symmetric. By the Borel-Cantelli lemma, and monotonicity of the averages, it suffices to show that for some constant K, and all $\varepsilon > 0$,

$$\sum_n \mathbb{P}_X \left\{ \mathbb{E}_\xi \left\| \sum_{i=1}^{2^n} \xi_i X_i \right\| > K(M+\varepsilon)2^{n/2} \right\} < \infty,$$

or, by definition of M, that

$$\sum_n \mathbb{P}_X \left\{ \mathbb{E}_\xi \left\| \sum_{i=1}^{2^n} \xi_i X_i \right\| > K\mathbb{E} \left\| \sum_{i=1}^{2^n} \xi_i X_i \right\| + \varepsilon 2^{n/2} \right\} < \infty.$$

To show this, we use of the isoperimetric approach developed in Section 6.3. Since $\mathbb{E}\|X\|^2 < \infty$, by Lemmas 7.6 and 7.8, there exists a sequence (k_n) of integers such that $\sum_n 2^{-k_n} < \infty$ and

$$\sum_n \mathbb{P} \left\{ \sum_{i=1}^{k_n} \|X_i\|^* > \varepsilon 2^{n/2} \right\} < \infty$$

where $(\|X_i\|^*)$ is the non-increasing rearrangement of $(\|X_i\|)_{i\leq 2^n}$. Now, Remark 6.18 applies similarly to the averages in the symmetric sequence (ξ_i). Note that $\mathbb{E}|\xi| \leq 1$. Hence, by (6.23) adapted to (ξ_i) with $q = 2K_0$ and $k = k_n$ ($\geq q$ for n large enough),

$$\mathbb{P}_X \left\{ \mathbb{E}_\xi \left\| \sum_{i=1}^{2^n} \xi_i X_i \right\| > 2q\mathbb{E} \left\| \sum_{i=1}^{2^n} \xi_i X_i \right\| + \varepsilon 2^{n/2} \right\}$$

$$\leq 2^{-k_n} + \mathbb{P} \left\{ \sum_{i=1}^{k_n} \|X_i\|^* > \varepsilon 2^{n/2} \right\}.$$

Letting $K = 2q = 4K_0$, the proof of Lemma 10.15 is complete.

We conclude the proof of the theorem. Since X satisfies the CLT, for every $k \geq 1$, there exists a finite dimensional subspace F_k of B such that if $T_k = T_{F_k} : B \rightarrow B/F_k$

$$CLT(T_k(X)) \leq \frac{1}{2Kk\|\xi\|_{2,1}} .$$

Recall from Proposition 10.4 that $CLT(\xi X) \leq 2\|\xi\|_{2,1} CLT(X)$. If we now apply Lemma 10.15 to $T_k(X)$ for each k, there exists Ω_k with $\mathbb{P}(\Omega_k) = 1$ such that for every ω in Ω_k

$$\limsup_{n \rightarrow \infty} \frac{1}{\sqrt{n}} \, \mathbb{E}_\xi \Big\| \sum_{i=1}^{n} \xi_i T_k (X_i(\omega)) \Big\| \leq \frac{1}{k} .$$

Let also Ω_0 be the set of full probability obtained when Lemma 10.15 is applied to X itself. Let $\Omega^0 = \bigcap_{k \geq 0} \Omega_k$; $\mathbb{P}(\Omega^0) = 1$. Let now $\omega \in \Omega^0$. For each $\varepsilon > 0$, there exists a finite dimensional subspace F of B such that if $T = T_F$,

$$\limsup_{n \rightarrow \infty} \frac{1}{\sqrt{n}} \, \mathbb{E}_\xi \Big\| \sum_{i=1}^{n} \xi_i T (X_i(\omega)) \Big\| < \varepsilon^2 .$$

Hence, if $n \geq n_0(\varepsilon)$,

$$\mathbb{P}_\xi \Big\{ \Big\| T \Big(\sum_{i=1}^{n} \xi_i X_i(\omega) / \sqrt{n} \Big) \Big\| > \varepsilon \Big\} \leq \varepsilon .$$

It follows that the sequence $(\sum_{i=1}^{n} \xi_i X_i(\omega) / \sqrt{n})$ is tight (it is bounded in probability since $\omega \in \Omega_0$). We conclude the proof by identifying the limit. Using basically that $\sum_{i=1}^{n} f^2(X_i)/n \rightarrow \mathbb{E} f^2(X)$ almost surely, it is not difficult to see, by the Lindeberg CLT (cf. e.g. [Ar-G2]) for example, that, for every f in B', there exists a set Ω_f of probability one such that for all $\omega \in \Omega_f$, $(\sum_{i=1}^{n} \xi_i f(X_i(\omega))/\sqrt{n})$ converges in distribution to a normal variable with variance $\mathbb{E} f^2(X)$. The proof is easily completed by considering a weakly dense countable subset of B'. Theorem 10.14 is established.

Notes and References

This chapter only concentrate on the classical central limit theorem ((CLT)) for sums of identically distributed independent random variables under the normalization $\sqrt{n}$. We would like to refer to the book of A. Araujo and E. Giné [Ar-G2] for a more complete account on the general CLT for real and Banach space valued random variables, as well as for precise and detailed historical references. See also the paper [A-A-G]. We also mention the recent book by V. Paulauskas and A. Rachkauskas [P-R2] on rates of convergence in the vector valued CLT, a topic not covered here. We note that some further results on the CLT, using empirical process methods, will be presented in Chapter 14.

Starting with Donsker's invariance principle [Do], the study of the CLT for Banach space valued random variables was initiated by E. Mourier [Mo] and R. Fortet [F-M1], [F-M2], S. R. S. Varadhan [Var], R. Dudley and V. Strassen [D-S], L. Le Cam [LC]. The proof that we present of the necessity of $\mathbb{E}X^2 < \infty$ on the line and similarly of $\mathbb{E}\|X\|^2 < \infty$ in cotype 2 spaces is due to N. C. Jain [Ja1] who also showed the necessity of $\|X\|_{2,\infty} < \infty$ in any Banach space [Ja1], [Ja2]. The improved Lemma 10.1 was then noticed independently in [A-A-G] and [P-Z]. Corollary 10.2 was observed in [Pi3]. Proposition 10.3 is due to G. Pisier and J. Zinn [P-Z]. The randomization property of Proposition 10.4 has been known for some time independently by X. Fernique and G. Pisier and was put forward in the paper [G-Z2]. Our proof is Pisier's and the best possibility of $L_{2,1}$ was shown in [L-T1].

The extension to Hilbert spaces of the classical CLT was obtained by S. R. S. Varadhan [Var]. A further extension in some smooth spaces, anticipating spaces of type 2, is described in [F-M1]. Examples of bounded random variables in $C[0,1]$ failing the CLT were provided in [D-S], and in L_p, $1 \leq p < 2$, by R. Dudley (cf. [Kue2]). A decisive step was accomplished by J. Hoffmann-Jørgensen and G. Pisier [HJ-P] with Theorem 10.5 and N. C. Jain [Ja2] with Theorem 10.7 and Proposition 10.8. This proposition is due independently to D. Aldous [Ald]. See also [Pi3], [HJ3], [HJ4]. Rosenthal's inequality appeared in [Ros]. Its interest in the study of the CLT in L_p-spaces was put forward in [G-M-Z] and further by J. Zinn in [Zi2] where Theorem 10.10 is explained. An attempt of a systematic study of Rosenthal's inequality for vector valued random variables is undertaken in [Led4]. The characterization of the CLT in L_p-spaces (and more general Banach lattices) goes back to [P-Z] (cf. also [G-Z1]). That the best possible necessary conditions for the CLT are not sufficient in $\ell_2(B)$ when B is not of cotype 2 is due to J. Zinn [Zi2], [G-Z1]. Some further CLT's in $\ell_p(\ell_q)$ are investigated in this last paper [G-Z1]. The example on the CLT and bounded CLT in ℓ_p, $2 < p < \infty$, is taken from [P-Z].

Theorem 10.11 is due to G. Pisier and J. Zinn [P-Z] thanks to several early results on the CLT and the LIL in L_p-spaces, $2 \leq p < \infty$. G. Pisier [Pi3] established Theorem 10.12 assuming a strong second moment with a proof containing the essential step of the general case. The final result is due independently to V. Goodman, J. Kuelbs, J. Zinn [G-K-Z] and B. Heinkel [He2]. The comments on the implication CLT $\Rightarrow$ LIL are taken from [Pi3].

The small ball criterion (Theorem 10.13) was obtained in [L-T3]. On the line, the result was noticed in [J-O]. We learned how to use Kanter's inequality (Proposition 6.4) in this study from X. Fernique. The proof of Theorem 10.13 with the isoperimetric approach is new. Theorem 10.14 is due to J. Zinn and the authors and also appeared in [L-T3] (with a different proof). Its proof, in particular Lemma 10.15, has been used recently in bootstrapping of empirical measures by E. Giné and J. Zinn [G-Z4].

11. Regularity of Random Processes

In Chapter 9 we described how certain conditions on Banach spaces can ensure the existence and the tightness of some probability measures. For example, if (x_i) is a sequence in a type 2 Banach space B such that $\sum_i \|x_i\|^2 < \infty$, then the series $\sum_i g_i x_i$ converges almost surely and defines a Gaussian Radon random variable with values in B. These conditions were further used in Chapters 9 and 10 to establish tightness properties of sums of independent random variables, especially in the context of central limit theorems.

In this chapter, another approach to the existence and tightness of certain measures is taken in the framework of random functions and processes. Given a random process $X = (X_t)_{t \in T}$ indexed by some set T, we investigate sufficient conditions for the almost sure boundedness or continuity of the sample paths of X in terms of the "geometry" (in the metric sense) of T. By geometry, we mean some metric entropy or majorizing measure condition which estimates the size of T in function of some parameters related to X. The setting of this study has its roots in a celebrated theorem of Kolmogorov which gives sufficient conditions for the continuity of processes X indexed by a compact subset of $\mathbb{R}$ in terms of a Lipschitz condition on the *increments* $X_s - X_t$ of the processes. Under this type of incremental conditions on the processes, this result was extended to processes indexed by regular subsets T of $\mathbb{R}^N$ and then further to abstract index sets T. In this chapter, we present several results in this general abstract setting. The first section deals with the metric entropy condition. The results, which naturally extend the more classical ones, are rather easy to prove and to use but nevertheless can be shown to be sharp in many respects. The second paragraph investigates majorizing measure conditions which are more precise than entropy conditions as they take more into account the local geometry of the index set. That majorizing measures are a key notion will be shown in the next chapter on Gaussian processes. These sufficient entropy or majorizing measure conditions for sample boundedness or continuity are used in the proofs in rather a similar manner: the main idea is indeed based on the rather classical covering technique and chaining argument already contained in Kolmogorov's theorem. In Section 11.3, we present important examples of applications to Gaussian, Rademacher and chaos processes.

Common to this chapter is the datum of a random process $X = (X_t)_{t \in T}$, that is a collection (X_t) of real valued random variables, indexed by some parameter set T which we assume to be a metric or pseudo-metric space. By pseudo-metric recall that we mean that T is equipped with a distance d which does not necessarily separate points ($d(s,t) = 0$ does not always imply $s = t$). Our main objective is to find sufficient conditions in order for X to be almost surely bounded or continuous, or to possess a version with these properties. We usually work with processes $X = (X_t)_{t \in T}$ which are in L_p, $1 \le p < \infty$, or, more generaly, in some Orlicz space L_ψ, i.e. $\|X_t\|_\psi < \infty$ for every t. Concerning almost sure boundedness, and according to what was described in Chapter 2, we therefore simply understand supremum like $\sup_{t \in T} X_t$, $\sup_{t \in T} |X_t|$, $\sup_{s,t \in T} |X_s - X_t|$... as lattice supremum in L_ψ; for example,

$$\mathbb{E} \sup_{s,t \in T} |X_s - X_t| = \sup\{\mathbb{E} \sup_{s,t \in F} |X_s - X_t| \,; \; F \text{ finite in } T\}.$$

We avoid in this way the usual measurability questions and moreover reduce the estimates which we will establish to the case of a *finite* parameter set T. Of course, we could also use separable versions and we do that anyway in the study of the sample continuity.

One more word before we turn to the object of this chapter. In the theorems below, we usually bound the quantity $\mathbb{E} \sup_{s,t \in T} |X_s - X_t|$ (for T finite for simplicity). Of course

$$\mathbb{E} \sup_{s,t \in T} |X_s - X_t| = \mathbb{E} \sup_{s,t \in T} (X_s - X_t)$$

which is also equal to $2\mathbb{E} \sup_{t \in T} X_t$ if the process is *symmetric* (i.e., the distribution in $\mathbb{R}^T$ of $-X$ and X are the same; for example Gaussian processes are symmetric). We also have, for every t_0 in T,

$$\mathbb{E} \sup_{t \in T} X_t \le \mathbb{E} \sup_{t \in T} |X_t| \le \mathbb{E}|X_{t_0}| + \mathbb{E} \sup_{s,t \in T} |X_s - X_t|$$

and, when X is symmetric,

$$\mathbb{E} \sup_{t \in T} X_t \le \mathbb{E} \sup_{t \in T} |X_t| \le \mathbb{E}|X_{t_0}| + 2\mathbb{E} \sup_{t \in T} X_t \,.$$

These inequalities are freely used below. The example of T being reduced to one point shows that the estimates we establish do not hold in general for $\mathbb{E} \sup_{t \in T} |X_t|$ rather than $\mathbb{E} \sup_{t \in T} X_t$ or $\mathbb{E} \sup_{s,t \in T} |X_s - X_t|$. The supremum notation will also often be shortened in $\sup_T$ or $\sup_t$, $\sup_{s,t}$ etc.

Recall that a Young function ψ is a convex increasing function on $\mathbb{R}_+$ such that $\lim_{t \to \infty} \psi(t) = \infty$ and $\psi(0) = 0$. The Orlicz space $L_\psi = L_\psi(\Omega, \mathcal{A}, \mathbb{P})$ associated to ψ is defined as the space of all real valued random variables Z on $(\Omega, \mathcal{A}, \mathbb{P})$ such that $\mathbb{E}\psi(|Z|/c) < \infty$ for some $c > 0$. Recall that it is a Banach space for the norm

$$\|Z\|_\psi = \inf\{c > 0 \,; \; \mathbb{E}\psi(|Z|/c) \le 1\}.$$

The general question that we study in the first two sections of this chapter is the following: given a Young function ψ and a random process $X = (X_t)_{t \in T}$ indexed by (T, d) and in L_ψ (i.e. $\|X_t\|_\psi < \infty$ for every t) satisfying the Lipschitz conditions in L_ψ

$$(11.1) \qquad \|X_s - X_t\|_\psi \leq d(s, t) \quad \text{for all} \ \ s, t \in T ,$$

find then estimates of $\sup_t X_t$ and sufficient conditions for the sample boundedness or continuity of X in terms of "the geometry of $(T, d; \psi)$". By this we mean the size of T measured in terms of d and ψ. Note that we could take as pseudo-metric d the one given by the process itself $d(s, t) = \|X_s - X_t\|_\psi$ so that T may be measured in terms of X. The main idea will be to convey, via the incremental conditions (11.1), boundedness and continuity of X in the function space L_ψ to the corresponding almost sure properties. This will be accomplished with a chaining argument.

Our main geometric measures of $(T, d; \psi)$ are the metric entropy condition and the majorizing measure condition. The first section develops the results under the concept of entropy which we already encountered in some of the preceding chapters in necessary results.

Let us note before we turn to these results that the study of the continuity (actually uniform continuity) will always follow rather easily from the various bounds established for the boundedness that thus appears as the main question in this investigation. This situation is rather classical and we already met it for example in Chapters 7-9 in the study of limit theorems. Let us also mention that the fact that we are working with pseudo-metrics rather than metrics is not really important since we can always identify two points s and t in T such that $d(s, t) = 0$; under (11.1), $X_s = X_t$ almost surely. Furthermore, all the conditions that we will use, entropy or majorizing measures, imply that (T, d) is totally bounded. For simplicity, one can therefore reduce everything, if one wishes it, to the case of a compact metric space (T, d).

11.1 Regularity of Random Processes Under Metric Entropy Conditions

Let (T, d) be a pseudo-metric space. Recall the entropy numbers $N(T, d; \varepsilon)$. That is, for each $\varepsilon > 0$, denote by $N(T, d; \varepsilon)$ the smallest number of open balls of radius $\varepsilon > 0$ in the pseudo-metric d which form a covering of T. (Recall that we could work equivalently with closed balls.) T is totally bounded for d if and only if $N(T, d; \varepsilon) < \infty$ for every $\varepsilon > 0$, a property which will always be satisfied under all the conditions we will deal with. Denote further by $D = D(T)$ the diameter of (T, d) i.e.,

$$D = \sup\{d(s, t); \ s, t \in T\}$$

(finite or infinite).

The following theorem is the main regularity result under metric entropy conditions for processes with increments satisfying (11.1). It only concerns so far boundedness but continuity will be achieved similarly later on. ψ^{-1} denotes the inverse function of ψ.

Theorem 11.1. *Let $X = (X_t)_{t \in T}$ be a random process in L_ψ such that for all s,t in T*

$$\|X_s - X_t\|_\psi \le d(s,t).$$

Then, if

$$\int_0^D \psi^{-1}\big(N(T,d;\varepsilon)\big)\,d\varepsilon < \infty,$$

X is almost surely bounded and we actually have

$$\mathbb{E}\sup_{s,t \in T} |X_s - X_t| \le 8 \int_0^D \psi^{-1}\big(N(T,d;\varepsilon)\big)\,d\varepsilon.$$

It is clear that the convergence of the entropy integral is understood when $\varepsilon \to 0$. The numerical constant 8 is without any special meaning.

We will actually prove a somewhat better result the proof of which is not more difficult.

Theorem 11.2. *Let ψ be a Young function and let $X = (X_t)_{t \in T}$ be a random process in $L_1 = L_1(\Omega, \mathcal{A}, \mathbb{P})$ such that for all measurable sets A in Ω and all s,t in (T,d),*

$$\int_A |X_s - X_t|\,d\mathbb{P} \le d(s,t)\mathbb{P}(A)\psi^{-1}\left(\frac{1}{\mathbb{P}(A)}\right).$$

Then, for every A,

$$\int_A \sup_{s,t \in T} |X_s - X_t|\,d\mathbb{P} \le 8\mathbb{P}(A)\int_0^D \psi^{-1}\big(\mathbb{P}(A)^{-1}N(T,d;\varepsilon)\big)\,d\varepsilon.$$

Note that this statement does not really concern ψ but rather the function $u\psi^{-1}(1/u)$, $0 < u \le 1$, and should perhaps be preferably stated in this way.

Before proving Theorem 11.2, let us explain the advantages of its formulation and how it includes Theorem 11.1. For the latter, simply note that when (11.1) holds, by convexity and Jensen's inequality

$$\int_A |X_s - X_t|\,d\mathbb{P} = d(s,t)\mathbb{P}(A)\int_A \psi^{-1} \circ \psi\left(\frac{|X_s - X_t|}{d(s,t)}\right)\frac{d\mathbb{P}}{\mathbb{P}(A)}$$

$$\le d(s,t)\mathbb{P}(A)\psi^{-1}\left(\frac{1}{\mathbb{P}(A)}\mathbb{E}\,\psi\left(\frac{|X_s - X_t|}{d(s,t)}\right)\right)$$

$$\le d(s,t)\mathbb{P}(A)\psi^{-1}\left(\frac{1}{\mathbb{P}(A)}\right).$$

Conversely, it should be noted that if Z is a positive random variable such that for every measurable set A

$$\int_A Z\,d\mathbb{P} \le \mathbb{P}(A)\psi^{-1}\left(\frac{1}{\mathbb{P}(A)}\right),$$

then, letting $A = \{Z > u\}$ and using Chebyshev's inequality, we have

$$\mathbb{P}\{Z > u\} \le \frac{1}{u}\int_{\{Z>u\}} Z\,d\mathbb{P} \le \frac{1}{u}\mathbb{P}\{Z > u\}\psi^{-1}\left(\frac{1}{\mathbb{P}\{Z > u\}}\right)$$

so that for every $u > 0$,

$$\mathbb{P}\{Z > u\} \le \frac{1}{\psi(u)}.$$

For Young functions of exponential type the latter is equivalent to saying that $\|Z\|_\psi < \infty$. However, for power type functions, it is less restrictive and this is why Theorem 11.2 is more general in its hypotheses. It includes for example conditions in weak L_p-spaces. Let $1 < p < \infty$ and assume indeed that we have a random process X such that for all s, t in T

(11.2) $$\|X_s - X_t\|_{p,\infty} \le d(s,t).$$

Then, as is easily seen by integration by parts,

$$\int_A |X_s - X_t|\,d\mathbb{P} \le q\,d(s,t)\mathbb{P}(A)^{1/q}$$

$$\le q\,d(s,t)\mathbb{P}(A)\left(\frac{1}{\mathbb{P}(A)}\right)^{1/p}$$

where $q = p/p - 1$ is the conjugate of p.

This for the advantages in the hypotheses. Concerning the conclusion, the formulation in Theorem 11.2 allows to obtain directly some quite sharp integrability and tail estimates of the sup-norm of the process X much in the spirit of those described in the first part of the book. If, for example, ψ is such that for some constant $C = C_\psi$ and all $x, y \ge 1$

$$\psi^{-1}(xy) \le C\psi^{-1}(x)\psi^{-1}(y)$$

(which is the case for example when $\psi(x) = x^p$, $1 \le p < \infty$), then the conclusion of Theorem 11.2 is that for all measurable sets A

$$\int_A \sup_{s,t}|X_s - X_t|\,d\mathbb{P} \le 8CE\mathbb{P}(A)\psi^{-1}\left(\frac{1}{\mathbb{P}(A)}\right)$$

where

$$E = E(T, d; \psi) = \int_0^D \psi^{-1}\big(N(T, d; \varepsilon)\big)\,d\varepsilon,$$

assumed of course to be finite. Then, by Chebyshev's inequality as before, for every $u > 0$,

(11.3) $$\mathbb{P}\left\{\sup_{s,t}|X_s - X_t| > u\right\} \le \left(\psi^{-1}\left(\frac{u}{8CE}\right)\right)^{-1}.$$

This applies in particular to the preceding setting of (11.2) for $\psi(x) = x^p$, $1 < p < \infty$, in which case we get

$$\|\sup_{s,t} |X_s - X_t|\|_{p,\infty} \leq 8qCE.$$

Hence, under the finiteness of the entropy integral, we conclude to a degree of integrability for the supremum which is exactly the same as the one we started with on the individuals $(X_s - X_t)$. In case we have $\|X_s - X_t\|_p \leq d(s,t)$ instead of (11.2) we end up with a small gap in this order of idea. This can however easily be repaired. Indeed, if Z is a positive random variable such that $\|Z\|_q = 1$ and if we set $\mathbb{Q}(A) = \int_A Z^q d\mathbb{P}$, then, from the assumption, for every measurable set A, and all s, t,

$$\int_A |X_s - X_t| d\mathbb{Q} \leq d(s,t) \mathbb{Q}(A)^{1/q}.$$

Theorem 11.2 with respect to $\mathbb{Q}$ yields

$$\int \sup_{s,t} |X_s - X_t| d\mathbb{Q} \leq 8CE$$

from which, by uniformity in $\mathbb{Q}$, it follows that

$$\|\sup_{s,t} |X_s - X_t|\|_p \leq 8CE.$$

If ψ is an exponential function $\psi_q(x) = \exp(x^q) - 1$, then (11.3) immediately implies that

$$\|\sup_{s,t} |X_s - X_t|\|_{\psi_q} \leq C_q E$$

for some constant C_q depending on q only. In this case actually, the general estimate (11.3) can be improved into a *deviation* inequality. Assume that ψ satisfies

$$\psi^{-1}(xy) \leq C(\psi^{-1}(x) + \psi^{-1}(y))$$

for all $x, y \geq 1$. The functions ψ_q satisfy this inequality. In this situation, Theorem 11.2 indicates that for every measurable set A,

$$\int_A \sup_{s,t} |X_s - X_t| d\mathbb{P} \leq 8C\mathbb{P}(A) \left(D\psi^{-1} \left(\frac{1}{\mathbb{P}(A)} \right) + E \right).$$

This easily implies that for every $u > 0$

(11.4) $$\mathbb{P}\{\sup_{s,t} |X_s - X_t| > 8C(E+u)\} \leq \left(\psi\left(\frac{u}{D}\right) \right)^{-1}.$$

This is an inequality of the type we obtained for bounded Gaussian or Rademacher processes in Chapters 3 and 4 with *two* parameters, E, the entropy integral, which measures some information on the sup-norm in L_p, $0 \leq p < \infty$, and the diameter D which may be assimilated to weak moments. For the purpose of comparison, note that, obviously,

$$E = \int_0^D \psi^{-1}\big(N(T, d; \varepsilon)\big)\,d\varepsilon \geq \psi^{-1}(1)D$$

($\psi^{-1}(1) > 0$ by convexity) and E is much bigger than D in general.

Now, we prove Theorem 11.2.

Proof of Theorem 11.2. It is enough to prove the inequality of the statement with T finite. Let ℓ_0 be the largest integer (in $\mathbb{Z}$) such that $2^{-\ell} \geq D$; let also ℓ_1 be the smallest integer ℓ such that the open balls $B(t, 2^{-\ell})$ with center t and radius $2^{-\ell}$ in the metric d are reduced to exactly one point. For each $\ell_0 \leq \ell \leq \ell_1$, let $T_\ell \subset T$ of cardinality $N(T, d; 2^{-\ell})$ such that the balls $\{B(t, 2^{-\ell}); \ t \in T_\ell\}$ form a covering of T. By induction, define maps $h_\ell : T_\ell \to T_{\ell-1}, \ \ell_0 < \ell \leq \ell_1$, such that $t \in B(h_\ell(t), 2^{-\ell+1})$. Set then $k_\ell : T \to T_\ell, \ k_\ell = h_{\ell+1} \circ \cdots \circ h_{\ell_1}, \ \ell_0 \leq \ell \leq \ell_1$ ($k_{\ell_1} = $ identity). Then, we can write the fundamental *chaining* identity which is at the basis of most of the results on boundedness and continuity of processes under conditions on increments. Since $2^{-\ell_0} \geq D$, T_{ℓ_0} is reduced to one point, call it t_0. Then, for every t in T,

$$X_t - X_{t_0} = \sum_{\ell=\ell_0+1}^{\ell_1} \big(X_{k_\ell(t)} - X_{k_{\ell-1}(t)}\big).$$

It follows that

$$\sup_{s,t \in T} |X_s - X_t| \leq 2 \sum_{\ell=\ell_0+1}^{\ell_1} \sup_{t \in T} |X_{k_\ell(t)} - X_{k_{\ell-1}(t)}|.$$

Now, for fixed ℓ, observe that

$$\mathrm{Card}\big\{ \big(X_{k_\ell(t)} - X_{k_{\ell-1}(t)}\big); \ t \in T \big\} \leq N\big(T, d; 2^{-\ell}\big).$$

Moreover, by construction, $d(k_\ell(t), k_{\ell-1}(t)) \leq 2^{-\ell+1}$ for every t. Hence, the hypothesis indicates that for every t, every $\ell_0 < \ell \leq \ell_1$ and every measurable set A,

$$\int_A |X_{k_\ell(t)} - X_{k_{\ell-1}(t)}|\,d\mathbb{P} \leq 2^{-\ell+1}\mathbb{P}(A)\psi^{-1}\left(\frac{1}{\mathbb{P}(A)}\right).$$

The conclusion will then easily follow from the following elementary lemma.

Lemma 11.3. *Let $(Z_i)_{i \leq N}$ be positive random variables on some probability space $(\Omega, \mathcal{A}, \mathbb{P})$ such that for all $i \leq N$ and all measurable sets A in Ω,*

$$\int_A Z_i \, d\mathbb{P} \leq \mathbb{P}(A)\psi^{-1}\left(\frac{1}{\mathbb{P}(A)}\right).$$

Then, for every measurable set A,

$$\int_A \max_{i \leq N} Z_i \, d\mathbb{P} \leq \mathbb{P}(A)\psi^{-1}\left(\frac{N}{\mathbb{P}(A)}\right).$$

Proof. Let $(A_i)_{i \leq N}$ be a (measurable) partition of A such that $Z_i = \max_{j \leq N} Z_j$ on A_i. Then

$$\int_A \max_{i \leq N} Z_i \, d\mathbb{P} = \sum_{i=1}^{N} \int_{A_i} Z_i \, d\mathbb{P} \leq \sum_{i=1}^{N} \mathbb{P}(A_i)\psi^{-1}\left(\frac{1}{\mathbb{P}(A_i)}\right)$$

$$\leq \mathbb{P}(A)\psi^{-1}\left(\frac{N}{\mathbb{P}(A)}\right)$$

where we have used in the last step that ψ^{-1} is concave and that $\sum_{i=1}^{N} \mathbb{P}(A_i) = \mathbb{P}(A)$. The lemma is proved. $\quad\square$

Note that the lemma applies when $\max_{i \leq N} \|Z_i\|_\psi \leq 1$. This lemma of course describes one of the key points of the entropy approach through the intervention of the cardinality N.

We can conclude the proof of Theorem 11.2. Together with the chaining inequality, the lemma implies that for any measurable set A,

$$\int_A \sup_{s,t} |X_s - X_t| d\mathbb{P} \leq 2 \sum_{\ell=\ell_0+1}^{\ell_1} \int_A \sup_t |X_{k_\ell(t)} - X_{k_{\ell-1}(t)}| d\mathbb{P}$$

$$\leq 4\mathbb{P}(A) \sum_{\ell > \ell_0} 2^{-\ell}\psi^{-1}\left(\mathbb{P}(A)^{-1}N(T,d;2^{-\ell})\right).$$

The conclusion follows from a simple comparison between series and integral:

$$\sum_{\ell > \ell_0} 2^{-\ell}\psi^{-1}\left(\mathbb{P}(A)^{-1}N(T,d;2^{-\ell})\right) \leq 2 \sum_{\ell > \ell_0} \int_{2^{-\ell-1}}^{2^{-\ell}} \psi^{-1}\left(\mathbb{P}(A)^{-1}N(T,d;\varepsilon)\right) d\varepsilon$$

$$\leq 2 \int_0^D \psi^{-1}\left(\mathbb{P}(A)^{-1}N(T,d;\varepsilon)\right) d\varepsilon$$

by definition of ℓ_0. (Note that some similar simple comparison between series and integral will be used frequently throughout this chapter and the next ones, usually without any further comments.) Theorem 11.2 is established.

Remark 11.4. What was thus used in this proof is the existence, for every ℓ, of a finite subset T_ℓ of T such that, for every t, there exists $t_\ell \in T_\ell$ with $d(t, t_\ell) \leq 2^{-\ell}$ and with the property

$$\sum_\ell 2^{-\ell}\psi^{-1}(\text{Card }T_\ell) < \infty.$$

It should be noted further that the preceding proof also applies to the random functions $X = (X_t)_{t \in T}$ with the following property: for every ℓ ($\geq \ell_0$), every (t, t_ℓ) as before and every measurable set A,

$$(11.5) \qquad \int_A |X_t - X_{t_\ell}| d\mathbb{P} \leq 2^{-\ell} \mathbb{P}(A) \psi^{-1} \left(\frac{1}{\mathbb{P}(A)} \right) + M_\ell \mathbb{P}(A)$$

where (M_ℓ) is a sequence of positive numbers satisfying $\sum_\ell M_\ell < \infty$. [In particular, this property is satisfied when, for some constant $C > 0$, for all measurable sets A and all s, t in T,

$$\int_A |X_s - X_t| d\mathbb{P} \leq d(s,t) \mathbb{P}(A) \left(\psi^{-1} \left(\frac{1}{\mathbb{P}(A)} \right) + C \right).]$$

The final bound involves then this quantity in the form of the inequality

$$\int_A \sup_{s,t \in T} |X_s - X_t| d\mathbb{P} \leq 8\mathbb{P}(A) \int_0^D \psi^{-1} \left(\mathbb{P}(A)^{-1} N(T, d; \varepsilon) \right) d\varepsilon + 2\mathbb{P}(A) \sum_{\ell \geq \ell_0} M_\ell$$

where we recall that ℓ_0 is the largest integer ℓ such that $2^{-\ell} \geq D$. The proof is straightforward. This simple observation can be useful as will be illustrated in Section 11.3.

Remark 11.5. Here, we collect some further easy observations which will be useful next. First note that in Theorems 11.1 and 11.2 (and Remark 11.4 too) we might have as well equipped T with the pseudo-metric $d(s,t) = \|X_s - X_t\|_\psi$ induced by X itself. Further, since the hypotheses and the conclusions of these theorems actually only involve the increments $X_s - X_t$ in absolute values, and since the only property used on them is that they satisfy the triangle inequality, the preceding results may be trivially extended to the setting of random distances; that is, random processes $(D(s,t))_{s,t \in T}$ on $T \times T$ such that for all s, t, u in T, $D(s,s) = 0 \leq D(s,t) = D(t,s) \leq D(s,u) + D(u,t)$ with probability one. These include $D(s,t) = |X_s - X_t|^\alpha$, $0 < \alpha \leq 1$, and $D(s,t) = \min(1, |X_s - X_t|)$ for example. Actually, we could also include in this way random processes with values in a Banach space by setting $D(s,t) = \|X_s - X_t\|$ (or some of the preceding variations). By random process with values in a Banach space, we simply mean here a collection $X = (X_t)_{t \in T}$ of random variables with values in a Banach space. An example will be discussed in Section 11.3. More extensions of this type can be obtained; we leave them to the interested reader.

Now, we present the continuity theorem in the preceding framework.

Theorem 11.6. *Let ψ be a Young function and let $X = (X_t)_{t \in T}$ be a random process in $L_1 = L_1(\Omega, \mathcal{A}, \mathbb{P})$ such that for all measurable sets A in Ω and all s, t in (T, d),*

$$\int_A |X_s - X_t| d\mathbb{P} \leq d(s,t) \mathbb{P}(A) \psi^{-1} \left(\frac{1}{\mathbb{P}(A)} \right).$$

Then, if

$$\int_0^D \psi^{-1}\big(N(T,d;\varepsilon)\big)d\varepsilon < \infty,$$

X admits a version $\widetilde{X}$ with almost all sample paths bounded and (uniformly) continuous on (T,d). Moreover, $\widetilde{X}$ satisfies the following property: for each $\varepsilon > 0$, there exists $\eta > 0$, depending only on ε and the finiteness of the entropy integral, but not on the process X itself, such that

$$\mathbb{E}\ \sup_{d(s,t)<\eta}\ |\widetilde{X}_s - \widetilde{X}_t| < \varepsilon.$$

Proof. We use the notation of the proof of Theorem 11.2. The main point is to show that, when T is finite, for every $\eta > 0$ and $\ell_0 \le \ell \le \ell_1$,

(11.6)
$$\mathbb{E}\ \sup_{d(s,t)<\eta}\ |X_s - X_t|$$
$$\le \eta\psi^{-1}\big(N(T,d;2^{-\ell})^2\big) + 8\sum_{m>\ell} 2^{-m}\psi^{-1}\big(N(T,d;2^{-m})\big).$$

Let η and ℓ be fixed. The proof of Theorem 11.2 indicates that the chaining identity

$$X_t - X_{k_\ell(t)} = \sum_{m=\ell+1}^{\ell_1} \big(X_{k_m(t)} - X_{k_{m-1}(t)}\big)$$

implies

$$\mathbb{E}\sup_{t\in T}|X_t - X_{k_\ell(t)}| \le 2\sum_{m>\ell} 2^{-m}\psi^{-1}\big(N(T,d;2^{-m})\big).$$

Let

$$U = \big\{(x,y) \in T_\ell \times T_\ell\,;$$
$$\exists\, u,v \text{ in } T \text{ such that } d(u,v) < \eta \text{ and } k_\ell(u) = x\,,\ k_\ell(v) = y\big\}.$$

If $(x,y) \in U$, we fix $u_{x,y}$, $v_{x,y}$ such that $k_\ell(u_{x,y}) = x$, $k_\ell(v_{x,y}) = y$ and $d(u_{x,y}, v_{x,y}) < \eta$. By Lemma 11.3,

$$\mathbb{E}\ \sup_{(x,y)\in U}\ |X_{u_{x,y}} - X_{v_{x,y}}| \le \eta\psi^{-1}(\text{Card } U)$$
$$\le \eta\psi^{-1}\big(N(T,d;2^{-\ell})^2\big).$$

Now, let s,t be arbitrary in T satisfying $d(s,t) < \eta$. Set $x = k_\ell(s)$, $y = k_\ell(t)$. Clearly $(x,y) \in U$. We can write by the triangle inequality

$$|X_s - X_t| \le |X_s - X_{k_\ell(s)}| + |X_{k_\ell(s)} - X_{u_{x,y}}| + |X_{u_{x,y}} - X_{v_{x,y}}|$$
$$+ |X_{v_{x,y}} - X_{k_\ell(t)}| + |X_{k_\ell(t)} - X_t|$$
$$\le \sup_{(x,y)\in U} |X_{u_{x,y}} - X_{v_{x,y}}| + 4\sup_{r\in T}|X_r - X_{k_\ell(r)}|$$

where we have used that $k_\ell(u_{x,y}) = k_\ell(s) = x$ and similarly for y. We have clearly (11.6).

We can conclude the proof of the theorem. We have obtained by (11.6) that, under the finiteness of the entropy integral, for each $\varepsilon > 0$ there exists $\eta > 0$ depending only on $\varepsilon > 0$ and T, d, ψ such that, for every finite and thus also countable subset S of T,

$$\mathbb{E} \sup_{\substack{s,t \in S \\ d(s,t) < \eta}} |X_s - X_t| < \varepsilon.$$

Since (T, d) is totally bounded, there exists S countable and dense in T. Then, set $\widetilde{X}_t = X_t$ if $t \in S$ and $\widetilde{X}_t = \lim X_s$ where this limit, in probability or in L_1, is taken for $s \to t$, $s \in S$. Then $(\widetilde{X}_t)_{t \in T}$ is clearly a version of X which satisfies all the required properties. To see in particular that $(\widetilde{X}_t)_{t \in T}$ has uniformly continuous sample paths on (T, d), let, for each n, $\eta_n > 0$ be such that

$$\mathbb{E} \sup_{d(s,t) < \eta_n} |\widetilde{X}_s - \widetilde{X}_t| < 4^{-n}.$$

Then, if $A_n = \{\sup_{d(s,t) < \eta_n} |\widetilde{X}_s - \widetilde{X}_t| > 2^{-n}\}$, $\sum_n \mathbb{P}(A_n) < \infty$ and the claim follows from the Borel-Cantelli lemma. The proof of Theorem 11.6 is complete.

It is plain that Remarks 11.4 and 11.5 also apply in the context of Theorem 11.6. Moreover, the dependence of $\eta > 0$ that we carefully describe in Theorem 11.6 has some easy consequences to tightness results. Assume that (T, d) is a compact metric space and denote by $C(T)$ the Banach space of all continuous functions on T equipped with the sup-norm. By Prokhorov's criterion and the Arzela-Ascoli characterization of compact sets in $C(T)$ (cf. e.g. [Bi]), it is easily seen that a family $\mathcal{X}$ of random variables $X = (X_t)_{t \in T}$ is relatively compact in the weak topology of probability distributions on $C(T)$ as soon as, for some t, $\{X_t; X \in \mathcal{X}\}$ is relatively compact as real valued random variables and, for each $\varepsilon > 0$, there is $\eta > 0$ such that for every X in $\mathcal{X}$

$$\mathbb{E} \sup_{d(s,t) < \eta} |X_s - X_t| < \varepsilon.$$

Now this last condition is exactly what is provided by Theorem 11.6 under the entropy condition. Thus, we have the following consequence which will be of interest in the study of the central limit theorem in Chapter 14.

Corollary 11.7. Let (T, d) be compact and let ψ be a Young function. Assume that

$$\int_0^D \psi^{-1}(N(T, d; \varepsilon)) d\varepsilon < \infty.$$

Let $\mathcal{X}$ be a family of separable random processes $X = (X_t)_{t \in T}$ in $L_1 = L_1(\Omega, \mathcal{A}, \mathbb{P})$ such that for all s, t in T and all measurable sets A,

$$\int_A |X_s - X_t| d\mathbb{P} \leq d(s,t) \mathbb{P}(A) \psi^{-1}\left(\frac{1}{\mathbb{P}(A)}\right).$$

Then, each element of $\mathcal{X}$ defines a tight probability distribution on $C(T)$ and $\mathcal{X}$ is weakly relatively compact if and only if, for some $t \in T$, $\{X_t; X \in \mathcal{X}\}$ is weakly relatively compact (as measures on $\mathbb{R}$).

As a first application, we would like, in particular, to indicate how the preceding results contain the continuity theorem of Kolmogorov. We state it in its classical and usual form although its various sharpenings are deduced similarly.

Corollary 11.8. *Let $X = (X_t)_{t \in [0,1]}$ be a separable random process indexed by $[0,1]$ such that for some $\alpha > 0$ and $p > 1$, and all s,t in $[0,1]$,*

$$\mathbb{E}|X_s - X_t|^\alpha \leq |s-t|^p.$$

Then X has almost surely bounded and continuous sample paths.

Proof. We apply Theorem 11.6 together with Remark 11.5. We distinguish between two cases. If $\alpha \geq p$, then

$$\|X_s - X_t\|_\alpha \leq d(s,t)$$

where d is the metric $d(s,t) = |s-t|^{p/\alpha}$ $(p/\alpha \leq 1)$. As is obvious, $N([0,1], d; \varepsilon)$ is of the order $\varepsilon^{-\alpha/p}$ and since $p > 1$ the corresponding entropy integral with $\psi(x) = x^\alpha$ in Theorem 11.6 is finite. The conclusion follows in this case. When $\alpha \leq p$, then, for all s,t in $[0,1]$,

$$\||X_s - X_t|^\gamma\|_p \leq d(s,t)$$

where $\gamma = \alpha/p \leq 1$ and here $d(s,t) = |s-t|$. Then, apply again Theorem 11.6 with this time Remark 11.5; indeed, since $\gamma \leq 1$, $|X_s - X_t|^\gamma$ defines a random distance and here $N([0,1], d; \varepsilon) \sim \varepsilon^{-1}$. The proof is complete.

While the preceding evaluations seem quite easy, they however appear to be sharp in various instances. The case of Gaussian processes treated in Section 11.3 and the next chapter is a first example. Other examples concerning processes indexed by regular subsets of $\mathbb{R}^N$ were treated in the literature (see e.g. [Ha], [H-K], [Pi9], [Ib], [Ta13] etc.). They basically indicate that, if *every* random process that satisfies some Lipschitz condition is almost surely bounded or continuous, then a corresponding integral is convergent. This is the natural formulation of the necessary results. They do not concern one single process but rather the whole family of processes satisfying the same incremental condition. Only in the Gaussian case the necessary conditions can concern one single process due to the comparison theorems. We shall come back to this in Chapter 12.

11.2 Regularity of Random Processes Under Majorizing Measure Conditions

One main feature (and weakness) of the entropy condition is that it gives some "weight" to *each* piece of T. This does not present any inconvenient if T is in some sense homogeneous; we will see how this is the case in Chapter 13 and how metric entropy is best possible (necessary) for some processes in such a homogeneous setting (cf. also the closing comments of Section 11.1). In general however, one has rather to think at some geometric measure of T which takes into account the possible lack of homogeneity of T. One way to handle this is the concept of majorizing measure.

Given a pseudo-metric space (T, d) and a Young function ψ as before, say that a probability measure m on T is a *majorizing measure* for $(T, d; \psi)$ if

$$(11.7) \qquad \gamma_m(T, d; \psi) = \sup_{t \in T} \int_0^D \psi^{-1}\left(\frac{1}{m(B(t, \varepsilon))}\right) d\varepsilon < \infty$$

where $B(t, \varepsilon)$ is the open ball in the d-metric with center t and radius $\varepsilon > 0$. (Again we could use essentially equivalently closed balls.) We thus call (11.7) a majorizing measure condition as opposed to the entropy condition studied in the previous section. Moreover, when we will speak of the existence of a majorizing measure on $(T, d; \psi)$, we mean the existence of a probability measure m on (T, d) such that $\gamma_m(T, d; \psi) < \infty$. This definition clearly gives a way to take more into account the local properties of the geometry of T. Our aim in this section will be to show how one can control random processes satisfying Lipschitz conditions under a majorizing measure condition as we did before with entropy and actually in a more efficient way.

We start with some remarks for a better understanding of condition (11.7) and for the comparison with the results of Section 11.1. If s and s' are two points in T with $d(s, s') = 2\eta > 0$, the open balls $B(s, \eta)$ and $B(s', \eta)$ are disjoint. Thus, if m is a probability measure on (T, d), one of these two balls, say $B(s, \eta)$, has a measure less than or equal to $1/2$. Therefore

$$\sup_{t \in T} \int_0^D \psi^{-1}\left(\frac{1}{m(B(t, \varepsilon))}\right) d\varepsilon \geq \int_0^\eta \psi^{-1}\left(\frac{1}{m(B(s, \varepsilon))}\right) d\varepsilon \geq \eta \psi^{-1}(2),$$

so that, as for the entropy condition, if $D = D(T)$ is the diameter of (T, d),

$$(11.8) \qquad \gamma_m(T, d; \psi) \geq \tfrac{1}{2}\psi^{-1}(2)D.$$

Also, if m is a majorizing measure satisfying (11.7), then (T, d) is totally bounded and actually

$$(11.9) \qquad \sup_{\varepsilon > 0} \varepsilon \psi^{-1}\left(N(T, d; \varepsilon)\right) \leq 2\gamma_m(T, d; \psi).$$

The proof of this easy fact is already instructive on the way to use majorizing measures. Let $N(T, d; \varepsilon) \geq N$. There exist $t_1, \ldots, t_N$ such that $d(t_i, t_j) \geq \varepsilon$ for all $i \neq j$. By definition of $\gamma_m(T, d; \psi) = \gamma$, for each $i \leq N$,

$$\frac{\varepsilon}{2}\psi^{-1}\left(\frac{1}{m\big(B(t_i,\varepsilon/2)\big)}\right)\leq\gamma,$$

that is $m(B(t_i,\varepsilon/2))\geq[\psi(2\gamma/\varepsilon)]^{-1}$. Since the balls $B(t_i,\varepsilon/2)$, $i\leq N$, are disjoint and since m is a probability, it follows that $\psi(2\gamma/\varepsilon)\geq N$ which is the desired result (11.9).

More important now is to observe that entropy conditions are stronger than majorizing measure conditions. That is, there is a probability measure m on T such that

$$(11.10)\quad \sup_{t\in T}\int_0^D\psi^{-1}\left(\frac{1}{m\big(B(t,\varepsilon)\big)}\right)d\varepsilon\leq K\int_0^D\psi^{-1}\big(N(T,d;\varepsilon)\big)d\varepsilon$$

where K is some numerical constant. This can be established in great generality (cf. [Ta13]) but we actually only prove it in a special case here. The general study of bounds on stochastic processes using majorizing measures indeed runs into many technicalities in which we decided not to enter here for the simplicity of the exposition. We will thus restrict this study to the case of a special class of Young functions ψ for which things become simpler. This restriction does not hide the main idea and interest of the majorizing measure technique. As we already mentioned it, this study can be conducted in rather a large generality and we refer to [Ta13] where this program is performed.

For the rest of this paragraph, we hence assume that ψ is a Young function such that for some constant C and all $x,y\geq 1$,

$$(11.11)\quad \psi^{-1}(xy)\leq C\big(\psi^{-1}(x)+\psi^{-1}(y)\big)\quad\text{and}\quad \int_0^\cdot\psi^{-1}(x^{-1})dx<\infty.$$

This class covers the main examples we have in mind, namely the exponential Young functions $\psi_q(x)=\exp(x^q)-1$, $1\leq q<\infty$ (and also $\psi_\infty(x)=\exp(\exp x)-e$). For simplicity in the notation, let us moreover assume that $C=1$ in (11.11) (which is actually easily seen not to be a restriction).

Let us show how, in this case, we may prove (11.10). Let, as usual, ℓ_0 be the largest integer with $2^{-\ell_0}\geq D$ where D is the diameter of (T,d). For every $\ell\geq\ell_0$, let $T_\ell\subset T$ denote the set of the centers of a minimal family such that the balls $B(t,2^{-\ell})$, $t\in T_\ell$, cover T. By definition, $\operatorname{Card}T_\ell=N(T,d;2^{-\ell})$. Consider the probability measure m on T given by

$$m=\sum_{\ell>\ell_0}2^{-\ell+\ell_0}N(T,d;2^{-\ell})^{-1}\sum_{t\in T_\ell}\delta_t$$

where δ_t is the Dirac measure at t. Clearly, for every t and $\ell>\ell_0$, $m(B(t,2^{-\ell}))\geq 2^{-\ell+\ell_0}N(T,d;2^{-\ell})^{-1}$. Hence,

$$\int_0^D\psi^{-1}\left(\frac{1}{m\big(B(t,\varepsilon)\big)}\right)d\varepsilon\leq\sum_{\ell>\ell_0}2^{-\ell}\psi^{-1}\left(\frac{1}{m\big(B(t,2^{-\ell})\big)}\right)$$

$$\leq\sum_{\ell>\ell_0}2^{-\ell}\psi^{-1}\big(2^{\ell-\ell_0}N(T,d;2^{-\ell})\big).$$

Using (11.11), this is estimated by

$$\sum_{\ell > \ell_0} 2^{-\ell} \psi^{-1}(2^{\ell - \ell_0}) + \sum_{\ell > \ell_0} 2^{-\ell} \psi^{-1}(N(T, d; 2^{-\ell}))$$

$$\leq 2^{-\ell_0 + 1} \int_0^{1/2} \psi(x^{-1}) dx + 2 \int_0^D \psi^{-1}(N(T, d; \varepsilon)) d\varepsilon$$

$$\leq 4D \int_0^{1/2} \psi(x^{-1}) dx + 2 \int_0^D \psi^{-1}(N(T, d; \varepsilon)) d\varepsilon$$

$$\leq 2 \left(1 + 2(\psi^{-1}(1))^{-1} \int_0^{1/2} \psi(x^{-1}) dx \right) \int_0^D \psi^{-1}(N(T, d; \varepsilon)) d\varepsilon$$

and the announced claim follows. Note that the constant however depends on ψ.

Let us note that the preceding proof actually shows that when $\int_0^D \psi^{-1}(N(T, d; \varepsilon)) d\varepsilon < \infty$, then m is a majorizing measure which satisfies (in addition to (11.10))

$$(11.12) \qquad \limsup_{\eta \to 0} \sup_{t \in T} \int_0^{\eta} \psi^{-1} \left(\frac{1}{m(B(t, \varepsilon))} \right) d\varepsilon = 0.$$

This condition is the one which enters to obtain continuity properties of stochastic processes as opposed to (11.7) which deals with boundedness. Therefore, we sometimes speak of bounded, resp. continuous, majorizing measure conditions. This is another advantage of majorizing measures upon entropy, to be able to give weaker sufficient conditions for the sample boundedness than for the sample continuity.

Let us now enter the heart of the matter and show how majorizing measures are used to control processes. Our approach will be to associate to each majorizing measure a (non-unique) ultrametric distance δ finer than the original distance d and for which there still exists a majorizing measure. This ultrametric structure is at the basis of the understanding of majorizing measures and will appear as the key notion in the study of the necessity in the next section. Alternatively, it can be seen as a way to *discretize* majorizing measures. This allows to use chaining arguments exactly as with entropy conditions. This program is accomplished in Proposition 11.10 and Corollary 11.12 below. We start with a simple lemma which allows us to conveniently reduce to a finite index set T.

Lemma 11.9. *Let (T, d) be a pseudo-metric space with diameter $D(T)$. Let m be a probability measure on (T, d) and recall that we set*

$$\gamma_m(T, d; \psi) = \sup_{t \in T} \int_0^{D(T)} \psi^{-1} \left(\frac{1}{m(B(t, \varepsilon))} \right) d\varepsilon .$$

Then, if A is a finite (or compact) subset of T, there is a probability measure μ on (A, d) such that

$$\gamma_\mu(A, d; \psi) = \sup_{t \in A} \int_0^{D(A)} \psi^{-1}\left(\frac{1}{\mu(B_A(t, \varepsilon))}\right) d\varepsilon$$

$$\leq 2 \sup_{t \in A} \int_0^{D(A)/2} \psi^{-1}\left(\frac{1}{m(B(t, \varepsilon))}\right) d\varepsilon$$

where $D(A)$ is the diameter of (A, d) and $B_A(t, \varepsilon)$ is the ball in A with center t and radius $\varepsilon > 0$. In particular, $\gamma_\mu(A, d; \psi) \leq 2\gamma_m(T, d; \psi)$.

Proof. For t in T, take $\varphi(t)$ in A with

$$d(t, \varphi(t)) = d(t, A) = \inf\{d(t, y); \ y \in A\}.$$

Set $\mu = \varphi(m)$, so that μ is supported by A. Fix x in A. For t in T, we have $d(t, \varphi(t)) = d(t, A) \leq d(t, x)$ and thus $d(x, \varphi(t)) \leq 2d(x, t)$. It follows that $\varphi(B(x, \varepsilon)) \subset B_A(x, 2\varepsilon)$ and thus $\mu(B_A(x, 2\varepsilon)) \geq m(B(x, \varepsilon))$. The proof is easily completed.

Recall that a metric space (U, δ) is ultrametric if δ satisfies the improved triangle inequality

$$\delta(u, v) \leq \max(\delta(u, w), \delta(w, v)), \quad u, v, w \in U.$$

The main feature of ultrametric spaces is that two balls of the same radius are either disjoint or equal. The next proposition deals with the nice functions ψ satisfying (11.11); we however recall that, at the expense of (severe) complications, a similar study can be driven in the general setting (cf. [Ta13]).

Proposition 11.10. *Let ψ satisfying (11.11) and let (T, d) be a finite metric space with diameter D. Let m be a probability measure on (T, d) and recall that we set*

$$\gamma_m(T, d; \psi) = \sup_{t \in T} \int_0^D \psi^{-1}\left(\frac{1}{m(B(t, \varepsilon))}\right) d\varepsilon.$$

There exist an ultrametric distance δ on T such that $d(s, t) \leq \delta(s, t)$ for all s, t in T and a probability measure μ on (T, δ) such that

$$\gamma_\mu(T, \delta; \psi) \leq K_\psi \gamma_m(T, d; \psi)$$

where $K_\psi > 0$ is a constant depending on ψ only.

Proof. Let ℓ_0 be the largest integer ℓ such that $4^{-\ell} \geq D$ and let ℓ_1 be the smallest one such that the balls $B(t, 4^{-\ell})$ of center t and radius $4^{-\ell}$ in the metric d are reduced to exactly one point. Assume $T = (t_i)$. Set $T_{\ell_1, i} = \{t_i\}$ for every i. For every $\ell = \ell_1 - 1, \ldots, \ell_0$, we construct by induction on $i \geq 1$, points $x_{\ell, i}$ and subsets $T_{\ell, i}$ of T as follows: setting $T_{\ell, 0} = \emptyset$,

$$m(B(x_{\ell, i}, 4^{-\ell})) = \max\{m(B(x, 4^{-\ell})); \ x \notin \bigcup_{j < i} T_{\ell, j}\},$$

$$T_{\ell, i} = \bigcup\{T_{\ell+1, k}; \ T_{\ell+1, k} \cap B(x_{\ell, i}, 4^{-\ell+1}) \neq \emptyset, \ \forall j < i, \ T_{\ell+1, k} \not\subset T_{\ell, j}\}.$$

Then, define $\delta(s,t) = 4^{-\ell+2}$ where ℓ is the largest integer such that s and t belong to the same $T_{\ell,i}$ for some i. δ is clearly an ultrametric distance. By decreasing induction on ℓ, it is easily verified that the diameter of each set $T_{\ell,i}$ is less than $4^{-\ell+2}$. It clearly follows by definition of δ that $d(s,t) \leq \delta(s,t)$ for all s,t in T.

For each ℓ, $(T_{\ell,i})$ forms the family of the δ-balls of radius $4^{-\ell+2}$. By construction, the balls $B(x_{\ell,i}, 4^{-\ell})$ when i varies are disjoint so that $\sum_i m(B(x_{\ell,i}, 4^{-\ell})) \leq 1$. Let $t_{\ell,i}$ be a fixed point in $T_{\ell,i}$. Consider

$$\mu' = \sum_{\ell=\ell_0}^{\ell_1} 4^{-\ell+\ell_0+1} \sum_i m(B(x_{\ell,i}, 4^{-\ell}))\delta_{t_{\ell,i}}$$

where δ_t is the Dirac measure at t. There is a probability measure $\mu \geq \mu'$. If $t \in T_{\ell,i}$, note that, by construction,

(11.13) $\qquad \mu(T_{\ell,i}) \geq 4^{-\ell+\ell_0-1}m(B(t,4^{-\ell})).$

We evaluate $\gamma_\mu(T,\delta;\psi)$ using (11.13) and the properties of ψ. Let t be fixed in $T_{\ell,i}$. By (11.13) and (11.11),

$$\sum_{\ell=\ell_0}^{\ell_1} 4^{-\ell}\psi^{-1}\left(\frac{1}{\mu(T_{\ell,i})}\right) \leq \sum_{\ell\geq\ell_0} 4^{-\ell}\psi^{-1}(4^{\ell-\ell_0+1}) + \sum_{\ell\geq\ell_0} 4^{-\ell}\psi^{-1}\left(\frac{1}{m(B(t,4^{-\ell}))}\right).$$

By definition of ℓ_0 (and (11.8)), this easily implies the conclusion. The proof is complete.

Remark 11.11. If (T,δ) is ultrametric and ψ satisfies (11.11), we may observe the following from the preceding proof: for every ℓ, let $\mathcal{B}_\ell$ be the family of the balls B of radius $2^{-\ell}$ (or $4^{-\ell}$ to agree with the proof of Proposition 11.10); then, perhaps more important than the probability measure μ we constructed (although it is actually equivalent), is the datum of a family of weights $\alpha(B,\ell) \geq 0$, $B \in \mathcal{B}_\ell$, such that $\sum_{B\in\mathcal{B}_\ell} \alpha(B,\ell) \leq 1$ (the measures $m(B(x_{\ell,i},4^{-\ell}))$ in the preceding proof). Furthermore, Proposition 11.10 may be expressed in the following "discretized" formulation. Denote by ℓ_0 the largest integer ℓ such that $2^{-\ell} \geq D$; then, there exists, for all $\ell \geq \ell_0$, finite sets T_ℓ in T and maps $\pi_\ell : T \to T_\ell$ such that $\pi_{\ell-1} \circ \pi_\ell = \pi_{\ell-1}$ and $d(t,\pi_\ell(t)) \leq 2^{-\ell}$ for every t and ℓ, and a *discrete* probability measure μ on $\{T_\ell; \ell \geq \ell_0\}$ satisfying

$$\sup_{t\in T}\sum_{\ell\geq\ell_0} 2^{-\ell}\psi^{-1}\left(\frac{1}{\mu(\{\pi_\ell(t)\})}\right) \leq K_\psi\gamma_m(T,d;\psi)$$

where K_ψ only depends on ψ. Indeed, if δ is the ultrametric structure obtained in Proposition 11.10, for every $\ell \geq \ell_0$, denote by $\mathcal{B}_\ell$ the family of the δ-balls of radius $2^{-\ell}$. For every $t \in T$, there is a unique element B of $\mathcal{B}_\ell$ with $t \in B$. Let $\pi_\ell(t)$ be one fixed point of B and let $\mu_\ell(\{\pi_\ell(t)\}) = \mu(B)$. The probability measure $\sum_{\ell\geq\ell_0} 2^{-\ell+\ell_0+1}\mu_\ell$ fulfills the conditions of the claim.

With the preceding results, we now present sufficient conditions in terms of majorizing measures for a random process to be almost surely bounded or continuous. The results are the analogs (actually improvements), in the setting of functions ψ satisfying (11.11), of Theorems 11.1, 11.2 and 11.6 dealing with entropy. We first establish the main result for a *general* Young function ψ in the case of an *ultrametric* index set, and then deduce the general case from Proposition 11.10 for a function ψ satisfying (11.11). As an alternate approach, one may use the preceding discretization (Remark 11.11) that, however, does not really clarify the steps in which the property (11.11) of ψ is used. We refer to [Ta13] for more details on this point and a more general study for arbitrary Young functions ψ.

Proposition 11.12. *Let ψ be an arbitrary Young function and let $X = (X_t)_{t \in T}$ be a random process in $L_1 = L_1(\Omega, \mathcal{A}, \mathbb{P})$ indexed by a finite ultrametric space (T, δ) such that, for all s, t in T and all measurable sets A in Ω,*

$$\int_A |X_s - X_t| d\mathbb{P} \le \delta(s,t) \mathbb{P}(A) \psi^{-1}\left(\frac{1}{\mathbb{P}(A)}\right).$$

Then, for any probability measure μ on (T, δ) and any measurable set A,

$$\int_A \sup_{s,t \in T} |X_s - X_t| d\mathbb{P} \le K\mathbb{P}(A) \sup_{t \in T} \int_0^D \psi^{-1}\left(\frac{1}{\mathbb{P}(A)\mu(B(t,\varepsilon))}\right) d\varepsilon$$

where $K > 0$ is a numerical constant and D is the diameter of (T, δ).

Proof. Set $T = (t_i)$ and let A be in $\mathcal{A}$. Let (A_i) be a measurable partition of A such that, on A_i,

$$\sup_{t \in T} |X_t - X_x| = |X_{t_i} - X_x|$$

where x is one (arbitrary) fixed point of T. Thus,

$$\int_A \sup_{t \in T} |X_t - X_x| d\mathbb{P} = \sum_i \int_{A_i} |X_{t_i} - X_x| d\mathbb{P}.$$

Let ℓ_0 be the largest integer ℓ such that $2^{-\ell} \ge D$ and, for $\ell \ge \ell_0$, denote by $\mathcal{B}_\ell$ the family of the balls of radius $2^{-\ell}$. For every B in $\mathcal{B}_\ell$, we fix $x(B) \in B$, and take $x(T) = x$. Further, we let $\pi_\ell(t) = x(B(t, 2^{-\ell}))$, $t \in T$, $\ell \ge \ell_0$. The usual chaining identity yields, for every t in T,

$$|X_t - X_x| \le \sum_{\ell > \ell_0} |X_{\pi_\ell(t)} - X_{\pi_{\ell-1}(t)}|.$$

For every B in $\mathcal{B}_\ell$, set $A_B = \bigcup\{A_i; \ t_i \in B\}$. Thus, we can write

$$\int_A \sup_{t \in T} |X_t - X_x| d\mathbb{P} \leq \sum_i \sum_{\ell > \ell_0} \int_{A_i} |X_{\pi_\ell(t_i)} - X_{\pi_{\ell-1}(t_i)}| d\mathbb{P}$$

$$= \sum_{\ell > \ell_0} \sum_{B \in \mathcal{B}_\ell} \sum_{t_i \in B} \int_{A_i} |X_{\pi_\ell(t_i)} - X_{\pi_{\ell-1}(t_i)}| d\mathbb{P}$$

$$= \sum_{\ell > \ell_0} \sum_{B \in \mathcal{B}_\ell} \int_{A_B} |X_{x(B)} - X_{x(\overline{B})}| d\mathbb{P}$$

where $B \subset \overline{B}$ and $\overline{B} \in \mathcal{B}_{\ell-1}$. Hence, from the hypothesis, since $\delta(x(B), x(\overline{B})) \leq 2^{-\ell+1}$,

$$\int_A \sup_{t \in T} |X_t - X_x| d\mathbb{P} \leq \sum_{\ell > \ell_0} \sum_{B \in \mathcal{B}_\ell} 2^{-\ell+1} \mathbb{P}(A_B) \psi^{-1}\left(\frac{1}{\mathbb{P}(A_B)}\right).$$

Now, let μ be a probability measure on (T, δ) and set

$$M = \sup_{t \in T} \int_0^D \psi^{-1}\left(\frac{1}{\mathbb{P}(A)\mu(B(t,\varepsilon))}\right) d\varepsilon$$

so that, for every t in T,

$$\sum_{\ell \geq \ell_0} 2^{-\ell} \psi^{-1}\left(\frac{1}{\mathbb{P}(A)\mu(B(t,2^{-\ell}))}\right) \leq 2M.$$

Integrating with respect to μ yields

$$\sum_{\ell \geq \ell_0} 2^{-\ell} \sum_{B \in \mathcal{B}_\ell} \mu(B) \psi^{-1}\left(\frac{1}{\mathbb{P}(A)\mu(B)}\right) \leq 2M,$$

while integrating with respect to the measure ν on T such that $\nu(\{t_i\}) = \mathbb{P}(A_i)$ yields

$$\sum_{\ell \geq \ell_0} \sum_{B \in \mathcal{B}_\ell} 2^{-\ell} \mathbb{P}(A_B) \psi^{-1}\left(\frac{1}{\mathbb{P}(A)\mu(B)}\right) \leq 2\mathbb{P}(A)M.$$

Next, we observe the following. If $\mathbb{P}(A_B) \leq \mathbb{P}(A)\mu(B)$,

$$\mathbb{P}(A_B) \psi^{-1}\left(\frac{1}{\mathbb{P}(A_B)}\right) \leq \mathbb{P}(A)\mu(B) \psi^{-1}\left(\frac{1}{\mathbb{P}(A)\mu(B)}\right)$$

since $\psi(u) \leq u\psi'(u)$ which shows that the function $u\psi^{-1}(1/u)$ is increasing. If $\mathbb{P}(A_B) \geq \mathbb{P}(A)\mu(B)$, we simply have

$$\mathbb{P}(A_B) \psi^{-1}\left(\frac{1}{\mathbb{P}(A_B)}\right) \leq \mathbb{P}(A_B) \psi^{-1}\left(\frac{1}{\mathbb{P}(A)\mu(B)}\right).$$

Assembling this observation with what we obtained previously yields

$$\int_A \sup_{t \in T} |X_t - X_x| d\mathbb{P} \leq 8\mathbb{P}(A)M$$

from which the conclusion follows. Proposition 11.12 is established.

Together with Lemma 11.9 and Proposition 11.10, the preceding basic result yields the following general theorem for functions ψ satisfying (11.11).

Theorem 11.13. *Let ψ be a Young function satisfying* (11.11) *and let $X = (X_t)_{t \in T}$ be a random process in $L_1 = L_1(\Omega, \mathcal{A}, \mathbb{P})$ indexed by the pseudo-metric space (T, d) such that, for all s, t in T and all measurable sets A in Ω,*

$$\int_A |X_s - X_t| d\mathbb{P} \leq d(s,t)\mathbb{P}(A)\psi^{-1}\left(\frac{1}{\mathbb{P}(A)}\right).$$

Then, for any probability measure m on (T, d) and any measurable set A,

$$\int_A \sup_{s,t \in T} |X_s - X_t| d\mathbb{P}$$

$$\leq K_\psi \mathbb{P}(A)\left(D\psi^{-1}\left(\frac{1}{\mathbb{P}(A)}\right) + \sup_{t \in T} \int_0^D \psi^{-1}\left(\frac{1}{m(B(t,\varepsilon))}\right) d\varepsilon\right)$$

where $D = D(T)$ is the diameter of (T, d) and $K_\psi > 0$ only depends on ψ. In particular,

$$\mathbb{E} \sup_{s,t \in T} |X_s - X_t| \leq K_\psi \sup_{t \in T} \int_0^D \psi^{-1}\left(\frac{1}{m(B(t,\varepsilon))}\right) d\varepsilon.$$

Let us mention that the various comments following Theorem 11.2 about integrability and tail behavior of the supremum of the processes under study can be repeated similarly from Theorem 11.13; simply replace the entropy integral by the corresponding majorizing measure integral. In particular, as an analog of (11.4), we have in the setting of Theorem 11.13 that, for every $u > 0$,

$$(11.14) \qquad \mathbb{P}\left\{\sup_{s,t \in T} |X_s - X_t| > K_\psi(\gamma + u)\right\} \leq \left(\psi\left(\frac{u}{D}\right)\right)^{-1}$$

where

$$\gamma = \gamma_m(T, d; \psi) = \sup_{t \in T} \int_0^D \psi^{-1}\left(\frac{1}{m(B(t,\varepsilon))}\right) d\varepsilon.$$

The next result concerns continuity of random processes under majorizing measure conditions. It is the analog of Theorem 11.6 and the proof actually simply needs to adapt appropriately the proof of Theorem 11.6. As announced, the majorizing measure condition has to be strengthened into (11.12).

Theorem 11.14. *Let ψ be a Young function satisfying* (11.11) *and let $X = (X_t)_{t \in T}$ be a random process in $L_1 = L_1(\Omega, \mathcal{A}, \mathbb{P})$ such that for all s, t in (T, d) and all measurable sets A in Ω,*

$$\int_A |X_s - X_t| d\mathbb{P} \le d(s, t) \mathbb{P}(A) \psi^{-1}\left(\frac{1}{\mathbb{P}(A)}\right).$$

Assume that there is a probability measure m on (T, d) such that

$$\lim_{\eta \to 0} \sup_{t \in T} \int_0^\eta \psi^{-1}\left(\frac{1}{m(B(t, \varepsilon))}\right) d\varepsilon = 0.$$

Then X admits a version $\widetilde{X}$ with almost all sample paths (uniformly) continuous on (T, d). Moreover, $\widetilde{X}$ satisfies the following property: for each $\varepsilon > 0$, there exists $\eta > 0$ depending on the preceding limit only, i.e. on T, d, ψ, m but not on X, such that

$$\mathbb{E} \sup_{d(s,t) < \eta} |\widetilde{X}_s - \widetilde{X}_t| < \varepsilon.$$

Proof. We simply sketch the steps of the proof of Theorem 11.6 in the majorizing measure setting. For each $\eta > 0$, set

$$\gamma(\eta) = \sup_{t \in T} \int_0^\eta \psi^{-1}\left(\frac{1}{m(B(t, \varepsilon))}\right) d\varepsilon$$

so that $\lim_{\eta \to 0} \gamma(\eta) = 0$. Fix $\eta > 0$. If A is a finite subset of T, we know from Lemma 11.9, or more precisely its proof, that there is a probability measure μ on (A, d) such that

$$\sup_{t \in T} \int_0^\eta \psi^{-1}\left(\frac{1}{\mu(B(t, \varepsilon))}\right) d\varepsilon \le 2\gamma(\eta/2) \le 2\gamma(\eta).$$

This observation allows us to assume that T is finite in what follows. Let ℓ be the largest integer such that $2^{-\ell} \ge \eta$. If (T, d) is ultrametric, the proof of Theorem 11.13 and its notation yield

$$\mathbb{E} \sup_{t \in T} |X_t - X_{\pi_\ell(t)}| \le K\gamma(\eta)$$

for some (numerical) constant K. Now, we can simply repeat in this case the argument leading to (11.6) in the proof of Theorem 11.6 to get that, for every $\tau > 0$,

(11.15) $$\mathbb{E} \sup_{d(s,t) < \tau} |X_s - X_t| \le \tau \psi^{-1}\left(N(T, d; 2^{-\ell})^2\right) + K\gamma(\eta).$$

Proposition 11.10, adapted to the present case, i.e. with η replacing the diameter, allows then to extend this property to the case of a general index set T, K depending then on ψ. Since (T, d) is totally bounded under the majorizing measure condition, the proof of Theorem 11.14 is completed exactly as the proof of Theorem 11.6.

Note that the precise dependence of η on ε in the last assertion of Theorem 11.14 can be made explicit from (11.15). This will not be required in the sequel so that we only gave the statement that will be sufficient in our applications. This will be in particular the case for the majorizing measure versions of Corollary 11.7 which we need not state since completely similar. Note further that a deviation inequality of the type (11.14) may be obtained for supremum over $d(s, t) < \eta$ from (11.15). We leave the details to the interested reader.

Various remarks developed in Section 11.1 in the context of entropy apply similarly in the setting of majorizing measure conditions. This is the case for example with Remark 11.5 which we need not repeat here; however we use it freely below. This is also the case for Remark 11.4 which might be worthwile to detail in this context. Here is its analog.

Remark 11.15. In the setting of Theorem 11.13, assume that the process $X = (X_t)_{t \in T}$ satisfies the following weaker assumption: for every $t \in T$ and every integer ℓ, and every measurable set A and every s in the ball with center t and radius $2^{-\ell}$,

$$(11.16) \qquad \int_A |X_s - X_t| d\mathbb{P} \leq 2^{-\ell} \mathbb{P}(A) \psi^{-1}\left(\frac{1}{\mathbb{P}(A)}\right) + M_\ell(t) \mathbb{P}(A)$$

where $(M_\ell(t))$ is a sequence of positive numbers such that

$$\sup_{t \in T} \sum_\ell M_\ell(t) < \infty.$$

Then, the conclusion of Theorem 11.13 holds similarly, i.e., under the majorizing measure condition, X is almost surely bounded. The quantitative bounds of course involve the preceding quantity. In case of Theorem 11.14 dealing with continuity, the condition on $(M_\ell(t))$ has to be strengthened into

$$\lim_{\ell_0 \to \infty} \sup_{t \in T} \sum_{\ell \geq \ell_0} M_\ell(t) = 0.$$

As for Remark 11.4, this extension to processes satisfying (11.16) follows directly from the proof of Theorem 11.13. We will see in the next paragraph how these simple observations can be rather useful in various applications.

11.3 Examples of Applications

In the last section of this chapter, we present some (rather important) examples for which the preceding results can be applied. They concern Gaussian and Rademacher processes and their corresponding chaos processes. In particular, the sufficient conditions which we described in order for a Gaussian process to be sample bounded or continuous may be considered as the first part of the study of the regularity of Gaussian processes. The second part devoted to necessity is the object of the next chapter.

Before we enter these examples, let us briefly indicate an elementary but convenient remark. We deal with the Young functions $\psi_q(x) = \exp(x^q) - 1$, $1 \leq q < \infty$. We have $\psi_q^{-1}(x) = (\log(1+x))^{1/q}$. The point of this observation is that we can deal equivalently, in the entropy or majorizing measure conditions, with the functions $(\log x)^{1/q}$, $x \geq 1$. For the entropy condition, it is clear that

$$\int_0^\infty \left(\log N(T, d; \varepsilon)\right)^{1/q} d\varepsilon = \int_0^D \left(\log N(T, d; \varepsilon)\right)^{1/q} d\varepsilon \geq (\log 2)^{1/q} D$$

and thus

(11.17) $$\int_0^D \psi_q^{-1}\left(N(T, d; \varepsilon)\right) d\varepsilon \leq 3 \int_0^\infty \left(\log N(T, d; \varepsilon)\right)^{1/q} d\varepsilon .$$

The reverse inequality (with constant 1) is obvious. Note that we can write, with the function $(\log x)^{1/q}$, the integral up to D or ∞ since $N(T, d; \varepsilon) = 1$ if $\varepsilon \geq D$. Similarly, for the majorizing measure condition, we have that for every probability measure m on T

(11.18)
$$\sup_{t \in T} \int_0^D \psi_q^{-1}\left(\frac{1}{m(B(t, \varepsilon))}\right) d\varepsilon$$
$$\leq 4 \sup_{t \in T} \int_0^\infty \left(\log \frac{1}{m(B(t, \varepsilon))}\right)^{1/q} d\varepsilon$$

and trivially also a reverse inequality. A similar property also holds for continuous majorizing measure conditions. We can further deal with $\psi_\infty(x) = \exp(\exp x) - e$ and replace $\psi_\infty^{-1}(x)$ by $\log(1 + \log x)$, or the more commonly used $\log^+ \log x$ (provided the diameter is taken in account in the inequalities). Accordingly, we freely use below either $\psi_q^{-1}(x)$ or $(\log x)^{1/q}$ depending on the context and/or historical references; *actually*, $(\log x)^{1/q}$ will be used most often.

Let $X = (X_t)_{t \in T}$ be a Gaussian process. Recall from Chapter 3 that by this we mean that the distribution of any finite dimensional random vector $(X_{t_1}, \ldots, X_{t_N})$, $t_i \in T$, is Gaussian. The distribution of X is therefore completely determined by its covariance structure $\mathbb{E} X_s X_t$, $s, t \in T$. The question raises to know under which condition(s) on its covariance structure, the Gaussian process X is (or admits a version which is) almost surely bounded or continuous.

Set
$$d_X(s, t) = \|X_s - X_t\|_2 , \quad s, t \in T .$$

The knowledge of the covariance structure implies a complete knowledge of this L_2-metric d_X. Therefore, d_X is a natural pseudo-metric on the index set T of the Gaussian process X. According to the previous sections, we may try to know how the "geometry" of (T, d_X) describes the boundedness or continuity properties of the sample paths of X. The results of the preceding sections provide rather a precise description of the situation (they will actually be shown to be best possible in the next chapter). To start with however, let us first mention that the comparison theorems of Section 3.3 can also be efficient in this study. Indeed, if X and Y are two Gaussian processes such that $d_Y(s,t) \leq d_X(s,t)$ for all s,t, and if X has nice regularity properties, boundedness or continuity of the sample paths, then, by the results of Section 3.3, these can be "transferred" to Y. This is clear for boundedness by the integrability properties of supremum of Gaussian processes. For continuity, we can use the following lemma.

Lemma 11.16. *Let* $Y = (Y_t)_{t \in T}$ *be a Gaussian process and let* d *be a metric on* T *such that* $d_Y(s,t) \leq d(s,t)$ *for all* s,t. *Then, for every* $\eta > 0$,

$$\mathbb{E} \sup_{d(s,t)<\eta} |Y_s - Y_t| \leq K \left(\sup_{t \in T} \mathbb{E} \sup_{d(s,t)<\eta} |Y_s - Y_t| + \eta \left(\log N(T,d;\eta) \right)^{1/2} \right)$$

where $K > 0$ *is a numerical constant.*

Proof. Fix $\eta > 0$ and let $N = N(T,d;\eta)$ (assumed to be finite and larger than 2). Let $U = (u_1, \ldots, u_N)$ in T be such that the d-balls of radius η and center u_i cover T. Clearly

$$\sup_{d(s,t)<\eta} |Y_s - Y_t| \leq 2 \max_{u \in U} \left(\sup_{d(t,u)<\eta} |Y_t - Y_u| \right) + \max_{\substack{u,v \in U \\ d(u,v)<3\eta}} |Y_u - Y_v|.$$

By (3.6) and the fact that $d_Y(t,u) \leq d(t,u)$, we have

$$\mathbb{E} \max_{u \in U} \left(\sup_{d(t,u)<\eta} |Y_t - Y_u| \right) \leq 2 \max_{u \in U} \mathbb{E} \sup_{d(t,u)<\eta} |Y_t - Y_u| + 3\eta (\log N)^{1/2}.$$

Similarly, by (3.13),

$$\mathbb{E} \max_{\substack{u,v \in U \\ d(u,v)<3\eta}} |Y_u - Y_v| \leq 3\eta (\log N^2)^{1/2}$$

and the lemma is proved.

The preceding claim follows immediately from this lemma when $d = d_X$ using Corollary 3.19. Note that the Gaussian comparison properties are only used in this approach through Corollary 3.19, that is Sudakov's minoration.

Therefore, let X be a Gaussian process with associated pseudo-metric d_X. The integrability properties of Gaussian variables indicate that, for all s,t in T,

$$\|X_s - X_t\|_{\psi_2} \leq 2d_X(s,t).$$

We are thus *immediately* in the setting of processes satisfying a Lipschitz condition as studied in the previous sections. The next two statements are then direct consequences of the results obtained there. The first statement which deals with entropy is known as Dudley's theorem. The numerical constant has no reason to be sharp.

Theorem 11.17. *Let* $X = (X_t)_{t \in T}$ *be a Gaussian process. Then*

$$\mathbb{E} \sup_{t \in T} X_t \leq 24 \int_0^\infty \big(\log N(T, d_X; \varepsilon) \big)^{1/2} d\varepsilon \,.$$

Furthermore, if this entropy integral converges, X has a version with almost all sample paths bounded and (uniformly) continuous on (T, d_X).

Theorem 11.18. *Let* $X = (X_t)_{t \in T}$ *be a Gaussian process. Then, for some numerical constant $K > 0$ and any probability measure m on (T, d_X),*

$$\mathbb{E} \sup_{t \in T} X_t \leq K \sup_{t \in T} \int_0^\infty \left(\log \frac{1}{m(B(t, \varepsilon))} \right)^{1/2} d\varepsilon \,.$$

Furthermore, if m satisfies

$$\lim_{\eta \to 0} \sup_{t \in T} \int_0^\eta \left(\log \frac{1}{m(B(t, \varepsilon))} \right)^{1/2} d\varepsilon = 0 \,,$$

X admits a version with almost all sample paths bounded and (uniformly) continuous on (T, d_X).

Note that if we are asked for the continuity properties of a Gaussian process X with respect to another metric d for which T is compact, we need simply assume in addition that d_X is continuous on (T, d), in other words that X is continuous in L_2 (or in probability). Actually, if (T, d) is any compact metric space, a Gaussian process $X = (X_t)_{t \in T}$ is continuous on (T, d) if and only if it is continuous on (T, d_X) and d_X is continuous on (T, d). Sufficiency is obvious. If X is d-continuous, so is d_X. For $\eta > 0$, let $A_\eta = \{(s, t) \in T \times T ; d_X(s, t) \leq \eta\}$. This is a closed set in $T \times T$ and $\bigcap_{\eta > 0} A_\eta = A_0$. Fix $\varepsilon > 0$. By compactness, there exist $\eta > 0$ and a finite set $A' \subset A_0$ such that, whenever $(s, t) \in A_\eta$, there exists $(s', t') \in A'$ with $d(s, s')$, $d(t, t') \leq \varepsilon$. We have by the triangle inequality

$$|X_s - X_t| \leq |X_s - X_{s'}| + |X_{s'} - X_{t'}| + |X_{t'} - X_t| \,.$$

Since $(s', t') \in A_0$, $X_{s'} = X_{t'}$ with probability one. It follows that

$$\mathbb{E} \sup_{d_X(s,t) \leq \eta} |X_s - X_t| \leq 2\mathbb{E} \sup_{d(s,t) \leq \varepsilon} |X_s - X_t| \,.$$

By the integrability properties of Gaussian random vectors, the right hand side of this inequality goes to 0 with ε, and thus the left hand side with η. It follows that X is d_X-uniformly continuous.

Recall that Theorem 11.18 is more general than Theorem 11.17 ((11.10)). It is remarkable that these two theorems drawn from the rather general results of the previous sections, which apply to large classes of processes, are sharp in this Gaussian setting. As will be discussed in Chapter 12, Theorem 11.17 may indeed be compared to Sudakov's minoration (Theorem 3.18) and, actually, the existence of a majorizing measure in Theorem 11.18 will be shown to be *necessary* for X to be bounded or continuous.

Closely related to Gaussian processes are Rademacher processes. Following Chapter 4, we say that a process $X = (X_t)_{t \in T}$ is a Rademacher process if there exists a sequence $(x_i(t))$ of functions on T such that for every t, $X_t = \sum_i \varepsilon_i x_i(t)$ assumed to converge almost surely (i.e. $\sum_i x_i(t)^2 < \infty$). Recall that (ε_i) denotes a Rademacher sequence, i.e. a sequence of independent random variables taking the values ± 1 with probability $1/2$. The basic observation is that, according to the subgaussian inequality (4.1), as in the Gaussian case,

$$\|X_s - X_t\|_{\psi_2} \le 5\|X_s - X_t\|_2$$

for all s, t in T. The preceding Theorems 11.17 and 11.18 therefore also apply to Rademacher processes. In particular, for any probability measure m on T equipped with the pseudo-metric $d(s, t) = (\sum_i |x_i(s) - x_i(t)|^2)^{1/2}$, we have

$$(11.19) \qquad \mathbb{E} \sup_t \sum_i \varepsilon_i x_i(t) \le K \sup_t \int_0^\infty \left(\log \frac{1}{m(B(t, \varepsilon))} \right)^{1/2} d\varepsilon$$

for some numerical constant $K > 0$.

The Gaussian results actually apply to the general class of the so-called subgaussian processes. A centered process $X = (X_t)_{t \in T}$ is said to be *subgaussian* with respect to a metric or pseudo-metric d on T if, for all s, t in T and every λ in $\mathbb{R}$,

$$(11.20) \qquad \mathbb{E} \exp(\lambda(X_s - X_t)) \le \exp\left(\frac{\lambda^2}{2} d(s, t)^2 \right).$$

Gaussian and Rademacher processes are subgaussian with respect to (a multiple of) their associated L_2-metric. If X is subgaussian with respect to d, by Chebyshev's inequality, for every $\lambda, u > 0$,

$$\mathbb{P}\{|X_s - X_t| > u\} \le 2 \exp\left(-\lambda u + \frac{\lambda^2}{2} d(s, t)^2 \right).$$

Minimizing over λ ($\lambda = u/d(s, t)^2$) yields

$$\mathbb{P}\{|X_s - X_t| > u\} \le 2 \exp(-u^2/2d(s, t)^2)$$

for all $u > 0$. Hence, for all s, t in T,

$$\|X_s - X_t\|_{\psi_2} \le 5d(s, t).$$

This is the property which we use on subgaussian processes. Actually, elementary computations using the expansion of the exponential function shows that if Z is a real valued mean zero random variable such that $\|Z\|_{\psi_2} \leq 1$, then, for all $\lambda \in \mathbb{R}$,

$$\mathbb{E}\exp(\lambda Z) \leq \exp(C^2\lambda^2)$$

where C is a numerical constant. Therefore, changing d by some multiple of it shows that the subgaussian definition (11.20) is equivalent to saying that $\|X_s - X_t\|_{\psi_2} \leq d(s,t)$ for all s,t, or $\mathbb{P}\{|X_s - X_t| > u\} \leq C \exp(-u^2/Cd(s,t)^2)$ for some constant C and all $u > 0$. We freely use this below.

As announced, Theorem 11.17 and 11.18 apply similarly to subgaussian processes. What we will actually use in applications concerning subgaussian processes (in Chapter 14) is the majorizing measure version of Corollary 11.7 for families of subgaussian processes. Let us record at this stage the following statement for further reference.

Proposition 11.19. *Assume that there is a probability measure m on (T,d) such that*

$$\lim_{\eta \to 0} \sup_{t \in T} \int_0^\eta \left(\log \frac{1}{m(B(t,\varepsilon))}\right)^{1/2} d\varepsilon = 0\,.$$

Then, for each $\varepsilon > 0$, there exists $\eta > 0$ such that for every (separable) process $X = (X_t)_{t \in T}$ which is subgaussian with respect to d,

$$\mathbb{E} \sup_{d(s,t)<\eta} |X_s - X_t| < \varepsilon\,.$$

Turning back to Rademacher processes $X = (X_t)_{t \in T}$, $X_t = \sum_i \varepsilon_i x_i(t)$, we have that $\|X_s - X_t\|_2 = (\sum_i |x_i(s) - x_i(t)|^2)^{1/2}$. In Section 4.1, we learned estimates of $\|X_s - X_t\|_{\psi_q}$, $2 < q \leq \infty$, for other metrics than this ℓ_2-metric, namely $\ell_{p,\infty}$-metrics $\|(x_i(s) - x_i(t))\|_{p,\infty}$ where p is the conjugate of q. These results yield further entropy or majorizing measure bounds of Rademacher processes in terms of ψ_q and these metrics. In particular, we have the following statement. Since $\|(x_i(s) - x_i(t))\|_{p,\infty}$ need not be distances in general, the proof is actually carried over with the true metrics $\|X_s - X_t\|_{\psi_q}$.

Lemma 11.20. *Let $X = (X_t)_{t \in T}$ be a Rademacher process, $X_t = \sum_i \varepsilon_i x_i(t)$, $t \in T$. For $1 \leq p < 2$, let $d_{p,\infty}(s,t) = \|(x_i(s) - x_i(t))\|_{p,\infty}$, s,t in T. Then, for any probability measure m on $(T, d_{p,\infty})$,*

$$\mathbb{E}\sup_t \sum_i \varepsilon_i x_i(t) \leq K_p \sup_{t \in T} \int_0^\infty \left(\log \frac{1}{m(B(t,\varepsilon))}\right)^{1/q} d\varepsilon$$

where K_p only depends on p and $q = p/p - 1$ is the conjugate of $p > 1$; when $p = 1$,

$$\mathbb{E}\sup_t \sum_i \varepsilon_i x_i(t) \leq K \sup_{t \in T} \int_0^\infty \log\left(1 + \log \frac{1}{m(B(t,\varepsilon))}\right) d\varepsilon\,.$$

After these classical and important examples, we now investigate some more specialized ones. They concern Gaussian processes with vector values and Gaussian chaos. We say Gaussian but actually these applications are exactly the same for Rademacher processes with the corresponding results in Chapter 4 and the previous discussion. We leave it to the interested reader to translate the results to the Rademacher case.

One of the main interests of these applications is the use of Remarks 11.4 and 11.15 concerning processes satisfying (11.5) or (11.16). In order to put the results in a clearer perspective, we decided to present the first application to vector valued Gaussian processes using the tool of entropy and Remark 11.4, and the second using majorizing measures. It is indeed fruitful to first analyze the questions in terms of entropy. Of course, Theorem 11.21 below also holds under the corresponding majorizing measure conditions.

We do not seek the greatest generality in the definition of processes with vector values. Assume simply that we are given a separable Banach space B and a family $X = (X_t)_{t \in T}$ indexed by T of Borel random variables X_t with values in B. X is Gaussian if each finite sample $(X_{t_1}, \ldots, X_{t_N})$, $t_i \in T$, is Gaussian in B^N. We may ask similarly for the almost sure boundedness or continuity properties of the sample paths of $X = (X_t)_{t \in T}$ in B. As a first simple observation, set, for all s, t in T,

$$d_X(s,t) = \|X_s - X_t\|_2 = (\mathbb{E}\|X_s - X_t\|^2)^{1/2}.$$

From (3.5), we have

$$\|X_s - X_t\|_{\psi_2} \le 8 d_X(s,t).$$

Then, we can make use of Remark 11.5 to see that if

$$(11.21) \qquad \int_0^\infty \left(\log N(T, d_X; \varepsilon)\right)^{1/2} d\varepsilon < \infty,$$

then X has a version with almost all sample paths bounded and continuous on (T, d_X).

The metric d_X in (3.5) is however too "strong". This inequality (3.5) is indeed a consequence of the precise deviation inequalities for norms of Gaussian random vectors in the form of Lemma 3.1. These involve, besides what can be called the "strong" parameter d_X, a "weak" parameter. Let us set indeed,

$$\sigma_X(s,t) = \sigma(X_s - X_t) = \sup_{\|f\| \le 1} \left(\mathbb{E}f^2(X_s - X_t)\right)^{1/2},$$

$s, t \in T$. Then we know from Lemma 3.1 that, for all s, t in T, and for all $u > 0$,

$$\mathbb{P}\{\|X_s - X_t\| > 2 d_X(s,t) + u\sigma_X(s,t)\} \le \exp(-u^2/2).$$

This is (basically) equivalent to saying that

$$\left\|\left(\|X_s - X_t\| - 2 d_X(s,t)\right)^+\right\|_{\psi_2} \le 2\sigma_X(s,t)$$

from which it follows that for every measurable set A

$$\int_A \|X_s - X_t\| d\mathbb{P}$$

(11.22)

$$\leq 2\sigma_X(s,t)\mathbb{P}(A)\psi_2^{-1}\left(\frac{1}{\mathbb{P}(A)}\right) + 2d_X(s,t)\mathbb{P}(A).$$

Therefore, we are in a position to use the general setting developed in the previous sections, and in particular Remarks 11.4 and 11.15. Similarly, we obtain the following result; as announced, it is described in the setting of entropy for a somewhat clearer picture of the argument but also holds under the corresponding majorizing measure conditions. It improves upon (11.21).

Theorem 11.21. *Let* $X = (X_t)_{t \in T}$ *be a Gaussian process with values in a separable Banach space* B. *Recall the weak and strong distances* σ_X *and* d_X *on* T *introduced above. Then, if*

$$\int_0^\infty \left(\log N(T, \sigma_X; \varepsilon)\right)^{1/2} d\varepsilon < \infty \quad and \quad \int_0^\infty \log^+ \log N(T, d_X; \varepsilon) \, d\varepsilon < \infty,$$

X *has a version with almost surely bounded and continuous paths on* (T, d_X).

Proof. Let us first show that there exists a sequence (a_ℓ) of positive numbers such that $\sum_\ell a_\ell < \infty$ and

$$\sum_\ell 2^{-\ell}\left(\log N(T, d_X; a_\ell)\right)^{1/2} < \infty.$$

Set $b_k = (\log N(T, d_X; 2^{-k}))^{1/2}$ for every k. Since $\int_0^\infty \log^+ \log N(T, d_X; \varepsilon) d\varepsilon < \infty$, we have $\sum_k 2^{-k} \log^+ b_k < \infty$. Define $\ell_k = [\log_2^+ (2^k b_k)] + 1$ where $[\cdot]$ is integer part and where $\log_2$ is the logarithm of base 2. We let a_ℓ to be 2^{-k} for all ℓ with $\ell_k < \ell \leq \ell_{k+1}$. Clearly

$$\sum_\ell a_\ell \leq \sum_k 2^{-k} \ell_{k+1} < \infty.$$

Further

$$\sum_\ell 2^{-\ell}\left(\log N(T, d_X; a_\ell)\right)^{1/2} \leq \sum_k 2^{-\ell_k + 1} b_k$$

which is finite too by definition of ℓ_k.

According to this, let now, for each ℓ, A_ℓ be minimal in T such that the d_X-balls with centers in A_ℓ and radius a_ℓ cover T. If D is such a ball, let further $C_\ell(D)$ be minimal in D such that the σ_X-balls with centers in $C_\ell(D)$ and radius $2^{-\ell}$ cover D. Set then, for every ℓ, $T_\ell = \bigcup_D C_\ell(D)$. We have

$$\log \operatorname{Card} T_\ell \leq \log N(T, \sigma_X; 2^{-\ell}) + \log N(T, d_X; a_\ell).$$

Moreover, for every $t \in T$ and every ℓ, there exists, by construction, t_ℓ in T_ℓ such that $\sigma_X(t, t_\ell) \leq 2^{-\ell}$ and $d_X(t, t_\ell) \leq 2a_\ell$. If we use (11.22) and recall that $\sum_\ell a_\ell < \infty$, we see that we are exactly in the setting of Remark 11.4. The conclusion therefore follows since this remark applies similarly to continuity as we noticed it.

Our last application deals with chaos processes. Gaussian and Rademacher chaos were introduced in Chapters 3 and 4 respectively where their integrability and tail behavior properties were investigated. As in those chapters, we restrict here to chaos of order 2. We moreover deal only with real valued chaos; we indicate at the end how the application can be amplified to more general cases. As above finally, we only deal with the Gaussian case; with the corresponding results in Chapter 4, the theorem that we will obtain applies similarly to Rademacher chaos.

Recall that (g_i) denotes an orthogaussian sequence. $X = (X_t)_{t \in T}$ is a Gaussian chaos process of order 2 if there is a sequence $(x_{ij}(t))$ of (real valued) functions on T such that, for every t, $X_t = \sum_{i,j} g_i g_j x_{ij}(t)$ where the sum is almost surely convergent. Following Section 3.2, we introduce two distances d_1 and d_2 on T by setting $d_1(s,t) = \|X_s - X_t\|_2$ and

$$d_2(s,t) = \sup_{|h| \leq 1} \left| \sum_{i,j} h_i h_j \big(x_{ij}(s) - x_{ij}(t) \big) \right|$$

for s, t in T. With respect to Section 3.2, we do not consider the third parameter since we have seen there that, for real valued sequences, the associated decoupled chaos is equivalent to X (at least if the diagonal terms are zero). It follows from Lemma 3.8 and the comments introducing it that there is a numerical constant K such that, for all s, t, $d_2(s,t) \leq K d_1(s,t)$ and

$$(11.23) \quad \mathbb{P}\big\{|X_s - X_t| > u d_1(s,t) + u^2 d_2(s,t)\big\} \leq K \exp(-u^2/K)$$

for every $u > 0$. In particular (for some possibly different constant K),

$$\mathbb{P}\big\{|X_s - X_t| > u d_1(s,t)\big\} \leq K \exp(-u/K).$$

We could then apply the results of the preceding sections and show the boundedness and continuity of X in terms of the distance d_1 with respect to the Young function $\psi_1(x) = \exp(x) - 1$ only. However, as in the previous application, the incremental estimates (11.23) involving the two distances d_1 and d_2 are more precise and lead to sharper conditions. (11.23) is used in the context of Remarks 11.4 and 11.15. To this aim, note that it implies that if $a \geq d_2(s,t)$, for all $u > 0$,

$$\mathbb{P}\left\{|X_s - X_t| > \frac{1}{a} d_1^2(s,t) + 2au\right\} \leq K \exp(-u/K)$$

(since $a^{-1} d_1^2(s,t) + 2au \geq \sqrt{u} d_1(s,t) + u d_2(s,t)$). Hence, for some (possibly different) numerical constant K,

$$\left\| \left(|X_s - X_t| - \frac{1}{a} d_1^2(s,t) \right)^+ \right\|_{\psi_1} \leq Ka.$$

Therefore, if $a \geq d_2(s,t)$ and if A is a measurable set in Ω,

$$(11.24) \quad \int_A |X_s - X_t| d\mathbb{P} \leq Ka\mathbb{P}(A)\psi_1^{-1}\left(\frac{1}{\mathbb{P}(A)}\right) + \frac{1}{a} d_1^2(s,t)\mathbb{P}(A).$$

These relations put us into the right situation in order to apply Theorems 11.13 and 11.14 together with Remark 11.15.

Now, we can state our result on the almost sure boundedness and continuity of Gaussian chaos processes.

Theorem 11.22. Let $X_t = \sum_{i,j} g_i g_j x_{ij}(t)$, $t \in T$, be a Gaussian chaos process of order 2 as just described with associated metrics d_1 and d_2. Assume that there exist probability measures m_1 and m_2 on T such that respectively

$$\lim_{\eta \to 0} \sup_{t \in T} \int_0^\eta \left(\log \frac{1}{m_1(B_1(t, \varepsilon))} \right)^{1/2} d\varepsilon = 0$$

and

$$\lim_{\eta \to 0} \sup_{t \in T} \int_0^\eta \log \frac{1}{m_2(B_2(t, \varepsilon))} \, d\varepsilon = 0$$

where $B_i(t, \varepsilon)$ is the d_i-ball with center t and radius $\varepsilon > 0$ $(i = 1, 2)$. Then $X = (X_t)_{t \in T}$ admits a version with almost all sample paths bounded and continuous on (T, d_1).

Proof. We only show that if T is finite and if M is a number such that

$$\sup_{t \in T} \int_0^\infty \left(\log \frac{1}{m_i(B_i(t, \varepsilon))} \right)^{i/2} d\varepsilon \leq M, \quad i = 1, 2,$$

then

$$\mathbb{E} \sup_{s, t \in T} |X_s - X_t| \leq KM$$

for some numerical constant K. From this and the material discussed in the preceding section, it is not difficult to deduce the full conclusion of the statement.

Thus, let T be finite and M be as before. According to Proposition 11.10, we may and do assume that d_1 and d_2 are ultrametric. For every t and ℓ set

$$\gamma(t, \ell) = 2^{-\ell} \left(\log \frac{1}{m_1(B_1(t, 2^{-\ell}))} \right)^{1/2}$$

so that $\sum_\ell \gamma(t, \ell) \leq KM$. Denote by $k(t, \ell)$ the largest integer k such that $2^{2j} \gamma(t, \ell + j) \leq 1$ for all $j \leq k$. We may observe that

$$(11.25) \qquad \sum_\ell 2^{-2k(t, \ell)} \leq 8KM .$$

To show this, let $L_n = \{\ell; \ell + k(t, \ell) = n\}$. We note that if $\ell \neq \ell'$ are elements of L_n, then $k(t, \ell) \neq k(t, \ell')$ from which it follows that $\sum_{\ell \in L_n} 2^{-2k(t, \ell)} \leq 2^{-2k(t, \ell_0) + 1}$ where $k(t, \ell_0) = \min\{k(t, \ell); \ell \in L_n\}$. Then, by definition of $k(t, \ell_0)$, $2^{-2k(t, \ell_0)} \leq 4\gamma(t, \ell_0 + k(t, \ell_0) + 1) = 4\gamma(t, n + 1)$. (11.25) clearly follows and implies in particular that

(11.26)
$$\sum_{\ell} \gamma\big(t, \ell + k(t,\ell)\big) \le 8KM \,.$$

As another property of the integers $k(t,\ell)$, note, as is easily checked, that $\ell + k(t,\ell)$ is increasing in ℓ. For every t and ℓ, consider the subset $H(t, 2^{-2\ell})$ of T consisting of the balls C of radius $2^{-\ell-k(t,\ell)-1}$ included in $B_1(t, 2^{-\ell-k(t,\ell)})$ such that $k(s,\ell) = k(t,\ell)$ for all s in C. From the definition of $k(t,\ell)$, the latter property only depends on C, and not on s in C. Since $t + k(t,\ell)$ is increasing in ℓ, $H(t, 2^{-2\ell-2}) \subset H(t, 2^{-2\ell})$. Furthermore, the subsets $H(t, 2^{-2\ell})$ when $t \in T$ form the family of the balls of radius $2^{-2\ell}$ for the (ultrametric) distance d' given by

$$d'(s,t) = 2^{-2\ell} \quad \text{where} \quad \ell = \sup\{j : \exists u \text{ such that } s, t \in H(u, 2^{-2j})\} \,.$$

Recall now from the assumptions that, for every t,

(11.27)
$$\sum_{\ell} 2^{-2\ell} \log \frac{1}{m_2\big(B_2(t, 2^{-2\ell})\big)} \le KM \,.$$

Let $B(t, 2^{-2\ell})$ be the ball with center t and radius $2^{-2\ell}$ for the distance $d = \max(d', d_2)$. Such a ball is the intersection of a d'-ball of radius $2^{-2\ell}$ and a d_2-ball of radius $2^{-2\ell}$. To this ball we can associate the weight

$$2^{-k(t,\ell)+k_0-1} m_1\big(B_1(t, 2^{-\ell-k(t,\ell)})\big) m_2\big(B_2(t, 2^{-2\ell})\big)$$

where k_0 is the smallest possible value for $k(t,\ell)$, $t \in T$, $\ell \in \mathbb{Z}$. We obtain in this way a family of weights as described in Remark 11.11. One can then construct a probability measure m on (T, d) such that, by (11.26) and (11.27), for all t in T,

$$\sum_{\ell} 2^{-2\ell} \log \frac{1}{m\big(B(t, 2^{-2\ell})\big)} \le KM$$

for some numerical constant K. We are in a position to conclude. Let s and t be such that $d(s,t) \le 2^{-2\ell}$. Then, by construction, $d_1(s,t) \le 2^{-\ell-k(t,\ell)}$ and $d_2(s,t) \le 2^{-2\ell}$. Hence, from (11.24), for every measurable set A,

$$\int_A |X_s - X_t| d\mathbb{P} \le K 2^{-2\ell} \mathbb{P}(A) \psi^{-1}\left(\frac{1}{\mathbb{P}(A)}\right) + 2^{-2k(t,\ell)} \mathbb{P}(A) \,.$$

Therefore, we see that we are exactly in the situation described by Remark 11.15 and (11.16) since (11.25) holds. We thus conclude the proof of Theorem 11.22 in this way.

Let us mention, to conclude this chapter, that the previous theorem might be extended to vector valued chaos processes, that is processes $X_t = \sum_{i,j} g_i g_j x_{ij}(t)$ where the functions $x_{ij}(t)$ take their values in a Banach space. According to the study of Section 3.2, three distances would then be involved with different entropy or majorizing measure conditions on each of them ($d+1$ distances for chaos of order d !). We do not pursue in this direction.

Notes and References

Various references have presented during the past years the theory of random processes on abstract index sets and their regularity properties as developed in this chapter. Expositions on Gaussian processes have been given in particular in the celebrated course [Fer4] by X. Fernique, and also by R. Dudley [Du2], V. N. Sudakov [Su4] and N. C. Jain and M. B. Marcus [J-M3]. General sufficient conditions for non-Gaussian processes satisfying incremental Orlicz conditions to be almost surely bounded or continuous are presented in the notes [Pi13] by G. Pisier and emphasized in [Fer7] and [We]. We refer to these authors for more accurate references and in particular to [Du2] for a careful historical description of the early developments of the theory of Gaussian (and non-Gaussian too) processes. Our exposition is based on [Pi13] and the recent work [Ta13].

The study of random processes under metric entropy conditions actually started with the Gaussian results of Section 11.3. The notion of ε-entropy goes back to A. N. Kolmogorov. The landmark paper [Du1] by R. Dudley, where Theorem 11.17 is established, introduced this fundamental abstract framework in the field. Credit for the introduction of ε-entropy applied to regularity of processes goes to V. Strassen (see [Du1]) and V. N. Sudakov [Su1]. It was only slowly realized after that the Gaussian structure of this result relies on the appropriate integrability properties of the increments $X_s - X_t$ only, and that the technique could be extended to large classes of non-Gaussian processes. On the basis of Kolmogorov's continuity theorem (which already contains the fundamental chaining argument) and this observation, several authors investigated sufficient conditions for the boundedness and continuity of processes whose increments $X_s - X_t$ are nicely controlled. Among the various articles, let us mention (see also [Du2]) [De], [Bou], [J-M1] (on subgaussian processes, see also [J-M3]), [Ha], [H-K], [N-N], [Ib] and [Kô] and [Pi9] on the important case of increments in L_p. The general Theorem 11.1 is due to G. Pisier [Pi13] (on the basis of [Pi9] thanks to an observation of X. Fernique). Its refined version Theorem 11.2 is equivalent to the (perhaps somewhat unorthodox) formulation of [Fer7]. The tail behaviors deduced from this statement were precisely analyzed in some cases in the context of empirical processes in [Ale1] (see also [Fer7], [We]). The uniform continuity and compactness results (Theorem 11.6 and Corollary 11.7) simplify prior proofs by X. Fernique [Fer11]. Kolmogorov's theorem (Corollary 11.8) may be found e.g. in [Slu], [Ne1], [Bi].

We refer to the survey paper [He3] for a history of majorizing measures (in view of the recent developments of the theory – see the next chapter [Ta18] –, the name of "minorizing measures" could seem more appropriate). In [G-R-R], A. M. Garsia, E. Rodemich and H. Rumsey establish a real variable lemma using integral bounds involving majorizing measures and apply it to the regularity of random processes. This lemma was further used and refined in [Gar] and [G-R] and usually provides interesting moduli of continuity. Let us note that our approach to majorizing measures is not completely similar to this real variable lemma. Our concerns go more to integrability and tail behaviors rather than moduli of continuity. More precisely, the technique of [G-R-R], refined by C. Preston [Pr1], [Pr2] and B. Heinkel [He1], [He3], allows for example to show the following *non-random* property. Let f be a real valued (continuous) function on some metric space (T, d) and let ψ be a Young function. Given a probability measure m on (T, d), denote by $\|\tilde{f}\|_{L_\psi(m \times m)}$ the Orlicz norm with respect to ψ of

$$\tilde{f}(s,t) = \frac{|f(s) - f(t)|}{d(s,t)} I_{\{d(s,t) \neq 0\}}$$

in the product space $(T \times T, \ m \times m)$. Then one can show (cf. [He3]) that for all s, t in T,

$$|f(s) - f(t)| \leq 20\|\tilde{f}\|_{L_\psi(m \times m)} \sup_{u \in T} \int_0^{d(s,t)/2} \psi^{-1}\left(\frac{1}{m\big(B(u,\varepsilon)\big)^2}\right) d\varepsilon.$$

Note the square of $m(B(u, \varepsilon))$ in the majorizing measure integral. (While this square is irrelevant when ψ has an exponential growth, this is no longer the case in general, and a main concern of the paper [Ta13] is to remove this square when it is not needed.) In concrete situations, the evaluation of the entropy integral (usually for the Lebesgue measure on some compact subset of $\mathbb{R}^N$) therefore yields various moduli of continuity and actually allows to study bounds on $\sup_{s,t} |X_s - X_t|/d(s,t)^\alpha$, $\alpha > 0$. These arguments have been proved useful in stochastic calculus by several authors (see e.g. [S-V], [Yo], [B-Y] etc., and in particular the recent connections between regularity of (stationary) Gaussian processes and regularity of local times of Lévy processes put forward by M. T. Barlow [Bar1], [Bar2], [B-H], [M-R]). On the basis of the seminal result of [G-R-R], C. Preston [Pr1], [Pr2] developed the concept of majorizing measure and basically obtained, in [Pr1], Theorem 11.18. However, in his main statement, C. Preston unnecessarily restricts his hypotheses. He was apparently not aware of the power of the present formulation which was put forward by X. Fernique who completely established Theorem 11.18 [Fer4]. X. Fernique developed in the meantime a somewhat different point of view based on the duality of Orlicz norms (cf. [Fer3], [Fer4]). Our exposition is taken from [Ta13] to which we actually refer for a more complete exposition involving general Young functions ψ. The ultrametric structure and discretization procedure (Proposition 11.10 and Corollary 11.12), implicit in [Ta5], are described in [A-G-O-Z], [Ta7], and [Ta13].

As we mentioned it, Theorem 11.17 is due to R. Dudley [Du1], [Du2] after some early observations and contributions by V. Strassen and V. N. Sudakov, while Theorem 11.18 is due to X. Fernique [Fer4]. Various comments on regularity of Gaussian processes are taken from these references as well as from [M-S1], [M-S2] (where Slepian's lemma is introduced in this study), [J-M3], [Ta5], [Ad]. Lemma 11.16 is taken from [Fer8] to show a result of [M-S2]. Let us mention here a volumic approach to regularity of Gaussian processes in the paper [Mi-P], [Pi18] where the local theory of Banach spaces is used to prove a conjecture of [Du1] on volume numbers of subsets of Hilbert space. Sub-gaussian processes have been emphasized in their relation to the central limit theorem in [J-M1], [He1], [J-M3] (cf. Chapter 14). Lemma 11.20 comes from [M-P2] (see also [M-P3]) and will be crucial in Chapter 13. The new technique of using simultaneously different majorizing measure conditions for different metrics in Theorems 11.21 and 11.22 is due to the second author. It becomes very natural in the new presentation of majorizing measures introduced in [Ta18]; see also [Ta20]. Theorem 11.22 about the regularity of Gaussian and Rademacher chaos processes improves observations of [Bo6].

12. Regularity of Gaussian and Stable Processes

In the preceding chapter, we presented some *sufficient* metric entropy and majorizing measure conditions for the sample boundedness and continuity of random processes satisfying incremental conditions. In particular, these results were applied to Gaussian random processes in Section 11.3. The main concern of this chapter is *necessity*. We will see indeed, as one of the main results, that the sufficient majorizing measure condition for a Gaussian process to be almost surely bounded or continuous is actually also necessary. This characterization thus provides a complete understanding of the regularity properties of Gaussian paths. The arguments of proof rely heavily on the basic ultrametric structure which lies behind a majorizing measure condition.

This characterization is performed in the first section which is completed by some equivalent formulations of the main result. Let us mention at this point that the study of necessity for non-Gaussian processes in this framework involves, rather than one given process, the whole family of processes satisfying some incremental condition with respect to the same Young function ψ and metric d. In the Gaussian case, Slepian's lemma and the comparison theorems (Section 3.3), which appear as a cornerstone in this study of necessity, confound those two situations and things become simpler. We refer the interested reader to [Ta13] for a study of necessity for general processes in this setting. A noticeable exception is however the case of stable processes where the series representation and conditional use of Gaussian techniques allow to describe necessary conditions as for the Gaussian processes. This extension to p-stable processes, $1 \leq p < 2$, is the subject of Section 12.2; as we will see, it is sufficiency which appears to be more difficult in the stable case. This chapter is completed with some applications to subgaussian processes and some remarkable type properties of the injection map $\mathrm{Lip}(T) \to C(T)$ when there is a Gaussian or stable majorizing measure on (T, d). The difficult subject of Rademacher processes is discussed with some conjectures in the very last part.

12.1 Regularity of Gaussian Processes

Recall that a random process $X = (X_t)_{t \in T}$ is said to be Gaussian if each finite linear combination $\sum_i \alpha_i X_{t_i}$, $\alpha_i \in \mathbb{R}$, $t_i \in T$, is a real valued Gaussian variable. The distribution of the Gaussian process X is therefore completely determined by its covariance structure $\mathbb{E} X_s X_t$, $s, t \in T$. As we know, to study the regularity properties of X, it is fruitful to analyze the geometry of T for the induced L_2-pseudo-metric $d_X(s, t) = \|X_s - X_t\|_2$, $s, t \in T$.

We have seen in Theorem 11.18 that, for *any* probability measure m on (T, d_X),

$$(12.1) \qquad \mathbb{E} \sup_{t \in T} X_t \leq K \sup_{t \in T} \int_0^\infty \left(\log \frac{1}{m(B(t, \varepsilon))} \right)^{1/2} d\varepsilon$$

where $B(t, \varepsilon)$ is the open ball with center t and radius $\varepsilon > 0$ in the pseudo-metric d_X and X has almost surely bounded sample paths if, for some probability measure m, the majorizing measure integral on the right hand side of (12.1) is finite. There is a similar result about continuity. Recall that $\mathbb{E} \sup_{t \in T} X_t$ is simply understood here as

$$\mathbb{E} \sup_{t \in T} X_t = \sup \left\{ \mathbb{E} \sup_{t \in F} X_t ; \ F \text{ finite in } T \right\}.$$

K is further some numerical constant which may vary from line to line in what follows. By (11.10), (12.1) contains the familiar entropic bound (Theorem 11.17)

$$(12.2) \qquad \mathbb{E} \sup_{t \in T} X_t \leq K \int_0^\infty \left(\log N(T, d_X; \varepsilon) \right)^{1/2} d\varepsilon.$$

Now, we described in Theorem 3.18 a lower bound in terms of entropy numbers which indicates that

$$(12.3) \qquad \sup_{\varepsilon > 0} \varepsilon \left(\log N(T, d_X; \varepsilon) \right)^{1/2} \leq K \mathbb{E} \sup_{t \in T} X_t.$$

These two bounds (12.2) and (12.3) appear to be rather close from each other. However, there is a small gap which may be put forward by the example of an independent Gaussian sequence (Y_n) such that $\|Y_n\|_2 = (\log(n+1))^{-1/2}$ for every n. This sequence, which defines an almost surely bounded process by (3.7), shows that boundedness of Gaussian processes cannot be characterized by the metric entropy integral in (12.2). The reason for this failure is due to the possible lack of homogeneity of T for the metric d_X. We will see in the next chapter that in, a *homogeneous* setting, the metric entropy condition *does characterize* the almost sure boundedness and continuity of Gaussian processes.

As we know, majorizing measures provide a way to take into account the possible lack of homogeneity of (T, d_X). Furthermore, by (11.9) and (11.10), the majorizing measure integral of (12.1) appears as an intermediate condition between the two entropic bounds (12.2) and (12.3). As a main result, we will show that the minoration (12.3) can be improved into

$$
(12.4) \qquad \sup_{t \in T} \int_0^\infty \left(\log \frac{1}{m(B(t,\varepsilon))} \right)^{1/2} d\varepsilon \le K \mathbb{E} \sup_{t \in T} X_t
$$

for *some* probability measure m on T. Hence, together with (12.1), the existence of a majorizing measure for the function ψ_2^{-1}, or $(\log(1/x))^{1/2}$ (cf. (11.18)), completely characterizes the boundedness of Gaussian processes X in terms *only* of their associated L_2-metric d_X. There is a similar result for continuity.

The proof of this result, to which we now turn, requires rather an involved study of majorizing measures. This will be accomplished in the abstract setting which we now describe.

The majorizing measure conditions which will be included in this study concern in particular those associated to the Young functions $\psi_q(x) = \exp(x^q) - 1$. To unify the exposition, let us thus consider a strictly decreasing function h on $(0,1]$ with $h(1) = 0$ and $\lim_{x \to 0} h(x) = \infty$. We assume that for all x, y in $(0,1]$

$$
(12.5) \qquad\qquad h(xy) \le h(x) + h(y).
$$

[This condition may be weakened into $h(xy) \le h(x) + ch(y)$ for some positive c and the reader can verified that the subsequent arguments go through in this case, the numerical constant of Theorem 12.5 depending then on h.] As announced, the main examples which we have in mind with this function h are the examples of $h_q(x) = (\log(1/x))^{1/q}$, $1 \le q < \infty$, and of $h_\infty(x) = \log(1 + \log(1/x))$.

Let us mention that we never attempt to find sharp numerical constants, but always use crude, but simple, bounds.

If (T, d) is a metric space, recall that we denote by $D = D(T)$ its diameter, i.e.

$$
D = D(T) = \sup\{d(s,t); \; s, t \in T\}.
$$

Given a probability measure m on the metric space (T, d), let

$$
\gamma_m(T) = \gamma_m(T, d) = \sup_{t \in T} \int_0^\infty h(m(B, (t, \varepsilon))) d\varepsilon
$$

where $B(t, \varepsilon)$ is the open ball with center t and radius $\varepsilon > 0$ in (T, d). Here, and throughout this study, when no ambiguity arises, we adopt the convention that $B(t, \varepsilon)$ denotes the ball for the distance on the space which contains t. We also let

$$
\gamma(T) = \gamma(T, d) = \inf \gamma_m(T, d) = \inf \gamma_m(T)
$$

where the infimum is taken over all probability measures m on T. For a subspace A of T, $\gamma(A) = \gamma(A, d)$ refers to the quantity associated to the metric space (A, d), i.e. $\gamma(A) = \inf \gamma_m(A)$ where the infimum is taken over the probability measures supported by A.

Recall that a metric space (U, δ) is called *ultrametric* if for u, v, w in U we have

$$\delta(u, w) \leq \max\big(\delta(u, v), \delta(v, w)\big).$$

The nice feature of ultrametric spaces is that two balls of the same radius are either identical or disjoint.

From now on, and until further notice, we assume that all the metric spaces are *finite*. Given a metric space (T, d) and an ultrametric space (U, δ), say that a map φ from U onto T is a contraction if $d(\varphi(u), \varphi(v)) \leq \delta(u, v)$ for u, v in U. Define the functional $\alpha(T) = \alpha(T, d)$ by

$$\alpha(T) = \inf\{\gamma(U); U \text{ is ultrametric and } T \text{ is the image of } U \text{ by a contraction}\}.$$

Although $\gamma(T)$ comes first in mind as a way to measure the size of T, the quantity $\alpha(T)$ is easier to manipulate and yields stronger results. First, we collect some simple facts.

Lemma 12.1. *The following hold under the preceding notations.*

(i) $\gamma(T) \leq \alpha(T)$.

(ii) *If* $A \subset T$, *then* $\gamma(A) \leq 2\gamma(T)$.

(iii) *If* U *is ultrametric and* $A \subset U$, *then* $\gamma(A) \leq \gamma(U)$.

(iv) *If* $A \subset T$, *then* $\alpha(A) \leq \alpha(T)$.

(v) $\alpha(T) = \inf\{\gamma(U); U$ *is ultrametric,* $D(U) \leq D(T)$ *and* T *is the image of* U *by a contraction*$\}$ *and this infimum is attained.*

(vi) $D(T) \leq 2\big[h(1/2)\big]^{-1}\gamma(T)$.

Proof. (i) Let φ be a contraction from U onto T, μ be a probability measure on U and $m = \varphi(\mu)$. For u in U, $\varepsilon > 0$, we have, since φ is a contraction, that $\varphi^{-1}(B(\varphi(u), \varepsilon)) \supset B(u, \varepsilon)$, so $m(B(\varphi(u), \varepsilon)) \geq \mu(B(u, \varepsilon))$. Since h is decreasing and φ onto, we get $\gamma_m(T) \leq \gamma_\mu(U)$, so $\gamma(T) \leq \gamma(U)$ since μ is arbitrary; therefore $\gamma(T) \leq \alpha(T)$ since U and φ are arbitrary. (ii) This was already shown in Lemma 11.9 but let us briefly recall the argument. For t in T, take $a(t)$ in A with $d(t, a(t)) = d(t, A)$. Let m be a probability measure on T and let $\mu = a(m)$ so that μ is supported by A. Fix x in A. For t in T, we have $d(t, A) \leq d(t, x)$, so $d(t, a(t)) \leq d(t, x)$ and $d(x, a(t)) \leq 2d(x, t)$. Since $\mu = a(m)$, it follows that $\mu(B(x, 2\varepsilon)) \geq m(B(x, \varepsilon))$. Hence, by a change of variables, $\gamma_\mu(A) \leq 2\gamma_m(T)$ which gives the results. (iii) With the notation of the proof of (ii), the ultrametricity gives $d(x, a(t)) \leq \max(d(x, t), d(t, a(t))) \leq d(x, t)$ and thus $\gamma(A) \leq \gamma(U)$ in this case. (iv) Let U be ultrametric and let φ be a contraction from U onto T. By (iii), we get $\alpha(A) \leq \alpha(\varphi^{-1}(A)) \leq \gamma(U)$ and thus $\alpha(A) \leq \alpha(T)$. (v) If (U, δ) is ultrametric and φ is a contraction from U onto T, consider the distance δ_1 on U given by $\delta_1(u, v) = \min(\delta(u, v), D(T))$. Then (U, δ_1) is ultrametric and φ is still a contraction from (U, δ_1) onto T.

By the argument of (i), $\gamma(U, \delta_1) \leq \gamma(U, \delta)$. The last assertion follows by a standard compactness argument. (vi) Take two points s and t in T and let $\eta = d(s, t)$. The balls $B(s, \eta/2)$ and $B(t, \eta/2)$ are disjoint so that if m is a probability measure on T, one of these balls, say the first, has a measure less than $1/2$. Therefore

$$\gamma_m(T) \geq \int_0^{\eta/2} h\big(m\big(B(s, \varepsilon)\big)\big) d\varepsilon \geq \frac{\eta}{2} h\Big(\frac{1}{2}\Big)$$

from which the result follows since m, s, t are arbitrary and $h(1/2) > 0$ by (12.5). The proof of the lemma is complete.

The next lemma is one of the key tools of this investigation. It exhibits a behavior of α that resembles a strong form of subadditivy.

Lemma 12.2. *Let T be a finite metric space with diameter $D = D(T)$. Suppose that we have a finite covering $A_1, \ldots, A_n$ of T. Then, for every positive numbers $a_1, \ldots, a_n$ with $\sum_{i=1}^n a_i \leq 1$,*

$$\alpha(T) \leq \max_{i \leq n} \big[\alpha(A_i) + D(T)h(a_i)\big].$$

Proof. From Lemma 12.1 (v), for every $i = 1, \ldots, n$, there exist an ultrametric space (U_i, δ_i) of diameter less than D, a contraction φ_i from U_i onto A_i and a probability measure μ_i on U_i such that $\alpha(A_i) = \gamma_{\mu_i}(U_i)$ (or arbitrarily close). Let U be the disjoint sum of the spaces $(U_i)_{i \leq n}$. Define the distance δ on U by $\delta(u, v) = \delta_i(u, v)$ whenever u, v belong to the same U_i, and $\delta(u, v) = D$ otherwise. Then (U, δ) is an ultrametric space and the map φ from U onto T given by $\varphi(u) = \varphi_i(u)$ for u in U_i is a contraction. Consider the positive measure μ' on U given by $\mu' = \sum_{i=1}^n a_i \mu_i$. Since $|\mu'| \leq 1$, there is a probability μ on U with $\mu \geq \mu'$. Take then u in U and let i be such that $u \in U_i$. By (12.5),

$$h\big(\mu\big(B(u, \varepsilon)\big)\big) \leq h\big(\mu'\big(B(u, \varepsilon)\big)\big)$$
$$\leq h\big(a_i \mu_i \big(B(u, \varepsilon)\big)\big)$$
$$\leq h\big(\mu_i \big(B(u, \varepsilon)\big)\big) + h(a_i).$$

It follows that

$$\int_0^\infty h\big(\mu\big(B(u, \varepsilon)\big)\big) d\varepsilon = \int_0^D h\big(\mu\big(B(u, \varepsilon)\big)\big) d\varepsilon$$
$$\leq \int_0^\infty h\big(\mu_i \big(B(u, \varepsilon)\big)\big) d\varepsilon + Dh(a_i)$$
$$\leq \alpha(A_i) + Dh(a_i).$$

Therefore

$$\alpha(U) \leq \gamma_\mu(U) \leq \max_{i \leq n} \big[\alpha(A_i) + Dh(a_i)\big]$$

from which the conclusion follows.

The next lemma is the basic step in the subsequent construction. If (T, d) is a finite metric space, let, for every integer k (in $\mathbb{Z}$),

$$\beta_k(T) = \alpha(T) - \sup_{x \in T} \alpha\big(B(x, 6^{-k})\big).$$

Lemma 12.3. *Let (T, d) be a finite metric space of diameter less than 6^{-k}. We are necessarily in one of the following two cases:*

(i) either there exists a subset S of T of diameter less than 6^{-k-1} satisfying

$$(12.6) \qquad \beta_{k+2}(S) \geq 2\big(\alpha(T) - \alpha(S)\big);$$

(ii) or there exist balls $(B_i)_{1 \leq i \leq N}$ of radius 6^{-k-2} with centers at mutual distance larger than $3 \cdot 6^{-k-2}$ such that

$$(12.7) \qquad \text{for every } i, \quad \alpha(B_i) \geq \alpha(T) - 6^{-k+1} h\left(\frac{1}{1+N}\right).$$

Proof. Suppose that (i) does not hold. By induction, we construct points x_i, $i \geq 1$, in T in the following manner: $\alpha(B(x_1, 6^{-k-2}))$ is maximal, and, if $x_1, \ldots, x_{i-1}$ have been constructed, we take x_i such that

$$\alpha\big(B(x_i, 6^{-k-2})\big) = \max\big\{\alpha\big(B(x, 6^{-k-2})\big), \ \forall j < i, \ d(x, x_j) \geq 3 \cdot 6^{-k-2}\big\}.$$

For every i, set then

$$S_i = B\big(x_i, 3 \cdot 6^{-k-2}\big) \backslash \bigcup_{j<i} B\big(x_j, 3 \cdot 6^{-k-2}\big).$$

Since (i) does not hold, necessarily, for any i, $\beta_{k+2}(S_i) \leq 2(\alpha(T) - \alpha(S_i))$. By construction $\beta_{k+2}(S_i) = \alpha(S_i) - \alpha(B(x_i, 6^{-k-2}))$. Thus, for every i,

$$(12.8) \qquad \begin{aligned} \alpha\big(B(x_i, 6^{-k-2})\big) &= \alpha(S_i) - \beta_{k+2}(S_i) \\ &\geq \alpha(S_i) - 2\big(\alpha(T) - \alpha(S_i)\big) \\ &\geq \alpha(T) - 3\big(\alpha(T) - \alpha(S_i)\big). \end{aligned}$$

The union of the S_i's covers T. They can be assumed to be ordered so that the sequence $(\alpha(S_i))$ is decreasing. If we let $a_i = (i+1)^{-2}$, we have $\sum_{i \geq 1} a_i \leq 1$. Therefore, by Lemma 12.2, there exists $i_0 \geq 1$ such that

$$(12.9) \qquad \alpha(S_{i_0}) \geq \alpha(T) - 6^{-k} h\big((i_0 + 1)^{-2}\big).$$

By (12.5), $h((i_0 + 1)^{-2}) \leq 2h((i_0 + 1)^{-1})$. Hence, if $I = \{1, \ldots, i_0\}$, since $(\alpha(S_i))$ is decreasing, we see from (12.9) that for every i in I,

$$(12.10) \qquad \alpha(S_i) \geq \alpha(T) - 2 \cdot 6^{-k} h\left(\frac{1}{1 + \operatorname{Card} I}\right).$$

Combining (12.10) with (12.8) yields that for every i in I,

$$\alpha\big(B(x_i, 6^{-k-2})\big) \geq \alpha(T) - 6^{-k+1} h\left(\frac{1}{1 + \operatorname{Card} I}\right).$$

Letting $B_i = B(x_i, 6^{-k-2})$ for i in I and $N = \operatorname{Card} I$ shows that we are in the case (ii) of the statement and that (12.7) is satisfied. Lemma 12.3 is established.

Now, we perform the main construction. Given a metric space T, we exhaust it with the alternative of Lemma 12.3 and construct in this way subsets (actually balls) which are well separated and whose α-functionals are big enough and carry enough information on $\alpha(T)$ itself. Iterative use of this proposition gives rise to a "tree" and an ultrametric structure. For two subsets A, B of a metric space (T, d), let $d(A, B) = \inf\{d(a, b); a \in A, \ b \in B\}$.

Proposition 12.4. *Let (T, d) be a finite metric space of diameter less than 6^{-k}. There exist an integer $\ell \geq k$ and subsets $(B_i)_{1 \leq i \leq N}$ of T of diameter less than $6^{-\ell-1}$ such that $d(B_i, B_j) \geq 6^{-\ell-2}$ for $i \neq j$, the diameter of $\bigcup_{i \leq N} B_i$ is $\leq 6^{-\ell+1}$ and*

$$(12.11) \qquad \textit{for every } i, \ \alpha(B_i) \geq \alpha(T) - 2 \cdot 6^{-k+1} h\left(\frac{1}{1 + N}\right).$$

Proof. By induction, we construct a decreasing sequence (T_m) of subsets of T such that $T_0 = T$, $D(T_m) \leq 6^{-k-m}$ and

$$(12.12) \qquad \beta_{k+m+1}(T_m) \geq 2\big(\alpha(T_{m-1}) - \alpha(T_m)\big)$$

for all $m \geq 1$. The construction stops (since T is finite) for some $m = n$ and in T_n we are necessarily in case (ii) of Lemma 12.3 (since if not we would be able to continue the exhaustion). First, we note that, for every $m \leq n$,

$$(12.13) \qquad \alpha(T) - \alpha(T_m) \leq \beta_{k+m+1}(T_m).$$

Indeed, this is clearly the case for $m = 1$ (since we even have in this case that $\beta_{k+2}(T_1) \geq 2(\alpha(T) - \alpha(T_1))$). Assume then that (12.13) is satisfied for m and let us show it for $m + 1$. We have by (12.12)

$$\beta_{k-m+2}(T_{m+1}) \geq 2\big(\alpha(T_m) - \alpha(T_{m+1})\big)$$
$$\geq \alpha(T_m) - \alpha(T_{m+1}) + \beta_{k+m+1}(T_m)$$

since $\alpha(T_m) - \alpha(T_{m+1}) \geq \beta_{k+m+1}(T_m)$ by definition of the functional β. Thus, by the induction hypothesis,

$$\beta_{k+m+2}(T_{m+1}) \geq \alpha(T) - \alpha(T_{m+1})$$

and (12.13) indeed holds.

Now, set $\ell = k + n$. Since, in T_n, we are in the case (ii) of Lemma 12.3, we can find balls $(B'_i)_{1 \leq i \leq N}$ (of T_n) of radius $6^{-\ell-2}$ with centers at mutual distance $\geq 3 \cdot 6^{-\ell-2}$ such that, for every i,

$$(12.14) \qquad \alpha(B'_i \cap T_n) \geq \alpha(T_n) - 6^{-k+1}h\left(\frac{1}{1+N}\right).$$

Since $\alpha(B'_i \cap T_n) \leq \alpha(T_n) - \beta_{\ell+1}(T)$, combining (12.13) and (12.14) yields that

$$\alpha(B'_i \cap T_n) \geq \alpha(T) - 2 \cdot 6^{-k+1}h\left(\frac{1}{1+N}\right)$$

and Proposition 12.4 holds with $B_i = B'_i \cap T_n$.

Let U be an ultrametric space. For x in U and $k \in \mathbf{Z}$, let $N_k(x)$ be the number of disjoint balls of radius 6^{-k-1} which are contained in $B(x, 6^{-k})$. Define

$$\xi_x(U) = \sum_{k \in \mathbf{Z}} 6^{-k}h\big(1/N_k(x)\big),$$

$$\xi(U) = \inf_{x \in U} \xi_x(U).$$

We note that if $D(U) \leq 6^{-k_0}$ and $B(x, 6^{-k_1}) = \{x\}$ for every x, we have

$$\xi_x(U) = \sum_{k_0 \leq k < k_1} 6^{-k}h\big(1/N_k(x)\big).$$

We can state and prove the main conclusion of the preceding construction.

Theorem 12.5. *There is a numerical constant K with the following property: for each function h satisfying (12.5) and each finite metric space (T, d), there exist an ultrametric space (U, δ) and a map $\phi : U \to T$ such that the following conditions hold:*

$$\alpha(T) \leq K\xi(U);$$

$$\text{for all } u, v \text{ in } U, \ \delta(u, v) \leq d\big(\phi(u), \phi(v)\big) \leq K\delta(u, v).$$

Proof. Let k_0 be the largest integer (in $\mathbf{Z}$) with $6^{-k_0} \geq D(T)$. Consider two points u, v of T with $d(u, v) = D(T)$. The space $U = (\{u, v\}, d)$ is ultrametric and the canonical injection ϕ from U in T satisfies $6^{-k_0-1} \leq d\big(\phi(u), \phi(v)\big) \leq 6^{-k_0}$. The balls $B(u, 6^{-k_0-1})$, $B(v, 6^{-k_0-1})$ are disjoint; so we have

$$\xi(U) \geq 6^{-k_0-1}h(1/2) \geq 6^{-1}h(1/2)D(T).$$

We intend to prove the theorem with $K = 4 \cdot 6^3$. By the preceding, the result holds unless $\alpha(T) \geq 4 \cdot 6^2 h(1/2)D(T)$. It thus remains to prove the theorem in that case only. By induction over $k \geq k_0$, we construct a family $\mathcal{B}$ of subsets A of T in the following way. The construction starts with $A = T$ and each step is performed by an application of Proposition 12.4 to each element of $\mathcal{B}$ obtained at the step before. That is, if A is an element of $\mathcal{B}$

and has a diameter less than or equal to 6^{-k}, there exist integers $k(A) \geq k$ and $N(A) \geq 1$, subsets $(B_i(A))_{1 \leq i \leq N(A)}$ of A of diameter $\leq 6^{-k(A)-1}$ and such that $d(B_i(A), B_j(A)) \geq 6^{-k(A)-2}$ and the diameter of $\bigcup_{i \leq N(A)} B_i(A)$ is $\leq 6^{-k(A)+1}$, with the following property: for every i,

$$(12.15) \qquad \alpha\big(B_i(A)\big) \geq \alpha(A) - 2 \cdot 6^{-k(A)+1} h\left(\frac{1}{1+N(A)}\right).$$

The construction stops when each element A of $\mathcal{B}$ is reduced to exactly one point and we denote by U the collection of points of T obtained in this way. For u, v in U, there exists A in $\mathcal{B}$ such that for two different $B_i(A)$, $B_j(A)$, $u \in B_i(A)$ and $v \in B_j(A)$. We set $\delta(u, v) = 6^{-k(A)-2}$. δ is an ultrametric distance on U. Moreover, if ϕ is the canonical injection map from U into T, we have by construction that $\delta(u, v) \leq d(\phi(u), \phi(v)) \leq 6^3 \delta(u, v)$. Fix x in U. Denote by $(A_\ell)_{\ell \geq 1}$ the decreasing sequence of the elements of $\mathcal{B}$ that contain x; $A_1 = T$. By (12.15), for every $\ell \geq 1$,

$$\alpha(A_{\ell+1}) \geq \alpha(A_\ell) - 2 \cdot 6^{-k(A_\ell)+1} h\left(\frac{1}{1+N(A_\ell)}\right).$$

Since $\alpha(\{x\}) = 0$, summation of these inequalities yields

$$(12.16) \qquad \alpha(T) \leq 12 \sum_{\ell \geq 1} 6^{-k(A_\ell)} h\left(\frac{1}{1+N(A_\ell)}\right).$$

For $k \geq k_0 + 3$, let $B(x, 6^{-k})$ be the δ-ball of U with center x and radius 6^{-k}. By definition of δ, if $k = k(A_\ell) + 2$ for some ℓ, then $N_k(x) = N(A_\ell)$, while if there is no such ℓ, $N_k(x) = 1$. Hence, from (12.16), and (12.5) since $1 + N(A_\ell) \leq 2N(A_\ell)$, we get

$$\alpha(T) \leq 12\left(h(1/2) \sum_{k > k_0} 6^{-k} + 6^2 \xi_x(U)\right)$$
$$\leq 12\big(6h(1/2)D(T) + 6^2 \xi_x(U)\big).$$

Since we are in the case $D(T) \leq \alpha(T)/4 \cdot 6^2 h(1/2)$, it follows that $\alpha(T) \leq 4 \cdot 6^3 \xi_x(U)$. Since x is arbitrary in U, we have $\alpha(T) \leq 4 \cdot 6^3 \xi(U)$ which is the announced claim.

With the abstract Theorem 12.5, we can now prove the existence of majorizing measures for bounded Gaussian processes (at least, to start with, indexed by a finite set). *In the rest of this section, $h(x) = (\log(1/x))^{1/2}$.*

Theorem 12.6. *Let $X = (X_t)_{t \in T}$ be a Gaussian process indexed by a finite set T, and provide T with the canonical distance $d_X(s, t) = \|X_s - X_t\|_2$. Then*

$$\alpha(T, d_X) \leq K \mathbb{E} \sup_{t \in T} X_t$$

where $K > 0$ is a numerical constant.

The fact here that d_X does not possibly separate all points of T is no problem: simply identify s and t such that $d_X(s,t) = 0$ and the new index set $\widetilde{T}$ obtained in this way is such that $\alpha(\widetilde{T}, d_X) = \alpha(T, d_X)$ and $\mathbb{E}\sup_{t\in\widetilde{T}} X_t = \mathbb{E}\sup_{t\in T} X_t$.

Let U, ϕ be as given by the application of Theorem 12.5 to the space (T, d_X) (or $(\widetilde{T}, d_X)$). It is thus enough to show that $\xi(U) \leq K\mathbb{E}\sup_{u\in U} X_{\phi(u)}$. We note that for u, v in U, $d_X(\phi(u), \phi(v)) \geq \delta(u, v)$ so that the theorem is a consequence of the following result that we single out for future reference. It is at this point, *actually the unique place in this study*, that the Gaussian structure via the comparison theorems based on Slepian's lemma plays its key rôle.(Recently [Ta18], the use of Slepian's lemma in this step has been replaced by the use of Sudakov's minoration and concentration properties.)

Proposition 12.7. *Let (U, δ) be a finite ultrametric space. Then, for each Gaussian process $X = (X_u)_{u\in U}$ such that $d_X(u, v) \geq \delta(u, v)$ whenever $u, v \in U$, we have $\xi(U) \leq K\mathbb{E}\sup_{u\in U} X_u$ where K is a numerical constant.*

Proof. Let k_0 be the largest integer such that $6^{-k_0} \geq D(U)$. For $k > k_0$, let $\mathcal{B}_k$ be the collection of the balls of radius 6^{-k}. Let $\mathcal{B} = \bigcup_{k>k_0} \mathcal{B}_k$. Consider an independent family $(g_B)_{B\in\mathcal{B}}$ of standard normal variables. For u in U, $k > k_0$, we simply write $g_{u,k} = g_{B(u,6^{-k})}$. We let $Z_u = \sum_{k>k_0} 6^{-k} g_{u,k}$. Let u, v in U and let ℓ be the largest such that $\delta(u, v) \leq 6^{-\ell}$. Then $B(u, 6^{-k}) = B(v, 6^{-k})$ for $k \leq \ell$, so $Z_u - Z_v = \sum_{k>\ell} 6^{-k}(g_{u,k} - g_{v,k})$. It follows that

$$\|Z_u - Z_v\|_2 \leq \sqrt{2}\sum_{k>\ell} 6^{-k} \leq 2\cdot 6^{-\ell-1} \leq 2\delta(u, v) \leq 2d_X(u, v).$$

Corollary 3.14 shows that it is enough to establish that $\xi(U) \leq A\mathbb{E}\sup_{u\in U} Z_u$ for some constant A. According to (3.14), we take A such that $(\log N)^{1/2} \leq A\mathbb{E}\max_{i\leq N} g_i$ for all N's where (g_i) is an orthogaussian sequence. By induction over n, we establish the following statement:

(H_n) If U has a diameter $\leq 6^{-k_0}$ and if, for each x in U, $B(x, 6^{-k}) = \{x\}$ with $k - k_0 \leq n$, then $\xi(U) \leq A\mathbb{E}\sup_{u\in U} Z_u$.

For $n = 0$, U contains one point only so that $\xi(U) = 0$ and (H_0) holds. Let us assume that (H_n) holds and let us prove (H_{n+1}). We enumerate $\mathcal{B}_{k_0+1}$ as $\{B_1, \ldots, B_q\}$. For $i \leq q$, let $\Omega_i = \{\forall p \leq q, p \neq i, g_{B_p} < g_{B_i}\}$. For u in U, define

$$Z'_u = \sum_{k>k_0+1} 6^{-k} g_{u,k} = Z_u - 6^{-k_0-1} g_{u,k_0+1}.$$

For $i \leq q$, consider a measurable map τ_i from Ω to B_i that satisfies $Z'_{\tau_i} = \sup_{u\in B_i} Z'_u$. Now define a measurable map τ from Ω to U by $\tau(\omega) = \tau_i(\omega)$ for ω in Ω_i. We have

$$\mathbb{E}\sup_{u\in U} Z_u \geq \mathbb{E}Z_\tau = \sum_{i\leq q} \mathbb{E}(I_{\Omega_i} Z_{\tau_i})$$

$$\geq \sum_{i\leq q} \mathbb{E}\big(I_{\Omega_i}(6^{-k_0-1}g_{B_i} + Z'_{\tau_i})\big)$$

$$= 6^{-k_0-1}\sum_{i\leq q} \mathbb{E}(I_{\Omega_i}g_{B_i}) + \sum_{i\leq q} \mathbb{E}(I_{\Omega_i}Z'_{\tau_i}).$$

Now

$$\sum_{i\leq q}\mathbb{E}(I_{\Omega_i}g_{B_i}) = \mathbb{E}\max_{i\leq q} g_{B_i} \geq A^{-1}(\log q)^{1/2}.$$

Further, the independence of the variables $(g_B)_{B\in\mathcal{B}}$ shows that I_{Ω_i} and Z'_{τ_i} are independent, and thus

$$\mathbb{E}(I_{\Omega_i}Z'_{\tau_i}) = \mathbb{P}(\Omega_i)\mathbb{E}Z'_{\tau_i} = \frac{1}{q}\mathbb{E}Z'_{\tau_i}.$$

By the induction hypothesis, for every i,

$$A\mathbb{E}Z'_{\tau_i} = A\mathbb{E}\sup_{u\in B_i} Z'_u \geq \xi(B_i).$$

Since the definition of ξ makes it clear that for each i

$$\xi(B_i) + 6^{-k_0-1}(\log q)^{1/2} \geq \xi(U),$$

the proof is complete.

Theorem 12.6 proves the existence of a majorizing measure for a Gaussian process when the index set is finite since $\gamma(T) \leq \alpha(T)$ (Lemma 12.1). Now, we deduce from this finite case the existence of a majorizing measure for almost surely bounded general Gaussian processes. The use of the functional α actually yields a seemingly *stronger*, but equivalent (see Remark 12.11), statement.

Anticipating on the next section, let us mention that the following two theorems, as well as their consequences, on necessary conditions for the boundedness and continuity of the sample paths of Gaussian processes actually hold similarly for other processes *once* a statement for T finite analogous to Theorem 12.6 can be established. This is the procedure which will indeed be followed for stable processes in the next section. Therefore, we write the *proofs* below with a general function h.

Metric spaces are no longer always finite.

Theorem 12.8. *Consider a bounded Gaussian process* $X = (X_t)_{t\in T}$. *Then, there exists a probability measure* m *on* (T, d_X) *such that*

$$\sup_{t\in T}\int_0^\infty \left(\log\frac{1}{\sup\{m(\{s\}); d_X(s,t) < \varepsilon\}}\right)^{1/2} d\varepsilon \leq K\mathbb{E}\sup_{t\in T} X_t$$

where $K > 0$ *is a numerical constant.*

Proof. Theorem 12.6 shows that for each finite subset F of T,

$$\alpha(F) \leq K\mathbb{E} \sup_{t \in F} X_t \leq K\mathbb{E} \sup_{t \in T} X_t .$$

It is hence enough to show that if $\alpha = \sup\{\alpha(F) ; \ F \subset T, \ F \text{ finite}\}$, there is a probability measure m on T such that the left hand side of the inequality of the theorem is less than $K\alpha$. Denote by k_0 the largest integer with $2^{-k_0} \geq D(T)$. Since X is almost surely bounded, $N(T, d_X; \varepsilon) < \infty$ for every $\varepsilon > 0$ (Theorem 3.18). For $\ell \geq k_0$, let T_ℓ be a finite subset of T such that each point of T is within a distance $\leq 2^{-\ell}$ of a point of T_ℓ. Consider a map a_ℓ from T to T_ℓ such that $d_X(t, a_\ell(t)) \leq 2^{-\ell}$. For each k, we know that $\alpha(T_\ell) \leq \alpha$. So there exist an ultrametric space (U_ℓ, δ_ℓ), a contraction φ_ℓ from U_ℓ on T_ℓ, and a probability measure μ_ℓ on U_ℓ such that $\gamma_{\mu_\ell}(U_\ell) \leq \alpha$. This implies that, for every u in U_ℓ and every ℓ,

$$(12.17) \qquad \sum_{k>k_0} 2^{-k} h\big(\mu_\ell(B(u, 2^{-k}))\big) \leq 2\alpha .$$

To each ball B in an ultrametric space U, we associate a point $v(B)$ of B. Let $\mathcal{B}_k$ be the family of the balls of radius 2^{-k} of U_ℓ. Denote by m_k^ℓ the probability measure on T that, for each B in $\mathcal{B}_k$, assigns mass $\mu_\ell(B)$ to the point $a_k(\varphi_\ell(v(B)))$. We note that m_k^ℓ is supported by T_k. Fix t in T. Choose t' in T_ℓ with $d_X(t, t') \leq 2^{-\ell}$. Take u in U_ℓ such that $\varphi_\ell(u) = t'$. For each k, we have $\delta_\ell(u, v(B(u, 2^{-k}))) \leq 2^{-k}$ so that $d_X(t', \varphi_\ell(v(B(u, 2^{-k})))) \leq 2^{-k}$ and

$$d_X\left(t, a_k\left(\varphi_\ell(v(B(u, 2^{-k})))\right)\right) \leq 2^{-k+1} + 2^{-\ell} .$$

We set $t_k^\ell = a_k(\varphi_\ell(v(B(u, 2^{-k}))))$ so $d_X(t, t_k^\ell) \leq 2^{-k+1} + 2^{-\ell}$ and $m_k^\ell(\{t_k^\ell\}) \geq \mu_\ell(B(u, 2^{-k}))$. It follows from (12.17) that we have

$$\sum_{k>k_0} 2^{-k} h(m_k^\ell(\{t_k^\ell\})) \leq 2\alpha .$$

Let $\mathcal{U}$ be a ultrafilter in $\mathbb{N}$. Since t_k^ℓ belongs to the finite set T_k, the limit $t_k = \lim_{\ell \to \mathcal{U}} t_k^\ell$ exists, and $d_X(t, t_k) \leq 2^{-k+1}$. Since m_k^ℓ is supported by the finite set T_k, the limit $m_k = \lim_{\ell \to \mathcal{U}} m_k^\ell$ exists and thus, for each t in U,

$$\sum_{k>k_0} 2^{-k} h(m_k(\{t_k\})) \leq 2\alpha .$$

Let $m = \sum_{k>k_0} 2^{k_0-k} m_k$, so m is a probability on T. We note that, by (12.5),

$$h(m(\{t_k\})) \leq h(2^{k_0-k} m_k(\{t_k\}))$$
$$\leq h(2^{k_0-k}) + h(m_k(\{t_k\})) .$$

It follows that

$$\sum_{k>k_0} 2^{-k} h\big(m(\{t_k\})\big) \le \sum_{k>k_0} 2^{-k} h(2^{k_0-k}) + 2\alpha$$

$$\le KD(T) + 2\alpha$$

where we have used (12.5) and $2^{-k_0} \ge D(T)$. Recall from Lemma 12.1 that $D(T) \le K\alpha$. The conclusion then follows since, clearly, for $\varepsilon \ge 2^{-k+1}$, $\sup\{m(\{s\}); d_X(s,t) < \varepsilon\} \ge m(\{t_k\})$.

Now, we present the necessary majorizing measure condition for the almost sure continuity of Gaussian processes.

Theorem 12.9. *Consider a Gaussian process* $X = (X_t)_{t\in T}$ *that is almost surely bounded and continuous (or that admits a version which is almost surely bounded and continuous) on* (T, d_X). *Then, there exists a probability measure* m *on* (T, d_X) *such that*

$$\limsup_{\eta \to 0}{}_{t\in T} \int_0^\eta \left(\log \frac{1}{\sup\{m(\{s\}); \ d_X(s,t) < \varepsilon\}} \right)^{1/2} d\varepsilon = 0.$$

Proof. For $n \ge 1$, let $a_n = 2^{-n}D$ where $D = D(T)$. Consider a family $B_{n,1}, \ldots, B_{n,p(n)}$ of d_X-balls of radius a_n that covers T where $p(n) = N(T, d_X; a_n)$. So, for $i \le p(n)$, if we denote by $t(n,i)$ the center of $B_{n,i}$, we have

$$\mathbb{E} \sup_{t\in B_{n,i}} X_t = \mathbb{E} \sup_{t\in B_{n,i}} \big(X_t - X_{t(n,i)}\big) \le \mathbb{E} \sup_{t\in B_{n,i}} |X_t - X_{t(n,i)}| \le \beta(a_n)$$

where we have set $\beta(\eta) = \sup_{t\in T} \mathbb{E} \sup_{d_X(s,t)<\eta} |X_s - X_t|$. Denote by $d_{n,i}$ the diameter of $B_{n,i}$ (that can be smaller than $2a_n$). Theorem 12.8 shows that there is a probability measure $m_{n,i}$ on $B_{n,i}$ such that for each t in $B_{n,i}$ we have

(12.18) $$\int_0^{d_{n,i}} h\big(\sup\{m_{n,i}(\{s\}); d_X(s,t) < \varepsilon\}\big) d\varepsilon \le K\beta(a_n).$$

Let $m'_{n,i} = \frac{1}{2}(\delta_{t(n,i)} + m_{n,i})$ and let

$$m' = \sum_{n\ge 1} \sum_{i\le p(n)} n^{-2} p(n)^{-1} m'_{n,i},$$

so $|m'| \le 1$. There is a probability measure m on T such that $m \ge m'$. Fix t in T and $0 < \eta \le D$. Let n be the smallest integer with $a_n \le \eta$, so $\eta \le 2a_n$. We note that if $t \in B_{n,i}$,

$$\sup\{m(\{s\}); d_X(s,t) < \varepsilon\} \ge \frac{1}{2n^2 p(n)} \sup\{m'_{n,i}(\{s\}); d_X(s,t) < \varepsilon\}.$$

Also, for $\varepsilon \ge \min(d_{n,i}, a_n)$,

$$\sup\{m(\{s\}); d_X(s,t) < \varepsilon\} \geq \frac{1}{2n^2 p(n)}.$$

From (12.18) therefore, if $t \in B_{n,i}$,

$$\int_0^\eta h\Big(\sup\{m(\{s\}); d_X(s,t) < \varepsilon\}\Big) d\varepsilon$$

$$\leq \int_0^{d_{n,i}} h\Big(\sup\{m_{n,i}(\{s\}); d_X(s,t) < \varepsilon\}\Big) d\varepsilon + \eta h\Big(\frac{1}{2n^2 p(n)}\Big)$$

$$\leq K\beta(a_n) + \eta h\Big(\frac{1}{p(n)}\Big) h\Big(\frac{1}{2n^2}\Big)$$

$$\leq K\beta(\eta) + \eta h\Big(N\Big(T, d_X; \frac{\eta}{2}\Big)^{-1}\Big) + \eta h\Big(\frac{1}{2}(\log 2)^2 \Big(\log \frac{D}{\eta}\Big)^{-2}\Big)$$

since $a_n \leq \eta \leq 2a_n$. Now, if the Gaussian process X is continuous on (T, d_X), $\lim_{\eta \to 0} \beta(\eta) = 0$ by the integrability properties of Gaussian random vectors. Furthermore, if $h(x) = (\log(1/x))^{1/2}$, we see from Corollary 3.19 that

(12.19) $$\lim_{\eta \to 0} \eta h\big(N(T, d_X; \eta)^{-1}\big) = 0.$$

These observations conclude the proof of Theorem 12.9.

The two preceding theorems describe the necessary majorizing measure conditions for a Gaussian process to have bounded or continuous sample paths that, together with Theorem 11.18, thus provide a complete description of the regularity properties of Gaussian processes. In the last part of this section, we describe a consequence of this result in terms of a convex hull representation of Gaussian processes.

Theorem 12.10. *Let $X = (X_t)_{t \in T}$ be a bounded Gaussian process. Let $\sigma = \sup_{t \in T}(\mathbb{E}|X_t|^2)^{1/2}$ and $M = \mathbb{E}\sup_{t \in T}|X_t|$. Then, there exists a Gaussian sequence $(Y_n)_{n \geq 1}$ with $\|Y_n\|_2 \leq KM(\log n + \sigma^{-2}M^2)^{-1/2}$ such that, for each t in T, we can write*

$$X_t = \sum_{n \geq 1} \alpha_n(t) Y_n$$

where $\alpha_n(t) \geq 0$, $\sum_{n \geq 1} \alpha_n(t) \leq 1$ and the series converges almost surely and in L_2. Moreover, each Y_n is a linear combination of at most two variables of the type X_t. If (and only if) X is continuous, (Y_n) can be chosen such that $\lim_{n \to \infty} (\log n)^{1/2} \|Y_n\|_2 = 0$.

Before the proof, let us mention that if (Y_n) is as in the theorem, by the Borel-Cantelli lemma and the Gaussian tail, it defines an almost surely bounded sequence. We even have that, for some numerical constant K_1,

$$\mathbb{P}\Big\{\sup_{n>1}|Y_n| > K_1(M + u\sigma)\Big\} \leq K_1 \exp(-u^2)$$

for all $u > 0$. Indeed, if K is such that $\|Y_n\|_2 \leq KM(\log n + \sigma^{-2}M^2)^{-1/2}$ in Theorem 12.10, we have

$$\mathbb{P}\Big\{\sup_{n\geq 1}|Y_n| > 2K(M + u\sigma)\Big\} \leq 2\sum_{n\geq 1}\exp\big(-4K^2(M + u\sigma)^2/2\|Y_n\|_2^2\big)$$

$$\leq 2\exp(-u^2)\sum_{n\geq 1}n^{-2}.$$

Since $\sup_{t\in T}|X_t| \leq \sup_{n\geq 1}|Y_n|$, we note, in particular, that the majorizing measure theorem contains, at least qualitatively, the tail behavior of isoperimetric nature of norms of Gaussian random vectors (cf. Section 3.1). We shall partially come back to this at the end of the section. The representation of Theorem 12.10 also implies, with a little more effort, that X is continuous when $\lim_{n\to\infty}(\log n)^{1/2}\|Y_n\|_2 = 0$.

Proof. We only show the assertion concerning boundedness, the continuous case being obtained with some easy modifications on the basis of Theorem 12.9. Let k_0 be the largest integer with $2^{-k_0} \geq D(T)$. It follows from Theorem 12.8 that there is a probability measure m on T such that for each t in T, we have

$$(12.20) \qquad \sum_{k\geq k_0}2^{-k}h\big(\sup\{m(\{s\}); d_X(s,t) < 2^{-k}\}\big) \leq KM.$$

For t in T, $k \geq k_0$, we pick t_k such that $d_X(t, t_k) < 2^{-k}$ and

$$m(\{t_k\}) = \sup\{m(\{s\}); d_X(s,t) < 2^{-k}\}.$$

We can assume that t_{k_0} does not depend on t. From (12.20), we see in particular that $2^{-k}h(m(\{t_k\})) \leq KM$. Thus, for each t in T, t_k belongs to the finite set

$$A_k = \{s \in T; \ m(\{s\}) \geq h^{-1}(2^k KM)\}.$$

Let $b_k = 2^{-k+k_0-1}h^{-1}(M/D)$. Using the fact that $D \leq 2\sigma \leq (2\pi)^{1/2}M$, it follows from (12.5) and (12.20) that

$$(12.21) \qquad \sum_{k\geq k_0}2^{-k}h\big(b_k m(\{t_k\})\big) \leq K_1 M$$

for some constant K_1. For each t in T, $k \geq k_0$, we define

$$a_{t,k} = 2^{-k}\big(h\big(b_k m(\{t_k\})\big) + h\big(b_{k+1}m(\{t_{k+1}\})\big)\big).$$

From (12.21), $\sum_{k\geq k_0}a_{t,k} \leq 3K_1M$. Define then, for $k \geq k_0$,

$$z_{t,k} = 6K_1 M a_{t,k}^{-1}(X_{t_{k+1}} - X_{t_k}).$$

Let Z_k be the set of all $z_{t,k}$ for t in T. Since t_k and t_{k+1} belong to the finite set A_{k+1}, Z_k is finite. Let Z be the union of the sets Z_k for $k > k_0$. Fix $\varepsilon > 0$. We note that

$$\|X_{t_{k+1}} - X_{t_k}\|_2 \le d_X(t, t_{k+1}) + d_X(t, t_k) \le 3 \cdot 2^{-k-1}$$

so, if $\|z_{t,k}\|_2 \ge \varepsilon$, we have $a_{t,k} \le 9 \cdot 2^{-k} K_1 M/\varepsilon$ and thus

$$h\big(b_k m(\{t_k\})\big) + h\big(b_{k+1} m(\{t_{k+1}\})\big) \le 9K_1 M/\varepsilon.$$

This implies that

$$m(\{t_k\}), \, m(\{t_{k+1}\}) \ge 2^{k-k_0+1} \left(h^{-1}\left(\frac{M}{D}\right) \right)^{-1} h^{-1}\left(\frac{9K_1 M}{\varepsilon}\right).$$

Since m is a probability, this shows that there are at most

$$2^{-k+k_0-1} h^{-1}\left(\frac{M}{D}\right) \left(h^{-1}\left(\frac{9K_1 M}{\varepsilon}\right) \right)^{-1}$$

possible choices for either t_k or t_{k+1} when $\|z_{t,k}\|_2 \ge \varepsilon$. Therefore

$$\text{Card}\{z \in Z_k; \|z\|_2 \ge \varepsilon\} \le 2^{-2k+2k_0-2} \left[h^{-1}\left(\frac{M}{D}\right) \left(h^{-1}\left(\frac{9K_1 M}{\varepsilon}\right) \right)^{-1} \right]^2$$

and

$$\text{Card}\{z \in Z; \|z\|_2 \ge \varepsilon\} \le \left[h^{-1}\left(\frac{M}{D}\right) \left(h^{-1}\left(\frac{9K_1 M}{\varepsilon}\right) \right)^{-1} \right]^2.$$

Thus, we can index Z as a sequence $(Y_n)_{n\ge 1}$ such that $\|Y_n\|_2$ does not increase. For each n,

$$n \le \text{Card}\{z \in Z; \|z\|_2 \ge \|Y_n\|_2\} \le \left[h^{-1}\left(\frac{M}{D}\right) \left(h^{-1}\left(\frac{9K_1 M}{\|Y_n\|_2}\right) \right)^{-1} \right]^2$$

so that

(12.22) $$\|Y_n\|_2 \le 9K_1 M \left(\frac{M}{D} + h\left(\frac{1}{\sqrt{n}}\right) \right)^{-1}.$$

Since $D \le 2\sigma$, for $h(x) = (\log(1/x))^{1/2}$, (12.22) implies, for all $n \ge 1$, $\|Y_n\|_2 \le 18K_1 M(\log n + \sigma^{-2} M^2)^{-1/2}$. In particular, by the Borel-Cantelli lemma, the Gaussian sequence (Y_n) is bounded almost surely. For each t in T, we have

$$X_t - X_{t_{k_0}} = \sum_{k \ge k_0} (X_{t_{k+1}} - X_{t_k}) = \sum_{k \ge k_0} (6K_1 M)^{-1} a_{t,k} z_{t,k}.$$

Since $\sum_{k \ge k_0} a_{t,k} \le 3K_1 M$, this implies that $X_t - X_{t_{k_0}} = \sum_{n \ge 1} \alpha_n(t) Y_n$ where $\sum_{n \ge 1} \alpha_n(t) \le 1/2$ and where the series converges almost surely since (Y_n) is bounded. Since $\|X_{t_{k_0}}\|_2 \le \sigma$, the proof of Theorem 12.10 is complete.

Remark 12.11. This remark aims to show that several of the techniques developed in the proofs of the preceding theorems go beyond the Gaussian case and apply to rather a general setting. As we noted, the Gaussian structure and Slepian's lemma were actually basic only in the proof of Proposition 12.7. For simplicity, let us consider the family of Young functions ψ_q, $1 \leq q \leq \infty$ (and associated functions $h_q(x) = (\log(1/x))^{1/q}$) although some more general functions may be imagined. If (T, d) is a metric space and m is a probability measure on (T, d), set, for $1 \leq q \leq \infty$,

$$\gamma_m^{(q)}(T) = \gamma_m^{(q)}(T, d) = \sup_{t \in T} \int_0^\infty h_q\big(m\big(B(t, \varepsilon)\big)\big) d\varepsilon$$

and

$$\gamma^{(q)}(T) = \gamma^{(q)}(T, d) = \inf \gamma_m^{(q)}(T, d)$$

where the infimum is taken over all probability measures m on T. (It might be useful to recall (11.18) at this point and the easy comparison between ψ_q^{-1} and h_q). K_q denotes below a constant depending only on q and not necessarily the same at each occurence. Our first observation, which we deduce from the proof of Theorem 12.8, is that there is a probability measure m on (T, d) such that

$$\sup_{t \in T} \int_0^\infty h_q\big(\sup\{m(\{s\}); d(s, t) < \varepsilon\}\big) d\varepsilon \leq K_q \gamma^{(q)}(T, d).$$

If T is finite, Proposition 11.10 indeed indicates that $\alpha(T, d) \leq K_q \gamma^{(q)}(T, d)$ (where α is defined with $h = h_q$). The proof of Theorem 12.8, which is given with a general function h, then implies the result. There is a similar result about continuous majorizing measures. From this observation, let us mention further the following one. Consider a stochastic process $X = (X_t)_{t \in T}$ continuous in L_1 (say) on (T, d), more precisely such that $\|X_s - X_t\|_1 \leq d(s, t)$ for all s, t in T. Assume again that $\gamma^{(q)}(T, d) < \infty$. Then, from the preceding, the proof of Theorem 12.10 shows that there is a sequence (Y_n) of random variables such that, for every n,

$$\|Y_n\|_1 \leq K_q \gamma^{(q)}(T, d) \left(\frac{\gamma^{(q)}(T, d)}{D(T)} + h_q\left(\frac{1}{\sqrt{n}}\right) \right)^{-1}$$

and such that, for every t, $X_t = \sum_n \alpha_n(t) Y_n$ where $\alpha_n(t) \geq 0$, $\sum_n \alpha_n(t) \leq 1$ and the series converges almost surely and in L_1. The main point in the Gaussian case is of course that $\gamma^{(2)}(T, d_X) \leq K \mathbb{E} \sup_{t \in T} X_t$.

It is interesting to interpret Theorem 12.10 (and similar comments may be given in the context of the observations of Remark 12.11) as a result about subsets of Hilbert space associated to bounded Gaussian processes. Let $X = (X_t)_{t \in T}$ be a bounded Gaussian process on $(\Omega, \mathcal{A}, \mathbb{P})$. Let $H = L_2(\Omega, \mathcal{A}, \mathbb{P})$ and identify T with the subset of H consisting of the family $(X_t)_{t \in T}$. Then, Theorem 12.10 proves the existence of a sequence (y^n) in H such that, for some $M > 0$ and every n, $|y^n| \leq M(\log(n + 1))^{-1/2}$, and $T \subset \text{Conv}(y^n)$. Let us rewrite this observation as a perhaps more geometric statement about the finite dimensional Hilbert space. Consider H of dimension N and denote by σ the normalized measure on its unit ball. For a subset T of H, consider

$$V(T) = \int \sup_{y \in T} |\langle x, y \rangle| d\sigma(x).$$

This quantity has been studied in Geometry under the name of mixed volume and plays an important rôle in the local theory of Banach spaces. Fix an orthonormal basis $(e_i)_{i \leq N}$ of H. For t in H, let $X_t = \sum_{i=1}^{N} g_i \langle t, e_i \rangle$ where (g_i) is a standard Gaussian sequence. Since the distribution of $(g_i)_{i \leq N}$ is rotation invariant,

$$\ell(T) = \mathbb{E} \sup_{t \in T} |X_t| = \mathbb{E} \left(\sum_{i=1}^{N} g_i^2 \right)^{1/2} \int \sup_{y \in T} |\langle x, y \rangle| d\sigma(x)$$

and thus $K^{-1} N^{1/2} V(T) \leq \ell(T) \leq K N^{1/2} V(T)$ for some numerical constant K. Now, define

$$A(T) = \inf \left\{ a > 0; \ \exists (y^n) \text{ in } H, \ |y^n| \leq a \left(\log(n+1) \right)^{-1/2}, \ T \subset \text{Conv}(y^n) \right\}.$$

Then Theorem 12.10 can be reformulated as

(12.23) $\qquad K^{-1} N^{-1/2} A(T) \leq V(T) \leq K N^{-1/2} A(T).$

As a closing remark, note that if we go back to the tail estimate (11.14) for the Young function $\psi = \psi_2$ associated to Gaussian processes, we see that the process techniques and the existence of majorizing measures (12.4) yield some deviation inequalities with the two parameters similar to those obtained from the isoperimetric considerations in Chapter 3. Such inequalities can also be deduced, as we have seen, from Theorem 12.10 and some elementary considerations on Gaussian sequences.

12.2 Necessary Conditions for the Boundedness and Continuity of Stable Processes

Let $X = (X_t)_{t \in T}$ be a p-stable random process, $0 < p < 2$. Recall that by this we mean that every finite linear combination $\sum_i \alpha_i X_{t_i}$, $\alpha_i \in \mathbb{R}$, $t_i \in T$, is a p-stable real valued random variable. As in the Gaussian case, we may consider the associated pseudo-metric d_X on T given by

$$d_X(s, t) = \sigma(X_s - X_t), \quad s, t \in T,$$

where $\sigma(X_s - X_t)$ is the parameter of the stable random variable $X_s - X_t$ (cf. Chapter 5).

Contrary to the Gaussian case, the pseudo-metric d_X does not entirely determine the distribution and therefore the regularity properties of a p-stable process X when $0 < p < 2$. The distribution of X is however determined by its spectral measure (Theorem 5.2) and, as we already noted in Chapter 5 with the representation, the existence and finiteness of a spectral measure for a p-stable process with $0 < p < 1$ already ensures its almost surely boundedness and continuity. This is no longer the case when $1 \le p < 2$. While a characterization in terms of d_X is therefore hopeless, nevertheless a best possible necessary majorizing measure condition for the almost sure boundedness and continuity of p-stable processes with $1 \le p < 2$ exists. This result extends the Gaussian results and forms the matter of this section.

As we mentioned it, the difference with the Gaussian setting is that this necessary condition is far from being sufficient. Sufficient conditions for the sample boundedness or continuity of p-stable processes may be obtained from Theorem 11.2 at some weak level; for example, if $1 < p < 2$, since

$$\|X_s - X_t\|_{p,\infty} \le C_p d_X(s,t), \quad s,t \in T,$$

the finiteness of the entropy integral

$$\int_0^D \left(N(T, d_X; \varepsilon) \right)^{1/p} d\varepsilon$$

implies that X has a version with almost all paths bounded and continuous on (T, d_X). The difference however between the necessary condition described below in Theorem 12.12 and this sufficient condition is huge. Trying to characterize regularity properties of p-stable processes with $1 \le p < 2$ seems to be a difficult question, still under study. The paper [Ta12] for example reflects some of the main problems.

For $1 \le p \le 2$, let q be the conjugate of p. For $p > 1$, $0 < x \le 1$, recall that we set $h_q(x) = (\log(1/x))^{1/q}$; further $h_\infty(x) = \log(1 + \log(1/x))$. These functions satisfy (12.5) and (12.6) and enter the setting of the abstract analysis of the preceding section. The main result of this section is the extension to p-stable processes with $1 \le p < 2$ of Theorems 12.8 and 12.9.

Theorem 12.12. *Let* $1 \le p < 2$. *Let* $X = (X_t)_{t \in T}$ *be a sample bounded p-stable process. Then there is a probability measure m on (T, d_X) such that*

$$\sup_{t \in T} \int_0^\infty h_q \left(\sup\{m(\{s\}); d_X(s,t) < \varepsilon\} \right) d\varepsilon \le K_p \| \sup_{s,t \in T} |X_s - X_t| \|_{p,\infty}$$

where $K_p > 0$ only depends on p. (In particular, we have $\gamma_m^{(q)}(T, d_X) \le K_p \| \sup_{s,t \in T} |X_s - X_t| \|_{p,\infty}$.) If, moreover, X has (or admits a version with) almost surely continuous sample paths on (T, d_X), then there exists a probability measure m on (T, d_X) such that

$$\lim_{\eta \to 0} \sup_{t \in T} \int_0^\eta h_q \left(\sup\{m(\{s\}); d_X(s,t) < \varepsilon\} \right) d\varepsilon = 0.$$

As we noted (Remark 12.11), with such a result, Theorem 12.10 has an extension to the stable case. This observation puts into light the gap between this necessary result and the known sufficient conditions for stable processes to be bounded or continuous simply because if (θ_n) is a standard p-stable sequence with $1 < p < 2$ (say), $(\theta_n/(\log n)^{1/q})$ is not almost surely bounded. For some special class of stationary p-stable processes, strongly stationary or harmonizable processes, we will however see in Chapter 13 that the necessary conditions of Theorem 12.12 are also sufficient.

We only give the proof of Theorem 12.12 for $p > 1$. While the proof when $p = 1$ uses related ideas and arguments, this case however seems to be of a different and much deeper nature (cf. [Ta18]). We restrict ourselves for clarity to the case $p > 1$, refering to [Ta8] for the complete result. In the following therefore, $1 < p < 2$. To put Theorem 12.12 in perspective, let us go back to Chapter 5, Section 5.3. There we saw how for a p-stable process X (Theorem 5.10),

$$\sup_{\varepsilon > 0} \varepsilon \big(\log N(T, d_X; \varepsilon)\big)^{1/q} \leq K_p \| \sup_{t \in T} |X_t| \|_{p,\infty} \, .$$

(By (11.9), Theorem 12.12 improves upon this result.) The main idea of the proof of this Sudakov type minoration was to realize the stable process $X = (X_t)_{t \in T}$ as conditionally Gaussian by the series representation of stable variables. We need not go back to all the details here and refer to the proof of Theorem 5.10 for this point. It was described there how $(X_t)_{t \in T}$ has the same distribution as $(X_t^\omega)_{t \in T}$ defined on $\Omega \times \Omega'$ where, for each ω in Ω, $(X_t^\omega)_{t \in T}$ is Gaussian. If d_ω is the canonical metric associated to the Gaussian process $(X_t^\omega)_{t \in T}$, the main tool was the comparison between the random distances d_ω and d_X given by (cf. (5.20))

(12.24) $$\mathbb{P}\big\{\omega; d_\omega(s, t) \leq \varepsilon d_X(s, t)\big\} \leq \exp(-c_\alpha \varepsilon^{-\alpha})$$

for all s, t in T and $\varepsilon > 0$, where $1/\alpha = 1/p - 1/2$ and c_α only depends on α. From (12.24), the idea was to "transfer" the Gaussian minoration inequalities of Sudakov's type on the random distances d_ω into similar ones for d_X. The idea of the proof of Theorem 12.12 is similar, trying to transfer the stronger minoration results of Section 12.1. Unfortunately, it turns out that it does not seem possible to use mixtures of majorizing measures. Instead, we are going to use the machinery of ultrametric structures of the last section to reduce the proof to the simpler, yet non-trivial, following property.

Theorem 12.13. *Let $1 < p < 2$ and let $X = (X_t)_{t \in T}$ be a p-stable process indexed by a finite set T. Equip T with the canonical distance d_X. Then*

$$\alpha(T, d_X) \leq K_p \| \sup_{s, t \in T} |X_s - X_t| \|_{p,\infty}$$

where $K_p > 0$ only depends on p.

This result is the analog of the Gaussian Theorem 12.6 and the functional α is the one used in Section 12.1 with $h = h_q$. We noted there that, with similar proofs than in the Gaussian case, Theorem 12.12 follows from Theorem 12.13 (for continuity, note that (12.19) holds with $h = h_q$ by (5.21)). Therefore, we concentrate on the proof of Theorem 12.13. Recall that if (U, δ) is ultrametric, we denote by $\mathcal{B}_k$ the family of the balls of U of radius 6^{-k} $(k \in \mathbb{Z})$. Let k_0 be the largest such that Card $\mathcal{B}_{k_0} = 1$. Let us set for simplicity $\Delta = 6^{-k_0}$. For x in U, we denote by $N_k(x) = N(x, k)$ the number of disjoint balls of $\mathcal{B}_{k+1}$ which are contained in $B(x, 6^{-k})$. We set

$$\xi_x^{(q)}(U) = \sum_{k \geq k_0} 6^{-k} \big(\log N(x, k)\big)^{1/q},$$

$$\xi^{(q)}(U) = \inf_{x \in U} \xi_x^{(q)}(U).$$

By Theorem 12.5 applied to (T, d_X) and $h = h_q$, we see that in order to establish Theorem 12.13, it is enough to prove the analog of Proposition 12.7 in this stable case. That is, if (U, δ) is a finite ultrametric space and $X = (X_u)_{u \in U}$ a p-stable process with $d_X \geq \delta$, then

$$(12.25) \qquad \xi^{(q)}(U) \leq K_p \big\| \sup_{u,v \in U} |X_u - X_v| \big\|_{p,\infty}.$$

Since the arguments of the proof of (12.25) have already proved their usefulness in some other contexts, we will try to detail (at least the first steps of) this study in a possible general context. Let us set

$$M = \big\| \sup_{u,v \in U} |X_u - X_v| \big\|_{p,\infty}.$$

X is conditionally Gaussian and we denote by $X^\omega = (X_u^\omega)_{u \in U}$, for every ω in Ω, the conditional Gaussian processes (see the proof of Theorem 5.10). By Fubini's theorem and the definition of M, there exists a set $\Omega_1 \subset \Omega$ with $\mathbb{P}(\Omega_1) \geq 1/2$ such that for ω in Ω_1

$$\mathbb{P}\big\{ \sup_{u,v \in U} |X_u^\omega - X_v^\omega| > 4M \big\} \leq 1/2.$$

By the integrability properties of norms of Gaussian processes (Corollary 3.2) and Theorem 12.6 we know the following: for ω in Ω_1, there exists a probability μ_ω on (U, d_ω) where $d_\omega(u, v) = \|X_u^\omega - X_v^\omega\|_2$, such that

$$(12.26) \qquad \sup_{x \in U} \int_0^\infty h_2\big(\mu_\omega\big(y \in U \,;\, d_\omega(x, y) < \varepsilon\big)\big)d\varepsilon \leq KM.$$

where K is a numerical constant.

The function $h_2(t) = (\log(1/t))^{1/2}$ is convex for $t \leq e^{-1/2}$ but not for $0 < t \leq 1$. For that reason, it will be more convenient to use the function $h_2'(t) = h_2(t/3)$ (the choice of 3 is rather arbitrary) which is convex on $(0, 1]$. To prove (12.25), and therefore Theorem 12.13, we will exhibit a subset Ω_2 of Ω with $\mathbb{P}(\Omega_2) \geq 3/4$ (for example), and constants K_1, K_2, K_3 depending on p only, such that for each ω in Ω_2 and each probability measure μ on U,

$$(12.27) \;\; \xi^{(q)}(U) \leq K_1 \sup_{x \in U} \int_0^{K_2 \Delta} h_2'\big(\mu\big(y \in U \,;\, d_\omega(x, y) < \varepsilon\big)\big)d\varepsilon + K_3\Delta.$$

We observe that $h_2'(t) \leq h_2(t) + (\log 3)^{1/2}$; so combining with (12.26) since $\mathbb{P}(\Omega_1 \cap \Omega_2) > 0$, we get $\xi^{(q)}(U) \leq K_1(KM + K_2(\log 3)^{1/2}\Delta) + K_3\Delta$. It is easily seen that $\Delta \leq K_4 M$ so that (12.25) holds.

Thus, we establish (12.27). The philosophy of the approach is that a large value of $\xi^{(q)}(U)$ means that (U, δ) is big in an appropriate sense. Since (12.24) means that $d_\omega(u, v)$ is, most of the time, not much smaller than $d_X(u, v) \geq \delta(u, v)$, we can expect that U will be big with respect to d_ω for most values of ω. The construction is made rather delicate however by the following feature: while (12.24) tells us precisely how $d_\omega(u, v)$ behaves compared to $d_X(u, v)$, if we take another couple (u', v'), we have no information about the joint behavior of $(d_\omega(u, v), d_\omega(u', v'))$.

As announced, we will try to develop the proof of (12.25) in a possible general framework. Let us therefore assume that we have a family (d_ω) of random distances on (U, δ) such that for some strictly increasing function θ on $\mathbb{R}_+$ with $\lim_{\varepsilon \to 0} \theta(\varepsilon) = 0$, and all u, v in U and $\varepsilon > 0$,

$$(12.28) \qquad \mathbb{P}\{\omega ; \ d_\omega(u, v) \leq \varepsilon\delta(u, v)\} \leq \theta(\varepsilon).$$

By (12.24), the stable case corresponds to $\theta(\varepsilon) = \exp(-c_\alpha \varepsilon^{-\alpha})$ where $1/\alpha = 1/p - 1/2$.

To show (12.27), it will be enough to show that for some probability measure λ on U

$$
\begin{aligned}
(12.29) \quad & K_5^{-1} \int_U d\lambda(x) \xi_x^{(q)}(U) \\
& \leq K_1 \int_U d\lambda(x) \int_0^{K_2\Delta} h_2'\big(\mu\big(y \in U ; \ d_\omega(x, y) < \varepsilon\big)\big) d\varepsilon + K_3\Delta .
\end{aligned}
$$

We choose as a convenient probability measure λ the following: λ is homogeneous in the sense that the mass of any ball of radius 6^{-k} is divided evenly among all the balls of radius 6^{-k-1} that it contains. Equivalently, for any x in U, $\lambda(\{x\}) = \big(\prod_{k \geq k_0} N(x, k)\big)^{-1}$. Let now $k \geq k_0$ be fixed. Let $B_1 \neq B_2$ in $\mathcal{B}_k$. Further, let $b, c > 0$ and define

$$
\begin{aligned}
& A(B_1, B_2, b, c) \\
& = \big\{\omega \in \Omega ; \ \lambda \otimes \lambda\big((x, y) \in B_1 \times B_2 ; \ d_\omega(x, y) \leq b 6^{-k}\big) \geq c\lambda(B_1)\lambda(B_2)\big\} .
\end{aligned}
$$

Under (12.28), one immediately checks by Fubini's theorem and Chebyshev's inequality that

$$(12.30) \qquad \mathbb{P}\big(A(B_1, B_2, b, c)\big) \leq \frac{\theta(b)}{c} .$$

For $B \in \mathcal{B}_k$ and $i \leq k$, there is a unique $B' \in \mathcal{B}_i$ that contains B. We denote by $N(B, i)$ the number of elements of $\mathcal{B}_{i+1}$ contained in B', so $N(B, i) = N(x, i)$ whenever x belongs to B. In particular, $N(B, k)$ is the number of elements of $\mathcal{B}_{k+1}$ that are contained in B. Also, if $i \leq k' \leq k$ and $B \in \mathcal{B}_k$, $B' \in \mathcal{B}_{k'}$, $B \subset B'$, we have $N(B, i) = N(B', i)$. We denote by $\mathcal{B}_k'$ the subset of $\mathcal{B}_k$ which consists of the balls B in $\mathcal{B}_k$ for which

$$N(B,k) > 2 \prod_{i<k} N(B,i).$$

Let B in $\mathcal{B}'_k$, so that in particular $N(B,k) > 1$. Therefore, there exist two balls B_1, B_2 in $\mathcal{B}_{k+1}$, $B_1, B_2 \subset B$, $B_1 \neq B_2$. We consider the event

$$C(B, B_1, B_2) = A\left(B_1, B_2, b, \left(2N(B,k)\right)^{-2}\right)$$

where $b = b(B, k)$ is chosen as

(12.31) $$b = \theta^{-1}\left(\left(2N(B,k)\right)^{-6}\right).$$

It follows from (12.30) that $\mathbb{P}(C(B, B_1, B_2)) \leq (2N(B,k))^{-4}$.

For D in $\mathcal{B}_k$, we consider the event

$$\overline{A}(D) = \bigcup C(B, B_1, B_2)$$

where the union is taken over all choices of $j \geq k$, B in $\mathcal{B}'_j$, $B \subset D$ and B_1, B_2 in $\mathcal{B}_{j+1}$, $B_1 \neq B_2$, $B_1, B_2 \subset B$. (If no such choice is possible, we set $\overline{A}(D) = \emptyset$.)

Lemma 12.14. *Under the previous notation,*

$$\mathbb{P}\left(\overline{A}(D)\right) \leq \left(2 \prod_{i<k} N(D,i)\right)^{-2}.$$

Proof. The proof goes by decreasing induction over k. If k is large enough so that D has only one point, then $\overline{A}(D) = \emptyset$ and the lemma holds. Now, assume that we have proved the lemma for $k+1$ and let D in $\mathcal{B}_k$. Assume first that $D \in \mathcal{B}'_k$, so $N(D,k) \geq 2$. Let

$$\overline{A}_1 = \bigcup\{\overline{A}(D'); \ D' \in \mathcal{B}_{k+1}, \ D' \subset D\},$$
$$\overline{A}_2 = \bigcup\{C(D, B_1, B_2); \ B_1, B_2 \in \mathcal{B}_{k+1}, \ B_1 \neq B_2, \ B_1, B_2 \subset D\}.$$

We have $\overline{A}(D) = \overline{A}_1 \cup \overline{A}_2$ and by the induction hypothesis, since $N(D,k) \geq 2$,

$$\mathbb{P}(\overline{A}_1) \leq N(D,k)\left(2 \prod_{i\leq k} N(D,i)\right)^{-2} \leq \frac{1}{2}\left(2 \prod_{i<k} N(D,i)\right)^{-2}.$$

Using that $\mathbb{P}(C(D, B_1, B_2)) \leq (2N(B,k))^{-4}$,

$$\mathbb{P}(\overline{A}_2) \leq \frac{1}{2} N(D,k)^2 \cdot \left(2N(D,k)\right)^{-4}$$
$$\leq \frac{1}{2} N(D,k)^{-2} \leq \frac{1}{2}\left(2 \prod_{i<k} N(D,i)\right)^{-2}.$$

The result therefore follows in this case. If $D \notin \mathcal{B}'_k$, with the same notation as before, $\overline{A}(D) = \overline{A}_1$. Thus, by the induction hypothesis,

$$\mathbb{P}\big(\overline{A}(D)\big) \le N(D,k)\bigg(2\prod_{i\le k} N(D,i)\bigg)^{-2} \le \bigg(2\prod_{i<k} N(D,i)\bigg)^{-2}.$$

This completes the proof of Lemma 12.14.

Now, let $\Omega_2 = \Omega\backslash\overline{A}(U)$, so $\mathbb{P}(\Omega_2) \ge 3/4$. Let us fix furthermore ω in Ω_2. For B in $\mathcal{B}'_k$, we set

$$a(B,k) = 6^{-k-2}b(B,k)$$

where $b(B,k)$ is given by (12.31). For x in B, set

$$H_x = \big\{z \in U;\ d_\omega(x,z) \le a(B,k)\big\}.$$

Then, we have, for all y in U,

$$(12.32) \qquad\qquad \lambda(x \in B;\ y \in H_x) \le \frac{3\lambda(B)}{2N(B,k)}.$$

Indeed, suppose otherwise and let y be in U such that (12.32) does not hold. For D in $\mathcal{B}_{k+1}$, $D \subset B$, we have $\lambda(D) = \lambda(B)/N(B,k)$. It follows that there are at least two balls B_1, B_2 of $\mathcal{B}_{k+1}$, $B_1, B_2 \subset B$ such that, for $\ell = 1,2$,

$$\lambda(x \in B_\ell;\ y \in H_x) \ge \frac{\lambda(B_\ell)}{2N(B,k)}.$$

For x_1, x_2 in H_y, $d_\omega(x_1,x_2) \le 2a(B,k)$; so we have

$$\lambda \otimes \lambda\big((x_1,x_2) \in B_1 \times B_2;\ d_\omega(x_1,x_2) \le 2a(B,k)\big) \ge \frac{\lambda(B_1)\lambda(B_2)}{4N(B,k)^2}.$$

This, however, contradicts the fact that $\omega \notin C(B,B_1,B_2)$ (by definition of Ω_2) and thus shows (12.32).

Let μ be a probability measure on U. Since the function h'_2 is convex,

$$I(B) = \frac{1}{\lambda(B)}\int_B h'_2\big(\mu(H_x)\big)d\lambda(x) \ge h'_2\bigg(\int g\,d\mu\bigg)$$

where $g(y) = \lambda(x \in B;\ y \in H_x)/\lambda(B)$. It follows from (12.32) that $0 \le g \le 3/2N(B,k)$, so that, by definition of h'_2, $I(B) \ge (\log(2N(B,k)))^{1/2}$. In particular

$$(12.33) \qquad a(B,k)I(B) \ge 6^{-k-2}b(B,k)\big(\log\big(2N(B,k)\big)\big)^{1/2}.$$

For x in U, let us enumerate as $k_1(x) < \cdots < k_{\ell(x)}(x)$ the indexes k such that $B(x,6^{-k}) \in \mathcal{B}'_k$. Note that $k_1(x) = k_0$. For $\ell \le \ell(x)$, let

$$c(x,\ell) = a\big(B\big(x,6^{-k_\ell(x)}\big),\ k_\ell(x)\big).$$

We have

$$
\int_U \sum_{\ell \leq \ell(x)} c(x,\ell) h_2' \big(\mu(y \in U \ ; \ d_\omega(x,y) < c(x,\ell))\big) d\lambda(x)
$$

(12.34)

$$
= \sum \int_B a(B,k) h_2' \big(\mu(y \in U \ ; \ h_\omega(x,y) \leq a(B,k))\big) d\lambda(x)
$$

where the summation is taken over each value of k and each B in $\mathcal{B}_k'$. By (12.33), we see that the latter quantity (12.34) dominates

(12.35)
$$
\sum 6^{-k-2} b(B,k) \big(\log(2N(B,k))\big)^{1/2} \lambda(B)
$$

where the summation is over the same range. Observe that for $\ell < \ell(x)$, $c(x,\ell+1) \leq c(x,\ell)/6$ (by definition of $\mathcal{B}_k'$ and since θ is increasing). Also, $c(x,1) \leq \theta^{-1}(1/2)\Delta$. It follows that, for every x,

$$
\sum_{\ell \leq \ell(x)} c(x,\ell) h_2' \big(\mu(y \in U \ ; \ d_\omega(x,y) < c(x,\ell))\big)
$$

$$
\leq 2 \int_0^{\theta^{-1}(1/2)\Delta} h_2' \big(\mu(y \in U \ ; \ d_\omega(x,y) < \varepsilon)\big) d\varepsilon .
$$

Let us summarize in a statement what we have obtained so far in this general approach. This statement could possibly be of some use in a related context.

Proposition 12.15. *Let (U,δ) be an ultrametric space and (d_ω) be a family of random distances on (U,δ) such that for some increasing function θ on $\mathbb{R}_+$ such that $\lim_{\varepsilon \to 0} \theta(\varepsilon) = 0$ and all u,v in U and $\varepsilon > 0$,*

$$
\mathbb{P}\{\omega \ ; \ d_\omega(u,v) \leq \varepsilon\delta(u,v)\} \leq \theta(\varepsilon) .
$$

Then, there exists Ω_0 with $\mathbb{P}(\Omega_0) \geq 3/4$ such that for every ω in Ω_0 and every probability measure μ on U, in the previous notation,

$$
\int_U d\lambda(x) \int_0^{\theta^{-1}(1/2)\Delta} h_2' \big(\mu(y \in U \ ; \ d_\omega(x,y) < \varepsilon)\big) d\varepsilon
$$

$$
\geq 2^{-7} \sum 6^{-k} \theta^{-1}\big((2N(B,k))^{-6}\big) \big(\log(2N(B,k))\big)^{1/2} \lambda(B)
$$

where the summation is taken over each value of k and each B in $\mathcal{B}_k'$.

From this general result, we can now complete the proof of Theorem 12.13 with the choice of θ given by (12.24).

Proof of Theorem 12.13. By (12.24), we take $\theta(\varepsilon) = \exp(-c_\alpha \varepsilon^{-\alpha})$. Then, from (12.31),

$$
b(B,k) = \left(\frac{6}{c_\alpha} \log(2N(B,k)) \right)^{-1/\alpha} .
$$

Since $1/2 - 1/\alpha = 1/q$, the right hand side of the inequality of Proposition 12.15 is simply $2^{-7}(6/c_\alpha)^{-1/\alpha} \int_U \eta_x(U) d\lambda(x)$ where

$$\eta_x(U) = \sum_{\ell \leq \ell(x)} 6^{-k_\ell(x)} \big(\log\big(2N(x, k_\ell(x))\big)\big)^{1/q} .$$

Therefore, in order to establish (12.29) (with $K_2 = (c_\alpha/\log 2)^{1/\alpha}$), we need simply find the appropriate lower bound for $\eta_x(U)$. For x in U, $s < \ell(x)$ and

$$k_0 = k_1(x) < \cdots < k_s(x) \leq i < k_{s+1}(x) ,$$

we have

$$N(x, i) \leq 2^{2^{i-k_0}} N\big(x, k_s(x)\big)^{2^{i-k_s(x)}} \cdots N\big(x, k_1(x)\big)^{2^{i-k_1(x)}}$$

as is shown by an immediate induction over i. Therefore,

$$\big(\log N(x, i)\big)^{1/q} \leq (2^{i-k_0} \log 2)^{1/q} + \sum_{\ell=1}^{s} \big(2^{i-k_\ell(x)} \log N\big(x, k_\ell(x)\big)\big)^{1/q} .$$

A simple computation then shows that there are constants K_6, K_7 such that $\xi_x^{(q)}(U) \leq K_6 \eta_x(U) + K_7 \Delta$. This shows (12.29) and concludes the proof of Theorem 12.13.

12.3 Applications and Conjectures on Rademacher Processes

The first application deals with subgaussian processes. Recall from Section 11.3 ((11.20)) that a centered process $Y = (Y_t)_{t \in T}$ is subgaussian with respect to a metric or pseudometric d if for all s, t in T and λ in $\mathbb{R}$,

$$\mathbb{E} \exp\big(\lambda(Y_s - Y_t)\big) \leq \exp\Big(\frac{\lambda^2}{2} d^2(s, t)\Big) .$$

As a consequence of the general study of Chapter 11, we noted there that the sufficient conditions for a Gaussian process to have bounded or continuous sample paths also apply for subgaussian processes. That is, if m is a probability measure on (T, d), then

$$\mathbb{E} \sup_{t \in T} Y_t \leq K \sup_{t \in T} \int_0^\infty \Big(\log \frac{1}{m\big(B(t, \varepsilon)\big)}\Big)^{1/2} d\varepsilon$$

where K is a numerical constant. Now, as a consequence of the main result of Section 12.1 and in particular of Theorem 12.8, we see that if d is the pseudo-metric of a Gaussian process $X = (X_t)_{t \in T}$, then

$$\mathbb{E} \sup_{t \in T} Y_t \leq K \mathbb{E} \sup_{t \in T} X_t .$$

where $K > 0$ is a numerical constant. Thus, we have the following statement. Its second part is proved similarly.

Theorem 12.16. *Let $X = (X_t)_{t \in T}$ be a Gaussian process and let $Y = (Y_t)_{t \in T}$ be subgaussian with respect to the canonical distance d_X associated to X. Then, $\mathbb{E} \sup_{t \in T} Y_t \leq K \mathbb{E} \sup_{t \in T} X_t$ for some numerical constant K. In particular, Y is almost surely bounded if X is. Further, if X is continuous, Y has a version with almost all sample paths continuous.*

The second application concerns some remarkable type properties of the canonical injection map $j : \mathrm{Lip}(T) \to C(T)$ first considered by J. Zinn. Let (T, d) be any compact metric space with diameter $D = D(T)$. Denote by $C(T)$ the space of all continuous functions on T with the sup-norm $\| \cdot \|_\infty$ and by $\mathrm{Lip}(T)$ the space of Lipschitz functions f on (T, d) equipped with the norm

$$\|f\|_{\mathrm{Lip}} = D^{-1}\|f\|_\infty + \sup_{s \neq t} \frac{|f(s) - f(t)|}{d(s, t)} .$$

Consider the canonical injection map $j : \mathrm{Lip}(T) \to C(T)$. For every $1 \leq p \leq 2$, let (θ_i) be a p-stable standard sequence (if $p = 2$, $(\theta_i) = (g_i)$, the orthogaussian sequence). Denote by $T_p(j)$ the smallest constant C such that for every finite sequence (x_i) in $\mathrm{Lip}(T)$,

$$\left\| \left\| \sum_i \theta_i j(x_i) \right\|_\infty \right\|_{p,\infty} \leq C \left(\sum_i \|x_i\|_{\mathrm{Lip}}^p \right)^{1/p} .$$

Denote further by $T_p'(j)$, $1 \leq p \leq 2$, the smallest constant C such that

$$\mathbb{E} \left\| \sum_i \varepsilon_i j(x_i) \right\|_\infty \leq C \begin{cases} \left(\sum_i \|x_i\|_{\mathrm{Lip}}^2 \right)^{1/2} & \text{if } p = 2 \\ \| (\|x_i\|_{\mathrm{Lip}}) \|_{p,\infty} & \text{if } 1 \leq p < 2 . \end{cases}$$

The introduction of those two type constants is motivated by the two possible definitions of p-stable operators as described at the end of Section 9.2. From this section actually, it can be seen that $T_2(j)$ and $T_2'(j)$ are equivalent, that is, for some numerical constant K, $K^{-1}T_2(j) \leq T_2'(j) \leq KT_2(j)$. The second inequality is simply that Gaussian averages "dominate" the corresponding Rademacher averages, while the first inequality is obtained by partial integration and the equivalence of moments of the Gaussian averages. For the same reason, for $1 \leq p < 2$, $T_p(j) \leq K_p T_p'(j)$ (cf. one part of the equivalence (iii) in Proposition 9.12). In general however, the type constants T_p and T_p' of operators between Banach spaces are not equivalent when $1 \leq p < 2$.

What we will discover, however, is that, for this particular operator j, the constants $T_p(j)$ and $T_p'(j)$ are equivalent for every $1 \leq p \leq 2$, and their finiteness equivalent to the existence of a majorizing measure condition on (T, d) for $h_q(x) = (\log(1/x))^{1/q}$ where q is the conjugate of p ($h_\infty(x) =$

$\log(1 + \log(1/x))$ if $p = 1$). Recall that if m is a probability measure m on (T, d), we let

$$\gamma_m^{(q)}(T, d) = \sup_{t \in T} \int_0^\infty h_q\big(m(B(t, \varepsilon))\big) d\varepsilon \,,$$

and set

$$\gamma^{(q)}(T, d) = \inf \gamma_m^{(q)}(T, d)$$

where the infimum runs over all probability measures m on T. Then, we have the following theorem.

Theorem 12.17. *Let $1 \leq p \leq 2$. There is a constant $K_p > 0$ depending only on p such that*

$$K_p^{-1} T_p'(j) \leq \gamma^{(q)}(T, d) \leq K_p T_p(j) \,.$$

Proof. First, we prove the left hand side inequality for $p = 2$. Let (x_i) be a finite sequence in $\mathrm{Lip}(T)$ such that, by homogeneity, $\sum_i \|x_i\|_{\mathrm{Lip}}^2 = 1$. Let $X = (X_t)_{t \in T}$ where $X_t = \sum_i \varepsilon_i x_i(t)$, $t \in T$, and set $d_X(s, t) = \|X_s - X_t\|_2 = (\sum_i |x_i(s) - x_i(t)|^2)^{1/2}$. Then, since X is subgaussian with respect to d_X (cf. (11.19)), for some numerical constant K,

$$\mathbb{E}\Big\|\sum_i \varepsilon_i j(x_i)\Big\|_\infty = \mathbb{E}\sup_{t \in T}\Big|\sum_i \varepsilon_i x_i(t)\Big|$$

$$\leq \sup_{t \in T}\Big(\sum_i x_i(t)^2\Big)^{1/2} + K\gamma^{(2)}(T, d_X) \,.$$

Now, by definition of $\|\cdot\|_{\mathrm{Lip}}$, for every s, t in T, $(\sum_i x_i(t)^2)^{1/2} \leq D$ and $d_X(s, t) \leq d(s, t)$. Thus

$$\mathbb{E}\Big\|\sum_i \varepsilon_i j(x_i)\Big\|_\infty \leq D + K\gamma^{(2)}(T, d)$$

$$\leq K'\gamma^{(2)}(T, d)$$

where we have used Lemma 12.1 (vi) in the last step. The result follows by homogeneity. Using Lemma 11.20, the proof is entirely similar for $1 \leq p < 2$.

Let us now establish that $\gamma^{(q)}(T, d) \leq K_p T_p(j)$. It is easy to see that if A is a subset of T and j' is the canonical injection from $\mathrm{Lip}(A)$ into $C(A)$, then $T_p(j') \leq T_p(j)$. On the other hand, we have shown in the proof of Theorem 12.8 that

$$\gamma^{(q)}(T) \leq K_p \sup\{\alpha(A); \ A \subset T, \ A \text{ finite}\}$$

(where, for simpliciticity, we do not specify q in α). Hence, it is enough to show that, when T is finite, $\alpha(T) \leq K_p T_p(j)$. Here, K_p denotes some constant depending on p only and not necessarily the same in each occurence. We claim that it is enough to show that when (T, d) is finite and ultrametric, there is a finite family (x_i) in $\mathrm{Lip}(T)$ with $(\sum_i \|x_i\|_{\mathrm{Lip}}^p)^{1/p} \leq 128$ (for example!) and such that if, for t in T, $X_t = \sum_i \theta_i x_i(t)$, then the canonical distance d_X of the

p-stable process $X = (X_t)_{t \in T}$ satisfies $d_X \geq d$. Indeed, we would then have $128 T_p(j) \geq \| \sup_{t \in T} |X_t| \|_{p,\infty}$, and, by the conjunction of Theorem 12.5 and Proposition 12.7 for $p = 2$, or (12.25) for $1 \leq p < 2$, the result. (Here we use the fact that (12.25) also holds for $p = 1$, cf. [Ta8].)

Let k_0 be the largest such that $4^{-k_0} \geq D$. For $k \geq k_0$, denote by $\mathcal{B}_k$ the family of balls of T of radius 4^{-k}. Since T is finite, there exists k_1 such that $B(x, 4^{-k_1}) = \{x\}$ for every x in T. Let $\mathcal{B} = \bigcup_{k_0 \leq k \leq k_1} \mathcal{B}_k$. For $\varepsilon = (\varepsilon_B)_{B \in \mathcal{B}} \in \mathcal{E} = \{0, 1\}^{\mathcal{B}}$, we define

$$\varphi_\varepsilon = \sum_{k=k_0+1}^{k_1} \sum_{B \in \mathcal{B}_k} 4^{-k} \varepsilon_B I_B .$$

We note that $\|\varphi_\varepsilon\|_\infty \leq \sum_{k > k_0} 4^{-k} \leq 2D$. Consider now s, t in T and let ℓ be the largest such that $d(s, t) \leq 4^{-\ell}$. If $B \in \mathcal{B}_m$ for some $m \leq \ell$, we have $I_B(s) = I_B(t)$. It follows that

$$|\varphi_\varepsilon(s) - \varphi_\varepsilon(t)| \leq \sum_{k > \ell} 4^{-k} \leq 2d(s, t) .$$

This shows that $\|\varphi_\varepsilon\|_{\mathrm{Lip}} \leq 4$. The definition of ℓ shows that the two balls $B_1 = B(s, 4^{-\ell-1})$, $B_2 = B(t, 4^{-\ell-1})$ are different. Since they belong to $\mathcal{B}_{\ell+1}$, we have

$$|\varphi_\varepsilon(s) - \varphi_\varepsilon(t)| \geq 4^{-\ell-1} |\varepsilon_{B_1}(s) - \varepsilon_{B_2}(t)| - \sum_{k > \ell+1} 4^{-k}$$

$$\geq 4^{-\ell-1} |\varepsilon_{B_1}(s) - \varepsilon_{B_2}(t)| - \frac{1}{2} 4^{-\ell-1} .$$

Since $|\varepsilon_{B_1}(s) - \varepsilon_{B_2}(t)|$ is zero or one, we have

$$|\varphi_\varepsilon(s) - \varphi_\varepsilon(t)| \geq \frac{1}{2} 4^{-\ell-1} |\varepsilon_{B_1}(s) - \varepsilon_{B_2}(t)| .$$

Set $N = 2^{\mathrm{Card} \mathcal{B}}$. It follows from the preceding inequality that

$$\left(\frac{1}{N} \sum_{\varepsilon \in \mathcal{E}} |\varphi_\varepsilon(s) - \varphi_\varepsilon(t)|^p \right)^{1/p} \geq \frac{1}{2^{1/p}} \cdot \frac{1}{2} 4^{-\ell-1} \geq \frac{1}{32} d(s, t) .$$

Set $x_\varepsilon = 32 N^{-1/p} \varphi_\varepsilon$, $\varepsilon \in \mathcal{E}$. The family $(x_\varepsilon)_{\varepsilon \in \mathcal{E}}$ is a finite family of elements of $\mathrm{Lip}(T)$ such that $\left(\sum_\varepsilon \|x_\varepsilon\|_{\mathrm{Lip}}^p \right)^{1/p} \leq 128$. Furthermore, if $(\theta_\varepsilon)_{\varepsilon \in \mathcal{E}}$ is an independent family of standard p-stable random variables and if $X_t = \sum_{\varepsilon \in \mathcal{E}} \theta_\varepsilon x_\varepsilon(t)$, $t \in T$, then we have just shown that $d_X(s, t) \geq d(s, t)$ for all s, t in T. As announced, this concludes the proof of Theorem 12.17.

We conclude this chapter with some observations and conjectures on Rademacher processes. Recall first that the various results on Gaussian processes described in Section 12.1 can be formulated in the language of subsets of Hilbert space (as, for example, next to Theorem 12.10). For example, if

$X = (X_t)_{t \in T}$ is a Gaussian process given by $X_t = \sum_i g_i x_i(t)$, where $(x_i(t))$ is a sequence of functions on T, we may identify T with the subset of ℓ_2 consisting of the elements $(x_i(t))$, $t \in T$. The boundedness of the Gaussian process X is then characterized by the majorizing measure condition $\gamma^{(2)}(T)$ for the Hilbertian distance $|\cdot|$ on $T \subset \ell_2$.

By Rademacher process $X = (X_t)_{t \in T}$ indexed by a set T, we mean that for some sequence $(x_i(t))$ of functions on T, $X_t = \sum_i \varepsilon_i x_i(t)$, assumed to be almost surely convergent (i.e. $\sum_i x_i(t)^2 < \infty$). As in Chapter 4, let $r(T) = \mathbb{E} \sup_{t \in T} |X_t|$. A Rademacher process is subgaussian. If $d_X(s,t) = \|X_s - X_t\|_2$, s,t in T, we thus know that $r(T) < \infty$ whenever there is a probability measure m on T such that $\gamma_m^{(2)}(T, d_X) < \infty$. To present our observations, it is convenient (and fruitfull too) to identify as before T with a subset of ℓ_2. By (11.19), we can write that

$$r(T) \leq |T| + K \gamma^{(2)}(T)$$

where $|T| = \sup_{t \in T} |t|$ and $\gamma^{(2)}(T)$ is as before the majorizing measure integral on $T \subset \ell_2$ with respect to the canonical metric $|\cdot|$ on ℓ_2 and K some numerical constant.

From this result, one might wonder, as for Gaussian processes, for some possible necessary majorizing measure condition for a Rademacher process to be almost surely bounded (or continuous). Denote by B_1 the unit ball of $\ell_1 \subset \ell_2$. Since $r(T) \leq \sup_{t \in T} \sum_i |x_i(t)|$, a natural *conjecture* would be that, for some numerical constant K,

$$T \subset K r(T) B_1 + A$$

where A is a subset of ℓ_2 such that $\gamma^{(2)}(A) \leq K r(T)$. This result would completely characterize boundedness of Rademacher process. Theorem 4.15 supports this conjecture. By the convex hull representation of subsets A for which $\gamma^{(2)}(A)$ is controlled (Theorem 12.10), one may look equivalently for a subset A of ℓ_2 such that $A \subset \text{Conv}(y^n)$ where (y^n) is a sequence in ℓ_2 such that, for every n, $|y^n| \leq K r(T)(\log(n+1))^{-1/2}$.

Let us call *M-decomposable* a Rademacher process $X = (X_t)_{t \in T}$, or rather T identified with a subset of ℓ_2, such that, for some $M > 0$ and some A in ℓ_2,

(12.36) $$T \subset M B_1 + A \quad \text{and} \quad \gamma^{(2)}(A) \leq M.$$

In the last part of this chapter, we would like to briefly describe some examples of M-decomposable Rademacher processes which go in the direction of the preceding conjecture.

As a first example, let us consider the case of a subset T of ℓ_2 for which there is a probability measure m such that $\gamma_m^{(q)}(T, \|\cdot\|_{p,\infty}) < \infty$, where $1 \leq p < 2$ and q is the conjugate of p. As a consequence of Lemma 11.20 and Remark 12.11, there exists a sequence (z^n) in ℓ_2 such that, if $M = K_p \gamma^{(q)}(T, \|\cdot\|_{p,\infty})$ where K_p only depends on p, $\|z^n\|_{p,\infty} \leq M(\log(n+1))^{-1/q}$ ($M \log(1+\log n)$ if $q = \infty$) and $T \subset \text{Conv}(z^n)$. We show that there exists then a sequence (y^n) in ℓ_2 with $|y^n| \leq K_p' M(\log(n+1))^{-1/2}$ for each n, such that,

$$T \subset K'_p M B_1 + \operatorname{Conv}(y^n)$$

where K'_p only depends on p. That is, T is $K'_p M$-decomposable. To check this assertion, let us restrict for simplicity to the case $q < \infty$, the case $q = \infty$ being similar. For every n,

$$\|z^n\|_{p,\infty} = \sup_{i \geq 1} i^{1/p} z_i^{n*} \leq M \left(\log(n+1)\right)^{-1/q}$$

where (z_i^{n*}) denotes the non-increasing rearrangement of $(|z_i^n|)$. Let $i_0 = i_0(n)$ be the largest integer $i \geq 1$ such that $i \leq (M/q)^q \log(n+1)$ so that

$$(12.37) \qquad \sum_{i=1}^{i_0} z_i^{n*} \leq M.$$

Let, for each n, y^n be the element of $\mathbb{R}^{\mathbb{N}}$ defined by $y_i^n = z_i^n$ if $z_i^n \notin \{z_1^{n*}, \ldots, z_{i_0}^{n*}\}$, $y_i^n = 0$ otherwise. Clearly

$$|y^n|^2 = \sum_{i>i_0} (z_i^{n*})^2 \leq K''_p M^2 \left(\log(n+1)\right)^{-1}$$

and the claims follows by (12.37).

Hence, under a majorizing measure condition of the type $\gamma^{(q)}(T, \|\cdot\|_{p,\infty}) < \infty$, T is decomposable. The next proposition describes a more general result based on Lemma 4.9. It deals with a natural class of bounded Rademacher processes which, according to the conjecture, could possibly describe all Rademacher processes.

Proposition 12.18. *Let (a^n) be a sequence in ℓ_2 such that for some M*

$$\sum_n \mathbb{P} \left\{ \left| \sum_i \varepsilon_i a_i^n \right| > M \right\} \leq \frac{1}{8}.$$

Let $T \subset \operatorname{Conv}(a^n)$. Then T is KM-decomposable where K is a numerical constant. More precisely, one can find a sequence (y^n) in ℓ_2 such that, for all n, $|y^n| \leq KM(\log(n+1))^{-1/2}$ with the property that $T \subset KMB_1 + \operatorname{Conv}(y^n)$.

Proof. By Lemma 4.2, $(\sum_i (a_i^n)^2)^{1/2} \leq 2\sqrt{2}M$ for each n. Denote by $K_1 \geq 1$ the numerical constant of Lemma 4.9 and set $M_1 = 2\sqrt{2}K_1 M$. By this lemma, there exist, for each n, u^n and v^n in ℓ_2 such that $a^n = u^n + v^n$ and

$$\sum_i |u_i^n| \leq M_1 \quad \text{and} \quad \exp\left(-\frac{K_1 M_1^2}{|v^n|^2}\right) \leq 2\mathbb{P}\left\{ \left| \sum_i \varepsilon_i a_i^n \right| > M_1 \right\}.$$

If, for $\varepsilon > 0$, $N(\varepsilon) = \operatorname{Card}\{n \,;\, |v^n| \geq \varepsilon\}$, it follows from the second of these inequalities and the hypothesis that $N(\varepsilon) \geq \frac{1}{4}\exp(K_1 M_1^2/\varepsilon^2)$. Therefore, we can rearrange (v^n) as a sequence (y^n) such that $|y^n|$ does not increase. In particular, for every N,

$$N \le \text{Card}\{n \,;\, |v^n| \ge |y^N|\} \le \frac{1}{4} \exp\left(\frac{K_1 M_1^2}{|y^N|^2}\right).$$

Hence, for every N, $|y^N| \le K_1 M_1^2 / \log(4N)$. The conclusion immediately follows since $\text{Conv}(v^n) = \text{Conv}(y^n)$.

Notes and References

The main results of this chapter are taken from the two papers [Ta5] and [Ta8] where the existence of majorizing measures for bounded Gaussian and p-stable random processes, $1 \le p < 2$, was established. This chapter basically compiles, with some omissions, these articles. References on Gaussian processes (in the spirit of what is developed in this chapter) presenting the results at their stage of developments are the article by R. Dudley [Du2], the well-known notes by X. Fernique [Fer4] and the paper by N. C. Jain and M. B. Marcus [J-M3]. A recent account on both continuity and extrema is the set of notes by R. J. Adler [Ad].

The finiteness of Dudley's entropy condition for almost surely bounded stationary Gaussian processes is due to X. Fernique [Fer4]. The result corresponds to Theorems 12.8 and 12.9 in an homogeneous setting and will be detailed in the next chapter. (That the finiteness of the entropy integral is not necessary in general was known since the examples of [Du1].) X. Fernique also established in [Fer5] the a posteriori important case of the existence of a majorizing measure in the case of an ultrametric index set. He conjectured back in 1974 the validity of the general case. This result has thus been obtained in [Ta5]. Let us also mention a "volumic" approach to the regularity of Gaussian processes, actually control of subsets of Hilbert space, in the papers [Du1] and [Mi-P] (see also [Pi18]).

One limitation of Theorem 12.5 is that it can help to understand only processes which are essentially described by a single distance. This limits to the case of Gaussian or p-stable processes. In the recent work [Ta18], this theorem is extended to a case where the single distance is replaced by a family of distances. The ideas of this result, when restricted to the case of one single distance, yield the new proof of Theorem 12.5 presented here, that is different from the original proof of [Ta5], and that is somewhat simpler and more constructive. Another contribution of [Ta18] is a new method to replace, in the proof of Proposition 12.7, the use of Slepian's lemma by the use of Sudakov's minoration and of the concentration of measure (expressed in this case by the Gaussian isoperimetric inequality). The use of these tools allows in [Ta18] to extend Theorem 12.12 (properly reformulated) to a large class of infinitely divisible processes.

Theorem 12.12 comes from [Ta8]. It improves upon the homogoneous case as well as the Sudakov type minorations obtained previously by M. B. Marcus and G. Pisier [M-P2] (cf. Chapter 5). The somewhat more general study for the proof has already been used in empirical processes in [L-T4].

Theorem 12.16 was known for a long time to follow from the majorizing measure conjecture. Its proof is actually rather indirect and it would be desirable to find a direct argument. As indicated by G. Pisier [Pi6] and described by the results of [Ta13], its validity actually implies conversely the existence of majorizing measures. This would yield a new approach to the results of Section 12.1. The type 2 property of the canonical injection map $j : \text{Lip}(T) \to C(T)$ has been put into light by J. Zinn [Zi1] to connect the Jain-Marcus central limit theorem for Lipschitz processes [J-M1] to the general type 2 theory of the CLT (cf. Chapter 14). The majorizing measure version of Zinn's map for $p = 2$ was noticed in [He1] and in [Ju] for $1 \leq p < 2$ (cf. also [M-P2], [M-P3]). Theorem 12.17 was obtained in [Ta5].

Consider $1 \leq \alpha < \infty$, and an independent identically distributed sequence (ζ_i), where the law of ζ_i has a density $a_\alpha \exp(-|x|^\alpha)$ with respect to Lebesgue's measure (a_α is the normalizing factor). Building on the ideas of [Ta18], in [Ta20] Theorem 12.8 is (properly reformulated) extended to the processes $(X_t)_{t \in T}$, where $X_t = \sum_i \zeta_i x_i(t)$, and where $(x_i(t))$ is a sequence of functions on T such that $\sum_i x_i^2(t) < \infty$. The still open case of Rademacher processes would correspond to the case "$\alpha = \infty$".

13. Stationary Processes and Random Fourier Series

In Chapter 11, we evaluated random processes indexed by an arbitrary index set T. In this chapter, we take advantage of some homogeneity properties of T and we investigate in this setting, using the general conclusions of Chapters 11 and 12, the more concrete random Fourier series. The tools developed so far indeed lead to a definitive treatment of those processes with applications to Harmonic Analysis. Our main reference for this chapter is the work by M. B. Marcus and G. Pisier [M-P1], [M-P2] to which we refer for an historical background and accurate references and priorities.

In the first section, we briefly indicate how majorizing measure and entropy conditions coincide in an homogeneous setting. We can therefore deal next with the simpler minded entropy conditions only. Using the necessity of the majorizing measure (entropy) condition for the boundedness and continuity of (stationary) Gaussian processes, we investigate and characterize in this way, in the second section, almost sure boundedness and continuity of large classes of random Fourier series. The case of stable random Fourier series and strongly stationary processes is studied next with the conclusions of Section 12.2. We conclude the chapter with some results and comments on random Fourier series with vector valued coefficients.

Let us note that we usually deal in this chapter with complex Banach spaces and use, often without further notice, the trivial extensions to the complex case of various results (such as e.g. the contraction principle).

13.1 Stationarity and Entropy

In this paragraph, we show that for translation invariant metrics majorizing measure and entropy conditions are the same. We shall adopt (throughout this chapter) the following homogeneous setting. Let G be a locally compact Abelian group with unit element 0. Let $\lambda(\cdot) = |\cdot|$ be the normalized translation invariant (Haar) measure on G. Consider furthermore a metric or pseudo-metric d on G which is translation invariant in the sense that $d(u + s, u + t) = d(s, t)$ for all u, s, t in G. Let finally T be a compact subset of non-empty interior of G.

Recall that $N(T, d; \varepsilon)$ denotes the minimal number of open balls (with centers in T) of radius $\varepsilon > 0$ in the pseudo-metric d which are necessary to cover T. More generally, for subsets A, B of G, denote by $N(A, B)$ the minimal number of translates of B (by elements of A) which are necessary to cover A, i.e.

$$N(A, B) = \inf \left\{ N \geq 1; \; \exists t_1, \ldots, t_N \in A, \; A \subset \bigcup_{i=1}^{N}(t_i + B) \right\}.$$

Here $t + B = \{t + s; \; s \in B\}$. Similarly, we let, for subsets A, B of G, $A + B = \{s + t; \; s \in A, t \in B\}$ and we define in the same way $A - B$, etc. In particular, we set $T' = T + T$ and $T'' = T + T' = T + T + T$.

The following lemma is an elementary statement about the preceding covering numbers which will be useful in various parts of this chapter. We set $B(0, \varepsilon) = \{t \in G; \; d(t, 0) < \varepsilon\}$.

Lemma 13.1. *Under the previous notation, we have:*

(i) if $B = T' \cap B(0, \varepsilon)$, $\varepsilon > 0$, then $N(T, d; \varepsilon) = N(T, B)$;

(ii) $N(T, d; 2\varepsilon) \leq N(T, B(0, \varepsilon) - B(0, \varepsilon))$;

(iii) if $A \subset G$, $N(T, A) \geq |T|/|A|$;

(iv) if $A \subset T'$, $N(T, A - A) \leq |T''|/|A|$.

Proof. (i) immediately follows from the definitions. (ii) If, for t in T, there exists t_i in T such that $t \in t_i + B(0, \varepsilon) - B(0, \varepsilon)$, this means that $t = t_i + u - v$ where $u, v \in B(0, \varepsilon)$. Hence, by the translation invariance,

$$d(t, t_i) = d(u - v, 0) \leq d(u, 0) + d(0, v) < 2\varepsilon$$

so that $N(T, d; 2\varepsilon) \leq N(T, B(0, \varepsilon) - B(0, \varepsilon))$. (iii) It is enough to consider the case $N(T, A) < \infty$. Then, if $T \subset \bigcup_{i=1}^{N}(t_i + A)$,

$$|T| \leq \sum_{i=1}^{N} |t_i + A| = N|A|$$

which gives the result. (iv) Assume that $|A| > 0$. Let $\{t_1, \ldots, t_M\}$ be maximal in T under the conditions $(t_i + A) \cap (t_j + A) = \emptyset$, $\forall i \neq j$. If $t \in T$, by maximality, $(t + A) \cap (t_i + A) \neq \emptyset$ for some $i = 1, \ldots, M$. Hence, $t + u = t_i + v$ for some u, v in A. Therefore $t \in t_i + A - A$ and $T \subset \bigcup_{i=1}^{M}(t_i + A - A)$. This implies that $M \geq N(T, A - A)$. Now the sets $(t_i + A)_{i \leq M}$ are disjoint in $T'' = T + T'$ and thus

$$|T''| \geq \sum_{i=1}^{M} |t_i + A| = M|A| \geq N(T, A - A)|A|.$$

The proof of Lemma 13.1 is complete.

With this lemma, we can compare majorizing measure and entropy conditions in translation invariant situations. The idea is simply that if a majorizing measure exists then the Haar measure is also a majorizing measure from which the conclusion follows by the preceding lemma.

Proposition 13.2. *Let ψ be a Young function. In the preceding notation, let m be a probability measure on $T \subset G$ and denote by $D = D(T)$ the d-diameter of T. Then*

$$\frac{1}{2} \int_0^D \psi^{-1}\left(\frac{|T|}{|T''|} N(T,d;\varepsilon)\right) d\varepsilon \leq \sup_{t \in T} \int_0^D \psi^{-1}\left(\frac{1}{m(B(t,\varepsilon))}\right) d\varepsilon .$$

Proof. Let us denote by M the right hand side of the inequality that we have to establish. Since $\psi^{-1}(1/x)$ is convex, by Jensen's inequality,

$$M \geq \int_0^D \frac{1}{|T|} \int_T \psi^{-1}\left(\frac{1}{m(B(t,\varepsilon))}\right) d\lambda(t) d\varepsilon$$

$$\geq \int_0^D \psi^{-1}\left(\frac{|T|}{\int_T m(B(t,\varepsilon)) d\lambda(t)}\right) d\varepsilon .$$

Now, by Fubini's theorem and the translation invariance,

$$\int_T m(B(t,\varepsilon)) d\lambda(t) = \int_T |T \cap B(s,\varepsilon)| dm(s) \leq |T' \cap B(0,\varepsilon)| .$$

Hence,

$$M \geq \int_0^D \psi^{-1}\left(\frac{|T|}{|T' \cap B(0,\varepsilon)|}\right) d\varepsilon .$$

To conclude, by (ii) and (iv) of Lemma 13.1, for every $\varepsilon > 0$,

$$N(T,d;2\varepsilon) \leq \frac{|T''|}{|T' \cap B(0,\varepsilon)|}$$

from which Proposition 13.2 follows.

Note that, conversely, if $\lambda_{T''}$ denotes restricted normalized Haar measure on $T'' = T + T + T$, then, for all $\eta > 0$,

$$(13.1) \quad \sup_{t \in T} \int_0^\eta \psi^{-1}\left(\frac{1}{\lambda_{T''}(B(t,\varepsilon))}\right) d\varepsilon \leq \int_0^\eta \psi^{-1}\left(\frac{|T''|}{|T|} N(T,d;\varepsilon)\right) d\varepsilon .$$

(Note the complete equivalence when $T = G$ is compact.) To show (13.1), observe that, if $t \in T$, $t + T' \cap B(0,\varepsilon) \subset T'' \cap B(t,\varepsilon)$. Hence, by the translation invariance and (i) and (iii) of Lemma 13.1,

$$\lambda_{T''}(B(t,\varepsilon)) \geq \frac{1}{|T''|} |T' \cap B(0,\varepsilon)| \geq \frac{|T|}{|T''|} N(T,d;\varepsilon) .$$

Proposition 13.2 allows to state in terms of entropy the characterization of the almost sure boundedness and continuity of stationary Gaussian processes.

Before we state the result, it is convenient to introduce some notation concerning entropy integrals. Let (T, d) be a pseudo-metric space with diameter D. For $1 \leq q < \infty$, we let

$$E^{(q)}(T, d) = \int_0^\infty \left(\log N(T, d; \varepsilon) \right)^{1/q} d\varepsilon .$$

As usual, the integral is taken from 0 to ∞ but of course stops at D. When $q = \infty$, we let

$$E^{(\infty)}(T, d) = \int_0^\infty \log\left(1 + \log N(T, d; \varepsilon) \right) d\varepsilon .$$

A process $X = (X_t)_{t \in G}$ indexed by G is stationary if, for every u in G, $(X_{u+t})_{t \in G}$ has the same distribution as X. If X is a stationary Gaussian process, its corresponding L_2-metric $d_X(s, t) = \|X_s - X_t\|_2$ is translation invariant. As a corollary to the majorizing measure characterization of the boundedness and continuity of Gaussian processes described in Chapter 12, and of Proposition 13.2, we can state:

Theorem 13.3. *Let G be a locally compact Abelian group and let T be a compact metrizable subset of G of non-empty interior. Let $X = (X_t)_{t \in G}$ be a stationary Gaussian process indexed by G. Then X has a version with almost all bounded and continuous sample paths on $T \subset G$ if and only if d_X is continuous on $G \times G$ and $E^{(2)}(T, d_X) < \infty$. Moreover, there is a numerical constant $K > 0$ such that*

$$K^{-1} \left(E^{(2)}(T, d_X) - L(T)^{1/2} D(T) \right) \leq \mathbb{E} \sup_{t \in T} X_t \leq K E^{(2)}(T, d_X)$$

where $D(T)$ is the diameter of (T, d_X) and $L(T) = \log(|T''|/|T|)$. Note that if $T = G$ is compact, $L(T) = 0$.

The sufficiency in this theorem is simply Dudley's majoration (Theorem 11.17) together with the continuity of d_X. The necessity and the left hand side inequality follow from Theorem 12.9 together with Proposition 13.2. (It might be useful to recall at this point the simple comparison (11.17).)

Note the following dichotomy contained in the statement of Theorem 13.3: stationary Gaussian processes are either almost surely continuous or almost surely unbounded (according to the finiteness or not of the entropy integral).

The choice of the compact set T does not affect the qualitative conclusion of Theorem 13.3. Indeed, let $(X_t)_{t \in G}$ be any stationary process. If T_1 and T_2 are two compact subsets of G with non-empty interiors, then $(X_t)_{t \in T_1}$ has a version with continuous sample paths if and only if $(X_t)_{t \in T_2}$ does. This is obvious by stationarity since each of the sets T_1 and T_2 can be covered by finitely many translates of the other. Consequently, if G is the union of a countable family of compact sets, then $(X_t)_{t \in T}$ has a version with continuous sample paths if and only if the entire process $(X_t)_{t \in G}$ does. This applies in the most important case $G = \mathbb{R}^n$. Boundedness (lattice-boundedness) of the stationary Gaussian process of Theorem 13.3 over *all* G holds if and only if

$$\int_0^\infty \left(\log N(G, d_X; \varepsilon)\right)^{1/2} d\varepsilon < \infty.$$

This can be shown by either repeating the arguments of the proof of Theorem 13.3 or by using the Bohr compactification of G and the fact that under the preceding entropy condition, X has a version which is an almost periodic function on G. The proof of this fact further indicates that, under this condition, a stationary Gaussian process $X = (X_t)_{t \in G}$ indexed by G admits an expansion as a series

$$X_t = \sum_n g_n \text{Re}\big(a_n \gamma_n(t)\big) + \sum_n g'_n \text{Im}\big(a_n \gamma_n(t)\big), \quad t \in G,$$

where (a_n) is a sequence of complex numbers in ℓ_2, (γ_n) is a sequence of continuous characters of G and (g_n), (g'_n) are independent standard Gaussian sequences. We refer to [M-P1, p. 134-138] for more details on these comments.

Theorem 13.3 is one of the main ingredients in the study of random Fourier series. Before we turn to this topic in the subsequent section, it is convenient for comparison and possible estimates in concrete situations to mention an equivalent of the entropy integral $E^{(q)}(T, d)$ for a translation invariant metric d. Assume for simplicity that T is a symmetric subset of G and let $\sigma(t) = d(t, 0)$, $t \in T'$. Consider the non-decreasing rearrangement $\bar{\sigma}$ of σ on $[0, |T'|]$ (with respect to T'); more precisely, for $0 < \delta \le |T'|$,

$$\bar{\sigma}(\delta) = \sup\{\varepsilon > 0; \ |T' \cap B(0, \varepsilon)| < \delta\}.$$

For $1 < p \le 2$, let

$$I^{(p)}(T, d) = \int_0^{|T'|} \frac{\bar{\sigma}(\varepsilon)}{\varepsilon \left(\log \frac{4|T'''|}{\varepsilon}\right)^{1/p}} d\varepsilon$$

where $T''' = T + T + T + T$. By Lemma 13.1 and elementary arguments, it can be shown that, if q is the conjugate of p, $E^{(q)}(T, d)$ and $I^{(p)}(T, d)$ are essentially of the same order. Namely, for some constant $K_p > 0$ depending on p only,

$$K_p^{-1}\big(D + I^{(p)}(T, d)\big) \le E^{(q)}(T, d) \le K_p\big(D + I^{(p)}(T, d)\big)$$

where D is the diameter of (T, d). A similar result holds for $p = 1$ (and $q = \infty$).

13.2 Random Fourier Series

In this section, we take advantage of Theorem 13.3 to study a class of random Fourier series and to develop some applications. The interested reader will find in the book [M-P1] a general and historical introduction to random Fourier series. Here, we more or less only concentrate on one typical situation. In particular, we only consider Abelian groups. We refer to [M-P1] for the non-Abelian case as well as for further aspects.

Throughout this section, we fix the following notation. G is a locally compact Abelian group with identity element 0 and Haar measure $\lambda(\cdot) = |\cdot|$. We denote by V a compact neighborhood of 0 (by the translation invariance, the results would apply to all compact sets with non-empty interior). Let Γ be the dual group of all characters γ of G (i.e. γ is a continuous complex valued function on G such that $|\gamma(t)| = 1$ and $\gamma(s)\gamma(t) = \gamma(s + t)$ for all s, t in G). Fix a countable subset A of Γ. Since the main interest will be the study of random Fourier series with spectrum in A, and since A generates a closed separable subgroup of Γ, we can assume without restricting the generality that Γ itself is separable. In particular, all the compact subsets of G are *metrizable*. As a concrete example of this setting, one may consider the compact group $\mathbb{T} = \mathbb{R}/\mathbb{Z}$ (identified with $[0, 2\pi]$) with $\mathbb{Z}$ as dual group.

$V \subset G$ being as before, we agree to denote by $\|\cdot\| = \sup_{t \in V} |\cdot|$ the sup-norm of the Banach space $C(V)$ of all continuous *complex* valued functions on V (or on G when $V = G$).

Let $(a_\gamma)_{\gamma \in A}$ be complex numbers such that $\sum_{\gamma \in A} |a_\gamma|^2 < \infty$. (We sometimes omit $\gamma \in A$ in the summation symbol only indicated then by $\sum_\gamma$ or just $\sum$.) Let also $(g_\gamma)_{\gamma \in A}$ be a standard Gaussian sequence. Following the comments at the end of Theorem 13.3, let us first consider the Gaussian process $X = (X_t)_{t \in G}$ given by

$$(13.2) \qquad\qquad X_t = \sum_{\gamma \in A} a_\gamma g_\gamma \gamma(t), \quad t \in G.$$

The question of course arises of the almost sure continuity or uniform convergence of the series X on V. Since (g_γ) is a symmetric sequence, these two properties are equivalent: if X is continuous (or admits a continuous version) on V, then by Itô-Nisio's theorem (Theorem 2.1), X converges uniformly (for any ordering of A). The process defined by (13.2) is a (complex valued) Gaussian process indexed by G with associated L_2-metric

$$d_X(s, t) = \left(\sum_{\gamma \in A} |a_\gamma|^2 |\gamma(s) - \gamma(t)|^2 \right)^{1/2}, \quad s, t \in G,$$

which is translation invariant. Since X is complex valued, to enter the setting of Theorem 13.3, we use the following. If we let

$$X'_t = \sum_{\gamma \in A} g_\gamma \mathrm{Re}\big(a_\gamma \gamma(t)\big) + \sum_{\gamma \in A} g'_\gamma \mathrm{Im}\big(a_\gamma \gamma(t)\big), \quad t \in G,$$

where (g'_γ) is an independent copy of (g_γ), $X' = (X'_t)_{t \in G}$ is a real valued stationary Gaussian process such that $d_{X'} = d_X$. The series X and X' converge uniformly almost surely simultaneously. As a consequence of Theorem 13.3, X' admits a version with almost all sample paths continuous on V and therefore converges uniformly almost surely on V if and only if

$E^{(2)}(V, d_{X'}) = E^{(2)}(V, d_X) < \infty$, and thus the same holds for X. That is, we have the following statement.

Theorem 13.4. *Let X be the random Fourier series of (13.2). The following are equivalent:*

(i) for some (all) ordering(s) of A, the series X converges almost surely uniformly on V;

(ii) $\sup_{F \subset A, F \text{ finite}} \mathbb{E} \| \sum_{\gamma \in F} a_\gamma g_\gamma \gamma \|^2 < \infty$;

(iii) $E^{(2)}(V, d_X) < \infty$.

Furthermore, for some numerical constant $K > 0$,

$$
K^{-1} \left(E^{(2)}(V, d_X) - L(V)^{1/2} \left(\sum_{\gamma \in A} |a_\gamma|^2 \right)^{1/2} \right)
$$

(13.3)

$$
\leq \left(\mathbb{E} \| X \|^2 \right)^{1/2} \leq K \left(E^{(2)}(V, d_X) + \left(\sum_{\gamma \in A} |a_\gamma|^2 \right)^{1/2} \right)
$$

where we recall that $L(V) = \log(|V''|/|V|)$ where $V'' = V + V + V$ and $\| X \| = \sup_{t \in V} |X_t|$.

Proof. If X converges uniformly in some ordering, then (ii) is satisfied by the integrability properties of Gaussian random vectors (Corollary 3.2) and conditional expectation. Let F be finite in A and denote by X^F the finite sum $\sum_{\gamma \in F} a_\gamma g_\gamma \gamma$. Then, as a consequence of Theorem 13.3 (considering as indicated before the natural associated real stationary Gaussian series), we have that (13.3) holds for X^F, with thus a numerical constant K independent of F finite in A. Then (iii) holds under (ii) by increasing F to A. Finally, and in the same way, if $E^{(2)}(V, d_X) < \infty$, for any ordering of A, X converges in L_2 with respect to the uniform norm on $C(V)$, and thus almost surely by Itô-Nisio's theorem: to see this, simply use a Cauchy argument in inequality (13.3) and dominated convergence in the entropy integral.

Note that we recover from this statement the equivalence between almost sure boundedness and continuity of Gaussian random Fourier series.

As the main question of this study, we shall be interested in similar results and estimates when the Gaussian sequence $(g_\gamma)_{\gamma \in A}$ is replaced by a Rademacher sequence $(\varepsilon_\gamma)_{\gamma \in A}$ or, more generally, by some *symmetric* sequence $(\xi_\gamma)_{\gamma \in A}$ of real valued random variables. While we are dealing with Fourier series with complex coefficients, for simplicity however we only consider real probabilistic strutures. The complex case can easily be deduced. Recall that, by the symmetry assumption, the sequence (ξ_γ) can be replaced by $(\varepsilon_\gamma \xi_\gamma)$ where (ε_γ) is an independent Rademacher sequence. This is the setting that we will adopt. By a standard symmetrization procedure, the results apply similarly to a sequence (ξ_γ) of independent mean zero random variables (cf. Lemma 6.3).

Assume therefore that we are given a sequence of complex numbers $(a_\gamma)_{\gamma \in A}$ and a sequence $(\xi_\gamma)_{\gamma \in A}$ of real valued random variables satisfying

$$\sum_{\gamma \in A} |a_\gamma|^2 \mathbb{E}|\xi_\gamma|^2 < \infty.$$

Consider the process $Y = (Y_t)_{t \in G}$ defined by

$$(13.4) \qquad Y_t = \sum_{\gamma \in A} a_\gamma \varepsilon_\gamma \xi_\gamma \gamma(t), \quad t \in G,$$

where, as above, (ε_γ) is a Rademacher sequence which is independent of (ξ_γ). We associate to this process the L_2-pseudo-metric

$$d_Y(s,t) = \left(\sum_{\gamma \in A} |a_\gamma|^2 \mathbb{E}|\xi_\gamma|^2 |\gamma(s) - \gamma(t)|^2 \right)^{1/2}, \quad s, t \in G,$$

which is translation invariant. If V is as usual a compact neighborhood of 0, we shall be interested, as previously in the Gaussian case, in the almost sure uniform convergence on V of the random Fourier series Y of (13.4). The main objective is to try to obtain bounds similar to those of the Gaussian random Fourier series (13.2). The basic idea consists in first writing the best possible entropy estimates conditionally on the sequence (ξ_γ) (that is, along the Rademacher sequence) and then in integrating and in making use of the translation invariant properties. The first step is simply the content of the following lemma which is the entropic (and complex) version of (11.19) and Lemma 11.20.

Lemma 13.5. *Let $x_i(t)$ be (complex) functions on a set T such that $\sum_i |x_i(t)|^2 < \infty$ for every t in T. Let $1 \le p \le 2$ and equip T with the pseudo-metric (functional) d_p defined by*

$$d_2(s,t) = \left(\sum_i |x_i(s) - x_i(t)|^2 \right)^{1/2}$$

if $p = 2$ and

$$d_p(s,t) = \| (x_i(s) - x_i(t)) \|_{p,\infty}$$

if $1 \le p < 2$. Then, for some constant K_p depending on p only,

$$\left(\mathbb{E} \sup_{t \in T} \left| \sum_i \varepsilon_i x_i(t) \right|^2 \right)^{1/2} \le K_p \left(R + E^{(q)}(T, d_p) \right)$$

where $R = \sup_{t \in T} \left(\sum_i |x_i(t)|^2 \right)^{1/2}$ or $\sup_{t \in T} \|(x_i(t))\|_{p,\infty}$ according as $p = 2$ or $p < 2$ and where $q = p/p - 1$ is the conjugate of p. Furthermore, if

$E^{(q)}(T, d_p) < \infty$, $\left(\sum_i \varepsilon_i x_i(t)\right)_{t \in T}$ *has a version with continuous paths on* (T, d_p).

As announced, this lemma applied conditionally together with the translation invariance property allows to evaluate random Fourier series of the type (13.4). The main result in this direction is the following theorem. Recall that for $V \subset G$ we set $L(V) = \log(|V''|/|V|)$ where $V'' = V + V + V$.

Theorem 13.6. *If* $E^{(2)}(V, d_Y) < \infty$, *the random Fourier series* Y *of* (13.4) *converges uniformly on* V *with probability one. Further, for some numerical constant* $K > 0$,

$$\left(\mathbb{E}\|Y\|^2\right)^{1/2} \leq K\left(\left(1 + L(V)^{1/2}\right)\left(\sum_{\gamma \in A} |a_\gamma|^2 \mathbb{E}|\xi_\gamma|^2\right)^{1/2} + E^{(2)}(V, d_Y)\right).$$

Proof. There is no loss of generality to assume by homogeneity that (for example)

$$\left(\sum_\gamma |a_\gamma|^2 \mathbb{E}|\xi_\gamma|^2\right)^{1/2} \leq \frac{1}{8}.$$

Denote by Ω_0 the set, of probability one, of all ω's for which $\sum_\gamma |a_\gamma|^2|\xi_\gamma(\omega)|^2 < \infty$. For ω in Ω_0, introduce the translation invariant pseudo-metric

$$d_\omega(s, t) = \left(\sum_\gamma |a_\gamma|^2|\xi_\gamma(\omega)|^2|\gamma(s) - \gamma(t)|^2\right)^{1/2}, \quad s, t \in G.$$

Clearly, for every s, t, $\mathbb{E}d_\omega^2(s, t) = d_Y^2(s, t)$ and

$$d_\omega(s, t) \leq R(\omega) = 2\left(\sum_\gamma |a_\gamma|^2|\xi_\gamma(\omega)|^2\right)^{1/2} < \infty.$$

We may as well assume that $R(\omega) > 0$ for all ω in Ω_0. Conditionally on the set Ω_0, which only depends on the sequence (ξ_γ), we now apply Lemma 13.5 (for $p = 2$). We get that, $\mathbb{E}_\varepsilon$ denoting as usual partial integration with respect to the Rademacher sequence (ε_γ),

(13.5)
$$\left(\mathbb{E}_\varepsilon \sup_{t \in V}\left|\sum_\gamma a_\gamma \varepsilon_\gamma \xi_\gamma(\omega)\gamma(t)\right|^2\right)^{1/2}$$
$$\leq K\left(R(\omega) + \int_0^{R(\omega)} \left(\log N(V, d_\omega; \varepsilon)\right)^{1/2} d\varepsilon\right).$$

The idea is now to integrate this inequality with respect to the sequence (ξ_γ) and to use the translation invariance properties to conclude. To this aim, let us set, for every integer $n \geq 1$,

$$W_n = \{t \in V'; \; d_Y(t,0) \le 2^{-n}\}$$

where $V' = V + V$. For all ω in Ω_0, define a sequence $(b_n(\omega))_{n \ge 0}$ of positive numbers by letting $b_0(\omega) = R(\omega)$ and, for $n \ge 1$,

$$b_n(\omega) = \max\left(2^{-n}, \; 4\int_{W_n} d_\omega(t,0)\frac{dt}{|W_n|}\right) \quad (2^{-n} \text{ if } |W_n| = 0).$$

Observe that, for every $n \ge 1$,

$$\left(\mathbb{E}b_n^2\right)^{1/2} \le 2^{-n} + 4\left(\int_{W_n} \mathbb{E}d_\omega^2(t,0)\frac{dt}{|W_n|}\right)^{1/2}$$

$$\le 2^{-n} + 4\left(\int_{W_n} d_Y^2(t,0)\frac{dt}{|W_n|}\right)^{1/2} \le 5 \cdot 2^{-n},$$

so that in particular $b_n(\cdot) \to 0$ almost surely. Let us denote by Ω_1 the set of full probability consisting of Ω_0 and of the set on which the sequence (b_n) converges to 0. For ω in Ω_1, and $n \ge 1$, set

$$B(n,\omega) = \{t \in W_n; \; d_\omega(t,0) \le b_n(\omega)/2\}.$$

Clearly, by Fubini's theorem and by the definition of $b_n(\omega)$,

$$(13.6) \qquad\qquad\qquad |B(n,\omega)| \ge \frac{|W_n|}{2}.$$

Now, we are ready to integrate with respect to ω inequality (13.5). To this aim, we use the following elementary lemma.

Lemma 13.7. *Let $f : (0, R] \to \mathbb{R}_+$ be decreasing. Let also (b_n) be a sequence of positive numbers with $b_0 = R$ and $b_n \to 0$. Then*

$$\int_0^R f(x)\,dx \le \sum_{n=0}^{\infty} b_n f(b_{n+1}).$$

By this lemma, for ω in Ω_1, we have

$$\int_0^{R(\omega)} \left(\log N(V, d_\omega; \varepsilon)\right)^{1/2} d\varepsilon \le \sum_{n=0}^{\infty} b_n(\omega)\left(\log N\left(V, d_\omega; b_{n+1}(\omega)\right)\right)^{1/2}.$$

By Lemma 13.1 and (13.6), for ω in Ω_1 and $n \ge 0$, we can write

$$N\left(V, d_\omega; b_{n+1}(\omega)\right) \le N\left(V, B(n+1,\omega) - B(n+1,\omega)\right)$$

$$\le \frac{|V''|}{|B(n+1,\omega)|}$$

$$\le 2\frac{|V''|}{|W_{n+1}|}$$

$$\le 2\frac{|V''|}{|V|}N(V, W_{n+1})$$

$$\le 2\frac{|V''|}{|V|}N\left(V, d_Y; 2^{-n-1}\right).$$

Hence (13.5) reads as

$$\left(\mathbb{E}_\varepsilon \sup_{t \in V} \left| \sum_\gamma a_\gamma \varepsilon_\gamma \xi_\gamma(\omega) \gamma(t) \right|^2 \right)^{1/2}$$

$$\leq K \left(R(\omega) + \sum_{n=0}^\infty b_n(\omega) \left(\log \left(2 \frac{|V''|}{|V|} N(V, d_Y; 2^{-n-1}) \right) \right)^{1/2} \right)$$

for every ω in Ω_1. Since $(\mathbb{E}b_n^2)^{1/2} \leq 5 \cdot 2^{-n}$, we have simply to integrate with respect to ω and we get, by the triangle inequality,

$$\left(\mathbb{E} \sup_{t \in V} |Y_t|^2 \right)^{1/2} \leq K \left(\frac{1}{4} + 5 \sum_{n=0}^\infty 2^{-n} \left(\log \left(2 \frac{|V''|}{|V|} N(V, d_Y; 2^{-n-1}) \right) \right)^{1/2} \right)$$

which yields the inequality of the theorem. A Cauchy argument together with dominated convergence in the entropy integral and Itô-Nisio's theorem proves then the almost sure uniform convergence of Y on V. The proof of Theorem 13.6 is complete.

Theorem 13.6 expresses a bound for the random Fourier series Y very similar to the bound described previously for Gaussian Fourier series. Actually, the comparison between Theorem 13.6 and the lower bound in (13.3) indicates that, for some numerical constant $K > 0$,

(13.7)
$$\left(\mathbb{E} \left\| \sum_\gamma a_\gamma \varepsilon_\gamma \xi_\gamma \gamma \right\|^2 \right)^{1/2}$$
$$\leq K \left(1 + L(V)^{1/2} \right) \left(\mathbb{E} \left\| \sum_\gamma a_\gamma (\mathbb{E}|\xi_\gamma|^2)^{1/2} g_\gamma \gamma \right\|^2 \right)^{1/2}.$$

(What we actually get is that

$$\left(\mathbb{E} \left\| \sum_\gamma a_\gamma \varepsilon_\gamma \xi_\gamma \gamma \right\|^2 \right)^{1/2} \leq K \left(\left(1 + L(V)^{1/2} \right) \left(\sum_\gamma |a_\gamma|^2 \mathbb{E}|\xi_\gamma|^2 \right)^{1/2} \right.$$
$$\left. + \left(\mathbb{E} \left\| \sum_\gamma a_\gamma (\mathbb{E}|\xi_\gamma|^2)^{1/2} g_\gamma \gamma \right\|^2 \right)^{1/2} \right)$$

but we will not use this improved formulation in the sequel. Recall that $L(V) \geq 0$.) In particular, we see by the contraction principle that

(13.8)
$$\left(\mathbb{E} \left\| \sum_\gamma a_\gamma \varepsilon_\gamma \xi_\gamma \gamma \right\|^2 \right)^{1/2}$$
$$\leq K \left(1 + L(V)^{1/2} \right) \sup_\gamma (\mathbb{E}|\xi_\gamma|^2)^{1/2} \left(\mathbb{E} \left\| \sum_\gamma a_\gamma g_\gamma \gamma \right\|^2 \right)^{1/2}.$$

One might wonder at this stage when an inequality such as (13.8) can be reversed. By the contraction principle in the form of Lemma 4.5,

$$\left(\mathbb{E}\left\|\sum_\gamma a_\gamma \varepsilon_\gamma \xi_\gamma \gamma\right\|^2\right)^{1/2} \geq \inf_\gamma \mathbb{E}|\xi_\gamma| \left(\mathbb{E}\left\|\sum_\gamma a_\gamma \varepsilon_\gamma \gamma\right\|^2\right)^{1/2}.$$

If we try to exploit this information to reverse inequality (13.8), we are faced with the question of knowing whether the Rademacher series $\sum_\gamma a_\gamma \varepsilon_\gamma \gamma$ dominates in the sup-norm the corresponding Gaussian series $\sum_\gamma a_\gamma g_\gamma \gamma$. We know ((4.8)) that

$$(13.9) \qquad \left(\mathbb{E}\left\|\sum_\gamma a_\gamma \varepsilon_\gamma \gamma\right\|^2\right)^{1/2} \leq K\left(\mathbb{E}\left\|\sum_\gamma a_\gamma g_\gamma \gamma\right\|^2\right)^{1/2}$$

(with $K = (\pi/2)^{1/2}$) but the converse inequality does not hold in general, unless we deal with a Banach space of finite cotype (cf. (4.9) and Proposition 9.14). Although $C(V)$ is of no finite cotype, the more general estimates (13.7) or (13.8) that we have obtained actually show that the inequality we are looking for is satisfied here due to the particular structure of the vector valued coefficients $a_\gamma \gamma$. The proof of this result is similar to the argument used in the proof of Proposition 9.25 showing that the Rademacher and Gaussian cotype 2 definitions are equivalent. Further, this property is indeed related to some cotype 2 Banach space of remarkable interest which we discuss next. The following proposition is the announced result of the equivalence of Gaussian and Rademacher random Fourier series.

Proposition 13.8. *For some numerical constant $K > 0$,*

$$\left(\mathbb{E}\left\|\sum_\gamma a_\gamma g_\gamma \gamma\right\|^2\right)^{1/2} \leq K\left(1 + L(V)^{1/2}\right)\left(\mathbb{E}\left\|\sum_\gamma a_\gamma \varepsilon_\gamma \gamma\right\|^2\right)^{1/2}.$$

(Recall that by (13.9) the converse inequality is satisfied with a constant independent of V.) Furthermore, $\sum_\gamma a_\gamma g_\gamma \gamma$ and $\sum_\gamma a_\gamma \varepsilon_\gamma \gamma$ converge uniformly almost surely simultaneously.

Proof. The second assertion follows from a Cauchy argument and the integrability properties of both Gaussian and Rademacher series. Thus, we can assume that we deal with finite sums. Let $c > 0$ to be specified. By the triangle inequality,

$$\left(\mathbb{E}\left\|\sum_\gamma a_\gamma g_\gamma \gamma\right\|^2\right)^{1/2}$$

$$\leq \left(\mathbb{E}\left\|\sum_\gamma a_\gamma g_\gamma I_{\{|g_\gamma|\leq c\}}\gamma\right\|^2\right)^{1/2} + \left(\mathbb{E}\left\|\sum_\gamma a_\gamma g_\gamma I_{\{|g_\gamma|>c\}}\gamma\right\|^2\right)^{1/2}.$$

By the contraction principle, the first term on the right hand side of this inequality is smaller than

$$c\left(\mathbb{E}\left\|\sum_\gamma a_\gamma \varepsilon_\gamma \gamma\right\|^2\right)^{1/2}.$$

To the second, we apply (13.7) to see that it is bounded by

$$K\left(1 + L(V)^{1/2}\right)\left(\mathbb{E}|g|^2 I_{\{|g|>c\}}\right)^{1/2}\left(\mathbb{E}\left\|\sum_\gamma a_\gamma g_\gamma \gamma\right\|^2\right)^{1/2}$$

where g is a standard normal variable. Let us then choose $c > 0$ such that

$$K\left(1 + L(V)^{1/2}\right)\left(\mathbb{E}|g|^2 I_{\{|g|>c\}}\right)^{1/2} \leq 1/2.$$

This can be achieved with c of the order of $1 + L(V)^{1/2}$ and the proposition follows. (Actually, a smaller function of $L(V)$ would suffice but we need not be concerned with this here.)

As a corollary of Theorem 13.7 and Proposition 13.8, we can now summarize the results on random Fourier series $Y = (Y_t)_{t \in G}$ of the type

$$Y_t = \sum_{\gamma \in A} a_\gamma \xi_\gamma \gamma(t), \quad t \in G,$$

where $\sum_\gamma |a_\gamma|^2 < \infty$ and where (ξ_γ) is a symmetric sequence of real valued random variables such that $\sup_\gamma \mathbb{E}|\xi_\gamma|^2 < \infty$ and $\inf_\gamma \mathbb{E}|\xi_\gamma| > 0$. Recall that by the symmetry assumption, (ξ_γ) has the same distribution as $(\varepsilon_\gamma \xi_\gamma)$ where (ε_γ) is an independent Rademacher sequence.

Corollary 13.9. *Let V be as usual a compact symmetric neighborhood of the unit of G. Let $Y = (Y_t)_{t \in G}$ be as just defined with associated metric*

$$d(s,t) = \left(\sum_\gamma |a_\gamma|^2 |\gamma(s) - \gamma(t)|^2\right)^{1/2}, \quad s,t \in G.$$

Then Y converges uniformly on V almost surely if and only if the entropy integral $E^{(2)}(V,d)$ is finite. Furthermore, for some numerical constant $K > 0$,

$$\left(K\left(1 + L(V)^{1/2}\right)\right)^{-1} \inf_\gamma \mathbb{E}|\xi_\gamma| \left(\left(\sum_\gamma |a_\gamma|^2\right)^{1/2} + E^{(2)}(V,d)\right)$$

$$\leq \left(\mathbb{E}\|Y\|^2\right)^{1/2}$$

$$\leq K\left(1 + L(V)^{1/2}\right)\sup_\gamma\left(\mathbb{E}|\xi_\gamma|^2\right)^{1/2}\left(\left(\sum_\gamma |a_\gamma|^2\right)^{1/2} + E^{(2)}(V,d)\right).$$

In particular, note that Y converges uniformly almost surely if and only if the associated Gaussian Fourier series $\sum_\gamma a_\gamma g_\gamma \gamma$ does. Let us mention further that several of the comments following Theorem 13.3 apply similarly in the context of Corollary 13.9 and that the equivalences of Theorem 13.4 hold in the same way. In particular, boundedness and continuity are equivalent.

The following is a consequence of Corollary 13.9.

Corollary 13.10. *Let Y be as in Corollary 13.9 with $\mathbb{E}|\xi_\gamma|^2 = 1$ for every γ in A. Then, $(Y_t)_{t \in V}$ satisfies the central limit theorem in $C(V)$ if and only if $E^{(2)}(V, d) < \infty$.*

Proof. $(\sum_\gamma a_\gamma g_\gamma \gamma(t))_{t \in V}$ is a Gaussian process with the same covariance structure as $(Y_t)_{t \in V}$. If $(Y_t)_{t \in V}$ satisfies the CLT, it is necessarily pregaussian so that $E^{(2)}(V, d) < \infty$ by (13.3). To prove the sufficiency, consider independent copies of $(Y_t)_{t \in V}$ associated to independent copies (ξ_γ^i) of the sequence (ξ_γ). Since, for every γ,

$$\frac{1}{\sqrt{n}}\left(\mathbb{E}\left|\sum_{i=1}^n \xi_\gamma^i\right|^2\right)^{1/2} = \left(\mathbb{E}|\xi_\gamma|^2\right)^{1/2} = 1,$$

the right hand side of the inequality of Corollary 13.9 together with an approximation argument shows that $(Y_t)_{t \in V}$ satisfies the CLT in $C(V)$ (cf. (10.4)).

Turning back to Proposition 13.8 and its proof, let us now explicit a remarkable Banach algebra of cotype 2 whose cotype 2 property actually basically amounts to the inequality (13.8) and the conclusion of Corollary 13.9. Let us assume here for simplicity that G is (Abelian and) *compact*. Denote by Γ its discrete dual group. Introduce the space $C_{\text{a.s.}} = C_{\text{a.s.}}(G)$ of all sequences of complex numbers $a = (a_\gamma)_{\gamma \in \Gamma}$ in ℓ_2 such that the Gaussian Fourier series $\sum_{\gamma \in \Gamma} a_\gamma g_\gamma \gamma(t)$, $t \in G$, converges uniformly on G almost surely. By what was described so far, we know that we get the same definition when the orthogaussian sequence (g_γ) is replaced by a Rademacher sequence (ε_γ), or even some more general symmetric sequence which enters the framework of Corollary 13.9. Alternatively, $C_{\text{a.s.}}$ is characterized by $E^{(2)}(G, d) < \infty$ where $d(s, t) = \left(\sum_\gamma |a_\gamma|^2 |\gamma(s) - \gamma(t)|^2\right)^{1/2}$, $s, t \in G$, which thus provides a *metric* description of $C_{\text{a.s.}}$ as opposed to the preceding *probabilistic* definition. Equip now the space $C_{\text{a.s.}}$ with the norm

$$[a] = \left(\mathbb{E}\left\|\sum_{\gamma \in \Gamma} a_\gamma g_\gamma \gamma\right\|^2\right)^{1/2}$$

for which it becomes a Banach space. By (13.9) and Proposition 13.8, an equivalent norm is obtained when (g_γ) is replaced by (ε_γ). Similarly, the equivalence of moments of both Gaussian and Rademacher series allow to consider L_p-norms for $1 \leq p \neq 2 < \infty$. Convenient also is to observe that

(13.10) $\frac{1}{4}\left([\text{Re}(a)] + [\text{Im}(a)]\right) \leq [a] \leq 2\left([\text{Re}(a)] + [\text{Im}(a)]\right)$

where $\text{Re}(a) = (\text{Re}(a_\gamma))_\gamma$ and $\text{Im}(a) = (\text{Im}(a_\gamma))_\gamma$. The right hand side inequality is obvious by the triangle inequality. The left hand side follows from the contraction principle. Indeed, if (b_γ) and (α_γ) are complex numbers such that $|\alpha_\gamma| = 1$ for all γ, then

$$\left(\mathbb{E} \left\| \sum_\gamma \alpha_\gamma b_\gamma g_\gamma \gamma \right\|^2 \right)^{1/2} \leq 2 \left(\mathbb{E} \left\| \sum_\gamma b_\gamma g_\gamma \gamma \right\|^2 \right)^{1/2}.$$

Replacing α_γ by $\bar{\alpha}_\gamma$ and b_γ by $\alpha_\gamma b_\gamma$, we get, since $|\alpha_\gamma| = 1$,

$$\left(\mathbb{E} \left\| \sum_\gamma b_\gamma g_\gamma \gamma \right\|^2 \right)^{1/2} \leq 2 \left(\mathbb{E} \left\| \sum_\gamma \alpha_\gamma b_\gamma g_\gamma \gamma \right\|^2 \right)^{1/2}.$$

For $\alpha_\gamma = a_\gamma / |a_\gamma|$ and $b_\gamma = \text{Re}(a_\gamma)$ or $\text{Im}(a_\gamma)$, (13.10) easily follows using one more time the contraction principle.

It is remarkable that the space $C_{\text{a.s.}}$ which arises from a sup-norm has nice cotype properties. This is the content of the following proposition which, as announced, basically amounts to inequality (13.8).

Proposition 13.11. *The space $C_{\text{a.s.}}$ is of cotype 2.*

Proof. We need to show that there is some constant C such that if $a^1, \ldots, a^N$ are elements of $C_{\text{a.s.}}$, then

$$\sum_{i=1}^{N} [\![a^i]\!]^2 \leq C \mathbb{E} \left[\!\!\left[\sum_{i=1}^{N} \varepsilon_i a^i \right]\!\!\right]^2.$$

By (13.10), we need only prove this for real elements $a^1, \ldots, a^N$. Consider the element a of $C_{\text{a.s.}}$ defined, for every γ in Γ, by $a_\gamma = \left(\sum_{i=1}^{N} |a_\gamma^i|^2 \right)^{1/2}$. By Jensen's inequality and (4.3), it is clear that

$$\mathbb{E} \left[\!\!\left[\sum_{i=1}^{N} \varepsilon_i a^i \right]\!\!\right]^2 \geq \frac{1}{2} [\![a]\!]^2$$

so that we have only to show $\sum_{i=1}^{N} [\![a^i]\!]^2 \leq C [\![a]\!]^2$. We simply deduce this from (13.8). Independently of the basic orthogaussian sequence (g_γ), let $A^1, \ldots, A^N$ be disjoint sets with equal probability $1/N$. Let, for γ in Γ,

$$\xi_\gamma = \sqrt{N} \sum_{i=1}^{N} \frac{a_\gamma^i}{a_\gamma} I_{A^i}.$$

Clearly

$$\sum_{i=1}^{N} [\![a^i]\!]^2 = \mathbb{E} \left\| \sum_\gamma a_\gamma \xi_\gamma g_\gamma \gamma \right\|^2.$$

Since $\mathbb{E}|\xi_\gamma|^2 = 1$ for every γ, the conclusion follows from (13.8). Proposition 13.11 is established.

It is interesting to present some remarkable subspaces of the Banach space $C_{\text{a.s.}} = C_{\text{a.s.}}(G)$. Let G be as before a compact Abelian group. A subset Λ of the dual group Γ of G is called a *Sidon* set if there is a constant C such that for every finite sequence (α_γ) of complex numbers

$$\sum_{\gamma \in \Lambda} |\alpha_\gamma| \le C \left\| \sum_{\gamma \in \Lambda} \alpha_\gamma \gamma \right\|.$$

Since the converse inequality (with $C = 1$) is clearly satisfied, $\{\gamma;\ \gamma \in \Lambda\}$ generates, when Λ is a Sidon set, a subspace of $C(G)$ isomorphic to ℓ_1. A typical example is provided by a lacunary sequence (in the sense of Hadamard) in $\mathbb{Z}$, the dual group of the torus group. As we have seen in Chapter 4, Section 4.1, the Rademacher sequence in the Cantor group is another example of Sidon set. If $a = (a_\gamma)_{\gamma \in \Gamma}$ is a sequence of complex numbers vanishing outside some Sidon set Λ of Γ, then the norm $[a]$ is equivalent to the ℓ_1-norm $\sum_\gamma |a_\gamma|$. ℓ_1 is of cotype 2 and it is remarkable that the norm $[\cdot]$ preserves this property. On the other hand, $C_{\text{a.s.}}$ is of no type $p > 1$ (since this is not the case for ℓ_1).

The consideration of the space $C_{\text{a.s.}}$ gives rise to another interesting observation. Let G be compact Abelian. Any function f in $L_2(G)$ admits a Fourier expansion $\sum_{\gamma \in \Gamma} \hat{f}(\gamma)\gamma$ which converges to f in $L_2(G)$. We denote by $[\hat{f}]$ the norm in $C_{\text{a.s.}}$ of the Fourier transform $\hat{f} = (\hat{f}(\gamma))_{\gamma \in \Gamma}$ of f. First, we note the following. Let F be a complex valued function on $\mathbb{C}$ such that

$$|F(x) - F(y)| \le |x - y| \quad \text{and} \quad |F(x)| \le 1 \quad \text{for all} \ \ x, y \in \mathbb{C}.$$

Let f be a function in $C(G)$ such that $\hat{f}$ belongs to $C_{\text{a.s.}}$. Then, $h = F \circ f$ belongs to $C_{\text{a.s.}}$ and for some numerical constant $K_1 \ge 1$,

(13.11) $$[\hat{h}] \le K_1 (\|f\| + [\hat{f}])$$

where we recall that $\|\cdot\|$ is the sup-norm (on G). This property is an easy consequence of the comparison theorems for Gaussian processes. For any t in G, set $f_t(x) = f(t + x)$. If (X_t) is the Gaussian process $X_t = \sum_\gamma \hat{f}(\gamma)g_\gamma\gamma(t)$, for all s, t in G,

$$\|X_s - X_t\|_2 = \|f_s - f_t\|_2$$

where the L_2-norm on the left is understood with respect to the Gaussian sequence (g_γ) and the one on the right with respect to the Haar measure on G. To establish (13.11), since F is 1-Lipschitz, we have

$$\|h_s - h_t\|_2 \le \|f_s - f_t\|_2 \quad \text{and} \quad \|h\|_2 \le 1$$

and the inequality then follows from (for example) Corollary 3.14 (in some complex formulation).

Let B be the space of the functions f in $C(G)$ whose Fourier transform $\hat{f}$ belongs to $C_{\text{a.s.}} = C_{\text{a.s.}}(G)$. Equip B with the norm

$$\|f\| = K_1\big(3\|f\| + [\hat{f}]\big),$$

where K_1 is the numerical constant which appears in (13.11), for which it becomes a Banach space. The trigonometric polynomials are dense in B (cf. Theorem 3.4). As a result, B is actually a Banach algebra for the pointwise product such that $\|fh\| \le \|f\| \|h\|$ for all f, h in B. This can be proved directly or as a consequence of (13.11). Let F be defined as

$$F(x) = \begin{cases} x^2/2 & \text{if } |x| \le 1, \\ x^2/2|x|^2 & \text{if } |x| > 1. \end{cases}$$

Then, if f, h are in B with $\|f\|, \|h\| \le 1$, since $4fh = (f+h)^2 - (f-h)^2$, as a corollary of (13.11),

$$[\widehat{fh}] \le 2K_1\big(2 + [\hat{f}] + [\hat{h}]\big).$$

The inequality $\|fh\| \le \|f\| \|h\|$ follows from the definition of $\|\cdot\|$.

Let $A(G)$ be the algebra of the elements of $C(G)$ whose Fourier transform is absolutely summable. The preceding algebra B is an example of a (strongly homogeneous) Banach algebra with $A(G) \not\subset B \not\subset C(G)$ on which all Lipschitz functions of order 1 operate. One might wonder for a *minimal* algebra with these properties. In this setting the following algebra might be of some interest. Let $\tilde{B}$ be the space of the functions f in $C(G)$ such that

$$\sup\left\{ E\sup_{t\in G}\Big|\sum_{i=1}^{n} a_i g_i f_t(x_i)\Big| \, ; \, n \ge 1, \; x_1, \dots, x_n \in G, \; \sum_{i=1}^{n} a_i^2 \le 1 \right\} < \infty.$$

It is not difficult to see, as above, that this quantity (at the exception perhaps of a numerical factor) defines a norm on $\tilde{B}$ for which $\tilde{B}$ is a Banach algebra with $A(B) \not\subset \tilde{B} \not\subset C(G)$ on which all 1-Lipschitz functions operate. Further, $\tilde{B}$ is smaller than B. To see this, let (Z_i) be independent random variables with values in G and common distribution λ (the normalized Haar measure on G). Then, if $f \in \tilde{B}$,

$$\sup_n \frac{1}{\sqrt{n}} \, \mathbb{E}\sup_{t\in G}\Big|\sum_{i=1}^{n} g_i f_t(Z_i)\Big| < \infty.$$

Therefore, by the central limit theorem in finite dimension, it follows that the Gaussian process with L_2-metric

$$\big(\mathbb{E}|f_s(Z_1) - f_t(Z_1)|^2\big)^{1/2} = \|f_s - f_t\|_2, \quad s, t \in G,$$

is almost surely bounded. Now (Theorem 13.4), f belongs to $C_{\text{a.s.}}$ which proves the desired claim.

A deeper analysis of the Banach algebra $\tilde{B}$ is still to be done. Is it, in particular, the smallest algebra on which the 1-Lipschitz functions operate?

13.3 Stable Random Fourier Series
and Strongly Stationary Processes

In the preceding sections, we investigated stationary Gaussian processes and random Fourier series of the type $\sum_\gamma a_\gamma \xi_\gamma \gamma$ where (ξ_γ) is a symmetric sequence of real valued random variables satisfying basically $\sup_\gamma \mathbb{E}|\xi_\gamma|^2 < \infty$. Here, we shall be interested in possible extensions to the stable case and to random Fourier series as above when the sequence (ξ_γ) satisfies some weaker moment assumptions (a typical example of both topics being given by the choice, for (ξ_γ), of a standard p-stable sequence).

Throughout this section, G denotes a locally compact Abelian group with dual group Γ, identity 0 and Haar measure $\lambda(\cdot) = |\cdot|$. If $X = (X_t)_{t \in G}$ is a *stationary* Gaussian process continuous in L_2, by Bochner's theorem, there is a measure m in Γ which represents the covariance of X in the sense that for all finite sequences of real numbers (α_j) and (t_j) in G,

$$\mathbb{E}\left|\sum_j \alpha_j X_{t_j}\right|^2 = \int_\Gamma \left|\sum_j \alpha_j \gamma(t_j)\right|^2 dm(\gamma).$$

Let $0 < p \le 2$. Say that a p-stable process $X = (X_t)_{t \in G}$ indexed by G is *strongly stationary*, or *harmonizable*, if there is a finite positive Radon measure m concentrated on Γ such that, for all finite sequences (α_j) of real numbers and (t_j) of elements of G,

$$(13.12) \quad \mathbb{E}\exp\left(i\sum_j \alpha_j X_{t_j}\right) = \exp\left(-\frac{1}{2}\int_\Gamma \left|\sum_j \alpha_j \gamma(t_j)\right|^p dm(\gamma)\right).$$

Going back to the spectral representation of stable processes (Theorem 5.2), we thus assume in this definition of strongly stationary stable processes a special property of the spectral measure m, namely to be concentrated on the dual group Γ of G. This property is motivated by several facts, some of which will become clear in the subsequent developments. In particular, strongly stationary stable processes are stationary in the usual sense; however, and refering to [M-P2], contrary to the Gaussian case not all stationary stable processes are strongly stationary.

The following example is noteworthy. Let $\tilde{\theta}$ be a complex stable random variable such that, as a variable in $\mathbb{R}^2$, $\tilde{\theta}$ has for spectral measure the uniform distribution on the unit circle. That is, if $\tilde{\theta} = \theta_1 + i\theta_2 = (\theta_1, \theta_2)$ and $\alpha = \alpha_1 + i\alpha_2 = (\alpha_1, \alpha_2) \in \mathbb{C}$,

$$(13.13) \quad \mathbb{E}\exp\big(i\mathrm{Re}(\overline{\alpha}\tilde{\theta})\big) = \mathbb{E}\exp\big(i(\alpha_1\theta_1 + \alpha_2\theta_2)\big) = \exp\big(-\tfrac{1}{2}c_p'{}^p|\alpha|^p\big)$$

where $c_p' = \left(\int_0^{2\pi} |\cos x|^p dx/2\pi\right)^{1/p}$. (Only in the Gaussian case $p = 2$, θ_1 and θ_2 are necessarily independent.) This definition is one 2-dimensional extension of the real valued stable variables for which spectral measures are concentrated on $\{-1, +1\}$. Let A be a countable subset of Γ. Let further $(\tilde{\theta}_\gamma)_{\gamma \in A}$ be a sequence of independent variables distributed as $\tilde{\theta}$ and $(a_\gamma)_{\gamma \in A}$ be complex numbers with $\sum_\gamma |a_\gamma|^p < \infty$. Then

$$X_t = \sum_{\gamma \in A} a_\gamma \tilde{\theta}_\gamma \gamma(t), \quad t \in G,$$

defines a *complex* strongly stationary p-stable process. Its real and imaginary parts are real valued strongly stationary p-stable processes in the sense of definition (13.12). The spectral measure is discrete in this case. Strongly stationary stable processes therefore include random Fourier series with stable coefficients. Since, up to a constant depending on p only, $\mathrm{Re}\tilde{\theta}$ and $\mathrm{Im}\tilde{\theta}$ are standard p-stable real valued variables, this study can be shown to include similarly random Fourier series of the type $\sum_\gamma a_\gamma \theta_\gamma \gamma$ where (θ_γ) is a standard p-stable sequence (real valued). We shall come back to this from a somewhat simpler point of view later.

In the first part of this section, we extend to strongly stationary p-stable processes, $1 \leq p < 2$, the Gaussian characterization of Theorem 13.3. Necessity will follow easily from Section 12.2 and Proposition 13.2. Sufficiency uses the series representation of stable processes (Corollary 5.3) together with some ideas developed in the preceding section on random Fourier series (conditional estimates and integration in presense of translation invariant metrics). As usual, we exclude from such a study the case $0 < p < 1$ since the only property that m is finite already ensures that a p-stable process, $0 < p < 1$, has a version with bounded and continuous paths. Thus, it is assumed henceforth that $1 \leq p \leq 2$, and actually also that $p < 2$ since the case $p = 2$ has already been studied.

Let $X = (X_t)_{t \in G}$ be a strongly stationary p-stable process with $1 \leq p < 2$ and with spectral measure m (in (13.12)). For s, t in G, denote by $d_X(s, t)$ the parameter of the real p-stable variable $X_s - X_t$, that is

$$d_X(s, t) = \left(\int_\Gamma |\gamma(s) - \gamma(t)|^p dm(\gamma) \right)^{1/p}.$$

d_X defines a translation invariant pseudo-metric. Let V be a fixed compact neighborhood of the unit element 0 of G. We shall always assume that V is metrizable. d_X is then continuous on $V \times V$ (by dominated convergence) (and thus also on $G \times G$). We know from Theorem 12.12 a general necessary condition for the boundedness of p-stable processes. Together with Proposition 13.2, it yields that if X has a version with almost all sample paths bounded on V, then $E^{(q)}(V, d_X) < \infty$ where $q = p/p-1$ is the conjugate of p. Furthermore, for some constant K_p depending only on p,

(13.14) $$\left\| \sup_{t \in V} |X_t| \right\|_{p,\infty} \geq K_p^{-1} \left(E^{(q)}(V, d_X) - L(V)^{1/q} D(V) \right)$$

where $D(V)$ is the diameter of (V, d_X) and $L(V) = \log(|V''|/|V|)$ ($V'' = V + V + V$). (When $q = \infty$ we agree that $L(V)^{1/q} = \log(1 + \log(|V''|/|V|))$.) In the following, $\left\| \sup_{t \in V} |X_t| \right\|_{p,\infty}$ will be denoted for simplicity by $\|X\|_{p,\infty}$.

This settles the necessary part of this study of the almost sure boundedness and continuity of strongly stationary p-stable processes. Now, we turn to the sufficiency and show, as a main result, that the preceding necessary entropy condition is also sufficient. As announced, the proof uses various arguments developed in Section 13.2. We only prove the result for $1 < p < 2$. The case $p = 1$ can basically be obtained similarly with however some more care. We refer to [Ta14] for the case $p = 1$.

Theorem 13.12. *Let $X = (X_t)_{t \in G}$ be a strongly stationary p-stable process, $1 \leq p < 2$. Let $q = p/p-1$. Then X has a version with almost all sample paths bounded and continuous on $V \subset G$ if and only if $E^{(q)}(V, d_X) < \infty$. Moreover, there is a constant $K_p > 0$ depending on p only such that*

$$K_p^{-1}\big(|m|^{1/p} + E^{(q)}(V, d_X) - L(V)^{1/q}D(V)\big)$$
$$\leq \|X\|_{p,\infty} \leq K_p\big((1 + L(V)^{1/q})|m|^{1/p} + E^{(q)}(V, d_X)\big)$$

where m is the spectral measure of X and where $D(V)$ the diameter of (V, d_X).

Proof. Necessity and the left hand side inequality have been discussed above. Recall the series representation of Corollary 5.3. Let Y_j be independent random variables distributed as $m/|m|$. Let further w_j be independent complex random variables all of them uniformly distributed on the unit circle of $\mathbb{R}^2$. Assume that the sequences (ε_j), (Γ_j), (w_j), (Y_j) are independent. Then, by Corollary 5.3 and (13.12), X has the same distribution as

$$c_p(c_p')^{-1}|m|^{1/p} \sum_{j=1}^{\infty} \Gamma_j^{-1/p}\varepsilon_j \mathrm{Re}\big(w_j Y_j(t)\big), \quad t \in G,$$

where c_p' appears in (13.13). For simplicity in the notation, we denote again by X this representation. Under $E^{(q)}(V, d_X) < \infty$, we will show, *exactly* as in the proof of Theorem 13.6, that this series has a version with almost all sample paths bounded and continuous, satisfying moreover the inequality of the theorem. By homogeneity, let us assume that $|m| = 1$. Conditionally on (Γ_j), (w_j) and (Y_j), we apply Lemma 13.5. For every ω of the probability space supporting these sequences, denote by $d_\omega(s, t)$ the translation invariant functional

$$d_\omega(s, t) = \|\big(\Gamma_j^{-1/p}(\omega)\mathrm{Re}\big(w_j(\omega)\big(Y_j(\omega, s) - Y_j(\omega, t)\big)\big)\big)\|_{p,\infty}.$$

Since the Y_j's take their values in Γ, the dual group of G, it is easy to verify that d_ω is, for almost all ω, continuous on $V \times V$. Indeed, if τ is a metric for which V is compact, we deduce from (5.8) and Corollary 5.9 that

$$\mathbb{E} \sup_{\substack{\tau(s,t)<\varepsilon \\ s,t \in V}} d_\omega(s, t) \leq K_p \mathbb{E} \sup_{\substack{\tau(s,t)<\varepsilon \\ s,t \in V}} |Y_1(s) - Y_1(t)|$$

for all $\varepsilon > 0$. (K_p denotes here and below some constant depending on $1 < p < 2$ only.) The claim thus follows. Moreover, since $\|\mathrm{Re}(w_1(Y_1(s) - Y_1(t)))\|_p = c_p' d_X(s,t)$, we have from exactly the same tools that

$$(13.15) \qquad \mathbb{E}d_\omega(s,t) \leq K_p d_X(s,t)$$

for all s,t. Once this has been observed, the rest of the proof is entirely similar to the proof of Theorem 13.6. By Lemma 13.5, letting $R(\omega) = 2\|(\Gamma_j^{-1/p}(\omega))\|_{p,\infty}$,

$$
\begin{aligned}
(13.16) \qquad & \mathbb{E}_\varepsilon \sup_{t \in V} \left| \sum_{j=1}^\infty \Gamma_j^{-1/p}(\omega) \varepsilon_j \mathrm{Re}(w_j(\omega) Y_j(\omega,t)) \right| \\
& \leq K_p \left(R(\omega) + \int_0^{R(\omega)} (\log N(V, d_\omega; \varepsilon))^{1/q} d\varepsilon \right)
\end{aligned}
$$

and, if this entropy integral is finite, $\left(\sum_{j=1}^\infty \Gamma_j^{-1/p}(\omega) \varepsilon_j \mathrm{Re}(w_j(\omega) Y_j(\omega,t))\right)_{t \in V}$ has a version with almost all (with respect to (ε_j)) sample paths continuous (since d_ω is continuous on $V \times V$). W_n and $b_n(\omega)$ being as in the proof of Theorem 13.6, we have from (13.15) that $\mathbb{E}b_n \leq K_p 2^{-n}$. Furthermore, from Lemma 13.7,

$$\int_0^{R(\omega)} (\log N(V, d_\omega; \varepsilon))^{1/q} d\varepsilon \leq \sum_{n=0}^\infty b_n(\omega) \left(\log \left(2\frac{|V''|}{|V|} N(V, d_X; 2^{-n-1}) \right) \right)^{1/q}.$$

Integrating with respect to ω, it follows that, when $E^{(q)}(V, d_X) < \infty$, for almost all ω the entropy integral on the left is finite and thus, by the preceding, $\left(\sum_{j=1}^\infty \Gamma_j^{-1/p}(\omega) \varepsilon_j \mathrm{Re}(w_j(\omega) Y_j(\omega,t))\right)_{t \in T}$ has a version with continuous sample paths with respect to (ε_j). Therefore, by Fubini's theorem and the representation, X has a version with continuous paths on V. Furthermore, integrating (13.16) together with the fact that $\mathbb{E}b_n \leq K_p 2^{-n}$ yields

$$\mathbb{E}\|X\| \leq K_p\big(1 + L(V)^{1/q} + E^{(q)}(V, d_X)\big).$$

Since $\mathbb{E}\|X\|$ is equivalent to $\|X\|_{p,\infty}$ (Proposition 5.6), the proof of Theorem 13.12 is complete (recall that we have assumed by homogeneity that $|m| = 1$).

Motivated by the previous result, we further investigate in the second part of this section random Fourier series with the objective of enlarging the conclusions of Theorem 13.6 or Corollary 13.9. In particular, we would like to study the case of a sequence (ξ_γ) there not necessarily in L_2. One typical example is a stable random Fourier series $\sum_\gamma a_\gamma \theta_\gamma \gamma$ where (θ_γ) is a standard p-stable sequence. We have seen that this example can be shown to enter the previous setting. Now, we present some natural extensions in the context of random Fourier series.

As in Section 13.2, G is a locally compact Abelian group with unit 0 and dual group Γ, V is a fixed compact symmetric neighborhood of 0 and A is a countable subset of Γ. Let $1 \le p < 2$. Let $(a_\gamma)_{\gamma \in A}$ be a sequence of complex numbers such that $\sum_\gamma |a_\gamma|^p < \infty$ and let $(\xi_\gamma)_{\gamma \in A}$ be a sequence of *independent and symmetric* real valued random variables. We are interested in the almost sure uniform convergence of the random Fourier series $Y = (Y_t)_{t \in G}$ where

$$(13.17) \qquad Y_t = \sum_{\gamma \in A} a_\gamma \xi_\gamma \gamma(t), \quad t \in G,$$

in terms of the translation invariant pseudo-metric

$$d(s,t) = \left(\sum_{\gamma \in A} |a_\gamma|^p |\gamma(s) - \gamma(t)|^p \right)^{1/p}, \quad s, t \in G.$$

The technique of proof of Theorems 13.6 and 13.12 enables us to extend the results of Section 13.2 to random variables ξ_γ which do not have finite second moments.

Theorem 13.13. *Assume that* $\sup_\gamma \|\xi_\gamma\|_{p,\infty} < \infty$. *Then, if* $E^{(q)}(V,d) < \infty$ *where* $q = p/p - 1$ *and where* d *is defined above, the random Fourier series* Y *of* (13.17) *converges uniformly on* V *with probability one. Furthermore, for some constant* $K_p > 0$ *depending on* p *only,*

$$\|Y\|_{p,\infty} \le K_p \sup_\gamma \|\xi_\gamma\|_{p,\infty} \left((1 + L(V)^{1/q}) \left(\sum_\gamma |a_\gamma|^p \right)^{1/p} + E^{(q)}(V,d) \right)$$

(where $\|Y\|_{p,\infty} = \| \sup_{t \in V} |Y_t| \|_{p,\infty}$*).*

Proof. It is entirely similar to (actually somewhat simpler than) the proofs of Theorems 13.6 and 13.12 so that we only mention a few observations. By independence and symmetry, Y can be replaced by

$$\sum_{\gamma \in A} a_\gamma \varepsilon_\gamma \xi_\gamma \gamma(t), \quad t \in G,$$

where (ε_γ) is a Rademacher sequence which is independent of (ξ_γ). Then, we use Lemma 13.5 conditionally on (ξ_γ) with respect to the metric

$$d_\omega(s,t) = \| \left(|a_\gamma \xi_\gamma(\omega)|(\gamma(s) - \gamma(t)) \right)_{\gamma \in A} \|_{p,\infty}.$$

Since the ξ_γ's are *independent*, we can integrate with respect to ω and use Lemma 5.8 and the hypothesis $\sup_\gamma \|\xi_\gamma\|_{p,\infty} < \infty$. The proof is completed similarly.

Note that if the sequence (ξ_γ) is only a *symmetric* sequence, the preceding theorem holds similarly *but* with $\sup_\gamma \|\xi_\gamma\|_p$ instead of $L_{p,\infty}$-moments. The argument is similar but, to integrate d_ω, since the ξ_γ's need not be independent, we simply use that

$$d_\omega(s,t) \le \left(\sum_\gamma |a_\gamma|^p |\xi_\gamma(\omega)|^p |\gamma(s) - \gamma(t)|^p \right)^{1/p}.$$

If the tails of the random variables ξ_γ are close to the tail of a standard p-stable variable with $1 \leq p < 2$, then the entropy condition $E^{(q)}(V,d) < \infty$ is also necessary for the random Fourier series Y of (13.17) to be almost surely bounded. More precisely, assume that for some $u_0 > 0$ and some $\delta > 0$,

$$\mathbb{P}\{|\xi_\gamma| > u\} \geq \delta u^{-p}$$

for all $u > u_0$ and all γ. Hence, by (5.2), if (θ_γ) is a standard p-stable sequence,

$$\inf_\gamma \mathbb{P}\{|\xi_\gamma| > u\}/\mathbb{P}\{|\theta_\gamma| > u\}$$

is bounded below for u sufficiently large. Therefore, if $\sum_\gamma a_\gamma \xi_\gamma \gamma$ converges uniformly almost surely, the same holds for $\sum_\gamma a_\gamma \theta_\gamma \gamma$ by Lemma 4.6. Now, we deal with a stable process and the complex version of (13.14) yields the announced claim. This approach can also be used to deduce Theorem 13.13 from Theorem 13.12.

Finally, the various comments developed in the Gaussian setting following Theorem 13.3 also apply in this stable case. The same is true for the equivalences of Theorem 13.4 in the context of stable Fourier series $\sum_\gamma a_\gamma \theta_\gamma \gamma$.

13.4 Vector Valued Random Fourier Series

In the last part of this chapter, we present some applications of the previous results to vector valued stationary processes and random Fourier series. The results are still fragmentary and concern so far Gaussian variables only.

Let B be a separable Banach space with dual space B'. Recall that by a process $X = (X_t)_{t \in T}$ with values in B we simply mean a family $(X_t)_{t \in T}$ such that, for each t, X_t is a Borel random variable with values in B. X is Gaussian if for every $t_1, \ldots, t_N$ in T, $(X_{t_1}, \ldots, X_{t_N})$ is Gaussian (in B^N). As in the preceding sections, let G be a locally compact Abelian group with identity 0 and dual group of characters Γ. Let us fix also a compact metrizable neighborhood V of 0. A process $X = (X_t)_{t \in G}$ indexed by G and with values in B is said to be stationary if, as in the real case, for every u in G, $(X_{u+t})_{t \in G}$ has the same distribution (on B^G) as X. Since B is separable, this is equivalent to saying that for every f in B', the real valued process $(f(X_t))_{t \in G}$ is stationary.

Almost sure boundedness and continuity of vector valued stationary Gaussian processes may be characterized rather easily through the corresponding properties along linear functionals. This is the content of the following statement. While this result is in relation with tensorization of Gaussian measures studied in Section 3.3, it does not seem possible to deduce it from the comparison theorems based on Slepian's lemma. Instead, we use majorizing measures and the deep results of Chapter 12. Recall that $L(V) = \log(|V''|/|V|)$, $V'' = V + V + V$.

Theorem 13.14. *Let $X = (X_t)_{t \in G}$ be a stationary Gaussian process with values in B. Then, for some numerical constant $K > 0$,*

$$\frac{1}{2}\left(\mathbb{E}\|X_0\| + \sup_{\|f\|\leq 1} \mathbb{E}\sup_{t\in V}|f(X_t)|\right)$$

$$\leq \mathbb{E}\sup_{t\in V}\|X_t\|$$

$$\leq K\left(1 + L(V)^{1/2}\right)\left(\mathbb{E}\|X_0\| + \sup_{\|f\|\leq 1}\mathbb{E}\sup_{t\in V}|f(X_t)|\right).$$

Moreover, X has a version with almost all continuous paths on V if and only if $\|X_s - X_t\|_2$ is continuous on $V \times V$ and

$$\lim_{\eta\to 0} \sup_{\|f\|\leq 1} \mathbb{E}\sup_{t\in\eta V}|f(X_t)| = 0.$$

Note that, by the results of Section 13.1, X is continuous on V if and only if

$$\lim_{\eta\to 0} \sup_{\|f\|\leq 1} \int_0^\eta \left(\log N(V, d_{f(X)}; \varepsilon)\right)^{1/2} d\varepsilon = 0$$

(and $\|X_s - X_t\|_2$ is continuous on $V \times V$) where $d_{f(X)}(s,t) = \|f(X_s) - f(X_t)\|_2$, $f \in B'$, $s, t \in G$.

Proof. We only show the part concerning boundedness and the inequalities of the theorem. Continuity follows similarly together with the preceding observation. Let B_1' be the unit ball of B'. We consider X as a process indexed by $B_1' \times G$. By Theorem 12.9, the real valued Gaussian process $X_0 = (f(X_0))_{f\in B_1'}$ indexed by B_1' has a majorizing measure; that is, there exists a probability measure m on (B_1', d_{X_0}) such that

$$(13.18) \quad \sup_{f\in B_1'} \int_0^\infty \left(\log\frac{1}{m\big(B(f,\varepsilon)\big)}\right)^{1/2} d\varepsilon \leq K\mathbb{E}\sup_{f\in B_1'} f(X_0) = K\mathbb{E}\|X_0\|$$

where, as in Chapter 12, $B(f,\varepsilon)$ is the ball of radius ε with respect to the metric on the space which contains its center f, that is, $d_{X_0}(f,g) = \|f(X_0) - g(X_0)\|_2$, $f, g \in B_1'$. (We use further this convention about balls in metric spaces below.) We intend to use (13.1) so let $\lambda_{V''}$ be the restricted normalized Haar measure on $V'' \subset G$. If we bound X considered as a (real valued) process on $B_1' \times V$ with the majorizing measure integral for $m \times \lambda_{V''}$ (on $B_1' \times V'' \supset B_1' \times V$), we get from Theorem 11.18 and Lemma 11.9 (which similarly applies with the function $(\log(1/x))^{1/2}$),

$$\mathbb{E}\sup_{(f,t)\in B_1'\times V} f(X_t)$$

$$(13.19)$$

$$\leq K \sup_{(f,t)\in B_1'\times V} \int_0^\infty \left(\log\frac{1}{m \times \lambda_{V''}\big(B((f,t),\varepsilon)\big)}\right)^{1/2} d\varepsilon$$

where $B((f,t),\varepsilon)$ is the ball for the L_2-pseudo-metric of X on $B_1' \times G$, i.e.

$$d\big((f,t),(g,s)\big) = \|f(X_t) - g(X_s)\|_2,$$

$f, g \in B'$, $s, t \in G$. To control the integral on the right hand side of (13.19), note the following. By the triangle inequality and the stationarity,

$$d((f,t),(g,s)) \leq d_{X_0}(f,g) + d_{f(X)}(s,t)$$

where we recall that $d_{f(X)}(s,t) = \|f(X_s) - f(X_t)\|_2$. It follows that, for every (f,t) in $B'_1 \times V$ and all $\varepsilon > 0$,

$$m \times \lambda_{V''}(B((f,t), 2\varepsilon)) \geq m(B(f,\varepsilon))\lambda_{V''}(B_{d_{f(X)}}(t,\varepsilon))$$

where $B_{d_{f(X)}}(t,\varepsilon)$ is the ball with respect to the metric $d_{f(X)}$. Therefore,

$$\mathbb{E} \sup_{B'_1 \times V} f(X_t) \leq 2K \left(\sup_{B'_1} \int_0^\infty \left(\log \frac{1}{m(B(f,\varepsilon))} \right)^{1/2} d\varepsilon \right.$$
$$\left. + \sup_{B'_1 \times V} \int_0^\infty \left(\log \frac{1}{\lambda_{V''}(B_{d_{f(X)}}(t,\varepsilon))} \right)^{1/2} d\varepsilon \right).$$

The first term on the right of this inequality is controlled by (13.18). We use (13.1) (which applies too with the function $(\log(1/x))^{1/2}$) to see that the second term is smaller than or equal to

$$\sup_{f \in B'_1} \int_0^\infty \left(\log \left(\frac{|V''|}{|V|} N(V, d_{f(X)}; \varepsilon) \right) \right)^{1/2} d\varepsilon.$$

Summarizing, we get from Theorem 13.3 that for some numerical constant K,

$$\mathbb{E} \sup_{t \in V} \|X_t\| \leq K \left(\mathbb{E}\|X_0\| + \sup_{f \in B'_1} \mathbb{E} \sup_{t \in V} |f(X_t)| + L(V)^{1/2} \sup_{f \in B'_1} (\mathbb{E} f^2(X_0))^{1/2} \right).$$

This inequality is stronger than the upper bound of the theorem. The minoration inequality is obvious. The proof is, therefore, complete.

One interesting application of Theorem 13.14 concerns Gaussian random Fourier series with vector valued coefficients. Let A be a fixed countable subset of the dual group of characters Γ of G. Let $(g_\gamma)_{\gamma \in A}$ be an orthogaussian sequence and $(x_\gamma)_{\gamma \in A}$ be a sequence of elements of a Banach space B. We assume that B is a complex Banach space. Suppose that the series $\sum_\gamma g_\gamma x_\gamma$ converges. Then define (using the contraction principle) the Gaussian random Fourier series $X = (X_t)_{t \in G}$ by

$$(13.20) \qquad X_t = \sum_{\gamma \in A} g_\gamma x_\gamma \gamma(t), \quad t \in G.$$

As in the scalar case, one might wonder for the almost sure uniform convergence of the series (13.20) (in the sup-norm $\sup_{t \in V} \| \cdot \|$) or, equivalently (by Itô-Nisio's theorem), the almost sure continuity of the process X on V. Theorem 13.14 implies the following.

Corollary 13.15. *In the preceding notation, there is a numerical constant K such that*

$$\frac{1}{2}\left(\mathbb{E}\left\|\sum_\gamma g_\gamma x_\gamma\right\| + \sup_{\|f\|\leq 1}\mathbb{E}\sup_{t\in V}\left|\sum_\gamma f(x_\gamma)g_\gamma\gamma(t)\right|\right)$$

$$\leq \mathbb{E}\sup_{t\in V}\left\|\sum_\gamma g_\gamma x_\gamma\gamma(t)\right\|$$

$$\leq K\left(1+L(V)^{1/2}\right)\left(\mathbb{E}\left\|\sum_\gamma g_\gamma x_\gamma\right\| + \sup_{\|f\|\leq 1}\mathbb{E}\sup_{t\in V}\left|\sum_\gamma f(x_\gamma)g_\gamma\gamma(t)\right|\right).$$

Furthermore, $\sum_\gamma g_\gamma x_\gamma\gamma$ converges uniformly almost surely if and only if

$$\lim\sup_{\|f\|\leq 1}\mathbb{E}\sup_{t\in V}\left|\sum_{\gamma\in F^c} f(x_\gamma)g_\gamma\gamma(t)\right| = 0$$

where the limit is taken over the finite sets F increasing to A.

The scalar case investigation of Sections 13.2 and 13.3 invites to consider the same convergence question for vector valued random Fourier series of the type (13.20) when, for example, the Gaussian sequence (g_γ) is replaced by a Rademacher sequence (ε_γ) or a standard p-stable sequence (θ_γ), $1 \leq p < 2$. These questions are not yet answered. By the equivalence of scalar Gaussian and Rademacher Fourier series (Proposition 13.8), it is plain from Corollary 13.15 that a Rademacher series $\sum_\gamma \varepsilon_\gamma x_\gamma\gamma$ is characterized as the corresponding Gaussian series provided $\sum_\gamma \varepsilon_\gamma x_\gamma$ and $\sum_\gamma g_\gamma x_\gamma$ converge simultaneously. We know that this holds for all sequences (x_γ) in B if and only if B is of finite cotype (Theorem 9.16) but, in general, $\sum_\gamma \varepsilon_\gamma x_\gamma$ is only dominated by $\sum_\gamma g_\gamma x_\gamma$ ((4.8)). However, we conjecture that Corollary 13.15 and its inequality also hold when (g_γ) is replaced by (ε_γ) (of course, note that the left hand side inequality is trivial). This conjecture is supported by the fact that it holds for Rademacher processes which are $Kr(T)$-decomposable in the terminology of Section 12.3; this fact is checked immediately by reproducing the argument of the proof of Theorem 13.14. The case of a p-stable standard sequence, $1 \leq p < 2$, in Corollary 13.15 is also open.

Stationary real valued Gaussian (and strongly stationary p-stable) processes are either continuous or unbounded. This follows from the characterizations that we described and extends to the classes of random Fourier series studied there. To conclude, we analyze this dichotomy for general random Fourier series with vector valued coefficients.

Let A be a countable subset of Γ. Further, let $(x_\gamma)_{\gamma\in A}$ be a sequence in a complex Banach space B and let $(\xi_\gamma)_{\gamma\in A}$ be independent symmetric real valued random variables. Assuming that $\sum_\gamma \xi_\gamma x_\gamma$ converges almost surely for one (or, by symmetry, all) ordering of A, consider the random Fourier series $X = (X_t)_{t\in G}$ given by

$$(13.21) \qquad X_t = \sum_{\gamma\in A}\xi_\gamma x_\gamma\gamma(t), \quad t\in G.$$

Let V be as usual a compact neighborhood of 0 in G. The random Fourier series X is said to be almost surely uniformly bounded if, for some ordering of $A = \{\gamma_n; n \geq 1\}$, the partial sums $\sum_{n=1}^{N} \xi_{\gamma_n} x_{\gamma_n} \gamma_n$, $N \geq 1$, are almost surely uniformly bounded with respect to the norm $\sup_{t \in V} \| \cdot \|$. Since (ξ_γ) is a symmetric sequence, by Lévy's inequalities, the preceding definition is independent of the ordering of A and is equivalent to the boundedness of X as a process on V. Similarly, the random Fourier series X of (13.21) is almost surely uniformly convergent if for some (or all by Itô-Nisio's theorem) ordering $A = \{\gamma_n; n \geq 1\}$, the preceding partial sums converge uniformly almost surely. Equivalently, X defines an almost surely continuous process on V with values in B. The next theorem describes how these two properties are equivalent for scalar random Fourier series of the type (13.21) and similarly for B-valued coefficients if B does not contain an isomorphic copy of c_0. The proof is based on Theorem 9.29 (which is identical for complex Banach spaces).

Theorem 13.16. *For a Banach space B, the following are equivalent:*

(i) B does not contain subspaces isomorphic to c_0;

(ii) every almost surely uniformly bounded random Fourier series of the type (13.21) with coefficients in B converges almost surely uniformly.

Proof. If (ii) does not hold, there exists a series (13.21) such that, for some ordering $A = \{\gamma_n; n \geq 1\}$, $\sum_n \xi_{\gamma_n} x_{\gamma_n} \gamma_n$ is an almost surely bounded series which does not converge. Let us set for simplicity $\xi_n = \xi_{\gamma_n}$ and $x_n = x_{\gamma_n}$ for every n. By Remark 9.31, there exist ω and a sequence (n_k) such that $(\xi_{n_k}(\omega) x_{n_k} \gamma_{n_k})$ is equivalent in the norm $\sup_{t \in V} \| \cdot \|$ to the canonical basis of c_0. Note that $\|\xi_{n_k}(\omega) x_{n_k}\| = \sup_{t \in V} \|\xi_{n_k}(\omega) x_{n_k} \gamma_{n_k}(t)\|$ so that, in particular (since $\gamma_{n_k}(0) = 1$),

$$\inf_k \|\xi_{n_k}(\omega) x_{n_k}\| > 0 .$$

Similarly, for every finite sequence (α_k) of complex numbers with $|\alpha_k| \leq 1$,

$$\left\| \sum_k \alpha_k \xi_{n_k}(\omega) x_{n_k} \right\| \leq C$$

for some constant C. We can then apply Lemma 9.30 to extract a further subsequence from $(\xi_{n_k}(\omega) x_{n_k})$ which will be equivalent to the canonical basis of c_0. This shows that (i) $\Rightarrow$ (ii). To prove the converse implication, we exhibit a random Fourier series of the type (13.21) (actually Gaussian) with coefficients in c_0 which is bounded but not uniformly convergent. Let G be the compact Cantor group $\{-1, +1\}^{\mathbb{N}}$ and set $V = G$. The characters on G consist of the Rademacher coordinate maps $\varepsilon_n(t)$. On some probability space, let (g_n) be a standard Gaussian sequence. For every n, let $I(n)$ denote the set of integers $\{2^n + 1, \ldots, 2^{n+1}\}$. Define then $X = (X_t)_{t \in G}$ by

$$X_t = \sum_n 2^{-n} \left(\sum_{i \in I(n)} g_i \varepsilon_i(t) \right) e_n , \quad t \in G ,$$

where (e_n) is the canonical basis of c_0. X is a (Gaussian) random Fourier series with values in c_0. It is almost surely bounded. To see it, note that

$$\sup_{N} \sup_{t \in G} \left\| \sum_{n=1}^{N} 2^{-n} \left(\sum_{i \in I(n)} g_i \varepsilon_i(t) \right) e_n \right\| = \sup_{N} \sup_{t \in G} \sup_{n \leq N} 2^{-n} \left| \sum_{i \in I(n)} g_i \varepsilon_i(t) \right|$$

$$= \sup_{n} 2^{-n} \sum_{i \in I(n)} |g_i|$$

(where we have used that (ε_n) spans ℓ_1 in $L_\infty(G)$). Now, we claim that $\sup_n 2^{-n} \sum_{i \in I(n)} |g_i| < \infty$ almost surely. Indeed, $\sup_n 2^{-n} \sum_{i \in I(n)} \mathbb{E}|g_i| < \infty$, and, by Chebyshev's inequality,

$$\mathbb{P} \left\{ \left| \sum_{i \in I(n)} |g_i| - \mathbb{E}|g_i| \right| > 2^n \right\} \leq 2^{-n}.$$

Hence the claim by the Borel-Cantelli lemma. With the same argument, X does not converge uniformly almost surely. The proof of Theorem 13.16 is complete.

Notes and References

The main references for this chapter are the book [M-P1] by M. B. Marcus and G. Pisier and their paper [M-P2] (see also [M-P3]). Random Fourier series go back to Paley, Salem and Zygmund. Kahane's ideas [Ka1] significantly contributed to the neat achievements of [M-P1].

Theorem 13.3 is due to X. Fernique [Fer4] (with of course a direct entropic proof). (See also [J-M3] for an exposition of this result more in the setting of random Fourier series.) It is the translation invariant version of the results of Section 12.1 and the key point in the subsequent investigation of random Fourier series. The equivalence of the boundedness and continuity of stationary Gaussian processes was known previously as Belaev's dichotomy [Bel] (see also [J-M3]) (a similar result for random Fourier series was proved by P. Billard, see [Ka1]). The basic Theorem 13.6 (in the case $G = \mathbb{R}$) is due to M. B. Marcus [Ma1], extended later in [M-P1]. The proof that we present is somewhat different and simpler; it has been put forward in [Ta14]. It does not use non-decreasing rearrangements as presented in [M-P1] (see also [Fer6]). The equivalence between Gaussian and Rademacher random Fourier series was put forward in [Pi6] and [M-P1]. The remarkable Banach space $C_{\text{a.s.}}(G)$ and associated Banach algebra $C_{\text{a.s.}}(G) \cap C(G)$ have been investigated by G. Pisier [Pi6], [Pi7]. He further provided an Harmonic Analysis description of $C_{\text{a.s.}}$ as the predual of a space of Fourier multipliers. A Sidon set generates a subspace isomorphic to ℓ_1 in $C(G)$. As a remarkable result, it was shown conversely by J. Bourgain and V. D. Milman [B-M] that if a subset Λ of Γ is such that the subspace C_Λ of $C(G)$ of all functions whose Fourier transform is supported by Λ is of finite cotype (i.e. does not contain ℓ_∞^n's uniformly), then Λ must be a Sidon set. (A prior contribution assuming C_Λ of cotype 2 is due to G. Pisier [Pi5]). For further conclusions on random Fourier series, in particular in non-Abelian groups, examples and quantitative estimates, we refer to [M-P1].

The results of Section 13.3 are taken from [M-P2] for the case $1 < p < 2$. The picture is completed in [Ta14] with the case $p = 1$ (and with a proof which inspired the proofs presented here). The complex probabilistic structures are carefully described in [M-P2]. Extensions to random Fourier series with infinitely divisible coefficients and ξ-radial processes are studied in [Ma4]. Further extensions to very general random Fourier series and harmonic processes are obtained in [Ta17].

The study of stationary vector valued Gaussian processes was initiated by X. Fernique to whom Theorem 13.14 is due [Fer10] (see also [Fer12]). He further extended this result in [Fer13]. Since the conclusion does not involve majorizing measures, one might wonder for a proof that does not use this tool. Theorem 13.14 was recently used in [I-M-M-T-Z] and [Fer14]. Theorem 13.16 is perhaps new.

Finally, related to the results of this chapter, note the following. Various central limit theorems (Corollary 13.10) for random Fourier series can be established [M-P1] with applications of the techniques to the empirical characteristic function [Ma2]. A law of the iterated logarithm for the empirical characteristic function can also be proved [Led1], [La]. Gaussian and Rademacher random Fourier quadratic forms (chaos) are studied and characterized in [L-M] with the result of Sections 13.2 and 13.4. In particular, it is shown there how random Fourier quadratic forms with either Rademacher or standard Gaussian sequences converge simultaneously.

14. Empirical Process Methods
in Probability in Banach Spaces

The purpose of this chapter is to present applications of the random process techniques developed so far to infinite dimensional limit theorems, and in particular to the central limit theorem (CLT). More precisely, we will be interested for example in the CLT in the space $C(T)$ of continuous functions on a compact metric space T. Since $C(T)$ is not well behaved with respect to the type or cotype 2 properties, we will rather have to seek for nice classes of random variables in $C(T)$ for which a central limit property can be established. This point of view leads to enlarge this framework and to investigate limit theorems for empirical measures or processes. Random geometric descriptions of the CLT may then be produced via this approach, as well as complete descriptions for nice classes of functions (indicator functions of some sets) on which the empirical processes are indexed. While these random geometric descriptions do not solve the central limit problem in infinite dimension (and are probably of little use in applications), however, they clearly describe the main difficulties inherent to the problem from the empirical point of view.

We do not try to give a complete account about empirical processes and their limiting properties but rather concentrate on some useful methods and ideas related to the material already discussed in this book. The examples of techniques which we chose to present are borrowed from the works by R. Dudley [Du4], [Du5] and by E. Giné and J. Zinn [G-Z2], [G-Z3], and actually we refer the interested reader to these authors for a complete exposition. The first section of this chapter presents various results on the CLT for subgaussian and Lipschitz processes in $C(T)$ under metric entropy or majorizing measure conditions. In the second section, we introduce the language of empirical processes and discuss the effect of pregaussianness in two cases: the first case concerns uniformly bounded classes while the second provides a random geometric description of Donsker classes, i.e. classes for which the CLT holds. Vapnik-Chervonenkis classes of sets form the matter of Section 14.3 where it is shown how these classes satisfy the classical limit properties uniformly over all probability measures, and are actually characterized in this way.

14.1 The Central Limit Theorem for Lipschitz Processes

Let (T, d) be a compact metric space and let $C(T)$ be the separable Banach space of all continuous functions on T equipped with the sup-norm $\| \cdot \|_\infty$. A Borel random variable X with values in $C(T)$ may be denoted in the process notation as $X = (X_t)_{t \in T} = (X(t))_{t \in T}$ and $(X(t))_{t \in T}$ has all its sample paths continuous on (T, d). If X is a random variable, we denote as usual by (X_i) a sequence of independent copies of X and let $S_n = X_1 + \cdots + X_n$, $n \geq 1$.

A subset K of $C(T)$ is relatively compact if and only if it is bounded and uniformly equicontinuous (Arzela-Ascoli). Equivalently, this is the case if there exist t_0 in T and a finite number M such that $|x(t_0)| \leq M$ for all x in K, and if for all $\varepsilon > 0$ there exists $\eta = \eta(\varepsilon) > 0$ such that $|x(s) - x(t)| < \varepsilon$ for all x in K and all s, t in T with $d(s, t) < \eta$. Combining this property with Prokhorov's Theorem 2.1 and the finite dimensional CLT, it follows that a random variable $X = (X(t))_{t \in T}$ satisfies the CLT in $C(T)$ if and only if $\mathbb{E}X(t) = 0$ and $\mathbb{E}X(t)^2 < \infty$ for every t and if, for each $\varepsilon > 0$, there is $\eta = \eta(\varepsilon) > 0$ such that

$$(14.1) \qquad \limsup_{n \to \infty} \mathbb{P}\left\{ \sup_{d(s,t)<\eta} \left| \frac{S_n(s) - S_n(t)}{\sqrt{n}} \right| > \varepsilon \right\} < \varepsilon .$$

Since the space $C(T)$ has no non-trivial type or cotype, and does not satisfy any kind of Rosenthal's inequality (cf. Chapter 10), the results that we can expect on the CLT in $C(T)$ can only concern special classes of random variables. We concentrate on the classes of subgaussian and Lipschitz variables, the first of which naturally extends the class of Gaussian variables (which trivially satisfy the CLT).

Recall that a centered process $X = (X(t))_{t \in T}$ is said to be subgaussian with respect to a metric d on T if for all real numbers λ and all s, t in T,

$$\mathbb{E} \exp\big(\lambda\big(X(s) - X(t)\big)\big) \leq \exp\left(\frac{\lambda^2}{2} d(s,t)^2 \right) .$$

Changing if necessary d into a multiple of it, we may require equivalently that $\|X(s) - X(t)\|_{\psi_2} \leq d(s, t)$ for all s, t in T (or $\mathbb{P}\{|X(s) - X(t)| > u d(s, t)\} \leq C \exp(-u^2/C)$ for all $u > 0$ and some constant C). We have seen in Section 11.3 that if (T, d) satisfies the majorizing measure condition

$$\limsup_{\eta \to 0} \sup_{t \in T} \int_0^\eta \left(\log \frac{1}{m(B(t, \varepsilon))} \right)^{1/2} d\varepsilon = 0$$

for some probability measure m on T, then the subgaussian process X has a version with almost all sample paths continuous on (T, d). Therefore, it defines (actually its version which we denote in the same way) a Radon random variable in $C(T)$. Note that by the main result of Chapter 12, the preceding condition is (essentially) equivalent to the existence of a Gaussian random variable G in $C(T)$ such that $\|G(s) - G(t)\|_2 \geq d(s,t)$ for all s,t in T. Now, under one of these (equivalent) assumptions, it is easily seen that the subgaussian process X also satisfies the CLT in $C(T)$. Indeed, by the independence and identical distribution of the summands, $S_n/\sqrt{n}$, for every n, is also subgaussian with respect to d. Then, from Proposition 11.19, we deduce that for every $\varepsilon > 0$, one can find $\eta > 0$, depending on ε, T, d only, such that, uniformly in n,

$$\mathbb{E} \sup_{d(s,t)<\eta} \left| \frac{S_n(s) - S_n(t)}{\sqrt{n}} \right| < \varepsilon.$$

Hence, X satisfies the CLT by (14.1). Therefore, we have the following result.

Theorem 14.1. *Let X be a Borel random variable with values in $C(T)$ which is subgaussian with respect to d. Assume that there is a probability measure m on (T, d) such that*

$$\lim_{\eta \to 0} \sup_{t \in T} \int_0^\eta \left(\log \frac{1}{m(B(t,\varepsilon))} \right)^{1/2} d\varepsilon = 0.$$

Then X satisfies the CLT.

We turn to the second class of random variables in $C(T)$ which we will study here and which are the Lipschitz random variables. They will be shown to be conditionally subgaussian and therefore will satisfy the CLT under conditions similar to those used for subgaussian variables. One first and main result is the following theorem.

Theorem 14.2. *Let X be a Borel random variable with values in $C(T)$ such that $\mathbb{E}X(t) = 0$ and $\mathbb{E}X(t)^2 < \infty$ for all t in T. Assume that there is a positive random variable M in L_2 such that for all ω's and all s,t in T,*

$$|X(\omega, s) - X(\omega, t)| \leq M(\omega) d(s,t).$$

Then, if (T, d) satisfies the majorizing measure condition

$$\lim_{\eta \to 0} \sup_{t \in T} \int_0^\eta \left(\log \frac{1}{m(B(t,\varepsilon))} \right)^{1/2} d\varepsilon = 0$$

for some probability measure m on (T, d), X satisfies the CLT in $C(T)$.

Recall that we may assume equivalently (Theorem 12.9) that d is the L_2-pseudo-metric of a Gaussian random variable in $C(T)$. We would like to mention that the exposition of the proof of Theorem 14.2 that we give is slightly more complicated than it should be. It should actually be similar to the proof of Theorem 14.5 below. We chose this exposition so to include in the same pattern Theorem 14.3.

Proof. Let X, and (X_i), be defined on $(\Omega, \mathcal{A}, \mathbb{P})$. By Proposition 10.4, or a simple symmetrization argument, we may and do assume that X is symmetrically distributed. Thus, (X_i) has the same distribution as $(\varepsilon_i X_i)$ where (ε_i) is a Rademacher sequence constructed on some different probability space. Further, there is a sequence (M_i) of independent copies of M such that $|X_i(\omega, s) - X_i(\omega, t)| \leq M_i(\omega)d(s, t)$ for all i, all ω, and all s, t in T. By the subgaussian inequality (4.1), for every ω, every integer n and every $u > 0$, and all s, t in T,

$$\mathbb{P}_\varepsilon \left\{ \frac{1}{\sqrt{n}} \left| \sum_{i=1}^n \varepsilon_i \big(X_i(\omega, s) - X_i(\omega, t) \big) \right| > u \right\}$$

$$\leq 2 \exp \left(- \frac{u^2}{\frac{2}{n} \sum_{i=1}^n |X_i(\omega, s) - X_i(\omega, t)|^2} \right)$$

$$\leq 2 \exp \left(- \frac{u^2}{\frac{2}{n} d(s,t)^2 \sum_{i=1}^n M_i(\omega)^2} \right),$$

where $\mathbb{P}_\varepsilon$ is, as usual, integration with respect to (ε_i). Let $a > 0$ to be specified and set, for every integer n, and every t in T,

$$Y^n(t) = \frac{1}{\sqrt{n}} \sum_{i=1}^n \varepsilon_i X_i(t) I_{\left\{ \sum_{j=1}^n M_j^2 \leq a^2 n \right\}}.$$

From the preceding, it clearly follows that for all s, t in T, all n and all $u > 0$,

$$\mathbb{P} \big\{ |Y^n(s) - Y^n(t)| > a\, ud(s, t) \big\} \leq 2 \exp(-u^2/2).$$

That is to say, for some numerical constant K, the processes $((Ka)^{-1} Y^n(t))_{t \in T}$ are subgaussian with respect to d. Therefore, under the majorizing measure condition of the theorem, we know from Proposition 11.19 that, for all $\delta > 0$, there exists $\eta > 0$ depending on δ, T, d, m only such that, uniformly in n,

(14.2) $$\mathbb{E} \sup_{d(s,t)<\eta} |Y^n(s) - Y^n(t)| < a\delta.$$

It is easy to conclude the proof of Theorem 14.2. Fix $\varepsilon > 0$ and let $a = a(\varepsilon) > 0$ be such that $a^2 \geq 2\mathbb{E}M^2/\varepsilon$. Hence, $\mathbb{P}\{\sum_{j=1}^n M_j^2 > a^2 n\} \leq \varepsilon/2$ for all n. For all $\eta > 0$, we can write

$$\mathbb{P} \left\{ \sup_{d(s,t)<\eta} \left| \frac{S_n(s) - S_n(t)}{\sqrt{n}} \right| > \varepsilon \right\} \leq \frac{\varepsilon}{2} + \mathbb{P} \left\{ \sup_{d(s,t)<\eta} |Y^n(s) - Y^n(t)| > \varepsilon \right\}$$

$$\leq \frac{\varepsilon}{2} + \frac{1}{\varepsilon} \mathbb{E} \sup_{d(s,t)<\eta} |Y^n(s) - Y^n(t)|.$$

If we then choose $\eta = \eta(\varepsilon) > 0$ small enough such that (14.2) is satisfied with $\delta = \varepsilon^2/2a$, we find that X satisfies (14.1), and therefore the CLT. The proof of Theorem 14.3 is complete.

Note that if the continuous majorizing measure condition in Theorem 14.2 is weakened into the corresponding bounded condition, then we can only conclude in general to the bounded CLT for the Lipschitz variable X. That the continuous majorizing measure condition is necessary is clear with the example of the random variable $X = (\varepsilon_n/(\log(n+1))^{1/2})$ in $C(\mathbb{N} \cup \{\infty\})$ which is Lipschitz with respect to the distance of the bounded, but not continuous, Gaussian sequence $(g_n/(\log(n + 1))^{1/2})$.

Although $C(T)$ is not of type 2, it is interesting to mention that Theorem 14.2 on Lipschitz random variables can be related to the general results on the CLT in type 2 spaces of Chapter 10. Actually, it rather concerns *operators* of type 2 and more precisely the canonical injection map $j : \mathrm{Lip}(T) \to C(T)$ investigated in Section 12.3. Recall that we denote by $\mathrm{Lip}(T)$ the space of Lipschitz functions x on T equipped with the norm

$$\|x\|_{\mathrm{Lip}} = D^{-1}\|x\|_\infty + \sup_{s \neq t} \frac{|x(s) - x(t)|}{d(s,t)}$$

where $D = D(T)$ is the diameter of (T,d). We have seen in Theorem 12.17 that if there is a (bounded) majorizing measure on (T, d) for the function $(\log 1/u)^{1/2}$, then j is an operator of type 2, and that its type 2 constant $T_2(j)$ satisfies $T_2(j) \leq K\gamma^{(2)}(T, d)$ for some numerical constant K. Let now X be Lipschitz with respect to d as in Theorem 14.2. Then $\mathbb{E}\|X\|_{\mathrm{Lip}}^2 < \infty$ and since j is of type 2, one might wish to use the CLT result for operators of type 2 (Corollary 10.6). However, there is a small problem since $\mathrm{Lip}(T)$ need *not* be separable and X need not be a Radon random variable in this space. This difficulty can be turned around in several ways. For example, from Proposition 9.11 for operators, we already have that, for every n,

$$(14.3) \qquad \frac{1}{\sqrt{n}} \mathbb{E} \left\| \sum_{i=1}^n j(X_i) \right\|_\infty \leq 2T_2(j)\left(\mathbb{E}\|X\|_{\mathrm{Lip}}^2\right)^{1/2}.$$

In particular, X already satisfies the bounded CLT in $C(T)$. Now, if there is a probability measure m on (T, d) such that,

$$(14.4) \qquad \lim_{\eta \to 0} \sup_{t \in T} \int_0^\eta \left(\log \frac{1}{m(B(t,\varepsilon))}\right)^{1/2} d\varepsilon = 0,$$

it is not difficult to see that the proof of Theorem 12.17 can be modified to show that, for every $\varepsilon > 0$, there exists a finite dimensional subspace F of $C(T)$ such that if T_F is the quotient map $C(T) \to C(T)/F$, then $T_2(T_F \circ j) < \varepsilon$. Applying (14.3) to $T_F \circ j$ then easily yields the CLT. However, in this last step, this approach basically amounts to the original proof of Theorem 14.2. As an alternate, but also somewhat cumbersome argument, one can show that under (14.4) there exists a distance d' on T such that $d(s,t)/d'(s,t) \to 0$ when $d(s,t) \to 0$ and for which still $\gamma^{(2)}(T, d') < \infty$. Since the balls for the norm in $\mathrm{Lip}(T, d)$ are compact in $\mathrm{Lip}(T, d')$, the Lipschitz random variable X of

Theorem 14.2 takes its values in some separable subspace of $\text{Lip}(T, d')$. Then Corollary 10.6 can be applied.

Since a random variable X satisfying the CLT in a Banach space B does not necessarily verify $\mathbb{E}\|X\|^2 < \infty$, but rather $\lim_{t\to\infty} t^2 \mathbb{P}\{\|X\| > t\} = 0$ (cf. Lemma 10.1), it was conjectured for some time that the hypothesis M in L_2 in Theorem 14.1 could possibly be weakened into M in $L_{2,\infty}$, i.e. $\sup_{t>0} t^2 \mathbb{P}\{M > t\} < \infty$. The next result shows how this is indeed the case. It is assumed explicitly that X is pregaussian since this property does not follow anymore from the Lipschitz assumption when M is not in L_2. The proof relies on inequality (6.30) and Lemma 5.8.

Theorem 14.3. *Let X be a pregaussian random variable with values in $C(T)$. Assume that there is a positive random variable M in $L_{2,\infty}$ such that for all ω's and all s, t in T*

$$|X(\omega, s) - X(\omega, t)| \leq M(\omega) d(s, t).$$

Then, if there is a probability measure m on (T, d) such that

$$\limsup_{\eta \to 0} \sup_{t \in T} \int_0^\eta \left(\log \frac{1}{m(B(t, \varepsilon))} \right)^{1/2} d\varepsilon = 0,$$

X satisfies the CLT in $C(T)$.

Proof. First, we have to transform the (necessary) pregaussian property into a majorizing measure condition. There exists a Gaussian variable in $C(T)$ with L_2-metric $d_X(s, t) = \|X_s - X_t\|_2$. By the comments following Theorem 11.18, this Gaussian process is also continuous with respect to d_X and thus, by Theorem 12.9, there is a probability measure m' on (T, d_X) which satisfies the same majorizing measure condition as m on (T, d). We would like to have this property for the *maximum* of the two distances d and d_X. Clearly, the measure $\mu = m \times m'$ on $T \times T$ equipped with the metric $\tilde{d}((s, t), (s', t')) = \max(d(s, s'), d_X(t, t'))$ satisfies

$$\limsup_{\eta \to 0} \sup_{T \times T} \int_0^\eta \left(\log \frac{1}{\mu(B((s, t), \varepsilon))} \right)^{1/2} d\varepsilon = 0.$$

Now, we simply project on the diagonal. For each couple (s, t) in $T \times T$, one can find (by a compactness argument) a point $\varphi(s, t)$ in T such that $\tilde{d}((s, t), (\varphi(s, t), \varphi(s, t))) \leq 2\tilde{d}((s, t), (u, u))$ for all u in T. Then, if $d(s, u)$ and $d_X(t, u)$ are both $< \varepsilon$, it follows by definition of $\varphi(s, t)$ that $d(\varphi(s, t), u)$ and $d_X(\varphi(s, t), u)$ are $< 3\varepsilon$. Hence $B((u, u), \varepsilon) \subset \varphi^{-1}(B(u, 3\varepsilon))$ where $B(u, 3\varepsilon)$ is the ball in T with center u and radius 3ε for the metric $\max(d, d_X)$. Therefore, letting $\tilde{m} = \varphi(\mu)$, $\tilde{m}(B(u, 3\varepsilon)) \geq \mu(B((u, u), \varepsilon))$ and thus

$$\limsup_{\eta \to \infty} \sup_{t \in T} \int_0^\eta \left(\log \frac{1}{\tilde{m}(B(t, \varepsilon))} \right)^{1/2} d\varepsilon - 0.$$

It follows from this discussion that, replacing d by $\max(d, d_X)$, we may and do assume in the following that $d_X \leq d$.

We can turn to the proof of Theorem 14.2. Instead of refering to (6.30), it is simpler, since we will only be concerned with real valued random variables, to state and to again prove the inequality that we will need.

Lemma 14.4. *Let (Z_i) be a finite sequence of independent real valued symmetric random variables in L_2. Then, for all $u > 0$,*

$$\mathbb{P}\left\{\left|\sum_i Z_i\right| > u\left(4\|(Z_i)\|_{2,\infty} + \left(\sum_i \mathbb{E}Z_i^2\right)^{1/2}\right)\right\} \leq 4\exp(-u^2/6).$$

Proof. Set $\sigma = \left(\sum_i \mathbb{E}Z_i^2\right)^{1/2}$. For any random variable $A > 0$, we can write by the triangle inequality and the definition of $\|(Z_i)\|_{2,\infty}$ that

$$\left|\sum_i Z_i\right| \leq \left|\sum_i Z_i I_{\{|Z_i| \leq A\}}\right| + \sum_i |Z_i| I_{\{|Z_i| > A\}}$$

$$\leq \left|\sum_i Z_i I_{\{|Z_i| \leq A\}}\right| + \frac{2}{A}\|(Z_i)\|_{2,\infty}^2.$$

Therefore, if we let $A = u^{-1}\|(Z_i)\|_{2,\infty}$,

$$\mathbb{P}\left\{\left|\sum_i Z_i\right| > u(4\|(Z_i)\|_{2,\infty} + \sigma)\right\} \leq \mathbb{P}\left\{\left|\sum_i Z_i I_{\{|Z_i| \leq A\}}\right| > u(2Au + \sigma)\right\}.$$

Let us observe that on the set $\{|Z_i| \leq A\}$

$$\frac{|Z_i|}{2Au + \sigma} \leq \min\left(\frac{|Z_i|}{\sigma}, \frac{1}{2u}\right).$$

Hence, by the symmetry of the variables Z_i and by the contraction principle in the form of (4.7) (applied conditionally on the Z_i's),

$$\mathbb{P}\left\{\left|\sum_i Z_i\right| > u(4\|(Z_i)\|_{2,\infty} + \sigma)\right\} \leq 2\mathbb{P}\left\{\left|\sum_i \varepsilon_i \min\left(\frac{|Z_i|}{\sigma}, \frac{1}{2u}\right)\right| > u\right\}$$

where (ε_i) is an independent Rademacher sequence. We now need simply apply Kolmogorov's inequality (Lemma 1.6) to get the result. Lemma 14.4 is proved.

Provided with this lemma, the proof of Theorem 14.3 is very much as the proof of Theorem 14.2, substituing the inequality of Lemma 14.4 to the subgaussian inequality. Since M is in $L_{2,\infty}$, by Lemma 5.8, one can find, for each $\varepsilon > 0$, some $a = a(\varepsilon) > 0$ such that, for every n,

$$\mathbb{P}\{\|(M_j)_{j \leq n}\|_{2,\infty} > a\sqrt{n}\} \leq \frac{\varepsilon}{2}.$$

Let, for every n and every t in T,

$$Y^n(t) = \frac{1}{\sqrt{n}} \sum_{i=1}^{n} \varepsilon_i X_i(t) I_{\{\|(M_j)_{j \le n}\|_{2,\infty} \le a\sqrt{n}\}} \, .$$

Lemma 14.4 implies that for every s, t in T and every $u > 0$,

$$\mathbb{P}\{|Y^n(s) - Y^n(t)| > (4a + 1)ud(s, t)\} \le 4 \exp(-u^2/6)$$

since $|X_i(s) - X_i(t)| \le M_i d(s, t)$ and $\|X_i(s) - X_i(t)\|_2 \le d(s, t)$ for all i and s, t in T. From this result, the proof of Theorem 14.3 is completed exactly as the proof of Theorem 14.2 using the subgaussian results (Proposition 11.19).

We conclude this section with an analogous study of some spectral measures of p-stable random vectors in $C(T)$. We already know the close relationships between the Gaussian CLT and the question of the existence of a stable random vector with a given spectral measure. The next example is another instance of this observation.

Given a positive finite Radon measure ν on $C(T)$, we would like to determine conditions under which ν is the spectral measure of some p-stable random variable in $C(T)$ with $1 \le p < 2$ (recall that the case $p < 1$ is trivial, cf. Chapter 5). Since this seems a difficult task in general, we consider, as for the CLT, the particular case corresponding to Lipschitz processes. Assume for simplicity (and without any loss of generality) that ν is a probability measure so that it is the distribution of a random variable Y in $C(T)$.

Theorem 14.5. *Let $1 \le p < 2$ and let $q = p/p - 1$. Let $Y = (Y(t))_{t \in T}$ be a random variable in $C(T)$ such that $\mathbb{E}|Y(t)|^p < \infty$ for all t and such that for all ω's and all s, t in T,*

$$|Y(\omega, s) - Y(\omega, t)| \le M(\omega)d(s, t)$$

for some positive random variable M in L_p. Assume that there is a probability measure m on (T, d) such that

$$\lim_{\eta \to 0} \sup_{t \in T} \int_0^\eta \left(\log \frac{1}{m(B(t, \varepsilon))} \right)^{1/q} d\varepsilon = 0$$

(if $q < \infty$; if $q = \infty$, use the function $\log^+ \log$). Then, the distribution ν of Y is the spectral measure of a p-stable random variable with values in $C(T)$.

Proof. It is similar to the proof of Theorem 14.2. For notational convenience, we restrict ourselves to the case $q < \infty$. Recall the series representation of stable random vectors and processes (Corollary 5.3). Let (Y_j) be independent copies of Y, and let (ε_j) be a Rademacher sequence. Assume as usual that (Γ_j), (ε_j), (Y_j) are independent. For each t, since $\mathbb{E}|Y(t)|^p < \infty$, the series $\sum_{j=1}^\infty \Gamma_j^{-1/p} \varepsilon_j Y_j(t)$ converges almost surely (and defines a p-stable real valued random variable). It will be enough to show that for each $\varepsilon > 0$ one can find $\eta > 0$ such that

$$(14.5) \qquad \mathbb{E} \sup_{d(s,t)<\eta} \left| \sum_{j=1}^{\infty} \Gamma_j^{-1/p} \varepsilon_j \big(Y_j(s) - Y_j(t) \big) \right| < \varepsilon .$$

Then, the series $\sum_{j=1}^{\infty} \Gamma_j^{-1/p} \varepsilon_j Y_j$ converges almost surely and in L_1 in $C(T)$ (Itô-Nisio's theorem). By Corollary 5.5, $c_p^{-1} \sum_{j=1}^{\infty} \Gamma_j^{-1/p} \varepsilon_j Y_j$ therefore defines there a p-stable random variable with spectral measure ν.

To establish (14.5), we first note that by independence, the contraction principle and (5.8),

$$\mathbb{E} \sup_{d(s,t)<\eta} \left| \sum_{j=1}^{\infty} \Gamma_j^{-1/p} \varepsilon_j \big(Y_j(s) - Y_j(t) \big) \right|$$

$$\leq \mathbb{E} \sup_{j \geq 1} \left(\frac{j}{\Gamma_j} \right)^{1/p} \mathbb{E} \sup_{d(s,t)<\eta} \left| \sum_{j=1}^{\infty} j^{-1/p} \varepsilon_j \big(Y_j(s) - Y_j(t) \big) \right|$$

$$\leq K_p \mathbb{E} \sup_{d(s,t)<\eta} \left| \sum_{j=1}^{\infty} j^{-1/p} \varepsilon_j \big(Y_j(s) - Y_j(t) \big) \right|$$

where K_p only depends on p. Using Lemma 1.7, for every ω of the space supporting the sequence (Y_j), and every s, t in T and $u > 0$,

$$\mathbb{P}_\varepsilon \left\{ \left| \sum_{j=1}^{\infty} j^{-1/p} \varepsilon_j \big(Y_j(\omega, s) - Y_j(\omega, t) \big) \right| > u \right\}$$

$$\leq 2 \exp \left(- \frac{u^q}{C_q d(s,t)^q \| (j^{-1/p} M_j(\omega)) \|_{p,\infty}^q} \right)$$

where (M_j) is a sequence of independent copies of M and where we have used that $|Y_j(\omega, s) - Y_j(\omega, t)| \leq M_j(\omega) d(s,t)$. Under the majorizing measure condition of the statement, we deduce from Theorem 11.14 that, for each $\varepsilon > 0$, one can find $\eta > 0$ such that, uniformly in ω,

$$\mathbb{E}_\varepsilon \sup_{d(s,t)<\eta} \left| \sum_{j=1}^{\infty} j^{-1/p} \varepsilon_j \big(Y_j(\omega, s) - Y_j(\omega, t) \big) \right| \leq \varepsilon \| (j^{-1/p} M_j(\omega)) \|_{p,\infty} .$$

An integration with respect to ω using Corollary 5.9 implies (14.5) and, as announced, the conclusion.

14.2 Empirical Processes and Random Geometry

In this section, we examine the CLT by yet another angle, namely by empirical process methods. Actually, we only present a short overview of these empirical

techniques with, in particular, a random geometric characterization of classes
for which the central limit property holds. We refer to [Du5] and [G-Z3] for
some of the basics of the theory as well as for a more detailed investigation.

First, we introduce the empirical process language. Let $(S, \mathcal{S})$ be a measurable space. If P is a probability on $(S, \mathcal{S})$, (X_i) will denote here, unless
otherwise indicated, a sequence of independent random variables defined on
some probability space $(\Omega, \mathcal{A}, \mathbb{P})$ with values in S and with common law P.
We will also use randomizing sequences such as Rademacher or standard Gaus-
sian sequences (ε_i) or (g_i), and denote accordingly by $\mathbb{P}_\varepsilon$, $\mathbb{E}_\varepsilon$, $\mathbb{P}_g$, $\mathbb{E}_g$ partial
integration with respect to (ε_i) or (g_i). The *empirical measures* P_n associated
to P are defined as the random measures on S given by

$$P_n(\omega) = \frac{1}{n} \sum_{i=1}^{n} \delta_{X_i(\omega)}, \quad \omega \in \Omega, \ n \geq 1,$$

where δ_x is point mass at x (recall that the X_i's have the common law P).

In this section (and the next one), $L_p = L_p(P)$, $0 \leq p \leq \infty$, is understood
to be $L_p(S, \mathcal{S}, P; \mathbb{R})$ (we write L_p or $L_p(P)$ depending on the context and on
the necessity of specifying the underlying probability P). $\|f\|_p$ denotes the L_p-
norm $(1 \leq p \leq \infty)$ of the measurable function f on S, and $d_p(f, g) = \|f - g\|_p$
denotes its associated metric. If f is in $L_1 = L_1(P)$, we further denote $P(f) =$
$E(f) = \int f dP$. We need also consider the random spaces $L_p(P_n)$, $1 \leq p < \infty$,
with their norms

$$\|f\|_{n,p} = \left(\frac{1}{n} \sum_{i=1}^{n} |f(X_i)|^p \right)^{1/p}$$

where f is a function on S, and denote by $d_{n,p}$ the associated random distances.

By a class of functions on S, we will always mean a family $\mathcal{F}$ of (real val-
ued) measurable functions f on $(S, \mathcal{S})$ such that $\|f(x)\|_{\mathcal{F}} = \sup_{f \in \mathcal{F}} |f(x)| <$
∞ for all x in S. (For any family $(a(f))_{f \in \mathcal{F}}$ of numbers indexed by a class $\mathcal{F}$,
we set, with some abuse of notation, $\|a(f)\|_{\mathcal{F}} = \sup_{f \in \mathcal{F}} |a(f)|$.) Given P on
$(S, \mathcal{S})$, the (centered) *empirical processes* based on P and indexed by a class
$\mathcal{F} \subset L_1(P)$ are defined as

$$(P_n - P)(f) = \frac{1}{n} \sum_{i=1}^{n} (f(X_i) - P(f)), \quad f \in \mathcal{F}, \ n \geq 1.$$

As always in this book, we do not enter the various and possibly intricate
measurability questions that the study of empirical processes raises. In order
not to hide the main ideas which we intend to emphasize here, we shall assume all classes $\mathcal{F}$ to be *countable*. Instead, we could require a separability
assumption on the processes $((P_n - P)(f))_{f \in \mathcal{F}}$.

Since we are assuming that $\|f(x)\|_{\mathcal{F}} < \infty$ for every x in S, the maps
$f \to f(X_i)$, $i \in \mathbb{N}$, define random elements in the space $\ell_\infty(\mathcal{F})$ of all bounded
functions $\mathcal{F} \to \mathbb{R}$ equipped with the sup-norm $\|\cdot\|_{\mathcal{F}}$. In this study of empirical
processes, we are therefore dealing with random variables taking their values

in the non-separable (unless $\mathcal{F}$ is finite) Banach space $\ell_\infty(\mathcal{F})$ entering, since we are assuming $\mathcal{F}$ countable, our general setting of infinite dimensional random variables (cf. Section 2.3). Many results presented throughout this book therefore apply in this empirical setting.

Limit properties are of course the main topic in the study of empirical processes as a way to approximate a given law P by empirical data P_n. We have the following definitions. A class $\mathcal{F}$ as above is said to be a *Glivenko-Cantelli class for P*, or P satisfies the strong law of large numbers uniformly on $\mathcal{F}$, if, with probability one,

$$\lim_{n\to\infty} \|P_n(f) - P(f)\|_{\mathcal{F}} = 0 \, .$$

This definition extends the classical result due to Glivenko and Cantelli according to which the class $\mathcal{F}$ of the indicator functions of the intervals $[0,t]$, $0 \leq t \leq 1$, is a Glivenko-Cantelli class for every probability P on $[0,1]$. Since weak convergence is involved, the definition of the central limit property in this non-separable framework requires some more care. Write for convenience $\nu_n = \sqrt{n}(P_n - P)$, $n \geq 1$. Then, a class $\mathcal{F}$ of functions on S is said to be a *Donsker class for P*, or P satisfies the central limit theorem uniformly on $\mathcal{F}$, if there is a Gaussian Radon probability measure γ_P on $\ell_\infty(\mathcal{F})$ such that, for every real valued bounded continuous function φ on $\ell_\infty(\mathcal{F})$,

$$\lim_{n\to\infty} \int^* \varphi(\nu_n)d\mathbb{P} = \int \varphi \, d\gamma_P \, .$$

The use of the upper integral takes into account the measurability questions. By the finite dimensional CLT, the probability measure γ_P is the law of a Gaussian process G_P indexed by $\mathcal{F}$ with covariance given by

$$\mathbb{E}G_P(f)G_P(g) = P(fg) - P(f)P(g) \, , \quad f,g \in \mathcal{F} \, .$$

Furthermore, to say that γ_P is Radon on $\ell_\infty(\mathcal{F})$ is equivalent to saying that G_P admits a version with almost all sample paths bounded and continuous on $\mathcal{F}$ with respect to the metric $\|(f - P(f)) - (g - P(g))\|_2$, $f,g \in \mathcal{F}$ (cf. [G-Z3]). If this property is satisfied, the class $\mathcal{F}$ is said to be *P-pregaussian*, so that a P-Donsker class is of course P-pregaussian. As before, these definitions extend the classical Kolmogorov-Smirnov-Donsker theorem for the class $\mathcal{F}$ of the indicator functions of the intervals $[0,t]$, $0 \leq t \leq 1$; the Gaussian process G_P appears as a generalization of the Brownian bridge (with P the Lebesgue measure on $[0,1]$). We note for further purposes that if G_P is continuous in the previous sense and if $\|P(f)\|_{\mathcal{F}} < \infty$ there exists a Gaussian process W_P with L_2-metric given by $\mathbb{E}|W_P(f) - W_P(g)|^2 = \|f - g\|_2^2 = d_2(f,g)^2$, f,g in $\mathcal{F}$ (the analog of the Brownian motion), which is almost surely continuous on $(\mathcal{F}, d_2)$. We may simply take for example $W_P(f) = G_P(f) + \theta P(f)$ where θ is a standard normal variable which is independent of G_P. To conclude this set of definitions, we should introduce Strassen classes satisfying the law of the iterated logarithm. Since we will basically be concerned only with the CLT, we leave this to the interested reader (cf. e.g. [K-D], [Du5], [D-P]).

As for the CLT in the space of continuous functions (cf. the preceding section), a class $\mathcal{F}$ is a P-Donsker class if and only if the processes ν_n satisfy a Prokhorov type asymptotic equicontinuity condition. We refer to [Du4], [Du5], [G-Z2], [G-Z3] for a complete description and a proof of the following statement which extends (14.1) to this empirical framework. It is already expressed in its randomized version (cf. Proposition 10.4) which will be useful in the sequel. For every $\eta > 0$, we let $\mathcal{F}_\eta = \{f - g;\ f, g \in \mathcal{F},\ d_2(f, g) < \eta\}$.

Theorem 14.6. *Let $\mathcal{F}$ be a class of functions on $(S, \mathcal{S}, P)$ such that $\|P(f)\|_{\mathcal{F}} < \infty$. Then, $\mathcal{F}$ is a Donsker class for P if and only if $(\mathcal{F}, d_2)$ is totally bounded and, for every $\varepsilon > 0$, there exists $\eta > 0$ such that*

$$\limsup_{n \to \infty} \mathbb{P}\left\{\left\|\sum_{i=1}^{n} \varepsilon_i f(X_i)/\sqrt{n}\right\|_{\mathcal{F}_\eta} > \varepsilon\right\} < \varepsilon.$$

The equivalence holds similarly if the Rademacher sequence (ε_i) is replaced by an orthogaussian sequence (g_i).

From the integrability properties in the CLT, in the form for example of Corollary 10.2 and (10.2), note that if $\mathcal{F}$ is a P-Donsker class we also have

$$(14.6) \qquad \lim_{\eta \to 0} \limsup_{n \to \infty} \frac{1}{\sqrt{n}} \mathbb{E}\left\|\sum_{i=1}^{n} \varepsilon_i f(X_i)\right\|_{\mathcal{F}_\eta} = 0$$

and similarly with (g_i) in place of (ε_i).

With these definitions and observations, we now turn to the two results on Donsker classes that we intend to present. The first result describes the effect of pregaussianness on the equicontinuity condition of Theorem 14.6 for uniformly bounded classes of functions. It combines Sudakov's minoration with exponential bounds. For every $\varepsilon > 0$ and every integer n, $\mathcal{F}_{\varepsilon,n}$ denotes $\mathcal{F}_\eta$ for $\eta = (\varepsilon/\sqrt{n})^{1/2}$.

Theorem 14.7. *Let $\mathcal{F}$ be a uniformly bounded class of functions on $(S, \mathcal{S}, P)$. Then, $\mathcal{F}$ is a P-Donsker class if and only if it is P-pregaussian and, for some (or all) $\varepsilon > 0$,*

$$\left\|\sum_{i=1}^{n} \varepsilon_i f(X_i)/\sqrt{n}\right\|_{\mathcal{F}_{\varepsilon,n}} \to 0 \quad \text{in probability}.$$

Proof. Assume without loss of generality that $\|f\|_\infty \le 1$ for all f in $\mathcal{F}$. Only sufficiency requires a proof. Let $\varepsilon > 0$ be fixed. Since $\mathcal{F}$ is P-pregaussian, we know that W_P is a Gaussian process which has a continuous version on $(\mathcal{F}, d_2)$. Therefore, by Sudakov's minoration (Corollary 3.19), $\lim_{n \to \infty} c_n(\varepsilon) = 0$ where

$$c_n(\varepsilon) = \left(\frac{\varepsilon}{\sqrt{n}}\right)^{1/2} \left(\log N\left(\mathcal{F}, d_2; \frac{1}{2}\left(\frac{\varepsilon}{\sqrt{n}}\right)^{1/2}\right)\right)^{1/2}.$$

By the definition of the entropy numbers, there exists a class $\mathcal{G} = \mathcal{G}(\varepsilon, n)$ maximal in $\mathcal{F}$ with respect to the relations $d_2(f, g) \geq (\varepsilon/\sqrt{n})^{1/2}$ such that

$$(14.7) \qquad \text{Card}\,\mathcal{G} \leq \exp(c_n(\varepsilon)^2 \sqrt{n}/\varepsilon).$$

By maximality, for every f in $\mathcal{F}$ there exists g in $\mathcal{G}$ satisfying $d_2(f, g) < (\varepsilon/\sqrt{n})^{1/2}$. Therefore, for every $\eta > 0$, and all sufficiently large n's depending on η, we can write for all $\delta > 0$,

$$\mathbb{P}\left\{ \left\| \sum_{i=1}^{n} \varepsilon_i f(X_i)/\sqrt{n} \right\|_{\mathcal{F}_\eta} > 3\delta \right\}$$

$$\leq 2\mathbb{P}\left\{ \left\| \sum_{i=1}^{n} \varepsilon_i f(X_i)/\sqrt{n} \right\|_{\mathcal{F}_{\varepsilon,n}} > \delta \right\} + \mathbb{P}\left\{ \left\| \sum_{i=1}^{n} \varepsilon_i f(X_i)/\sqrt{n} \right\|_{\mathcal{G}_\eta} > \delta \right\}.$$

So, by the hypothesis, it is enough to show that for all $\delta > 0$,

$$(14.8) \qquad \lim_{\eta \to 0} \limsup_{n \to \infty} \mathbb{P}\left\{ \left\| \sum_{i=1}^{n} \varepsilon_i f(X_i)/\sqrt{n} \right\|_{\mathcal{G}_\eta} > \delta \right\} = 0.$$

Set now, for every n,

$$A(\varepsilon, n) = \left\{ \forall f \neq g \text{ in } \mathcal{G} = \mathcal{G}(\varepsilon, n),\ d_{n,2}(f, g) \leq 2d_2(f, g) \right\}$$

where we recall that we denote by $d_{n,2}(f, g)$ the random distances $\left(\sum_{i=1}^{n} (f - g)^2(X_i)/n \right)^{1/2}$. Let $h = f - g$, $f \neq g$ in $\mathcal{G}$; then $\|h\|_\infty \leq 2$ since $\mathcal{F}$ is uniformly bounded by 1 and $\|h\|_2 \geq (\varepsilon/\sqrt{n})^{1/2}$ by the definition of $\mathcal{G}$. By Lemma 1.6, for all large enough n's,

$$\mathbb{P}\left\{ \|h\|_{n,2} > 2\|h\|_2 \right\} \leq \mathbb{P}\left\{ \sum_{i=1}^{n} (h^2(X_i) - \mathbb{E}h^2(X_i)) > 3n\|h\|_2^2 \right\}$$

$$\leq \exp(-n\|h\|_2^2/50)$$

$$\leq \exp(-\varepsilon\sqrt{n}/50).$$

Hence, by (14.7),

$$(14.9) \quad \limsup_{n \to \infty} \mathbb{P}\left(A(\varepsilon, n)^c \right) \leq \limsup_{n \to \infty} \left(\text{Card}\,\mathcal{G}(\varepsilon, n) \right)^2 \exp(-\varepsilon\sqrt{n}/50) = 0.$$

For each n and for each ω in $A(\varepsilon, n)$, consider the Gaussian process

$$Z_{\omega,n}(f) = \frac{1}{\sqrt{n}} \sum_{i=1}^{n} g_i f(X_i(\omega)), \quad f \in \mathcal{G}.$$

Since $\omega \in A(\varepsilon, n)$, clearly $\mathbb{E}_g |Z_{\omega,n}(f) - Z_{\omega,n}(f')|^2 \leq 4d_2(f, f')$. Now W_P has d_2 as associated L_2-metric and possesses a continuous version on $(\mathcal{F}, d_2)$. It clearly follows, from Lemma 11.16 for example, that

$$\lim_{\eta \to 0} \mathbb{E}_g \|Z_{\omega,n}(f)\|_{\mathcal{G}_\eta} = 0$$

which therefore holds for every n and for every ω in $A(\varepsilon, n)$. Standard comparison of Rademacher averages with Gaussian averages combined with (14.9) then implies (14.8) and thus the conclusion. Theorem 14.7 is established.

The second result of this section further investigates the influence of pregaussianness in the study of Donsker classes $\mathcal{F}$ (no longer necessarily uniformly bounded). While we only used Sudakov's minoration before, we now take advantage of existence of majorizing measures (Chapter 12). The result that we present indicates precisely how the pregaussian property actually controls a whole "portion" of $\mathcal{F}$. In the remaining part, no cancellation (one of the main features of the study of sums of independent random variables) occurs. For clarity, we first give a quantitative rather than a qualitative statement. If P is a probability on $(S, \mathcal{S})$ and $\mathcal{F}$ is a class of functions in $L_2 = L_2(P)$, recall the Gaussian process $W_P = (W_P(f))_{f \in \mathcal{F}}$. For classes of functions $\mathcal{F}, \mathcal{F}_1, \mathcal{F}_2$, we write $\mathcal{F} \subset \mathcal{F}_1 + \mathcal{F}_2$ to signify that each f in $\mathcal{F}$ can be written as $f_1 + f_2$ where $f_1 \in \mathcal{F}_1$, $f_2 \in \mathcal{F}_2$.

Theorem 14.8. *There is a numerical constant K with the following property: for every P-pregaussian class $\mathcal{F}$ such that $\|f\|_{\mathcal{F}} \in L_1 = L_1(P)$ and for every n, there exist classes $\mathcal{F}_1^n, \mathcal{F}_2^n$ in $L_2 = L_2(P)$ such that $\mathcal{F} \subset \mathcal{F}_1^n + \mathcal{F}_2^n$ and*

$$(i\,) \ \mathbb{E}\left\|\sum_{i=1}^n |f(X_i)|/\sqrt{n}\right\|_{\mathcal{F}_1^n} \le K\left(\mathbb{E}\|W_P(f)\|_{\mathcal{F}} + \mathbb{E}\left\|\sum_{i=1}^n \varepsilon_i f(X_i)/\sqrt{n}\right\|_{\mathcal{F}}\right),$$

$$(ii) \ \mathbb{E}\left\|\sum_{i=1}^n g_i f(X_i)/\sqrt{n}\right\|_{\mathcal{F}_2^n} \le K\mathbb{E}\|W_P(f)\|_{\mathcal{F}}.$$

Proof. We may and do assume that $\mathcal{F}$ is a finite class. By Theorem 12.6, there exist an ultrametric distance $\delta \ge d_2$ on $\mathcal{F}$ and a probability measure m on $(\mathcal{F}, \delta)$ such that

$$(14.10) \qquad \sup_{f \in \mathcal{F}} \int_0^\infty \left(\log \frac{1}{m(B(f, \varepsilon))}\right)^{1/2} d\varepsilon \le K\mathbb{E}\|W_P(f)\|_{\mathcal{F}}$$

where $B(f, \varepsilon)$ are the balls for the metric δ. K is some numerical constant, possibly changing from line to line below, and eventually yielding the constant of the statement. We use (14.10) as in Proposition 11.10 and Remark 11.11. Denote by ℓ_0 the largest ℓ for which $2^{-\ell} \ge D$ where D is the d_2-diameter of $\mathcal{F}$. For every $\ell \ge \ell_0$, let $\mathcal{B}_\ell$ be the family of δ-balls of radius $2^{-\ell}$. For every f in $\mathcal{F}$, there is a unique element B of $\mathcal{B}_\ell$ with $f \in B$. Let then $\pi_\ell(f)$ be one fixed point of B and let $\mu_\ell(\{\pi_\ell(f)\}) = m(B)$. Let further $\mu = \sum_{\ell \ge \ell_0} 2^{-\ell+\ell_0+1}\mu_\ell$ which defines a probability measure. We note that $d_2(f, \pi_\ell(f)) \le 2^{-\ell}$ for all f and ℓ and that $\pi_{\ell-1} \circ \pi_\ell = \pi_{\ell-1}$. From (14.10),

(14.11) $$\sup_{f\in\mathcal{F}}\sum_{\ell\geq\ell_0}2^{-\ell}\left(\log\frac{2^{\ell-\ell_0}}{\mu(\{\pi_\ell(f)\})}\right)^{1/2}\leq K\mathbb{E}\|W_P(f)\|_{\mathcal{F}}$$

(where we have used the definition of ℓ_0). Now, let n be fixed (so that we do not specify it every time we should). For every f in $\mathcal{F}$ and $\ell > \ell_0$, set

$$a(f,\ell)=\sqrt{n}2^{-\ell}\left(\log\frac{2^{\ell-\ell_0}}{\mu(\{\pi_\ell(f)\})}\right)^{-1/2}.$$

Given f in $\mathcal{F}$ and $x \in S$, let

$$\ell(x,f)=\sup\{\ell;\forall j\leq\ell,\ |\pi_j(f)(x)-\pi_{j-1}(f)(x)|\leq a(f,j)\}.$$

Define then f_2 by $f_2(x) = \pi_{\ell(x,f)}(f)(x)$ and $f_1 = f - f_2$, and let $\mathcal{F}_1 = \mathcal{F}_1^n = \{f_1;\ f\in\mathcal{F}\}$, $\mathcal{F}_2 = \mathcal{F}_2^n = \{f_2;\ f\in\mathcal{F}\}$, with the obvious abuse in notation. The classes $\mathcal{F}_1^n$ and $\mathcal{F}_2^n$ are the classes of the desired decomposition and we thus would like to show that they satisfy (i) and (ii) respectively.

We start with (ii). Set $\mathcal{F}_2 - \mathcal{F}_2 = \{f_2 - f_2';\ f_2,f_2'\in\mathcal{F}_2\}$ and $u = \mathbb{E}\|W_P(f)\|_{\mathcal{F}}$. We work with $\mathcal{F}_2 - \mathcal{F}_2$ rather than $\mathcal{F}_2$ since the process bounds of Chapter 11 are usually stated in this way. By definition of u, this will make no difference. We evaluate, for every $t > 0$ (or only $t \geq t_0$ large enough), the probability

$$\mathbb{P}\left\{\mathbb{E}_g\left\|\sum_{i=1}^n g_if(X_i)/\sqrt{n}\right\|_{\mathcal{F}_2-\mathcal{F}_2}>tu\right\}.$$

In a first step, let us show that this probability is less than $\mathbb{P}(A(t)^c)$ where, for K_2 to be specified later,

$$A(t)=\{\forall\ell\geq\ell_0,\ \forall f\in\mathcal{F},$$
$$\|(\pi_\ell(f)-\pi_{\ell-1}(f))I_{\{|\pi_\ell(f)-\pi_{\ell-1}(f)|\leq a(f,\ell)\}}\|_{n,2}\leq K_2^{-1}t2^{-\ell}\}$$

(recall the random norms and distances $\|\cdot\|_{n,2}$, $d_{n,2}$). Let f, f' in $\mathcal{F}$ and denote by j the largest ℓ such that $\pi_\ell(f) = \pi_\ell(f')$. Then $\ell(x,f) \geq j$ if and only if $\ell(x,f') \geq j$. That is, we can write for every x in S that

$$f_2(x)-f_2'(x)=\left(\pi_{\ell(x,f)}(f)(x)-\pi_j(f)(x)\right)I_{\{\ell(\cdot,f)\geq j\}}(x)$$
$$-\left(\pi_{\ell(x,f')}(f')(x)-\pi_j(f')(x)\right)I_{\{\ell(\cdot,f')\geq j\}}(x).$$

It follows that

$$d_{n,2}(f_2,f_2')=\|f_2-f_2'\|_{n,2}$$
$$\leq\sum_{\ell\geq j}\|(\pi_\ell(f)-\pi_{\ell-1}(f))I_{\{|\pi_\ell(f)-\pi_{\ell-1}(f)|\leq a(f,\ell)\}}\|_{n,2}$$
$$+\sum_{\ell\geq j}\|(\pi_\ell(f')-\pi_{\ell-1}(f'))I_{\{|\pi_\ell(f')-\pi_{\ell-1}(f')|\leq a(f',\ell)\}}\|_{n,2}$$

and thus, on the set $A(t)$, for all f, f',

$$d_{n,2}(f_2, f_2') \le K_2^{-1}t2^{-j+2} \le 8K_2^{-1}t\delta(f, f')$$

(by definition of δ and j). From this property, it follows from the majorizing measure bound of Theorem 11.18 and (14.10) that, for all $t > 0$,

$$(14.12) \qquad \mathbb{P}\left\{ \mathbb{E}_g \left\| \sum_{i=1}^n g_i f(X_i)/\sqrt{n} \right\|_{\mathcal{F}_2 - \mathcal{F}_2} > tu \right\} \le \mathbb{P}\left(A(t)^c \right)$$

for K_2 well chosen from (14.10) and the constant of Theorem 11.18. We therefore have to evaluate $\mathbb{P}(A(t)^c)$. To this aim, we use some exponential inequalities in the form, for example, of Lemma 1.6. Note that $\|\pi_\ell(f) - \pi_{\ell-1}(f)\|_2 \le 3 \cdot 2^{-\ell}$. Recentering, we deduce from Lemma 1.6 that, for every f in $\mathcal{F}$, and every $\ell > \ell_0$,

$$\mathbb{P}\left\{ \left\| (\pi_\ell(f) - \pi_{\ell-1}(f)) I_{\{|\pi_\ell(f) - \pi_{\ell-1}(f)| \le a(f,\ell)\}} \right\|_{n,2} > K_2^{-1}t2^{-\ell} \right\}$$

$$\le \exp\left(-tn2^{-2\ell}a(f,\ell)^{-2} \right)$$

for all large enough $t \ge t_1$ (independent of n, f and ℓ). By the definition of $a(f, \ell)$, this probability is estimated by

$$\exp\left(-t \log \left(\frac{2^{\ell - \ell_0}}{\mu(\{\pi_\ell(f)\})} \right) \right).$$

If $c \ge 2$, $\exp(-t \log c) \le (ct^2)^{-1}$ as soon as $t \ge t_2$ where t_2 is numerical. Therefore, if $t \ge t_0 = \max(t_1, t_2)$, we have obtained

$$\mathbb{P}\left(A(t)^c \right) \le \frac{1}{t^2} \sum_{\ell > \ell_0} \sum_{\{\pi_\ell(f)\}} 2^{\ell - \ell_0} \mu(\{\pi_\ell(f)\}) \le \frac{1}{t^2}$$

where we have used that $\pi_{\ell-1} \circ \pi_\ell = \pi_{\ell-1}$ and that μ is a probability. By (14.12), integration by parts then yields

$$\mathbb{E}\left\| \sum_{i=1}^n g_i f(X_i)/\sqrt{n} \right\|_{\mathcal{F}_2 - \mathcal{F}_2} \le \left(t_0 + \frac{1}{t_0} \right) u$$

from which (ii) immediately follows since, for any f in $\mathcal{F}$, $\|f_2\|_2 \le \|f\|_2 + 2^{-\ell_0} \le 5u$.

The main observation to establish (i) is the following: since we have $|\pi_{\ell(x,f)+1}(f)(x) - \pi_{\ell(x,f)}(f)(x)| > a(f, \ell+1)$, for every f_1 in $\mathcal{F}_1$,

$$\|f_1\|_1 = E|f| \le \sum_{\ell \ge \ell_0} E\left(|f - \pi_\ell(f)| I_{\{|\pi_{\ell+1}(f) - \pi_\ell(f)| > a(f,\ell+1)\}} \right).$$

By Cauchy-Schwarz and Chebyshev's inequalities,

$$\|f_1\|_1 \leq \sum_{\ell \geq \ell_0} a(f, \ell+1)^{-1} \|f - \pi_\ell(f)\|_2 \|\pi_{\ell+1}(f) - \pi_\ell(f)\|_2$$

$$\leq \sum_{\ell \geq \ell_0} a(f, \ell+1)^{-1} 3 \cdot 2^{-2\ell-1}$$

$$\leq K_1 \frac{u}{\sqrt{n}}$$

for some numerical constant K_1, where the last inequality is (14.11) (recall that $u = \mathbb{E}\|W_P(f)\|_{\mathcal{F}}$). It is easy to conclude. Set

$$v = \mathbb{E}\left\|\sum_{i=1}^n \varepsilon_i f(X_i)/\sqrt{n}\right\|_{\mathcal{F}}.$$

Since $\mathcal{F}_1 \subset \mathcal{F} - \mathcal{F}_2$, we already know from (ii) that

$$\mathbb{E}\left\|\sum_{i=1}^n \varepsilon_i f(X_i)/\sqrt{n}\right\|_{\mathcal{F}_1} \leq v + Ku.$$

From the comparison properties for Rademacher averages (Theorem 4.12),

$$\mathbb{E}\left\|\sum_{i=1}^n \varepsilon_i |f(X_i)|/\sqrt{n}\right\|_{\mathcal{F}_1} \leq 2(v + Ku),$$

and further, by Lemma 6.3,

$$\mathbb{E}\left\|\sum_{i=1}^n (|f(X_i)| - \mathbb{E}|f(X_i)|)/\sqrt{n}\right\|_{\mathcal{F}_1} \leq 4(v + Ku).$$

Since $\|f_1\|_1 \leq K_1 u n^{-1/2}$ for every f_1 in $\mathcal{F}_1$, (i) immediately follows. Therefore, the proof of Theorem 14.8 is complete.

It is noteworthy that the proof of Theorem 14.8 actually yields more than its statement. We have shown that, with high probability, the class $\mathcal{F}_2^n$ equipped with the random distances $d_{n,2}$ is a Lipschitz image of $(\mathcal{F}, \delta)$ which is controlled by the pregaussian hypothesis. The class $\mathcal{F}_1^n$ is controlled in $L_1(P_n)$. In this sense, Theorem 14.8 may appear as a random geometric description of the central limit property. If $\mathcal{F}$ is P-Donsker, then $\mathcal{F}$ is decomposed in two classes, the first for which the random distances $d_{n,2}$ are controlled by the (necessary) P-pregaussian property, the second being controlled in the $\|\cdot\|_{n,1}$ random norms for which no cancellation occurs. Conversely, such a decomposition clearly contains the Donsker property. Note that the levels of truncation chosen in the proof of Theorem 14.8 correspond, when the class $\mathcal{F}$ is reduced to one point, to the classical level $\sqrt{n}$.

To draw a possible qualitative version of Theorem 14.8, let us state the following (see also [Ta4], [G-Z3]). Recall that $\mathcal{F}_\eta = \{f - g;\ f, g \in \mathcal{F},\ d_2(f, g) < \eta\}$.

Corollary 14.9. *Let $\mathcal{F}$ be a P-pregaussian class. Then, for all $\eta > 0$ and every integer n, one can find classes $\mathcal{F}_1^n(\eta)$, $\mathcal{F}_2^n(\eta)$ in $L_2(P)$ such that $\mathcal{F}_\eta \subset \mathcal{F}_1^n(\eta) + \mathcal{F}_2^n(\eta)$ and such that*

$$\lim_{\eta \to 0} \limsup_{n \to \infty} \mathbb{E} \left\| \sum_{i=1}^n g_i f(X_i)/\sqrt{n} \right\|_{\mathcal{F}_2^n(\eta)} = 0$$

and

$$\limsup_{\eta \to 0} \limsup_{n \to \infty} \mathbb{E} \left\| \sum_{i=1}^n |f(X_i)|/\sqrt{n} \right\|_{\mathcal{F}_1^n(\eta)}$$

$$\leq K \limsup_{\eta \to 0} \limsup_{n \to \infty} \mathbb{E} \left\| \sum_{i=1}^n \varepsilon_i f(X_i)/\sqrt{n} \right\|_{\mathcal{F}_\eta}$$

(where K is a numerical constant). In particular, $\mathcal{F}$ is P-Donsker if and only if

$$\lim_{\eta \to 0} \limsup_{n \to \infty} \mathbb{E} \left\| \sum_{i=1}^n |f(X_i)|/\sqrt{n} \right\|_{\mathcal{F}_1^n(\eta)} = 0 \,.$$

14.3 Vapnik-Chervonenkis Classes of Sets

While the previous section dealt with random characterizations of Donsker classes, this section is devoted to the study of nice classes of indicator functions for which the classical limit theorems can be established. These classes of sets are the so-called Vapnik-Chervonenkis classes which naturally extend the case of the intervals $[0, t]$, $0 \leq t \leq 1$, on $[0, 1]$. As we will see moreover, the limit properties of empirical processes indexed by Vapnik-Chervonenkis classes actually hold *uniformly* over all probability distributions.

Let S be a set and $\mathcal{C}$ be a class of subsets of S. Let A be a subset of S of cardinality k. Say that $\mathcal{C}$ *shatters* A if each subset of A is the trace of an element of $\mathcal{C}$, i.e. $\mathrm{Card}(\mathcal{C} \cap A) = 2^k$ where $\mathcal{C} \cap A = \{\mathcal{C} \cap A; \ \mathcal{C} \in \mathcal{C}\}$. Say that $\mathcal{C}$ is a *Vapnik-Chervonenkis class* (VC class in short) if there exists an integer $k \geq 1$ such that no subset A of S of cardinality k is shattered by $\mathcal{C}$, i.e. for every A in S with $\mathrm{Card}\,A = k$, we have $\mathrm{Card}(\mathcal{C} \cap A) < 2^k$. Denote by $v(\mathcal{C})$ the smallest k with this property. The class $\mathcal{C} = \{[0, t]; \ 0 \leq t \leq 1\}$ in $[0, 1]$ is a VC class with $v(\mathcal{C}) = 2$. The following result is the most striking fact about VC classes.

Proposition 14.10. *Let $\mathcal{C}$ be a VC class in S and let $v = v(\mathcal{C})$. Then, for any finite subset A of S,*

$$\mathrm{Card}(\mathcal{C} \cap A) \leq \mathrm{Card}\{B \subset A; \ \mathrm{Card}\,B < v\} \,.$$

In particular, if $\mathrm{Card}\,A = n$ and $n \geq v$,

$$\mathrm{Card}(\mathcal{C} \cap A) \leq \left(\frac{en}{v} \right)^v \,.$$

The second part of the proposition follows from the fact that $\text{Card}\{B \subset A;\ \text{Card}\,B < v\} = \sum_{j<v} \binom{n}{j}$ and from an easy estimate of the latter. In particular, it indicates that we pass from the a priori information that $\text{Card}(\mathcal{C} \cap A) < 2^n$ to a *polynomial* growth of this cardinality. This section is devoted to the applications to empirical processes indexed by VC classes of this basic property.

We may note that if $B \subset A$ is such that $\text{Card}(\mathcal{C} \cap A) = 2^{\text{Card}B}$, then $\text{Card}B < v$. Proposition 14.10 therefore follows from the more general following result (by letting $\mathcal{U} = \mathcal{C} \cap A$), the proof of which uses rearrangement techniques.

Proposition 14.11. *Let A be a finite set and $\mathcal{U}$ be a class of subsets of A. Then,*
$$\text{Card}\,\mathcal{U} \leq \text{Card}\{B \subset A;\ B \text{ is shattered by } \mathcal{U}\}.$$

Proof. The idea is to find a simple operation (symmetrization) that will make $\mathcal{U}$ more regular while at the same time not decreasing the number of sets shattered by $\mathcal{U}$. One then applies this operation until the set $\mathcal{U}$ is so regular that the result is obvious. Given x in A, we define $T_x(\mathcal{U}) = \{T_x(U);\ U \in \mathcal{U}\}$ where for U in $\mathcal{U}$, $T_x(U) = U\backslash\{x\}$ if $x \in U$ and $U\backslash\{x\} \notin \mathcal{U}$ and $T_x(U) = U$ otherwise. The first observation is that

$$(14.13) \qquad \text{Card}\,T_x(\mathcal{U}) = \text{Card}\,\mathcal{U}.$$

To show this, it suffices to establish that T_x is one-to-one on $\mathcal{U}$. Suppose that $T_x(U_1) = T_x(U_2)$ for U_1, U_2 in $\mathcal{U}$. Then, by the definition of the operation T_x, $U_1\backslash\{x\} = U_2\backslash\{x\}$. Since $U_1 = U_2$ when $x \in T_x(U_1)$, let us assume that $x \notin T_x(U_1)$ and that $U_1 \neq U_2$, and proceed to a contradiction. Suppose for example that $x \in U_1$, $x \notin U_2$. Then, since $U_1\backslash\{x\} = U_2 \in \mathcal{U}$, $T_x(U_1) = U_1$, so $x \in T_x(U_1)$ which is a contradiction. This shows (14.13). Let us now establish that if $T_x(\mathcal{U})$ shatters B, then $\mathcal{U}$ shatters B. If $x \notin B$, $T_x(\mathcal{U})$ and $\mathcal{U}$ have the same trace on B. If $x \in B$, for $B' \subset B\backslash\{x\}$, there is $T \in T_x(\mathcal{U})$ such that $T \cap B = B'\cup\{x\}$. T is of the form $T_x(U)$ for some U in $\mathcal{U}$. Since $x \in T$, both U and $U\backslash\{x\}$ belong to $\mathcal{U}$, so $\mathcal{U}$ shatters B. We can now conclude the proof of the proposition. Let $w(\mathcal{U}) = \sum_{U \in \mathcal{U}} \text{Card}\,U$. Let $\mathcal{U}'$ be such that $\mathcal{U}'$ is obtained from $\mathcal{U}$ by applications of some transformations T_x and such that $w(\mathcal{U}')$ is minimal. Then, for each U in $\mathcal{U}'$ and each x in U, we must have $U\backslash\{x\} \in \mathcal{U}'$ for, otherwise, $w(T_x(\mathcal{U}')) < w(\mathcal{U}')$. This means that $\mathcal{U}'$ is hereditary in the sense that if $B' \subset B \in \mathcal{U}'$, then $B' \in \mathcal{U}'$. In particular, $\mathcal{U}'$ shatters each set it contains so that the result of the proposition is obvious for $\mathcal{U}'$. Since by (14.13), $\text{Card}\,\mathcal{U}' = \text{Card}\,\mathcal{U}$ and since $\mathcal{U}$ shatters more sets than $\mathcal{U}'$, the proof is complete.

Let now $(S,\mathcal{S})$ be a measurable space and consider a class $\mathcal{C} \subset \mathcal{S}$. If Q is a probability measure on $(S,\mathcal{S})$, we let, for any A, B in $\mathcal{S}$,

$$d_Q(A, B) = \big(Q(A \triangle B)\big)^{1/2} = \|I_A - I_B\|_2$$

(where $A \triangle B = (A \triangle B^c) \cup (A^c \cap B)$ and where the norm $\| \cdot \|_2$ is understood with respect to Q). Recall the entropy numbers $N(\mathcal{C}, d_Q; \varepsilon)$. The next theorem is a consequence of Proposition 14.10 and appears as the fundamental property in the study of limit theorems for empirical processes indexed by VC classes.

Theorem 14.12. *Let $\mathcal{C} \subset \mathcal{S}$ be a VC class. There is a numerical constant K such that for all probability measures Q on $(S, \mathcal{S})$ and all $0 < \varepsilon \leq 1$,*

$$\log N(\mathcal{C}, d_Q; \varepsilon) \leq K v(\mathcal{C}) \left(1 + \log \frac{1}{\varepsilon} \right).$$

Proof. Let Q and ε be fixed. Let N be such that $N(\mathcal{C}, d_Q; \varepsilon) \geq N$. There exist $A_1, \ldots, A_N$ in $\mathcal{C}$ such that $d_Q(A_k, A_\ell) \geq \varepsilon$ for all $k \neq \ell$. If (X_i) are independent random variables distributed according to Q, $\mathbb{P}\{X_i \notin A_k \triangle A_\ell\} \leq 1 - \varepsilon^2$ and thus

$$\mathbb{P}\{(A_k \triangle A_\ell) \cap \{X_1, \ldots, X_M\} = \emptyset\} \leq (1 - \varepsilon^2)^M$$

for all $k \neq \ell$ and all integers M. Therefore, if M is chosen such that $N^2(1 - \varepsilon^2)^M < 1$,

$$\mathbb{P}\{\forall k \neq \ell, \ (A_k \triangle A_\ell) \cap \{X_1, \ldots, X_M\} \neq \emptyset\} > 0$$

and there exist points $x_1, \ldots, x_M$ in S such that, for $k \neq \ell$, $(A_k \triangle A_\ell) \cap \{x_1, \ldots, x_M\}) \neq \emptyset$. Then, Proposition 14.10 indicates that, necessarily,

$$N \leq \left(\frac{eM}{v} \right)^v$$

where $v = v(\mathcal{C})$, at least when $M \geq v$. Take $M = [2\varepsilon^{-2} \log N] + 1$ so that $N^2(1 - \varepsilon^2)^M < 1$. Assume first that $\log N \geq v$ (≥ 1) so that $v \leq M \leq 4\varepsilon^{-2} \log N$. Then, the inequality $N \leq (eM/v)^v$ yields

$$\log N \leq v \left(\log \frac{4e}{\varepsilon^2} + \log \left(\frac{\log N}{v} \right) \right)$$

$$\leq v \log \frac{4e}{\varepsilon^2} + \frac{1}{e} \log N$$

where we have used that $\log x \leq x/e$, $x > 0$. It follows that

$$\log N \leq 2v \log \frac{4e}{\varepsilon^2},$$

an inequality which is also satisfied when $\log N \leq v$ since $\varepsilon \leq 1$. Since $N \leq N(\mathcal{C}, d_Q; \varepsilon)$ is arbitrary, Theorem 14.12 is established.

To agree with the notation and language of the previous section, we identify a class $\mathcal{C} \subset \mathcal{S}$ with the class $\mathcal{F}$ of the indicator functions I_C, $C \in \mathcal{C}$, and thus write $\| \cdot \|_{\mathcal{C}}$ for $\| \cdot \|_{\mathcal{F}}$. We also assume that we only deal with countable classes $\mathcal{C}$ in order to avoid the usual measurability questions. This condition may be replaced by some appropriate separability assumption. The next statement is one of the main results on VC classes. It expresses that VC classes are Donsker classes for all underlying probability distributions.

Theorem 14.13. *Let $C \subset S$ be a VC class. Then C is a Donsker class for every probability measure P on (S, S).*

Proof. Let P be a probability measure on (S, S). As a first simple consequence of Theorem 14.12, C is totally bounded in $L_2 = L_2(P)$. By Theorem 14.6, it therefore suffices to show that

$$(14.14) \quad \lim_{\eta \to 0} \limsup_{n \to \infty} \mathbb{P} \left\{ \sup_{d_P(C,D) < \eta} \left| \sum_{i=1}^n \varepsilon_i (I_C - I_D)(X_i)/\sqrt{n} \right| > \varepsilon \right\} = 0$$

for all $\varepsilon > 0$ (where, of course, C, D belong to C). Recall the empirical measures $P_n(\omega) = \sum_{i=1}^n \delta_{X_i(\omega)}/n$, $\omega \in \Omega$, $n \geq 1$. For every ω and every n, the random (in the Rademacher sequence (ε_i)) process $\left(\sum_{i=1}^n \varepsilon_i I_C(X_i(\omega))/\sqrt{n} \right)_{C \in C}$ is subgaussian with respect to $d_{P_n(\omega)}$. Since Theorem 14.12 actually implies

$$\lim_{\eta \to 0} \sup_Q \int_0^\eta \left(\log N(C, d_Q; \varepsilon) \right)^{1/2} d\varepsilon = 0,$$

it is plain that Proposition 11.19 yields a conclusion which is *uniform* over the distances $d_{P_n(\omega)}$ in the sense that for every $\varepsilon > 0$, there exists $\eta = \eta(\varepsilon) > 0$ such that *for every n and every ω,*

$$(14.15) \quad \mathbb{E}_\varepsilon \sup_{d_{P_n(\omega)}(C,D) < \eta} \left| \frac{1}{\sqrt{n}} \sum_{i=1}^n \varepsilon_i (I_C - I_D)(X_i(\omega)) \right| < \varepsilon^2$$

where, as usual, $\mathbb{E}_\varepsilon$ is partial integration with respect to (ε_i). From this result, the proof will be completed if the random distances $d_{P_n(\omega)}$ can be replaced by d_P, at least on a large enough set of ω's. As for (14.15), if C is VC,

$$\sup_n \frac{1}{\sqrt{n}} \mathbb{E} \left\| \sum_{i=1}^n \varepsilon_i I_C(X_i) \right\|_C < \infty.$$

By Lemma 6.3 (actually the subsequent comment), we also have

$$\sup_n \frac{1}{\sqrt{n}} \mathbb{E} \left\| \sum_{i=1}^n (I_C(X_i) - P(C)) \right\|_C < \infty.$$

The same property holds for $C \triangle C = \{C \triangle C'; \ C, C' \in C\}$ since it is also VC by Proposition 14.10. Hence, given $\varepsilon > 0$ and $\eta = \eta(\varepsilon) > 0$ so that (14.15) is satisfied, there exists n_0 such that $\mathbb{P}(A(n)) \leq \varepsilon$ for all $n \geq n_0$ where

$$A(n) = \left\{ \|(P_n - P)(C \triangle C')\|_{C \triangle C} > \eta^2/4 \right\}.$$

We can easily conclude. For all $n \geq n_0$, using (14.15),

$$\mathbb{P}\left\{ \sup_{d_P(C,D)<\eta/2} \left| \sum_{i=1}^{n} \varepsilon_i (I_C - I_D)(X_i)/\sqrt{n} \right| > \varepsilon \right\}$$

$$\leq \varepsilon + \mathbb{P}\left\{ A(n)^c ; \sup_{d_{P_n}(C,D)<\eta} \left| \sum_{i=1}^{n} \varepsilon_i (I_C - I_D)(X_i)/\sqrt{n} \right| > \varepsilon \right\}$$

$$\leq 2\varepsilon .$$

This gives (14.14) so that Theorem 14.13 is established.

The preceding proof based on the key Theorem 14.12 actually carries more information than the actual statement of Theorem 14.13. It indeed indicates a uniformity property over all probability measures P. With the same argument leading to (14.15) via Theorem 14.12, and as we actually used it in the proof, there is a numerical constant $K > 0$ such that if $\mathcal{C}$ is VC,

$$(14.16) \qquad \sup_{n} \mathbb{E}\|\nu_n(\mathcal{C})\|_{\mathcal{C}} \leq K v(\mathcal{C})^{1/2}$$

for all probability measures P on $(S, \mathcal{S})$, where we recall that $\nu_n = \sqrt{n}(P_n - P)$. This property implies a uniform strong law of large numbers in the sense that for some sequence (a_n) of positive numbers decreasing to 0 and for any P

$$\sup_{n} \frac{1}{a_n} \|(P_n - P)(\mathcal{C})\|_{\mathcal{C}} < \infty \quad \text{almost surely} .$$

This may be obtained for example from Theorem 8.6 (or the "bounded" version of Theorem 10.12) which yields the best possible $a_n = (LLn/n)^{1/2}$. One may also invoke the SLLN results in the form of Theorem 7.9 for example. (Recall that since we are dealing with indicators, the corresponding random variables in $\ell_\infty(\mathcal{C})$ are uniformly bounded.)

It is remarkable that these uniform limit properties actually characterize VC classes. Following (14.16), suppose for example that we are given a class $\mathcal{C}$ in $\mathcal{S}$ such that, for some finite M,

$$(14.17) \qquad \sup_{n} \mathbb{E}\|\nu_n(C)\|_{\mathcal{C}} \leq M$$

for all probability measures P on $(S, \mathcal{S})$. If we then recall the Gaussian processes G_P with covariance $P(C \cap D) - P(C)P(D)$, $C, D \in \mathcal{C}$, introduced in Section 14.2, we also have, by the finite dimensional CLT, that

$$(14.18) \qquad \mathbb{E}\|G_P(C)\|_{\mathcal{C}} \leq M$$

for every P. In particular, if $P = \frac{1}{k}\sum_{i=1}^{k} \delta_{x_i}$ where $x_1, \ldots, x_k$ are points in S, G_P can be described as

$$G_P(C) = \frac{1}{\sqrt{k}} \sum_{i=1}^{k} g_i\big(\delta_{x_i}(C) - P(C)\big), \quad C \in \mathcal{C} ,$$

where (g_i) is a standard normal sequence. Therefore, from (14.18),

(14.19) $$\mathbb{E}\left\|\sum_{i=1}^{k} g_i \delta_{x_i}(C)\right\|_C \leq (M+1)\sqrt{k}.$$

Now, suppose that C is not a VC class. Then, for every k, there exists a subset $A = \{x_1, \ldots, x_k\}$ of S of cardinality k such that $\mathrm{Card}(C \cap A) = 2^k$. Therefore, as is obvious, for every $\alpha = (\alpha_1, \ldots, \alpha_k)$ in $\mathbb{R}^k$,

(14.20) $$\sum_{i=1}^{k} |\alpha_i| \leq 2\left\|\sum_{i=1}^{k} \alpha_i \delta_{x_i}(C)\right\|_C.$$

If we integrate this inequality along the orthogaussian sequence (g_i), we get a contradiction with (14.19) when k is large enough. Therefore C is necessarily a VC class under (14.18) and thus a fortiori under (14.17). (By an argument close in spirit to the closed graph theorem as in Proposition 14.14 below, it is actually enough to have that C is P-pregaussian for every P.)

The next proposition strengthens this conclusion, showing in particular that the conclusion is not restricted to the CLT.

Proposition 14.14. *Let C be a (countable) class in S. Assume that there exists a decreasing sequence of positive numbers (a_n) tending to 0 such that for every probability P on $(S, \mathcal{S})$, the sequence $(\|(P_n - P)(C)\|_C / a_n)$ is bounded in probability. Then C is a VC class.*

Proof. We may assume that $\sup_n (a_n \sqrt{n})^{-1} < \infty$. By Hoffmann-Jørgensen's inequalities (Proposition 6.8),

$$\sup_n \frac{1}{a_n} \mathbb{E}\|(P_n - P)(C)\|_C < \infty.$$

From Lemma 6.3, we see that, for every n,

$$\mathbb{E}\left\|\sum_{i=1}^{n} \varepsilon_i \big(I_C(X_i) - P(C)\big)\right\|_C \leq 2\mathbb{E}\left\|\sum_{i=1}^{n} \big(I_C(X_i) - P(C)\big)\right\|_C.$$

Therefore, if we set

$$\Phi(P) = \sup_n \frac{1}{na_n} \mathbb{E}\left\|\sum_{i=1}^{n} \varepsilon_i I_C(X_i)\right\|_C,$$

we clearly get

$$\Phi(P) \leq \sup_n \frac{1}{\sqrt{n}a_n} + 2\sup_n \frac{1}{a_n}\mathbb{E}\|(P_n - P)(C)\|_C.$$

Hence $\Phi(P) < \infty$ for every P. We would like to show that there is some finite constant M such that

(14.21) $\Phi(P) \leq M$ for all probability measures P.

To this aim, let us first show that if P^0 and P^1 are two probability measures on $(S, \mathcal{S})$, and if $P = \alpha P^0 + (1 - \alpha)P^1$, $0 \leq \alpha \leq 1$, then $\Phi(P) \geq \alpha \Phi(P^0)$. Let (X_i^0) (resp. (X_i^1)) be independent random variables with common distribution P^0 (resp. P^1). Let further (λ_i) be independent of everything that was introduced before and consisting of independent random variables with law $\mathbb{P}\{\lambda_i = 0\} = 1 - \mathbb{P}\{\lambda_i = 1\} = \alpha$. Then (X_i) has the same distribution as $(X_i^{\lambda_i})$. In particular, by the contraction principle,

$$\mathbb{E}\left\|\sum_{i=1}^{n} \varepsilon_i I_C(X_i)\right\|_C \geq \mathbb{E}\left\|\sum_{i=1}^{n} \varepsilon_i I_C(X_i^0) I_{\{\lambda_i=0\}}\right\|_C .$$

Jensen's inequality with respect to the sequence (λ_i) then yields

$$\mathbb{E}\left\|\sum_{i=1}^{n} \varepsilon_i I_C(X_i)\right\|_C \geq \alpha \mathbb{E}\left\|\sum_{i=1}^{n} \varepsilon_i I_C(X_i^0)\right\|_C$$

and thus the desired inequality $\Phi(P) \geq \alpha \Phi(P^0)$. This observation easily implies (14.21). Indeed, if it is not satisfied, one can find a sequence (P^k) of probability measures on $(S, \mathcal{S})$ such that $\Phi(P^k) \geq 4^k$ for every k. If we then let $P = \sum_{k=1}^{\infty} 2^{-k} P^k$, $\Phi(P) \geq 2^{-k} \Phi(P^k) \geq 2^k$ for every k, contradicting $\Phi(P) < \infty$.

Now, we can conclude the proof of Proposition 14.14. If C is not VC, there exists, for every k, $A = \{x_1, \ldots, x_k\}$ in S such that (14.20) holds. Take then $P = \frac{1}{k} \sum_{i=1}^{k} \delta_{x_i}$. Fix then n large enough so that $n > 4Mna_n$, which is possible since $a_n \to 0$. Consider $k > 2n^2$ such that $\mathbb{P}(\Omega_0) \geq 1/2$ where

$$\Omega_0 = \{\forall i \neq j \leq n, X_i \neq X_j\}.$$

Then, since $\Phi(P) \leq M$, we can write by (14.20)

$$Mna_n \geq \mathbb{E}\left\|\sum_{i=1}^{n} \varepsilon_i I_C(X_i)\right\|_C$$

$$\geq \int_{\Omega_0}\left\|\sum_{i=1}^{n} \varepsilon_i I_C(X_i)\right\|_C \geq \frac{n}{2}\mathbb{P}(\Omega_0) \geq \frac{n}{4}$$

which is a contradiction. The proof of Proposition 14.14 is complete.

As in Section 14.1, the nice limit properties of empirical processes indexed by VC classes may be related to the type property of a certain operator between Banach spaces. Denote by $M(S, \mathcal{S})$ the Banach space of all bounded measures μ on $(S, \mathcal{S})$ equipped with the norm $\|\mu\| = |\mu|(S)$. Let $C \subset \mathcal{S}$; consider the operator $j : M(S, \mathcal{S}) \to \ell_\infty(C)$ defined by $j(\mu) = (\mu(C))_{C \in C}$. We denote by $T_2(j)$ the type 2 constant of j, that is the smallest constant C such that for all finite sequences (μ_i) in $M(S, \mathcal{S})$,

$$\mathbb{E}\left\|\sum_{i} \varepsilon_i j(\mu_i)\right\|_C \leq C\left(\sum_{i} \|\mu_i\|^2\right)^{1/2}$$

(provided there exists one). We have the following result.

Theorem 14.15. *In the preceding notation, C is a VC class if and only if j is an operator of type 2. Moreover, for some numerical constant $K > 0$,*

$$K^{-1}(v(C))^{1/2} \leq T_2(j) \leq K(v(C))^{1/2}.$$

Proof. We establish that $T_2(j) \leq K(v(C))^{1/2}$. Let (μ_i) be a finite sequence in $M(S, \mathcal{S})$. To prove the type 2 inequality, we may assume that the measures μ_i are positive and, by homogeneity, that $\sum_i \|\mu_i\|^2 = 1$. Set $Q = \sum_i \|\mu_i\|\mu_i$. Then Q is a probability measure on $(S, \mathcal{S})$ and we clearly have

$$\left(\sum_i |\mu_i(C) - \mu_i(D)|^2\right)^{1/2} \leq d_Q(C, D)$$

for all C, D. Therefore, the entropic version of inequality (11.19) together with Theorem 14.12 applied to the process $\left(\sum_i \varepsilon_i \mu_i(C)\right)_{C \in \mathcal{C}}$ where $\mathcal{C} \subset \mathcal{S}$ is VC yields

$$\mathbb{E}\left\|\sum_i \varepsilon_i j(\mu_i)\right\|_C \leq 1 + K \int_0^\infty (\log N(\mathcal{C}, d_Q; \varepsilon))^{1/2} d\varepsilon$$

$$\leq 1 + K'(v(C))^{1/2}$$

where K, K' are numerical constants. Since $v(C) \geq 1$, the right hand side inequality of the theorem follows. To prove the converse inequality, set $v = v(C)$. Since $T_2(j) \geq 1$, we can assume that $v \geq 2$. By definition of v, there exists $A = \{x_1, \ldots, x_{v-1}\}$ in S such that $\mathrm{Card}(\mathcal{C} \cap A) = 2^{v-1}$. Then

$$\mathbb{E}\left\|\sum_{i=1}^{v-1} \varepsilon_i j(\delta_{x_i})\right\|_C \leq T_2(j)(v-1)^{1/2},$$

and, by (14.20),

$$\frac{1}{2}(v-1) \leq T_2(j)(v-1)^{1/2}.$$

Therefore, $T_2(j) \geq (v-1)^{1/2}/2 \geq v^{1/2}/4$, which completes the proof of Theorem 14.15. $\qquad\blacksquare$

From this result together with the limit theorems for sums of independent random variables with values in Banach spaces with some type (or for *operators* with some type) (cf. Chapter 7, 8, 9, 10), one can essentially again deduce Theorem 14.13 and the consequent strong limit theorems for empirical processes indexed by VC classes. Moreover, one can note that C is actually a VC class as soon as the associated operator j has *some* type $p > 1$, simply because if this holds, we are in a position to apply Proposition 14.14. Finally, the property (14.20) makes it clear that the notion of VC class is related to ℓ_1^n-spaces. The following statement is the Banach space theory formulation of the previous investigation.

Theorem 14.16. *Let* $x_1, \ldots, x_n$ *be functions on some set* T *taking the values* ± 1. *Let* $r(T) = \mathbb{E} \sup_{t \in T} \left| \sum_{i=1}^{n} \varepsilon_i x_i(t) \right|$. *There exists a numerical constant* K *such that for every* k *such that* $k \le r(T)^2 / Kn$, *one can find* $m_1 < m_2 < \cdots < m_k$ *in* $\{1, \ldots, n\}$ *such that the set of values* $\{x_{m_1}(t), \ldots, x_{m_k}(t)\}$, $t \in T$, *is exactly* $\{-1, +1\}^k$. *In other words, the subsequence* $x_{m_1}, \ldots, x_{m_k}$ *generates a subspace isometric to* ℓ_1^k *in* $\ell_\infty(T)$.

Proof. Let

$$M = \mathbb{E} \sup_{t \in T} \left| \sum_{i=1}^{n} \varepsilon_i \left(\frac{1 + x_i(t)}{2} \right) \right|$$

and consider the class $\mathcal{C}$ of subsets of $\{1, 2, \ldots, n\}$ of the form $\{i \in \{1, \ldots, n\}$; $x_i(t) = 1\}$, $t \in T$. Theorem 14.15 applied to this class yields

$$M \le T_2(j)\sqrt{n} \le K \left(n v(\mathcal{C}) \right)^{1/2}.$$

Therefore, by definition of $v(\mathcal{C})$, the conclusion of the theorem is fulfilled for all $k < M^2 / K^2 n$. Note that $r(T) \le 2M + \sqrt{n}$. Since we may assume that $M \ge \sqrt{n}$ (otherwise there is nothing to prove), we see that when $k \le r(T)^2 / K'n$ for some large enough K', we have that $k < M^2 / K^2 n$ in which case we already know that the conclusion holds.

Notes and References

As announced, this chapter only presents a few examples of the empirical process methods and their applications. Our framework is essentially the one put forward in the work of E. Giné and J. Zinn [G-Z2], [G-Z3], itself initiated by R. Dudley [Du4]. Some general references on empirical processes are the notes and books [Ga], [Du5], [Pol], [G-Z3]. The interested reader will complete appropriately this chapter and this short discussion with those references and the papers cited therein.

The central limit theorems for subgaussian and Lipschitz processes (Theorems 14.1 and 14.2) are due to N. C. Jain and M. B. Marcus [J-M1]. R. Dudley and V. Strassen [D-S] introduced entropy in this study and established Theorem 14.2 with M uniformly bounded. Another prior partial result may be found in [Gin] (where the technique of proof is actually close to the nowadays bracketing arguments – see below). These authors worked under the metric entropy condition; the majorizing measure version of these results was obtained in [He1]. In [Zi1], J. Zinn connects the Jain-Marcus CLT for Lipschitz processes with the type 2 property of the operator $j : \mathrm{Lip}(T) \to C(T)$. The analog of Theorem 14.2 for random variables in c_0 is studied in [Pau], [He4], [A-G-O-Z], [Ma-Pl]. A stronger version of Theorem 14.3 is established in [A-G-O-Z] where it is shown that the conclusion actually holds for *local* Lipschitz processes, namely processes X in $C(T)$ such that for every t in T and every $\varepsilon > 0$,

$$\| \sup_{d(s,t)<\varepsilon} |X_s - X_t| \|_{2,\infty} \leq \varepsilon .$$

The proof of this result relies on bracketing techniques in the context of empirical processes. (The arguments of proof are related to Theorem 14.8 and the truncations used there; a similar decomposition is developed from which the local Lipschitz conditions provide the control of the corresponding $L_1(P_n)$-portion.) Bracketing was initiated in [Du4]; further results are obtained in [Oss], [A-G-Z], [L-T4]. In those last two articles, analogous results for the LIL are discussed, improving upon [Led1]. The simple proof of the (weaker) Theorem 14.3 and the inequality of Lemma 14.4 are due to B. Heinkel [He7] (see also [He8]). Theorem 14.5 is taken from [M-P3].

The equicontinuity criterion for Donsker classes is due to R. Dudley [Du4]; its randomized version (as stated as Theorem 14.6) is taken from [G-Z2]. E. Giné and J. Zinn made clever use of the Gaussian randomization and the Gaussian process techniques together with exponential bounds to achieve remarkable progress in the understanding of the Donsker property. Theorem 14.7 is theirs [G-Z2]. Theorem 14.8 and the random geometric description of Donsker classes have been obtained in [Ta4], motivated by the investigation and the prior results in [G-Z2]. We refer to [G-Z3] for an alternate exposition and for more details. Further statements in this spirit appear in [L-T4]. Donsker classes of sets are investigated in [G-Z2], [G-Z3], [Ta6] (see also [V-C2]).

Vapnik-Chervonenkis classes of sets were introduced in [V-C1] and were shown there to satisfy uniform laws of large numbers. CLT, LIL and invariance principle for VC classes have been established in [Du4], [K-D], [D-P] respectively. Our exposition of this section is based on the observations by G. Pisier [Pi14]. Proposition 14.10 was established, independently, in [V-C1], [Sa], [Sh]. The proof based on Proposition 14.11 seems to be due to P. Frankl [Fr]. We learned it from V. D. Milman. The main Theorem 14.12 on the uniform entropy control of VC classes was observed by R. Dudley [Du4] to whom Theorem 14.13 is due. That VC classes are actually characterized by uniform limit properties of the empirical measures was noticed in a particular case (for the Donsker property) in [D-D] and completely understood via the map j and Theorem 14.15 by G. Pisier [Pi14]. Theorem 14.16 is also taken from [Pi14]. Universal Donsker classes of functions do not seem to have similar nice descriptions (see however [G-Z5]); for some results connected with type 2 maps, cf. [Zi4]. Let us also mention the extension of the VC definition to classes of functions (VC graphs) with, in particular, a characterization of the Donsker property [Ale2] in the spirit of the best possible conditions for the CLT in Banach spaces (cf. Chapter 10). See [A-T] for the corresponding LIL result.

15. Applications to Banach Space Theory

This last chapter emphasizes some applications of isoperimetric methods and of process techniques of Probability in Banach spaces to the local theory of Banach spaces. The applications which we present are only a sample of some of the recent developments in the local theory of Banach spaces (and we refer to the lists of references, and seminars and proceedings, for further main examples in the historical developments). They demonstrate the power of probabilistic ideas in this context. This chapter is organized along its subtitles of rather independent context. Several questions and conjectures are presented in addition, some with details as in Sections 15.2 and 15.6, the others in the last paragraph on miscellaneous problems.

15.1 Subspaces of Small Codimension

Before we turn to the object of this first section, it is convenient to briefly present a covering lemma in the spirit of Lemma 9.5 which will be of help here. We denote by $B_2 = B_2^N$ the Euclidean unit ball of $\mathbb{R}^N$.

Lemma 15.1. *There exists a subset H of $2B_2^N$ of cardinality at most 5^N such that $B_2^N \subset \operatorname{Conv} H$.*

Proof. It is similar to the proof of Lemma 9.5. Let N be fixed. For $0 < \delta < 1$, let $\widetilde{H}$ be maximal in B_2 such that $|x - y| > \delta$ for all x, y in $\widetilde{H}$. Then the balls of radius $\delta/2$ with centers in $\widetilde{H}$ are disjoint and contained in $(1 + \delta/2)B_2$. If we compare the volumes, we get

$$\operatorname{Card} \widetilde{H} \left(\frac{\delta}{2}\right)^N \operatorname{vol} B_2 \leq \left(1 + \frac{\delta}{2}\right)^N \operatorname{vol} B_2$$

so that $\operatorname{Card} \widetilde{H} \leq (1 + 2/\delta)^N$. By maximality of $\widetilde{H}$, it is easily seen that each x in B_2 can be written as

$$x = \sum_{k=0}^{\infty} \delta^k h_k$$

where $(h_k) \subset \widetilde{H}$. Then take $\delta = 1/2$, $H = 2\widetilde{H}$ and the lemma follows.

As a consequence of this lemma, note that $\widetilde{H} = \frac{1}{2}H$ in the preceding proof is such that $\widetilde{H} \subset B_2^N$, Card $\widetilde{H} \leq 5^N$ and

$$|x| \leq 2 \sup_{h \in \widetilde{H}} |\langle x, h \rangle|$$

for all x in $\mathbb{R}^N$.

Let T be a convex body in $\mathbb{R}^N$, that is T is a compact convex symmetric (about the origin) subset of $\mathbb{R}^N$ with non-empty interior. As usual, denote by (g_i) an orthonormal Gaussian sequence. We let (as in Chapter 3)

$$\ell(T) = \mathbb{E} \sup_{t \in T} \left| \sum_{i=1}^{N} g_i t_i \right| = \int_{\mathbb{R}^N} \sup_{t \in T} |\langle x, t \rangle| d\gamma_N(x)$$

where $t = (t_1, \ldots, t_N)$ in $\mathbb{R}^N$ and γ_N is the canonical Gaussian measure on $\mathbb{R}^N$. The result of this section describes a way of finding a subspace F of $\mathbb{R}^N$ whose intersection with T has a small diameter as soon as the codimension of F is large with respect to $\ell(T)$. This result is one of the main steps in the proof of the remarkable quotient of a subspace theorem of V. D. Milman [Mi2]. As for Dvoretzky's theorem, we present two proofs of the result, both of them based on Gaussian variables. The first proof uses isoperimetry and Sudakov's minoration, the second the Gaussian comparison theorems.

For g_{ij}, $1 \leq i \leq N$, $1 \leq j \leq k$, $k \leq N$, a family of independent standard normal variables, we denote by G the random operator $\mathbb{R}^N \to \mathbb{R}^k$ with matrix (g_{ij}). If S is a subset of $\mathbb{R}^N$, we set $\text{diam}(S) = \sup\{|x - y|; \ x, y \in S\}$. If F is a vector subspace of a vector space B, the *codimension* of F in B is the dimension of the quotient space B/F.

Theorem 15.2. *Let T be a convex body in $\mathbb{R}^N$. There exists a numerical constant $K > 0$ such that for all $k \leq N$*

$$\mathbb{P}\left\{ \text{diam}(T \cap \text{Ker}G) \leq K \frac{\ell(T)}{\sqrt{k}} \right\} \geq 1 - \exp(-k/K).$$

In particular, there exists a subspace F of $\mathbb{R}^N$ of codimension k (obtained as $F = \text{Ker}G(\omega)$ for some appropriate ω) such that

$$\text{diam}(T \cap F) \leq K \frac{\ell(T)}{\sqrt{k}}.$$

First proof of Theorem 15.2. We start with an elementary observation. For every x in $\mathbb{R}^N$ and all $u > 0$,

(15.1) $$\mathbb{P}\{|G(x)| < u|x|\} \leq \left(\frac{eu^2}{k} \right)^{k/2}.$$

We may assume by homogeneity that $|x| = 1$. Then $|G(x)|^2$ has the same distribution as $\sum_{i=1}^{k} g_i^2$. For every $\lambda > 0$,

$$\mathbb{E}\exp\left(-\lambda \sum_{i=1}^{k} g_i^2\right) = (\mathbb{E}\exp(-\lambda g_1^2))^k = \left(\frac{1}{1+2\lambda}\right)^{k/2}.$$

Therefore,

$$\mathbb{P}\{|G(x)|^2 < u^2\} = \mathbb{P}\left\{\exp\left(-\lambda \sum_{i=1}^{k} g_i^2\right) > \exp(-\lambda u^2)\right\}$$

$$\leq \left(\frac{1}{1+2\lambda}\right)^{k/2} \exp(\lambda u^2).$$

Choose then for example $\lambda = k/2u^2$ and (15.1) follows.

The next lemma is the main step of this proof. It is based on the isoperimetric type inequalities for Gaussian measures.

Lemma 15.3. *Under the previous notation, for every integer k and every $u > 0$,*

$$\mathbb{P}\left\{\sup_{x \in T} |G(x)| > 4\ell(T) + u\right\} \leq 5^k \exp\left(-\frac{u^2}{8R^2}\right)$$

where $R = \sup_{x \in T} |x|$.

Proof. By Lemma 15.1 (or rather by its immediate consequence), there exists H in B_2^k with cardinality less than or equal to 5^k such that

$$|G(x)| \leq 2 \sup_{h \in H} |\langle G(x), h\rangle|$$

for all x in $\mathbb{R}^N$. Therefore, it suffices to show that for every $|h| \leq 1$,

$$(15.2) \qquad \mathbb{P}\left\{\sup_{x \in T} |\langle G(x), h\rangle| > 2\ell(T) + \frac{u}{2}\right\} \leq \exp\left(-\frac{u^2}{8R^2}\right).$$

By the Gaussian rotational invariance and the definition of G, the process $(\langle G(x), h\rangle)_{x \in T}$ is distributed as $\left(|h| \sum_{i=1}^{N} g_i x_i\right)_{x \in T}$. Then (15.2) is a direct consequence of Lemma 3.1. (Of course, we need not be concerned here with sharp constants in the exponential function and the simpler inequality (3.2), for example, can be used equivalently.) Lemma 15.3 is established. $\square$

We can conclude this first proof of Theorem 15.2. By Sudakov's minoration (Theorem 3.18),

$$\sup_{\varepsilon > 0} \varepsilon \left(\log N(T, \varepsilon B_2^N)\right)^{1/2} \leq K_1 \ell(T)$$

for some numerical constant K_1. Therefore, there exists a subset S of T of cardinality less than $\exp(K_1^2 k)$ such that the Euclidean balls of radius $k^{-1/2}\ell(T)$

and with centers in S cover T. Let $\overline{T} = 2T \cap (k^{-1/2}\ell(T)B_2^N)$ and define the random variables

$$A = \sup_{x \in \overline{T}} |G(x)|, \quad B = \inf_{x \in S} \frac{|G(x)|}{|x|}.$$

By Lemma 15.3 applied to $\overline{T}$, for every $u > 0$,

$$\mathbb{P}\{A > (8+u)\ell(T)\} \leq 5^k \exp\left(-\frac{ku^2}{8}\right),$$

so that, if $u = 6$ for example ($k \geq 1$),

$$\mathbb{P}\{A > 14\ell(T)\} \leq \exp(-2k).$$

On the other hand, by (15.1),

$$\mathbb{P}\{B < u\} \leq \exp(K_1^2 k)\left(\frac{eu^2}{k}\right)^{k/2}.$$

If we choose $u = K_2^{-1}\sqrt{k}$ where $K_2 = \exp(K_1^2 + 3)$, it follows that

$$\mathbb{P}\left\{B < K_2^{-1}\sqrt{k}\right\} \leq \exp(-2k).$$

Therefore, the set $\{A \leq 14\ell(T), B \geq K_2^{-1}\sqrt{k}\}$ has a probability larger than $1 - 2\exp(-2k) \geq 1 - \exp(-k)$. On this set of ω's, take $x \in T \cap \mathrm{Ker}G(\omega)$. There exists y in S with $|x - y| \leq k^{-1/2}\ell(T)$. Since $G(\omega, x) = 0$, we can write

$$|x| \leq |y| + \frac{\ell(T)}{\sqrt{k}} \leq \frac{|G(\omega, y)|}{B(\omega)} + \frac{\ell(T)}{\sqrt{k}} = \frac{|G(\omega, x - y)|}{B(\omega)} + \frac{\ell(T)}{\sqrt{k}}$$

$$\leq \frac{A(\omega)}{B(\omega)} + \frac{\ell(T)}{\sqrt{k}} \leq (14K_2 + 1)\frac{\ell(T)}{\sqrt{k}}.$$

Hence $\mathrm{diam}(T \cap \mathrm{Ker}G(\omega)) \leq K\ell(T)/\sqrt{k}$ with $K = 2(14K_2+1)$. This completes the first proof of Theorem 15.2.

Second proof of Theorem 15.2. It is based on Corollary 3.13. Let S be a closed subset of the unit sphere S_2^{N-1} of $\mathbb{R}^N$. Let $(g_i), (g_i')$ be independent orthogaussian sequences. For x in S and y in the Euclidean unit sphere S_2^{k-1} of $\mathbb{R}^k$, define the two Gaussian processes

$$X_{x,y} = \langle G(x), y \rangle + g = \sum_{i=1}^{N} \sum_{j=1}^{k} g_{ij} x_i y_j + g$$

where g is a standard normal variable which is independent of (g_{ij}) and

$$Y_{x,y} = \sum_{i=1}^{N} g_i x_i + \sum_{j=1}^{k} g_j' y_j.$$

It is easy to see that for all x, x' in S and all y, y' in S_2^{k-1},

$$\mathbb{E}(X_{x,y}X_{x',y'}) - \mathbb{E}(Y_{x,y}Y_{x',y'}) = (1 - \langle x, x' \rangle)(1 - \langle y, y' \rangle)$$

so that this difference is always non-negative and equal to 0 when $x = x'$. By Corollary 3.13, for every $\lambda > 0$,

$$(15.3) \qquad \mathbb{P}\Big\{\inf_{x \in S} \sup_{y \in S_2^{k-1}} Y_{x,y} > \lambda\Big\} \leq \mathbb{P}\Big\{\inf_{x \in S} \sup_{y \in S_2^{k-1}} X_{x,y} > \lambda\Big\}.$$

By the definitions of $X_{x,y}$ and $Y_{x,y}$, it follows that the right hand side of the inequality is majorized by

$$\mathbb{P}\Big\{\inf_{x \in S} |G(x)| + g > \lambda\Big\} \leq \mathbb{P}\Big\{\inf_{x \in S} |G(x)| > 0\Big\} + \tfrac{1}{2}\exp(-\lambda^2/2)$$

while the left hand side is bounded below by

$$\mathbb{P}\Big\{\Big(\sum_{j=1}^{k} g'^2_j\Big)^{1/2} - \sup_{x \in S}\sum_{i=1}^{N} g_i x_i > \lambda\Big\} \geq \mathbb{P}\{Z_k > 2\lambda\} - \mathbb{P}\Big\{\sup_{x \in S}\sum_{i=1}^{N} g_i x_i > \lambda\Big\}$$

$$\geq \mathbb{P}\{Z_k > 2\lambda\} - \frac{1}{\lambda}\mathbb{E}\sup_{x \in S}\sum_{i=1}^{N} g_i x_i$$

where we have set $Z_k = \big(\sum_{j=1}^{k} g'^2_j\big)^{1/2}$. We can write

$$k = \mathbb{E}Z_k^2 \leq \frac{k}{10^2} + \int_{\{Z_k^2 \geq k/10^2\}} Z_k^2 \, d\mathbb{P}$$

$$\leq \frac{k}{10^2} + \big(\mathbb{E}Z_k^4\big)^{1/2}\big(\mathbb{P}\{Z_k^2 \geq k/10^2\}\big)^{1/2};$$

since $\mathbb{E}Z_k^4 \leq (k+1)^2$ (elementary computation), we get

$$\mathbb{P}\{Z_k^2 \geq k/10^2\} \geq \Big[\Big(1 - \frac{1}{10^2}\Big)\frac{k}{k+1}\Big] \geq \frac{1}{2}$$

at least if $k \geq 3$, something we may assume without any loss of generality (increasing if necessary the value of K in the conclusion of the theorem). Hence, by the preceeding, if $\lambda = \sqrt{k}/20$, the left hand side of (15.3) is larger than

$$\frac{1}{2} - \frac{20}{\sqrt{k}}\mathbb{E}\sup_{x \in S}\sum_{i=1}^{N} g_i x_i.$$

Then, let $S = r^{-1}T \cap S_2^{N-1}$ where $r > 0$. We have obtained from (15.3) (with $\lambda = \sqrt{k}/20$) that

$$\mathbb{P}\Big\{\inf_{x \in S} |G(x)| > 0\Big\} \geq \frac{1}{2} - \frac{20}{\sqrt{k}} \cdot \frac{\ell(T)}{r} - \frac{1}{2}\exp\Big(-\frac{k}{800}\Big).$$

Hence, if we choose $r = K\ell(T)/\sqrt{k}$ for a large enough numerical constant K, we see from the preceding inequality that $\mathbb{P}\{\inf_{x \in S} |G(x)| > 0\} > 0$. Therefore, there exists an ω such that $|G(\omega, x)| > 0$ for all x in S. That is, if we let $F = \mathrm{Ker}G(\omega)$ (which is of codimension k), $S \cap F = \emptyset$. By definition of $S = r^{-1}T \cap S_2^{N-1}$, this implies that, for every x in $T \cap F$, $|x| < r = K\ell(T)/\sqrt{k}$. The qualitative conclusion of the theorem already follows. To improve the preceding argument to the quantitative estimate, we simply need combine it with concentration properties. We improve the minoration of the left hand side of (15.3) in the following way. Let m and M be respectively medians of $Z_k = (\sum_{j=1}^{k} g_j'^2)^{1/2}$ and of $\sup_{x \in S} \sum_{i=1}^{N} g_i x_i$. For every $\lambda > 0$,

$$\mathbb{P}\left\{\inf_{x \in S} \sup_{y \in S_2^{k-1}} Y_{x,y} > \lambda\right\}$$

$$\geq \mathbb{P}\left\{Z_k - \sup_{x \in S} \sum_{i=1}^{N} g_i x_i > \lambda\right\}$$

$$\geq \mathbb{P}\{Z_k \geq m - \lambda\} - \mathbb{P}\left\{\sup_{x \in S} \sum_{i=1}^{N} g_i x_i > m - 2\lambda\right\}$$

$$\geq 1 - \mathbb{P}\{m - Z_k > \lambda\} - \mathbb{P}\left\{\sup_{x \in S} \sum_{i=1}^{N} g_i x_i > M + (m - M - 2\lambda)\right\}.$$

By Lemma 3.1, if $m - M - 2\lambda > 0$, this is larger than or equal to

$$1 - \tfrac{1}{2}\exp(-\lambda^2/2) - \tfrac{1}{2}\exp\left[-(m - M - 2\lambda)^2/2\right].$$

Then, let us choose $\lambda = (m - M)/3$ to get the lower bound

$$1 - \exp[-(m - M)^2/18]$$

and (15.3) thus reads as

$$\mathbb{P}\left\{\inf_{x \in S} |G(X)| > 0\right\} \geq 1 - \tfrac{3}{2}\exp\left[-(m - M)^2/18\right].$$

Previously, we saw that $m \geq \sqrt{k}/10$. On the other hand, if S is as above $r^{-1}T \cap S_2^{N-1}$,

$$M \leq 2\mathbb{E}\sup_{x \in S} \sum_{i=1}^{N} g_i x_i \leq \frac{2}{r}\ell(T)$$

so that

$$\mathbb{P}\{\mathrm{diam}(T \cap \mathrm{Ker}G) \leq 2r\} \geq \mathbb{P}\left\{\inf_{x \in S} |G(x)| > 0\right\}$$

$$\geq 1 - \frac{3}{2}\exp\left[-\frac{1}{18}\left(\frac{\sqrt{k}}{10} - \frac{2\ell(T)}{r}\right)^2\right].$$

The conclusion of the theorem follows. Again, let us observe that the simpler inequality (3.2) may be used instead of Lemma 3.1.

Note that this second proof may be rather simplified yielding moreover best constants in the statement of the theorem by the use of Theorem 3.16. Take $X_{x,y} = \langle G(x), y \rangle$ and

$$Y_{x,y} = \sum_{i=1}^{N} g_i x_i + |x| \sum_{j=1}^{k} g_j' y_j \, .$$

For all x, x' in S and all y, y' in S_2^{k-1}, we have

$$\mathbb{E}|Y_{x,y} - Y_{x',y'}|^2 - \mathbb{E}|X_{x,y} - X_{x',y'}|^2$$
$$= |x|^2 + |x'|^2 - 2|x|\,|x'|\langle y, y' \rangle - 2\langle x, x' \rangle \big(1 - \langle y, y' \rangle\big)$$
$$\geq |x|^2 + |x'|^2 - 2|x|\,|x'|\langle y, y' \rangle - 2|x|\,|x'|\big(1 - \langle y, y' \rangle\big)$$

so that this difference is always ≥ 0 and equal to 0 if $x = x'$. By Theorem 3.16, this implies

$$\mathbb{E} \inf_{x \in S} \sup_{y \in S_2^{k-1}} Y_{x,y} \leq \mathbb{E} \inf_{x \in S} \sup_{y \in S_2^{k-1}} X_{x,y}$$

and hence, by the definitions of $X_{x,y}$ and $Y_{x,y}$, and with $S = r^{-1}T \cap S_2^{N-1}$,

$$\mathbb{E} \inf_{x \in S} |G(x)| \geq a_k - \frac{1}{r}\ell(T)$$

where $a_k = \mathbb{E}Z_k = \mathbb{E}\big((\sum_{j=1}^{k} g_j'^2)^{1/2}\big)$. By the law of large numbers, a_k is of the order of $\sqrt{k}$. Then write

$$\mathbb{P}\Big\{ \inf_{x \in S} |G(x)| = 0 \Big\} \leq \mathbb{P}\Big\{ \inf_{x \in S} |G(x)| - \mathbb{E}\inf_{x \in S} |G(x)| \leq \frac{1}{r}\ell(T) - a_k \Big\}$$

which is majorized, using concentration ((1.6) e.g.), by

$$\exp\left[\frac{1}{2}\Big(a_k - \frac{\ell(T)}{r}\Big)^2\right].$$

As before, this yields the conclusion of the theorem, with, as announced, improved numerical constants.

15.2 Conjectures on Sudakov's Minoration for Chaos

Denote by $\ell_2(\mathbb{N} \times \mathbb{N})$ the space of all sequences $t = (t_{ij})$ indexed by $\mathbb{N} \times \mathbb{N}$ such that $|t| = \big(\sum_{i,j} t_{ij}^2\big)^{1/2} < \infty$. Let T be a subset of $\ell_2(\mathbb{N} \times \mathbb{N})$. Let further (g_i) be an orthogaussian sequence (on $(\Omega, \mathcal{A}, \mathbb{P})$) and consider the Gaussian chaos process $\big(\sum_{i,j} g_i g_j t_{ij}\big)_{t \in T}$. While we have studied in Chapter 3 the integrability properties and the tail behaviors of $\sup_{t \in T} |\sum_{i,j} g_i g_j t_{ij}|$, little seems to be known on the "metric-geometric" conditions on T equivalent to this supremum to be almost surely finite (if there are any?). The sufficient conditions of

Theorem 11.22 are too strong and by no way necessary. One approach could be to view the preceding chaos process, after decoupling, as a mixture of Gaussian processes, i.e. to study, given ω', $\left(\sum_i g_i (\sum_j g_j'(\omega')t_{ij})\right)_{t\in T}$ where (g_j') is a standard Gaussian sequence constructed on some different probability space $(\Omega', \mathcal{A}', \mathbb{P}')$. Unfortunately, this approach seems to be doomed to failure: the random distances

$$d_{\omega'}(s,t) = \left(\sum_i \left|\sum_j g_j'(\omega')(s_{ij} - t_{ij})\right|^2\right)^{1/2}$$

do not have the property that was essential in Section 12.2, i.e. that $\mathbb{P}'\{\omega'; d_{\omega'}(s,t) < \varepsilon\}$ is very small for $\varepsilon > 0$ small.

Let us reduce to decoupled chaos and set

$$\ell^{(2)}(T) = \mathbb{E}\sup_{t\in T}\left|\sum_{i,j} g_i g_j' t_{ij}\right|.$$

(With respect to the notation of Section 15.1, the 2 in $\ell^{(2)}(T)$ indicates that we are dealing with chaos of the second order.) By the study of Section 3.2, we know that, at least for symmetric (t_{ij}), this reduction to the decoupled setting is no loss of generality in the understanding of when T defines an almost surely bounded chaos process. A first step in this study would be an analog of Sudakov's minoration. As we discussed it in Section 11.3 on sufficiency, there are two natural distances to consider. The first distance is the usual L_2-metric $|s - t|$ and the second is the injective metric or norm given by

$$\|t\|_\vee = \sup_{|h|,|h'|\leq 1} |\langle th, h'\rangle| = \sup_{|h|\leq 1} |th|$$

where $th = \left(\sum_j h_j t_{ij}\right)_i$. Clearly $\|t\|_\vee \leq |t|$.

First, we investigate an instructive example. Denote, for n fixed, by $\ell_2(n \times n)$ the subset of $\ell_2(\mathbb{N} \times \mathbb{N})$ consisting of the elements (t_{ij}) for which $t_{ij} = 0$ if i or $j > n$. Let T be the unit ball of $\ell_2(n \times n)$ for the injective norm $\|\cdot\|_\vee$. Clearly

$$\sup_{t\in T}\left|\sum_{i,j=1}^n g_i g_j' t_{ij}\right| \leq \left(\sum_{i=1}^n g_i^2\right)^{1/2}\left(\sum_{j=1}^n g_j'^2\right)^{1/2}$$

so that $\ell^{(2)}(T) \leq n$.

Proposition 15.4. *Let T be as before, i.e. $T = \{t \in \ell_2(n \times n); \|t\|_\vee \leq 1\}$. There is a numerical constant $c > 0$ (independent of n) such that*

(i) $\left(\log N(T, |\cdot|; c\sqrt{n})\right)^{1/2} \geq cn$;

(ii) $\left(\log N(T, \|\cdot\|_\vee; \frac{1}{2})\right)^{1/2} \geq cn$.

Proof. (ii) is obvious by volume considerations. We give a simple probabilistic proof of (i). Let $(\varepsilon_{ij})_{1 \le i,j \le n}$ be a doubly-indexed Rademacher sequence. For $|h|, |h'| \le 1$, by the subgaussian inequality (4.1), for every $u > 0$,

$$\mathbb{P}\left\{ \left| \sum_{i,j} h_i h'_j \varepsilon_{ij} \right| > u \right\} \le 2 \exp\left(-\frac{u^2}{2} \right).$$

By Lemma 15.1, there exists $H \subset 2B_2^n$ with Card $H \le 5^n$ such that $B_2^n \subset$ Conv H. By the preceding, since Card $H \le 5^n$,

$$\mathbb{P}\left\{ \forall h, h' \in H \; ; \; \left| \sum_{i,j} h_i h'_j \varepsilon_{ij} \right| > 4\sqrt{n} \right\} \le \frac{1}{2}.$$

By definition of $\| \cdot \|_\vee$ and of H, it follows that $\mathbb{P}\{\|(\varepsilon_{ij})\|_\vee \le 4\sqrt{n}\} \ge 1/2$. Let η_{ij}, $1 \le i, j \le n$, be a family of ± 1. Then $(|\varepsilon_{ij} - \eta_{ij}|^2/4)$ is a sequence of random variables taking the values 0 and 1 with probability 1/2. Recentering, Lemma 1.6 implies that for some $c > 0$,

$$\mathbb{P}\left\{ \sum_{i,j} |\varepsilon_{ij} - \eta_{ij}|^2 \le n^2 \right\} \le \exp(-cn^2).$$

That is, $\mathbb{P}\{|(\varepsilon_{ij} - \eta_{ij})| \le n\} \le \exp(-cn^2)$. So one needs at least $\frac{1}{2}\exp(cn^2)$ balls of radius n in the Euclidean metric $| \cdot |$ to cover $\{\|(\varepsilon_{ij})\|_\vee \le 4\sqrt{n}\} \subset 4\sqrt{n}T$. Therefore, one needs at least $\frac{1}{2}\exp(cn^2)$ balls of radius $\sqrt{n}/4$ to cover T. The proof of Proposition 15.4 is complete.

It follows from this proposition that, for $\varepsilon = c\sqrt{n}$,

$$\varepsilon \left(\log N(T, | \cdot |; \varepsilon) \right)^{1/4} \ge c^{3/2} n \ge c^{3/2} \ell^{(2)}(T),$$

and, for $\varepsilon = 1/2$,

$$\varepsilon \left(\log N(T, \| \cdot \|_\vee; \varepsilon) \right)^{1/2} \ge \frac{c}{2} n \ge \frac{c}{2} \ell^{(2)}(T)$$

since, as we have seen, $\ell^{(2)}(T) \le n$. It is natural to *conjecture* that these inequalities are best possible, i.e. that for any subset T in $\ell_2(\mathbb{N} \times \mathbb{N})$,

(15.4)
$$\sup_{\varepsilon > 0} \varepsilon \left(\log N(T, | \cdot |; \varepsilon) \right)^{1/4} \le K \ell^{(2)}(T)$$

and

$$\sup_{\varepsilon > 0} \varepsilon \left(\log N(T, \| \cdot \|_\vee; \varepsilon) \right)^{1/2} \le K \ell^{(2)}(T)$$

for some numerical constant K. Recently, (15.4) has been proved in [Ta16] (relying on an appropriate version of Theorem 15.2), where it is also proved that for $\delta > 2$

$$\sup_{\varepsilon > 0} \varepsilon \left(\log N(T, \| \cdot \|_\vee; \varepsilon) \right)^{1/\delta} \le K(\delta) \ell^{(2)}(T).$$

Now, we would like to present a simple result that is in relation with these questions.

Proposition 15.5. *Let* (g_i), (g_j'), $(\tilde{g}_{ij})$ *be independent standard Gaussian sequences. Then, if* $(x_{ij})_{1 \le i,j \le n}$ *is a finite sequence in a Banach space,*

$$\mathbb{E}\Big\| \sum_{i,j=1}^{n} \tilde{g}_{ij} x_{ij} \Big\| \le \Big(\frac{\pi}{2}\Big)^{1/2} \sqrt{n}\, \mathbb{E}\Big\| \sum_{i,j=1}^{n} g_i g_j' x_{ij} \Big\|.$$

Proof. We simply write that

$$\sqrt{n}\, \mathbb{E}\Big\| \sum_{i,j=1}^{n} g_i g_j' x_{ij} \Big\| = \mathbb{E}\Big\| \sum_{i,j,k=1}^{n} \tilde{g}_{ik} g_j' x_{ij} \Big\|.$$

By symmetry, if (ε_i) is a Rademacher sequence which is independent of the sequences $(\tilde{g}_{ik})$ and (g_j'),

$$\mathbb{E}\Big\| \sum_{i,j,k=1}^{n} \tilde{g}_{ik} g_j' x_{ij} \Big\| = \mathbb{E}\Big\| \sum_{i,j,k=1}^{n} \varepsilon_k \tilde{g}_{ik} \varepsilon_j g_j' x_{ij} \Big\|.$$

Jensen's inequality with respect to partial integration in (ε_i) shows

$$\mathbb{E}\Big\| \sum_{i,j,k=1}^{n} \tilde{g}_{ik} g_j' x_{ij} \Big\| \ge \mathbb{E}\Big\| \sum_{i,j=1}^{n} \tilde{g}_{ik} g_j' x_{ij} \Big\|.$$

The contraction principle in (g_j') (cf. (4.8)) and symmetry imply the result.

15.3 An Inequality of J. Bourgain

In his remarkable recent work on $\Lambda(p)$-sets [Bour], J. Bourgain establishes the following inequality. Let (ξ_i) be independent random variables with common distribution

$$\mathbb{P}\{\xi_i = 1\} = 1 - \mathbb{P}\{\xi_i = 0\} = \delta, \quad 0 < \delta < 1.$$

Let T be a subset of $\mathbb{R}_+^N$ and set

$$E = \int_0^\infty \big(\log N(T, \varepsilon B_2^N)\big)^{1/2} d\varepsilon$$

(assumed to be finite). Set further $R = \sup_{t \in T} |t|$. An element t in $\mathbb{R}^N$ has coordinates $t = (t_1, \ldots, t_N)$.

Theorem 15.6. *Under the previous notation, there is a numerical constant* $K > 0$ *such that for all* $p \ge 1$ *and all* $1 \le m \le N$,

$$\left\| \sup_{t \in T} \max_{\mathrm{Card} I \leq m} \sum_{i \in I} \xi_i t_i \right\|_p$$

$$\leq KR \left[\sqrt{\delta m} + \sqrt{p} \left(\log \frac{1}{\delta} \right)^{-1/2} \right] + K \left(\log \frac{1}{\delta} \right)^{-1/2} E .$$

The inequality of the theorem is equivalent to saying (see Lemma 4.10) that for some constant K and all $u > 0$

(15.5)
$$\mathbb{P} \left\{ \sup_{t \in T} \max_{\mathrm{Card} I \leq m} \sum_{i \in I} \xi_i t_i > u + K \left[R \sqrt{\delta m} + \left(\log \frac{1}{\delta} \right)^{-1/2} E \right] \right\}$$

$$\leq K \exp \left(-\frac{u^2}{K R^2} \log \frac{1}{\delta} \right) .$$

We will establish (15.5) using the entropic bound (11.4) for the *vector valued* process $\sum_{i \in I} \xi_i t_i$, $I \subset \{1, \ldots, N\}$, $\mathrm{Card} I \leq m$, $t \in T$, with respect to the norm

$$\max_{\mathrm{Card} I \leq m} \sum_{i \in I} \xi_i t_i .$$

To this aim, the crucial next lemma will indicate that the increments of this process satisfy the appropriate regularity condition.

As usual, if $t = (t_1, \ldots, t_N) \in \mathbb{R}^N$, $\|(t_i)\|_{2,\infty}$ denotes the weak ℓ_2-norm of the sequence $(t_i)_{i \leq N}$. Recall that $\|(t_i)\|_{2,\infty} \leq |t|$.

Lemma 15.7. *Let t be in $\mathbb{R}_+^N$ satisfying $\|(t_i)\|_{2,\infty} \leq 1$ and let also $1 \leq m \leq N$. There is a numerical constant K such that for all $u \geq K \sqrt{\delta m}$,*

$$\mathbb{P} \left\{ \max_{\mathrm{Card} I \leq m} \sum_{i \in I} \xi_i t_i > u \right\} \leq K \exp \left(-\frac{u^2}{K} \log \frac{1}{\delta} \right) .$$

Proof. Let $(Z_i)_{i \leq N}$ be the non-increasing rearrangement of the sequence $(\xi_i t_i)_{i \leq N}$ so that

$$\max_{\mathrm{Card} I \leq m} \sum_{i \in I} \xi_i t_i = \sum_{i=1}^{m} Z_i :$$

Note that since $\|(t_i)\|_{2,\infty} \leq 1$, and $\xi_i = 0$ or 1, we also have that $\|(\xi_i t_i)\|_{2,\infty} \leq 1$, that is $Z_i \leq i^{-1/2}$, $i \geq 1$. By the binomial estimate (Lemma 2.5), for every i and every $u > 0$,

$$\mathbb{P}\{Z_i > u\} = \mathbb{P} \left\{ \sum_{j=1}^{N} I_{\{\xi_j t_j > u\}} \geq i \right\} \leq \left(\frac{e}{i} \sum_{j=1}^{N} \mathbb{P}\{\xi_j t_j > u\} \right)^i .$$

Since $\|(t_i)\|_{2,\infty} \leq 1$,

$$\sum_{j=1}^{N} \mathbb{P}\{\xi_j t_j > u\} \leq \sum_{j=1}^{N} \mathbb{P}\{\xi_1 > u\sqrt{j}\}$$

$$= \sum_{j=1}^{N} \mathbb{P}\{\xi_1 > u^2 j\} \leq \mathbb{E}\left(\frac{\xi_1}{u^2}\right) = \frac{\delta}{u^2}.$$

Therefore, for all integers $i \leq N$ and all $u > 0$,

(15.6) $$\mathbb{P}\{Z_i > u\} \leq \left(\frac{e\delta}{u^2 i}\right)^i.$$

Without loss of generality, we can assume that $m = 2^p$. Let $u > 0$ be fixed. Let $j \geq 1$ be the largest such that $u^2 \geq 2\mathrm{e}10^4 2^j$ and take $j = 0$ if $u^2 < 2\mathrm{e}10^4$. (We do not attempt to deal with sharp numerical constants.) We observe that, if $j \geq 1$,

$$\sum_{i=1}^{2^j} Z_i \leq \sum_{i=1}^{2^j} i^{-1/2} \leq 2 \cdot 2^{j/2} \leq \frac{u}{2}.$$

We also observe that

$$\sum_{i > 2^j} Z_i \leq \sum_{\ell=j}^{p} 2^{\ell} Z_{2^{\ell}}.$$

It follows from these two observations that, for all $j \geq 0$,

$$\mathbb{P}\left\{\sum_{i=1}^{m} Z_i > u\right\} \leq \mathbb{P}\left\{\sum_{\ell=j}^{p} 2^{\ell} Z_{2^{\ell}} > \frac{u}{2}\right\} \leq \sum_{\ell=j}^{p} \mathbb{P}\{2^{\ell} Z_{2^{\ell}} > u_{\ell}\}$$

whenever (u_{ℓ}) are positive numbers such that $\sum_{\ell=j}^{p} u_{\ell} \leq u/2$. Let $v_{\ell} = 10^{-2} u 2^{-(\ell-j)/2}$, $w_{\ell} = 10(\delta 2^{\ell})^{1/2}$, $j \leq \ell \leq p$, and set $u_{\ell} = \max(v_{\ell}, w_{\ell})$. It is clear that $\sum_{\ell \geq j} v_{\ell} \leq u/4$ while, when $u \geq K\sqrt{\delta m}$ and $K \geq 10^3$,

$$\sum_{\ell \leq p} w_{\ell} \leq 3 \cdot 10(\delta 2^{p+1})^{1/2} \leq \frac{u}{4}.$$

Hence, we have $\sum_{\ell=j}^{p} u_{\ell} \leq u/2$. By the binomial estimates (15.6),

(15.7) $$\mathbb{P}\left\{\sum_{i=1}^{m} Z_i > u\right\} \leq \sum_{\ell=j}^{p} \left(\frac{e\delta 2^{\ell}}{u_{\ell}^2}\right)^{2^{\ell}} \leq \sum_{\ell=j}^{p} \exp\left[-2^{\ell} \log\left(\frac{u_{\ell}^2}{e\delta 2^{\ell}}\right)\right].$$

Note that for $2\delta \leq x \leq 4$,

$$\log\left(\frac{x}{\delta}\right) \geq \frac{x}{4} \log \frac{1}{\delta}.$$

Therefore, the first term $(\ell = j)$ of the sum on the right hand side of (15.7) gives rise to an exponent of the form

$$2^j \log\left(\frac{u_j^2}{e\delta 2^j}\right) \geq 2^j \log\left(\frac{v_j^2}{e\delta 2^j}\right) = 2^j \log\left(\frac{u^2}{e10^4\delta 2^j}\right) \geq \frac{1}{4}\cdot\frac{u^2}{e10^4}\log\frac{1}{\delta}$$

provided that

$$2\delta \leq \frac{u^2}{e10^4 2^j} \leq 4.$$

This clearly holds by definition of $j \geq 1$ and also if $j = 0$ since $u \geq K\sqrt{\delta m}$ ($K \geq 10^3$). Hence,

$$(15.8) \qquad \exp\left[-2^j \log\left(\frac{u_j^2}{e\delta 2^j}\right)\right] \leq \exp\left(-\frac{u^2}{K}\log\frac{1}{\delta}\right)$$

for $K \geq 4e10^4$.

Now, we would like to show that for every $\ell > j$

$$(15.9) \qquad 2^\ell \log\left(\frac{u_\ell^2}{e\delta 2^\ell}\right) \geq \frac{6}{5}2^{\ell-1}\log\left(\frac{u_{\ell-1}^2}{e\delta 2^{\ell-1}}\right),$$

that is,

$$\log\left(\frac{u_\ell^2}{e\delta 2^\ell}\right) \geq \frac{3}{5}\log\left(\frac{u_{\ell-1}^2}{e\delta 2^{\ell-1}}\right).$$

If $u_{\ell-1} = w_{\ell-1}$, then $u_\ell = w_\ell$ and this inequality is trivially satisfied by the definition of w_ℓ. If $u_{\ell-1} = v_{\ell-1}$, we note the following:

$$\frac{u_\ell^2}{e\delta 2^\ell} \geq \frac{v_\ell^2}{e\delta 2^\ell} = \frac{1}{4}\cdot\frac{v_{\ell-1}^2}{e\delta 2^{\ell-1}} = \frac{1}{4}\cdot\frac{u_{\ell-1}^2}{e\delta 2^{\ell-1}} \geq \frac{1}{4}\cdot\frac{w_{\ell-1}^2}{e\delta 2^{\ell-1}} \geq 2^3.$$

Using that $\log(x/4) \geq (3\log x)/5$ when $x \geq 2^5$, (15.9) is thus satisfied.

We can conclude the proof of Lemma 15.7. By (15.7), (15.8) and (15.9), we have

$$\mathbb{P}\left\{\sum_{i=1}^m Z_i > u\right\} \leq \sum_{k\geq 0}\exp\left[-\left(\frac{6}{5}\right)^k\frac{u^2}{K}\log\frac{1}{\delta}\right]$$

$$\leq \sum_{k\geq 1}\exp\left(-\frac{k}{3}\cdot\frac{u^2}{K}\log\frac{1}{\delta}\right)$$

$$\leq 2\exp\left(-\frac{u^2}{3K}\log\frac{1}{\delta}\right)$$

at least if $u^2\log(1/\delta) \geq 3K$. The inequality of Lemma 15.7 follows with (for example) $K = 12e10^4$ since there is nothing to prove when $u^2\log(1/\delta) \leq 3K$.

Proof of Theorem 15.6. For s, t in $\mathbb{R}^N$, define the random distance

$$D(s,t) = \left(\log\frac{1}{\delta}\right)^{1/2}\max_{\mathrm{Card}I\leq m}\left|\sum_{i\in I}\xi_i(s_i - t_i)\right|.$$

Lemma 15.7 implies that for all s, t and all $u \geq K(m\delta\log\frac{1}{\delta})^{1/2}$,

$$\mathbb{P}\{D(s,t) > u|s - t|\} \leq K \exp\left(-\frac{u^2}{K}\right).$$

Here, K denotes a positive numerical constant which is not necessarily the same each time it appears. From the preceding, for all $u > 0$,

$$\mathbb{P}\left\{D(s,t) > \left(u + K\left(m\delta \log \frac{1}{\delta}\right)^{1/2}\right)|s - t|\right\} \leq K \exp\left(-\frac{u^2}{K}\right).$$

Hence, for all measurable sets A, and all s, t,

$$\int_A D(s,t)d\mathbb{P} \leq K\mathbb{P}(A)|s - t|\left(\psi_2^{-1}\left(\frac{1}{\mathbb{P}(A)}\right) + \left(m\delta \log \frac{1}{\delta}\right)^{1/2}\right)$$

where we recall that $\psi_2(x) = \exp(x^2) - 1$. Therefore, we are in a position to apply Remark 11.4 to the random distance $D(s,t)$ (cf. Remark 11.5). It follows that, for all measurable sets A,

$$\int_A \sup_{s,t\in T} D(s,t)d\mathbb{P} \leq K\mathbb{P}(A)\left(D\psi_2^{-1}\left(\frac{1}{\mathbb{P}(A)}\right) + R\left(m\delta \log \frac{1}{\delta}\right)^{1/2} + E\right)$$

and hence ((11.4)), for all $u > 0$,

$$\mathbb{P}\left\{\sup_{s,t\in T} D(s,t) > u + K\left(R\left(m\delta \log \frac{1}{\delta}\right)^{1/2} + E\right)\right\} \leq K \exp\left(-\frac{u^2}{KR^2}\right).$$

Since for every t, by Lemma 15.7,

$$\mathbb{P}\left\{D(0,t) > u + KR\left(m\delta \log \frac{1}{\delta}\right)^{1/2}\right\} \leq K \exp\left(-\frac{u^2}{KR^2}\right)$$

for all $u > 0$, the inequality (15.5) follows. The proof of Theorem 15.6 is thus complete.

Note of course that we can replace the entropy integral by sharper majorizing measure conditions.

15.4 Invertibility of Submatrices

This section is devoted to the exposition of one simple but significant result of the work [B-T1] (see also [B-T2]) by J. Bourgain and L. Tzafriri. The proof of this result uses several probabilistic ideas and arguments already encountered throughout this book.

Let A be a real $N \times N$ matrix considered as a linear operator on $\mathbb{R}^N$. Denote by $\|A\| = \|A\|_{2\to2}$ its norm as an operator $\ell_2^N \to \ell_2^N$. If σ is a subset of $\{1, \ldots, N\}$, denote by R_σ the restriction operator, that is the projection from $\mathbb{R}^N$ onto the span of the unit vectors $(e_i)_{i\in\sigma}$ where (e_i) is the canonical basis of $\mathbb{R}^N$. R_σ^t is the transpose operator. By restricted invertibility of A, we mean the existence of a subset σ of $\{1, \ldots, N\}$ with cardinality of the order of N such that $R_\sigma A R_\sigma^t$ is an isomorphism. If the diagonal of A is the identity matrix I, this will be achieved by simply constructing the set σ such that $\|R_\sigma (A - I) R_\sigma^t\| < 1/2$.

The following statement is the main result which we present.

Theorem 15.8. *There is a numerical constant K with the following property: for every $0 < \delta < K^{-1}$ and every $N \geq K/\delta$, whenever $A = (a_{ij})$ is an $N \times N$ matrix with $a_{ii} = 0$ for all i, there exists a subset σ of $\{1, \ldots, N\}$ such that $\operatorname{Card} \sigma \geq \delta N$ and such that*

$$\|R_\sigma A R_\sigma^t\| \leq K \sqrt{\delta} \|A\|.$$

The following restricted invertibility statement is an immediate consequence of Theorem 15.8.

Corollary 15.9. *There is a numerical constant K with the following property: for every $0 < \varepsilon \leq 1$, $c \geq 1$ and $N \geq K c \varepsilon^{-2}$, whenever A is an $N \times N$ matrix of norm $\|A\| \leq c$ with only 1's on the diagonal, there exists a subset σ of $\{1, \ldots, N\}$ of cardinality larger than or equal to $\varepsilon^2 N / Kc$ such that $R_\sigma A R_\sigma^t$ restricted to $R_\sigma \ell_2^N$ is invertible and its inverse satisfies*

$$\|(R_\sigma A R_\sigma^t)^{-1}\| < 1 + \varepsilon.$$

The following proposition, the proof of which is based on a random choice argument, is the main step in the proof of Theorem 15.8.

Proposition 15.10. *Let $0 < \delta < 1$ and $N \geq 8/\delta$. Then, for all $N \times N$ matrices $A = (a_{ij})$ with $a_{ii} = 0$, there exists $\sigma \subset \{1, \ldots, N\}$ such that $\operatorname{Card} \sigma \geq \delta N / 2$ and such that*

$$\|R_\sigma A R_\sigma^t\|_{2 \to 1} \leq 50 \delta \|A\| \sqrt{N}$$

where $\| \cdot \|_{2 \to 1}$ is the operator norm $\ell_2^M \to \ell_1^M$ $(M = \operatorname{Card} \sigma)$.

Before we prove this proposition, let us show how to deduce Theorem 15.8 from this result. We need the following second step which clarifies the rôle of the norm $\| \cdot \|_{2 \to 1}$.

Proposition 15.11. *Let $u : \ell_2^n \to \ell_1^m$. There exists a subset τ of $\{1, \ldots, m\}$ with $\operatorname{Card} \tau \geq m/2$ such that*

$$\|R_\tau u\|_{2 \to 2} \leq \left(\frac{\pi}{m}\right)^{1/2} \|u\|_{2 \to 1}.$$

Proof. Consider the dual operator $u^* : \ell_\infty^m \to \ell_2^n$, $\|u^*\|_{\infty \to 2} = \|u\|_{2 \to 1}$. By the "little Grothendieck theorem" (cf. [Pi15]), if $\pi_2(u^*)$ is the 2-summing norm of u^*,

$$\pi_2(u^*) \le (\pi/2)^{1/2} \|u^*\|_{\infty \to 2} .$$

Therefore, by Pietsch's factorization theorem (cf. e.g. [Pie], [Pi15]), there exists positive numbers $(\lambda_i)_{i \le m}$ with $\lambda_i > 0$, $\sum_{i=1}^m \lambda_i = 1$ such that, for all $x = (x_1, \ldots, x_m)$ in ℓ_∞^m,

$$|u^*(x)| \le \left(\frac{\pi}{2} \right)^{1/2} \|u^*\|_{\infty \to 2} \left(\sum_{i=1}^m \lambda_i x_i^2 \right)^{1/2} .$$

Let $\tau = \{i \le m ; \ \lambda_i \le 2/m\}$. Then $\operatorname{Card} \tau \ge m/2$ and, from the preceding,

$$\|u^* R_\tau^t\|_{2 \to 2} \le \left(\frac{\pi}{2} \right)^{1/2} \|u^*\|_{\infty \to 2}$$

from which Proposition 15.11 follows.

These two propositions clearly imply the theorem. If $N \ge 8/\delta$ and A is an $N \times N$ matrix with $a_{ii} = 0$, there exists, by Proposition 15.10, σ in $\{1, \ldots, N\}$ such that $\operatorname{Card} \sigma \ge \delta N/2$ and

$$\|R_\sigma A R_\sigma^t\|_{2 \to 1} \le 50 \delta \|A\| \sqrt{N} .$$

Then apply Proposition 15.11 to $u = R_\sigma A R_\sigma^t$. It follows that one can find $\tau \subset \sigma$ with $\operatorname{Card} \tau \ge \frac{1}{2} \operatorname{Card} \sigma \ge \delta N/4$ and such that

$$\|R_\tau A R_\tau^t\|_{2 \to 2} \le \left(\frac{\pi}{\operatorname{Card} \sigma} \right)^{1/2} 50 \delta \|A\| \sqrt{N} \le 50 (2\pi \delta)^{1/2} \|A\| .$$

Theorem 15.8 immediately follows.

Proof of Proposition 15.10. Let $0 < \delta < 1$ and let ξ_i be independent random variables with distribution $\mathbb{P}\{\xi_i = 1\} = 1 - \mathbb{P}\{\xi_i = 0\} = \delta$. Let $A(\omega)$ be the random operator $\mathbb{R}^N \to \mathbb{R}^N$ with matrix $(\xi_i(\omega)\xi_j(\omega)a_{ij})$. We use a decoupling argument in the following form. For $I \subset \{1, \ldots, N\}$, let

$$B_I = \mathbb{E} \left\| \sum_{i \notin I, j \in I} \xi_i \xi_j a_{ij} e_i \otimes e_j \right\|_{2 \to 1} .$$

Since

$$\frac{1}{2^N} \sum_I \sum_{i \notin I, j \in I} a_{ij} = \frac{4}{2^N} \sum_{i,j} a_{ij}$$

(recall that $a_{ii} = 0$), we have

(15.10) $$\mathbb{E}\|A(\omega)\|_{2 \to 1} \le \frac{4}{2^N} \sum_I B_I .$$

Therefore, we estimate B_I for each fixed $I \subset \{1, \ldots, N\}$. We have

$$\left\| \sum_{i \notin I, j \in I} \xi_i \xi_j a_{ij} e_i \otimes e_j \right\|_{2 \to 1} = \sup_{|h| \leq 1} \sum_{i \notin I} \xi_i \left| \sum_{j \in I} \xi_j h_j a_{ij} \right|.$$

For every i and every h, let $d_i(h) = \sum_{j \in I} \xi_j h_j a_{ij}$. Therefore,

$$B_I = \mathbb{E} \sup_{|h| \leq 1} \sum_{i \notin I} \xi_i |d_i(h)|.$$

Let (ξ'_i) be an independent copy of (ξ_i). Working conditionally on $\xi_j, \xi'_j, j \in I$, we get by centering and Jensen's inequality that

$$B_I \leq \mathbb{E} \sup_{|h| \leq 1} \left| \sum_{i \notin I} (\xi_i - \mathbb{E}\xi_i) |d_i(h)| \right| + \delta \sup_{|h| \leq 1} \sum_{i \notin I} |d_i(h)|$$

$$\leq \mathbb{E} \sup_{|h| \leq 1} \left| \sum_{i \notin I} (\xi_i - \xi'_i) |d_i(h)| \right| + \delta \sup_{|h| \leq 1} \sum_{i \notin I} |d_i(h)|.$$

By Cauchy-Schwarz,

$$\sup_{|h| \leq 1} \sum_{i \notin I} |d_i(h)| \leq \|A\| \sqrt{N}.$$

On the other hand, if (ε_i) is a Rademacher sequence which is independent of (ξ_i), by symmetry,

$$\mathbb{E} \sup_{|h| \leq 1} \left| \sum_{i \notin I} (\xi_i - \xi'_i) |d_i(h)| \right| \leq 2\mathbb{E} \sup_{|h| \leq 1} \left| \sum_{i \notin I} \varepsilon_i \xi_i |d_i(h)| \right|.$$

By the comparison properties of Rademacher averages (Theorem 4.12), we have further

$$\mathbb{E} \sup_{|h| \leq 1} \left| \sum_{i \notin I} \varepsilon_i \xi_i |d_i(h)| \right| \leq 2\mathbb{E} \sup_{|h| \leq 1} \left| \sum_{i \notin I} \varepsilon_i \xi_i d_i(h) \right|.$$

Summarizing, we have obtained so far

$$B_I \leq 4\mathbb{E} \sup_{|h| \leq 1} \left| \sum_{i \notin I} \varepsilon_i \xi_i d_i(h) \right| + \delta \|A\| \sqrt{N}.$$

Going back to the definition of $d_i(h)$, we see that

$$\sup_{|h| \leq 1} \left| \sum_{i \notin I} \varepsilon_i \xi_i d_i(h) \right|^2 = \sum_{j \in I} \xi_j \left| \sum_{i \notin I} \varepsilon_i \xi_i a_{ij} \right|^2.$$

If we integrate this expression with respect to the Rademacher sequence and then with respect to the variables ξ_i, $i \notin I$, we obtain

$$\delta \sum_{j \in I} \xi_j \left(\sum_{i \notin I} a_{ij}^2 \right).$$

Since $\sum_i a_{ij}^2 \le \|A\|^2$ for every j, we finally get

$$B_I \le 4\left(\mathbb{E}\sup_{|h|\le 1}\left|\sum_{i\notin I}\varepsilon_i\xi_i d_i(h)\right|^2\right)^{1/2} + \delta\|A\|\sqrt{N}$$

$$\le 4\left(\delta^2\|A\|^2 N\right)^{1/2} + \delta\|A\|\sqrt{N}$$

$$\le 5\delta\|A\|\sqrt{N}.$$

This estimate holds for any $I \subset \{1,\ldots,N\}$. Hence, by (15.10),

$$\mathbb{E}\|A(\omega)\|_{2\to 1} \le 20\delta\|A\|\sqrt{N}.$$

In particular,

$$\mathbb{P}\{\|A(\omega)\|_{2\to 1} \le 50\delta\|A\|\sqrt{N}\} > 1/2.$$

Since clearly

$$\mathbb{P}\left\{\sum_{i=1}^{N}\xi_i \ge \frac{\delta N}{2}\right\} > \frac{1}{2},$$

there exists an ω in the intersection of these two events; $\sigma = \{i \le N\,;\,\xi_i(\omega) = 1\}$ then fulfills the conclusion of Proposition 15.10. The proof is complete.

15.5 Embedding Subspaces of L_p into ℓ_p^N

This section is devoted to the following problem of the local theory of Banach spaces. Given an n-dimensional subspace E of $L_p = L_p([0,1],dt)$, $1 \le p < \infty$, and $\eta > 0$, what is the smallest integer $N = N(E,\eta)$ such that there is a subspace F of ℓ_p^N with $d(E,F) \le 1 + \eta$? Here $d(E,F)$ is the Banach-Mazur distance between Banach spaces defined as the infimum of $\|U\|\,\|U^{-1}\|$ over all isomorphisms $U : E \to F$ (more precisely, the logarithm of d is a distance); $d(E,F) = 1$ if E and F are isometric. Partial results in case $E = \ell_2^n$ or in case $E = \ell_p^n$, $1 \le p < 2$, follow from the general study about Dvoretzky's theorem and the stable type in Chapter 9.

The results which we present are taken from [B-L-M] and [Ta15]. At this point, they are different when $p = 1$ or $p > 1$ so that we distinguish these cases in two separate statements. In the first statement, we need the concept of the K-convexity constant of a Banach space [Pi11], [Pi16]. If E is a (finite dimensional) Banach space, denote by $K(E)$ the norm of the natural projection from $L_2(E) = L_2(\Omega,\mathcal{A},\mathbb{P};E)$ onto the span of the functions $\sum_i \varepsilon_i x_i$ where (ε_i) is a Rademacher sequence on $(\Omega,\mathcal{A},\mathbb{P})$. Thus, if $f \in L_2(E)$, we have

$$\left\|\sum_i \varepsilon_i \mathbb{E}(\varepsilon_i f)\right\|_{L_2(E)} \le K(E)\|f\|_{L_2(E)}.$$

It has been noticed in [TJ2] that then the same inequality holds when (ε_i) is replaced by a standard Gaussian sequence (g_i). We use this freely below. It is further known and due to G. Pisier [Pi8] that $K(E) \leq C(\log(n + 1))^{1/2}$ if E is a subspace of dimension n of L_1 (C numerical).

We can state the two main results.

Theorem 15.12. *There are numerical constants η_0 and C such that for any n-dimensional subspace E of L_1 and any $0 < \eta < \eta_0$, we have*

$$N(E, \eta) \leq CK(E)^2 \eta^{-2} n.$$

In particular,

$$N(E, \eta) \leq C\eta^{-2} n \log(n + 1).$$

Theorem 15.13. *Let $1 < p < \infty$. There are numerical constants η_0 and C such that for any n-dimensional subspace E of L_p and any $0 < \eta < \eta_0$, we have*

$$N(E, \eta) \leq Cp\eta^{-2} n^{p/2} (\log n)^2 \log(\eta^{-1} n)$$

if $p \geq 2$, and

$$N(E, \eta) \leq C\eta^{-2} n (\log n)^2 \max\left(\frac{1}{p - 1}, \log(\eta^{-1} n)\right)$$

if $1 < p \leq 2$.

Common to the proofs of these theorems is a basic random choice argument. First, we develop it in general.

We agree to denote below by $\| \cdot \|_p = \| \cdot \|_{L_p(\mu)}$, $1 \leq p < \infty$, the norm of $L_p(S, \Sigma, \mu)$ where μ is a σ-finite measure on (S, Σ) (hopefully with no confusion when (S, Σ, μ) varies). We first observe that, given $\eta > 0$, a finite dimensional subspace E of $L_p = L_p([0, 1], dt)$ is always at a distance (in the sense of Banach-Mazur) less than $1 + \eta$ of a subspace of ℓ_p^M for some (however large) M. This can be seen in a number of ways, e.g. using that $\|f - E^k f\|_p \to 0$ where E^k denotes the conditional expectation with respect to the k-th dyadic subalgebra of $[0, 1]$.

From this observation, we concentrate on subspaces E of dimension n of ℓ_p^M for some fixed M. To embed E into $\ell_p^{M_1}$ where M_1 is of the order $M/2$, we will, for each coordinate, flip a coin and disregard the coordinate if "head" comes up. More formally, consider a sequence $\varepsilon = (\varepsilon_i)$ of Rademacher random variables and consider the operator $U_\varepsilon : \ell_p^M \to \ell_p^M$ given by: if $x = (x_i)_{i \leq M} \in \ell_p^M$, $U_\varepsilon(x) = ((1 + \varepsilon_i)^{1/p} x_i)_{i \leq M}$. We will give conditions under which U_ε is, with large probability, almost an isometry when restricted to E. We note that $U_\varepsilon(\ell_p^M)$ is isometric to $\ell_p^{M_1}$ where $M_1 = \mathrm{Card}\{i \leq M ; \varepsilon_i = 1\}$ and, with probability $\geq 1/2$, we have $M_1 \leq M/2$. For x in ℓ_p^M, consider the random variable

$$Z_x = \|U_\varepsilon(x)\|_p^p - \|x\|_p^p.$$

We have

$$Z_x = \sum_{i=1}^{M}(1 + \varepsilon_i)|x_i|^p - \sum_{i=1}^{M}|x_i|^p = \sum_{i=1}^{M}\varepsilon_i|x_i|^p .$$

Let E_1 denote the unit ball of E and set

$$A_E = \sup_{x \in E_1}\left|\sum_{i=1}^{M}\varepsilon_i|x_i|^p\right|.$$

The restriction R_ε of U_ε to E satisfies $\|R_\varepsilon\| \le (1 + A_E)^{1/p}$ and $\|R_\varepsilon^{-1}\| \le (1 - A_E)^{1/p}$ so that, when $A_E \le 1/2$,

$$d\big(E, U_\varepsilon(E)\big) \le 1 + \frac{12}{p}A_E .$$

Since $\mathbb{P}\{A_E \le 3\mathbb{E}A_E\} \ge 2/3$, we have obtained:

Proposition 15.14. *Let E be a subspace of ℓ_p^M, $1 \le p < \infty$, of dimension n and let A_E be as above. If $\mathbb{E}A_E \le 1/6$, there exist $M_1 \le M/2$ and a subspace H of $\ell_p^{M_1}$ such that*

$$d(E, H) \le 1 + \frac{36}{p}\,\mathbb{E}A_E .$$

To apply successfully this result, one therefore must ensure that $\mathbb{E}A_E$ is small. The next lemma is one of the useful tools towards this goal. It is an immediate consequence of a result by D. Lewis [Lew].

Lemma 15.15. *Let E be an n-dimensional subspace of ℓ_p^M, $1 \le p < \infty$. Then, one can find a probability measure μ on $\{1, \ldots, M\}$ and a subspace F of $L_p(\mu)$ isometric to E which admits a basis $(\psi_j)_{j \le n}$ orthogonal in $L_2(\mu)$ such that $\sum_{j=1}^{n}\psi_j^2 = 1$ and such that $\|\psi_j\|_2 = n^{-1/2}$ for all $j \le n$.*

In this lemma, if we split each atom of μ of mass $a \ge 2/M$ in $[aM/2] + 1$ equal pieces, we can assume that each atom of μ has mass $\le 2/M$ and that μ is now supported by $\{1, \ldots, M'\}$ where M' is an integer less than $3M/2$. Also we can assume that $\lambda_i = \mu(\{i\}) > 0$ for all $i \le M'$. *We always use Lemma 15.15 in this convenient form below.*

Let F be as before and let F_1 be the unit ball of $F \subset L_p(\mu)$. Our main task in the proof of both theorems will be to show that

(15.11)
$$\Lambda_E = \mathbb{E}\sup_{x \in F_1}\left|\sum_{i=1}^{M'}\lambda_i\varepsilon_i|x(i)|^p\right|$$

can be nicely estimated (recall that $\lambda_i = \mu(\{i\})$). If we denote by $G \subset \ell_p^{M'}$ the image of F by the map

$$L_p(\mu) \to \ell_p^{M'}$$

$$x = (x(i))_{i \le M'} \to (\lambda_i^{1/p}x(i))_{i \le M'},$$

then $\mathbb{E}A_G = \Lambda_E$ and G is isometric to F. Applying Proposition 15.14 to G in $\ell_p^{M'}$, we see that if $\Lambda_E \leq 1/6$, there exists a subspace H of $\ell_p^{M_1}$ where $M_1 \leq M'/2 \leq 3M/4$ such that

$$(15.12) \qquad d(E,H) = d(F,H) = d(G,H) \leq 1 + \frac{36}{p}\Lambda_E .$$

This is the procedure which we will use to establish Theorems 15.12 and 15.13. When $p = 1$, we estimate Λ_E by the K-convexity constant $K(E)$ of E while, when $p > 1$, we use Dudley's majorizing result (Theorem 11.17).

We turn to the proof of Theorem 15.12.

Proof of Theorem 15.12. The main point is the following proposition. In this proof, $p = 1$.

Proposition 15.16. *Let Λ_E be as in* (15.11). *Then*

$$\Lambda_E \leq CK(E)\left(\frac{n}{M}\right)^{1/2}$$

for some numerical constant $C > 0$.

Proof. By an application of the comparison properties for Rademacher averages (Theorem 4.12) and by comparison with Gaussian averages (4.8),

$$\Lambda_E \leq 2\mathbb{E}\sup_{x \in F_1}\left|\sum_{i=1}^{M'}\lambda_i\varepsilon_i x(i)\right| \leq \sqrt{2\pi}\mathbb{E}\sup_{x \in F_1}\left|\sum_{i=1}^{M'}\lambda_i g_i x(i)\right|.$$

(Of course, we may also invoke the Gaussian comparison theorems; however, Theorem 4.12 is simpler.)

We denote by $\langle \cdot, \cdot \rangle$ the scalar product in $L_2(\mu)$. There exists f in $L_\infty(\Omega, \mathcal{A}, \mathbb{P}; F) = L_\infty(F)$ of norm 1 such that

$$\sup_{x \in F_1}\left\langle \sum_{j=1}^{n} g_j\psi_j, x \right\rangle = \left\langle \sum_{j=1}^{n} g_j\psi_j, f \right\rangle .$$

Thus

$$\mathbb{E}\sup_{x \in F_1}\left\langle \sum_{j=1}^{n} g_j\psi_j, x \right\rangle = \sum_{j=1}^{n}\langle \psi_j, \mathbb{E}(g_j f)\rangle .$$

If we set $y_j = \mathbb{E}(g_j f)$, we have by definition of $K(F) = K(E)$,

$$\left\|\sum_{j=1}^{n} g_j y_j\right\|_{L_2(F)} \leq K(E).$$

Now

$$\left\|\sum_{j=1}^{n} g_j y_j\right\|_{L_2(F)}^2 = \mathbb{E}\int\left|\sum_{j=1}^{n} g_j y_j(t)\right|^2 d\mu(t) = \int\left(\sum_{j=1}^{n} y_j(t)^2\right) d\mu(t).$$

Since $\sum_{j=1}^{n} \psi_j^2 = 1$, we can write by Cauchy-Schwarz

$$\sum_{j=1}^{n}\langle\psi_j, y_j\rangle = \int\left(\sum_{j=1}^{n}\psi_j(t)y_j(t)\right) d\mu(t)$$

$$\leq \int\left(\sum_{j=1}^{n} y_j(t)^2\right)^{1/2} d\mu(t) \leq K(E).$$

Hence, we have obtained

(15.13) $$\mathbb{E}\sup_{x\in F_1}\left\langle\sum_{j=1}^{n} g_j n^{1/2}\psi_j, x\right\rangle \leq n^{1/2}K(E).$$

Set $v_i = I_{\{i\}}$ so that $(\lambda_i^{-1/2} v_i)_{i\leq M'}$ is an orthonormal basis of $L_2(\mu)$ ($\lambda_i = \mu(\{i\})$). Since $(n^{1/2}\psi_j)_{j\leq n}$ is an orthonormal basis of $F \subset L_2(\mu)$, the rotational invariance of Gaussian distributions indicates that

(15.14) $$\mathbb{E}\sup_{x\in F_1}\left\langle\sum_{j=1}^{n} g_j n^{1/2}\psi_j, x\right\rangle = \mathbb{E}\sup_{x\in F_1}\left\langle\sum_{i=1}^{M'}\lambda_i^{-1/2} g_i v_i, x\right\rangle.$$

Hence, (15.13) reads as

$$\mathbb{E}\sup_{x\in F_1}\left\langle\sum_{i=1}^{M'}\lambda_i^{-1/2} g_i v_i, x\right\rangle \leq n^{1/2}K(E).$$

Finally, since $\lambda_i \leq 2/M$ for each i, by the contraction principle,

$$\mathbb{E}\sup_{x\in F_1}\left|\sum_{i=1}^{M'}\lambda_i g_i x(i)\right| = \mathbb{E}\sup_{x\in F_1}\left\langle\sum_{i=1}^{M'} g_i v_i, x\right\rangle$$

$$\leq \max_{i\leq M'}\lambda_i^{1/2}\mathbb{E}\sup_{x\in F_1}\left\langle\sum_{i=1}^{M'}\lambda_i^{-1/2} g_i v_i, x\right\rangle$$

$$\leq \left(\frac{2n}{M}\right)^{1/2} K(E).$$

Proposition 15.16 follows (with $C = 2\sqrt{\pi}$).

We establish Theorem 15.12 by a simple iteration of the preceding proposition. We show that, for $\eta \leq C^{-1}$, there exists a subspace H of ℓ_1^N such that $d(E, H) \leq 1 + C\eta$ and with $N \leq CK(E)^2\eta^{-2}n$ where $C > 0$ is a numerical constant. Note that for any Banach spaces X, Y, $K(Y) \leq d(X,Y)^2 K(X)$. As we have seen, given $\eta > 0$, there exist M and a subspace H^0 of ℓ_1^M with $d(E, H^0) \leq 1 + \eta$. In particular, $K(H^0) \leq (1+\eta)^2 K(E)$. Let C_1 be the constant of Proposition 15.16 and set $C_2 = 288 C_1$. Assume that $0 < \eta \leq 10^{-2}$. Further, we can assume that $M \geq C_2^2 K(E)^2 \eta^{-2} n$ otherwise we simply take $H = H^0$. By (15.12), we construct by induction a sequence of integers $(M_j)_{j \geq 0}$, $M_0 = M$, that satisfy $M_{j+1} \leq 3M_j/4$, and subspaces H^j, $j \geq 0$, of $\ell_1^{M_j}$ such that, for all $j \geq 0$,

$$d(H^j, H^{j+1}) \leq 1 + C_2 K(E) \left(\frac{n}{M_j} \right)^{1/2},$$

and we stop at j_0, the first integer for which $M_{j_0} \leq C_2^2 K(E)^2 \eta^{-2} n$. Indeed, suppose that $j < j_0$ and that $M_0, \ldots M_j$, $H^0, \ldots, H^j$ have been constructed. Note that

$$\prod_{\ell=0}^{j} \left(1 + C_2 K(E) \left(\frac{n}{M_\ell} \right)^{1/2} \right) \leq \prod_{\ell=0}^{j} \left(1 + C_2 K(E) \left(\frac{3}{4} \right)^{(j-\ell)/2} \left(\frac{n}{M_j} \right)^{1/2} \right)$$

$$(15.15)$$

$$\leq e^{10\eta} \leq 2$$

(since $\eta \leq 10^{-2}$). In particular, it follows that

$$K(H^j) \leq (1+\eta)^2 e^{10\eta} K(E) \leq 8K(E),$$

and thus, by Proposition 15.16,

$$\Lambda_{H^j} \leq C_1 K(H^j) \left(\frac{n}{M_j} \right)^{1/2} \leq 8 C_1 K(E) \left(\frac{n}{M_j} \right)^{1/2} \leq \frac{1}{6}$$

so that, by (15.12), there exist $M_{j+1} \leq 3M_j/4$ and a subspace H^{j+1} of $\ell_1^{M_{j+1}}$ with

$$d(H^j, H^{j+1}) \leq 1 + 36 C_1 K(H^j) \left(\frac{n}{M_j} \right)^{1/2}$$

$$\leq 1 + C_2 K(E) \left(\frac{n}{M_j} \right)^{1/2}.$$

This proves the induction procedure. The result follows. We have

$$d(H^0, H^{j_0}) \leq \prod_{j=0}^{j_0-1} \left(1 + C_2 K(E) \left(\frac{n}{M_j} \right)^{1/2} \right) \leq e^{10\eta}$$

(by (15.15)) and thus, $d(E, H^{j_0}) \leq (1+\eta)e^{10\eta}$. The proof of Theorem 15.12 is complete.

Now, we turn to the case $p > 1$ and to the proof of Theorem 15.13.

Proof of Theorem 15.13. We first need the following easy consequences of Lemma 15.15.

Lemma 15.17. *Let F be as given in Lemma 15.15. If $x \in F$, then*

$$\max_{i \leq M'} |x(i)| \leq n^{\max(1/p, 1/2)} \|x\|_p .$$

Furthermore, for all $r \geq 1$,

$$\mathbb{E} \Big\| \sum_{j=1}^{n} g_j \psi_j \Big\|_r \leq C r^{1/2}$$

where C is a numerical constant.

Proof. If $x = \sum_{j=1}^{n} \alpha_j \psi_j$, $\|x\|_2 = n^{-1/2} \big(\sum_{j=1}^{n} \alpha_j^2 \big)^{1/2}$, so that, for all $i \leq M'$,

$$(15.16) \qquad |x(i)| \leq \Big(\sum_{j=1}^{n} \alpha_j^2 \Big)^{1/2} \Big(\sum_{j=1}^{n} \psi_j(i)^2 \Big)^{1/2} = n^{1/2} \|x\|_2 .$$

This already gives the result for $p \geq 2$ since then $\|x\|_2 \leq \|x\|_p$. When $p \leq 2$,

$$\|x\|_2^2 \leq \max_i |x(i)|^{2-p} \|x\|_p^p \leq n^{1-p/2} \|x\|_2^{2-p} \|x\|_p^p$$

and thus $\|x\|_2 \leq n^{(1/p)-(1/2)} \|x\|_p$. By (15.16), the first claim in Lemma 15.17 is established in the case $p \leq 2$ as well. The second claim immediately follows from Jensen's inequality and the fact that $\sum_{j=1}^{n} \psi_j^2 = 1$.

Recall that $\lambda_i = \mu(\{i\}) > 0$, $\lambda_i \leq 2/M$, $i \leq M'$. Let $J = \{ i \leq M' ;$ $\lambda_i > 1/M^2 \}$. Then, by Lemma 15.18,

$$\mathbb{E} \sup_{x \in F_1} \Big| \sum_{i \notin J} \lambda_i \varepsilon_i |x(i)|^p \Big| \leq \sum_{i \notin J} \lambda_i \, n^{\max(1, p/2)} \leq \frac{3}{2M} \, n^{\max(1, p/2)} .$$

Hence, by the triangle inequality and the contraction principle,

$$(15.17) \quad \Lambda_E \leq \frac{3}{2M} \, n^{\max(1, p/2)} + \Big(\frac{2}{M} \Big)^{1/2} \mathbb{E} \sup_{x \in F_1} \Big| \sum_{i \in J} \lambda_i^{1/2} \varepsilon_i |x(i)|^p \Big| .$$

Our aim is to study the process $\sum_{i \in J} \lambda_i^{1/2} \varepsilon_i |x(i)|^p$, $x \in F_1$, with associated pseudo-metric

$$\delta(x, y) = \Big(\sum_{i \in J} \lambda_i \big(|x(i)|^p - |y(i)|^p \big)^2 \Big)^{1/2} ,$$

and use to Dudley's entropy integral to bound it appropriately. Therefore, we need to evaluate various entropy numbers. J being fixed as before, for $x = (x(i))_{i \leq M'}$ in $L_p(\mu)$, we set

$$\|x\|_J = \max_{i \in J} |x(i)|$$

and we agree to denote by B_J the corresponding unit ball. In this notation, let us first observe the following. If $x, y \in F$, $\|x\|_p, \|y\|_p \leq 1$, then

$$(15.18) \qquad \delta(x,y) \leq 2pn^{(p-2)/4}\|x-y\|_J \qquad \text{if } p \geq 2$$

and

$$(15.19) \qquad \delta(x,y) \leq 6\|x-y\|_J^{p/2} \qquad \text{if } p \leq 2.$$

For a proof, observe that if $a, b \geq 0$,

$$a^p - b^p \leq p(a^{p-1} + b^{p-1})|a-b|.$$

Thus

$$\delta(x,y)^2 \leq 2p^2 \sum_{i \in J} \lambda_i \max(|x(i)|^{2p-2}, |y(i)|^{2p-2})|x(i) - y(i)|^2.$$

If $p \geq 2$, we proceed as follows. For all i, $|x(i)|^{2p-2} \leq \max_j |x(j)|^{p-2}|x(i)|^p$ and, if $i \in J$, $|x(i) - y(i)|^2 \leq \|x-y\|_J^2$. Hence, since $\|x\|_p, \|y\|_p \leq 1$, using Lemma 15.17,

$$\delta(x,y)^2 \leq 2p^2 n^{(p-2)/2}\|x-y\|_J^2 \sum_{i \in J} \lambda_i(|x(i)|^p + |y(i)|^p)$$

and (15.18) is satisfied. When $p \leq 2$, Hölder's inequality with $\alpha = p/(2p-2)$ and $\beta = p/(2-p)$ shows that

$$\sum_{i \in J} \lambda_i \max(|x(i)|^{2p-2}, |y(i)|^{2p-2})|x(i) - y(i)|^{2-p}$$

$$\leq \left(\sum_{i \in J} \lambda_i \max(|x(i)|^p, |y(i)|^p) \right)^{(2p-2)/p} \left(\sum_{i \in J} \lambda_i |x(i) - y(i)|^p \right)^{(2-p)/p}$$

$$\leq (\|x\|_p^p + \|y\|_p^p)^{(2p-2)/p}\|x-y\|_p^{2-p} \leq 4.$$

This yields (15.19).

The following proposition is the basic entropy estimate which will be required for the case $p \geq 2$. It is based on the dual version of Sudakov's minoration ((3.15)) which will also be used for $p \leq 2$. For every $r \geq 1$, denote by $B_{F,r}$ the unit ball of F considered as a subspace of $L_r(\mu)$ (in particular $B_{F,p} = F_1$).

Proposition 15.18. *Let $p \geq 2$ and let $F \subset L_p(\mu)$ be as given by Lemma 15.15. Then, for every $u > 0$,*

$$\log N(B_{F,p}, uB_J) \leq \frac{C}{u^2} n \log M$$

where $C > 0$ is a numerical constant.

Proof. Let $r > 1$ and let r' be the conjugate of r. By (3.15), for some constant C and all $u > 0$,

$$u\big(\log N(B_{F,2}, uB_{F,r})\big)^{1/2} \leq C\mathbb{E} \sup_{x \in B_{F,r'}} \Big| \sum_{i=1}^{M'} \lambda_i^{-1/2} g_i x(i) \Big|.$$

As in (15.14), since $(n^{1/2}\psi_j)_{j\leq n}$ is an orthonormal basis of F, the rotational invariance of Gaussian measures implies that the expectation on the right hand side of the preceding inequality is equal to

$$\mathbb{E} \sup_{x \in B_{F,r'}} \Big\langle \sum_{j=1}^{n} g_j n^{1/2}\psi_j, x \Big\rangle = n^{1/2}\mathbb{E}\Big\| \sum_{j=1}^{n} g_j \psi_j \Big\|_r.$$

Therefore, the second assertion in Lemma 15.17 shows that, for some numerical constant C and all $r > 1$ and $u > 0$,

$$(15.20) \qquad \log N(B_{F,2}, uB_{F,r}) \leq \frac{C}{u^2} rn.$$

Then, it is easy to conclude the proof of Proposition 15.18. Since $p \geq 2$, $B_{F,p} \subset B_{F,2}$. On the other hand, if $x \in B_{F,r}$ for some r,

$$\sum_{i=1}^{M'} \lambda_i |x(i)|^r \leq 1$$

and thus $|x(i)|^r \leq \lambda_i^{-1} \leq M^2$ if $i \in J$. Therefore, $B_{F,r} \subset eB_J$ if $r = \log M^2$. Hence, for this choice of r,

$$N(B_{F,p}, uB_J) \leq N\Big(B_{F,2}, \frac{u}{e}B_{F,r}\Big)$$

and the conclusion follows from (15.20).

Let us show how to deduce Theorem 15.13 for $p \geq 2$. By (15.18),

$$\int_0^\infty \big(\log N(F_1, \delta; u)\big)^{1/2} du \leq 2pn^{(p-2)/4} \int_0^{2\sqrt{n}} \big(\log N(B_{F,p}, uB_J)\big)^{1/2} du.$$

Then note that since $n^{-1/2}B_{F,p} \subset B_J$ (Lemma 15.17), by simple volume considerations,

$$N(B_{F,p}, uB_J) \leq N\big(B_{F,p}, un^{-1/2}B_{F,p}\big) \leq \Big(1 + \frac{2\sqrt{n}}{u}\Big)^n.$$

Together with Proposition 15.18, we get

$$\int_0^{2\sqrt{n}} \big(\log N(B_{F,p}, uB_J)\big)^{1/2} du$$

$$\leq \int_0^1 \Big(n\log\Big(1 + \frac{2\sqrt{n}}{u}\Big)\Big)^{1/2} du + \int_1^{2\sqrt{n}} (Cn\log M)^{1/2}\frac{du}{u}.$$

It follows that for some numerical constant C,

$$\int_0^\infty \left(\log N(F_1, \delta; u)\right)^{1/2} du \le \left[Cp^2 n^{p/2} (\log n)^2 \log M\right]^{1/2}.$$

Therefore, by Dudley's majoration theorem (Theorem 11.17) and by (15.17), if $n \le M$,

$$\Lambda_E \le \left[Cp^2 \frac{n^{p/2}}{M} (\log n)^2 \log M\right]^{1/2}.$$

An iteration on the basis of (15.12) similar to the iteration performed in the proof of Theorem 15.12 (but simpler) then yields the conclusion for $p \ge 2$.

Next, we turn to the case $1 < p \le 2$. It relies on the following analog of Proposition 15.18 that will basically be obtained by duality (after an interpolation argument).

Proposition 15.19. *Let $1 < p \le 2$ and let $F \subset L_p(\mu)$ be as given in Lemma 15.15. Then, for every $0 < u \le 2n^{1/p}$,*

$$\log N(B_{F,p}, uB_J) \le \frac{C}{u^p} n \max\left(\frac{1}{p-1}, \log M\right)$$

where $C > 0$ is a numerical constant.

Proof. Let $q = p/p - 1$ be the conjugate of p. Let $v \le 2n^{(2-p)/2p}$. Let further $h \ge q$ and define θ, $0 \le \theta \le 1$, by

$$\frac{1}{q} = \frac{1-\theta}{2} + \frac{\theta}{h}.$$

For x, y in $B_{F,2}$, by Hölder's inequality,

$$\|x - y\|_q \le \|x - y\|_2^{1-\theta} \|x - y\|_h^\theta \le 2^{1-\theta} \|x - y\|_h^\theta$$

so that, if $\|x - y\|_h \le (v/2)^{1/\theta}$, then $\|x - y\|_q \le v$. Hence, (15.20) for $r = h$ yields

$$\log N(B_{F,2}, vB_{F,q}) \le \log N\left(B_{F,2}, \left(\frac{v}{2}\right)^{1/\theta} B_{F,h}\right) \le Chn\left(\frac{v}{2}\right)^{-2/\theta}.$$

Here and below, $C > 0$ denotes some numerical constant, not necessarily the same each time it appears. Since

$$\frac{1}{\theta} = \frac{p}{2-p} - \frac{2p}{(2-p)h},$$

it follows that (recall that $v \le 2n^{(2-p)/2p}$),

$$\log N(B_{F,2}, vB_{F,q}) \le Chn\, n^{2/h} \left(\frac{v}{2}\right)^{-2p/(2-p)}.$$

Then, let us choose

$$h = h(n) = \max(q, \log n)$$

so that

$$\log N(B_{F,2}, vB_{F,q}) \leq Chn\left(\frac{v}{2}\right)^{-2p/(2-p)}.$$

The proof of (3.16) of the equivalence of Sudakov and dual Sudakov inequalities indicates that we have in the same way

(15.21) $$\log N(B_{F,p}, vB_{F,2}) \leq Chn\left(\frac{v}{2}\right)^{-2p/(2-p)}$$

where the constant C may be chosen independent of p.

As in the proof of Proposition 15.18, if $r = \log M^2$, then $B_{F,r} \subset eB_J$. Thus, using furthermore the properties of entropy numbers, for $u, v > 0$,

$$\log N(B_{F,p}, uB_J) \leq \log N(B_{F,p}, vB_{F,2}) + \log N\left(B_{F,2}, \frac{u}{ev}B_{F,r}\right).$$

Let us choose

$$v = 2^{p/2}\left(\frac{u}{e}\right)^{(2-p)/2}$$

(so that $v \leq 2n^{(2-p)/2p}$ since $u \leq 2n^{1/p}$). By (15.21) applied to the first term on the right hand side of the preceding inequality and (15.20) to the second (for $r = \log M^2$), we get that, for some numerical constant C (recall that $1 < p \leq 2$),

$$\log N(B_{F,p}, uB_J) \leq Cu^{-p}n(h + \log M).$$

Since we may assume that $n \leq M$, by definition of $h = h(n)$, the proof of Proposition 15.19 is complete.

The proof of Theorem 15.13 for $1 < p \leq 2$ is then completed exactly as above for $p \geq 2$. The preceding proposition together with Dudley's theorem ensures similarly that

$$\Lambda_E \leq \left[C\frac{n}{M}(\log n)^2 \max\left(\frac{1}{p-1}, \log M\right)\right]^{1/2}.$$

An iteration on the basis of (15.12) concludes the proof in the same way.

15.6 Majorizing Measures on Ellipsoids

Consider a sequence (y^n) in a Hilbert space H such that $|y^n| \leq (\log(n+1))^{-1/2}$ for all n and set $T = \{y^n; \ n \geq 1\}$. We know (cf. Section 12.3) that the canonical Gaussian process $X_t = \sum_i g_i t_i$, $t = (t_i) \in T \subset H$, is almost surely bounded. The same holds of course if X is indexed by Conv T. In particular, by

Theorem 12.8, there is a majorizing measure on $\operatorname{Conv} T$. However, no explicit construction is known. The difficulty is that this majorizing measure should depend on the relative position of the y^n's, not just on their lengths. More generally, we can ask: given a set A in H with a majorizing measure, construct a majorizing measure on $\operatorname{Conv} A$. Of a similar nature: given sets A_i, $i \le n$, with majorizing measures, construct a majorizing measure on $\frac{1}{n} \sum_{i=1}^{n} A_i$.

As an example, we construct in this section explicit majorizing measures on ellipsoids. The construction is based on the evaluation of the entropy integral for products of balls.

Let (a_i) be a sequence of positive numbers with $\sum_i a_i^2 \le 1$. In $H = \ell_2$, consider the ellipsoid

$$\mathcal{E} = \left\{ x \in H ; \ \sum_i \frac{x_i^2}{a_i^2} \le 1 \right\}.$$

Let (g_i) be an orthogaussian sequence. Since

$$\mathbb{E} \sup_{x \in \mathcal{E}} \left| \sum_i g_i x_i \right|^2 = \mathbb{E} \left(\sum_i g_i^2 a_i^2 \right) = \sum_i a_i^2 \le 1,$$

we know by Theorem 12.8 that there is a majorizing measure (for the function $(\log(1/x))^{1/2})$ on $\mathcal{E}$ with respect to the ℓ_2-metric (that can actually be made into a continuous majorizing measure). One may wonder for an explicit description of such a majorizing measure. This is the purpose of this section.

What will actually be explicit is the following. Let I_k, $k \ge 0$, be the disjoint sets of integers given by

$$I_k = \left\{ i ; \ 2^{-k-1} < a_i \le 2^{-k} \right\}.$$

Note that since $\sum_i a_i^2 \le 1$, we have that $\sum_k 2^{-2k} \operatorname{Card} I_k \le 4$. Then, consider the ellipsoid

$$\mathcal{E}' = \left\{ x \in H ; \ \sum_k 2^{2k} \sum_{i \in I_k} x_i^2 \le 1 \right\}.$$

We will exhibit a probability measure m on $\sqrt{3} \mathcal{E}'$ such that, for all x in $\mathcal{E}$,

$$(15.22) \qquad \int_0^\infty \left(\log \frac{1}{m(B(x,\varepsilon))} \right)^{1/2} d\varepsilon \le C$$

where $C > 0$ is a numerical constant and where $B(x, \varepsilon)$ is the (Hilbertian) ball with center x and radius $\varepsilon > 0$ in $\sqrt{3} \mathcal{E}'$. Since $\mathcal{E} \subset \mathcal{E}'$, we can use Lemma 11.9 to obtain a majorizing measure on $\mathcal{E}$ (however less explicit). Furthermore, it can be shown to be a continuous majorizing measure. We leave this to this interested reader.

Let U be the set of all sequences of integers $\bar{n} = (n_k)_{k \ge 0}$ such that $n_k \le k$ for all k and such that $\sum_k 2^{-n_k} \le 3$. For such a sequence $\bar{n} = (n_k)$, let $\Pi(\bar{n})$ be the product of balls

$$\Pi(\bar{n}) = \left\{ x \in H \,;\; \forall k, \; \sum_{i \in I_k} 2^{2k} x_i^2 \leq 2^{-n_k} \right\}.$$

The family $\{\Pi(\bar{n}) \,;\; \bar{n} \in U\}$ covers $\mathcal{E}'$. Indeed, given x in $\mathcal{E}'$ and k an integer, let m_k be the integer such that

$$2^{-m_k-1} < \sum_{i \in I_k} 2^{2k} x_i^2 \leq 2^{-m_k}.$$

Then $x \in \Pi(\bar{n})$ where $\bar{n} = (n_k)$, $n_k = \min(k, m_k)$. Note conversely that $\Pi(\bar{n}) \subset \sqrt{3}\mathcal{E}'$ for every $\bar{n}$ in U. The main step in this construction will be to show that the products of balls $\Pi(\bar{n})$ satisfy the entropy condition uniformly over all $\bar{n}$ in U; that is, for some constant C and all $\bar{n}$ in U,

(15.23)
$$\sum_p 2^{-p} \big(\log N(\Pi(\bar{n}), 2^{-p} B_2) \big)^{1/2} \leq C$$

(where B_2 is the unit ball of ℓ_2).

Therefore, let $\bar{n} = (n_k)$ be a fixed element of U. Set for all k, $\ell_k = n_k + 2k$, and consider the product of balls

$$\Pi = \Pi(\bar{n}) = \left\{ x \in H \,;\; \forall k, \; \sum_{i \in I_k} x_i^2 \leq 2^{-\ell_k} \right\}.$$

Note that since $\sum_k 2^{-2k} \operatorname{Card} I_k \leq 4$ and $\sum_k 2^{-n_k} \leq 3$, by Cauchy-Schwarz and the definition of ℓ_k,

(15.24)
$$\sum_k (2^{-\ell_k} \operatorname{Card} I_k)^{1/2} \leq 2\sqrt{3}.$$

In addition, we may assume that the sequence $(2^{\ell_k} \operatorname{Card} I_k)$ is *increasing*. This property will allow us to easily select the balls with small entropy in the product Π.

For any integer p, let $k(p)$ be the smallest such that

$$\sum_{k \geq k(p)} 2^{-\ell_k} \leq 2^{-2p-2}.$$

Then
$$N(\Pi, 2^{-p} B_2) \leq N(\Pi_p, 2^{-p-1} B_2)$$

where Π_p is the finite product $\Pi_{k<k(p)} B(k)$ of the finite dimensional Euclidean balls

$$B(k) = \left\{ (x_i)_{i \in I_k} \,;\; \sum_{i \in I_k} x_i^2 \leq 2^{-\ell_k} \right\}.$$

We agree to denote similarly by B_2 the unit ball of ℓ_2 and of the finite dimensional space ℓ_2^N for the corresponding dimension (that will be clear from the context). Let us now recall that, by volume considerations, for every $\varepsilon > 0$,

(15.25)
$$N\big(B(k),\varepsilon B_2\big) \le \left(1+\frac{2\cdot 2^{-\ell_k/2}}{\varepsilon}\right)^{\mathrm{Card}\, I_k}.$$

Furthermore, it is easily seen that if $(\varepsilon_k)_{k<p}$ are positive numbers with $\sum_{k<k(p)} \varepsilon_k^2 \le 2^{-2p-2}$, then

(15.26)
$$N(\Pi_p, 2^{-p-1}B_2) \le \prod_{k<k(p)} N\big(B(k),\varepsilon_k B_2\big).$$

Let us choose in this inequality
$$\varepsilon_k^2 = 2^{-\ell_k-4(p-j)-1}$$

where j is such that $k(j-1) \le k < k(j)$. We have

$$
\begin{aligned}
\sum_{k<k(p)} \varepsilon_k^2 &= \sum_{j\le p} 2^{-4(p-j)-1} \sum_{k(j-1)\le k<k(j)} 2^{-\ell_k} \\
&\le \sum_{j\le p} 2^{-4(p-j)-1} \sum_{k\ge k(j-1)} 2^{-\ell_k} \\
&\le \sum_{j\le p} 2^{2j-4p-1} \le 2^{-2p-2}
\end{aligned}
$$

by definition of $k(j-1)$. Thus, we are in a position to apply (15.26). Together with (15.25), it yields that

$$
\begin{aligned}
\log N(\Pi, 2^{-p}B_2) &\le \sum_{k<k(p)} \log N\big(B(k),\varepsilon_k B_2\big) \\
&\le \sum_{j\le p}\left(\sum_{k(j-1)\le k<k(j)} \mathrm{Card}\, I_k\right) \log\big(2^{2(p-j)+2}\big).
\end{aligned}
$$

It follows that for some numerical constant C

(15.27)
$$
\begin{aligned}
\sum_p 2^{-p}\big(\log N(\Pi,2^{-p}B_2)\big)^{1/2} \\
\le C\sum_j 2^{-j}\left(\sum_{k(j-1)\le k<k(j)} \mathrm{Card}\, I_k\right)^{1/2}.
\end{aligned}
$$

For every j, set
$$u_j = \sum_{k(j-1)\le k<k(j)} \big(2^{-\ell_k}\mathrm{Card}\, I_k\big)^{1/2}$$

so that $\sum_j u_j \le 2\sqrt{3}$ by (15.24). Since the sequence $(2^{\ell_k}\mathrm{Card}\, I_k)$ is increasing,

(15.28)
$$\sum_{k(j-1)\le k<k(j)} \mathrm{Card}\, I_k \le u_j\big(2^{\ell_{k(j)-1}}\mathrm{Card}\, I_{k(j)-1}\big)^{1/2}.$$

By definition of $k(j)$, $\sum_{k>k(j)-1} 2^{-\ell_k} > 2^{-2j-2}$. Hence,

$$\sum_{k(j)-1<k<k(j+1)} 2^{-\ell_k} > 2^{-2j-2} - 2^{-(2j+1)-2} = 2^{-2j-3}.$$

Therefore,

$$2^{-2j}\left(2^{\ell_{k(j)}-1}\operatorname{Card} I_{k(j)-1}\right)^{1/2} \leq 2^3\left(\sum_{k(j)-1<k<k(j+1)} 2^{-\ell_k}\right)\left(2^{\ell_{k(j)}-1}\operatorname{Card} I_{k(j)-1}\right)^{1/2}$$

$$\leq 2^3 \sum_{k(j)-1<k<k(j+1)}\left(2^{-\ell_k}\operatorname{Card} I_k\right)^{1/2}$$

$$\leq 2^3(u_j + u_{j+1})$$

where again we have used that $(2^{\ell_k}\operatorname{Card} I_k)$ is increasing. Therefore, by (15.28),

$$\sum_j 2^{-j}\left(\sum_{k(j-1)\leq k<k(j)} \operatorname{Card} I_k\right)^{1/2} \leq \sum_j 2^{-2j} u_j\left(2^{\ell_{k(j)}-1}\operatorname{Card} I_{k(j)-1}\right)^{1/2}$$

$$\leq 2^{3/2}\sum_j\left(u_j(u_j + u_{j+1})\right)^{1/2}.$$

Since $\sum_j u_j \leq 2\sqrt{3}$ by (15.24), the announced claim (15.23) follows from the Cauchy-Schwarz inequality and (15.27).

Now, we make use of this result (15.23) to exhibit a concrete majorizing measure satisfying (15.22). Recall that we denote by U the set of all sequences of integers $\bar{n} = (n_k)_{k\geq 0}$, $n_k \geq 0$, such that $n_k \leq k$ for all k and $\sum_k 2^{-n_k} \leq 3$. For every $j \geq 0$, let φ_j be the restriction map from $\mathbb{N}^{\mathbb{N}}$ onto $\mathbb{N}^{j+1}$, i.e. $\varphi_j((n_k)) = (n_k)_{k\leq j}$. Set $U_j = \varphi_j(U)$ and note that $\operatorname{Card} U_j \leq (j+1)!$. As for the elements of U, for every $\bar{n} = (n_k)_{k\leq j}$ in U_j, we denote by $\Pi(\bar{n})$ the (finite dimensional) product of balls

$$\Pi(\bar{n}) = \left\{ x \in H ; \forall k \leq j, \sum_{i\in I_k} 2^{2k} x_i^2 \leq 2^{-n_k}, \forall k > j, \forall i \in I_k, x_i = 0 \right\}.$$

For every j and every $\bar{n}$ in U_j, let $(x(j,\bar{n}))$ be a family of points in $\Pi(\bar{n})$ of cardinality $N(\Pi(\bar{n}), 2^{-j}B_2)$, such that the Euclidean balls with centers in $(x(j,\bar{n}))$ and radius 2^{-j} cover $\Pi(\bar{n})$. Consider

$$m = \sum_{j\geq 0}\sum_{\bar{n}\in U_j}\left(2^{j+1}(j+1)!N(\Pi(\bar{n}), 2^{-j}B_2)\right)^{-1}\delta_{x(j,\bar{n})}$$

where δ_x is the point mass at x. Since $\Pi(\bar{n}) \subset \sqrt{3}\mathcal{E}'$ for every $\bar{n}$ in U, m is a measure on $\sqrt{3}\mathcal{E}'$ such that, by construction, $|m| \leq 1$. Now, if x is in $\mathcal{E} \subset \mathcal{E}'$, there exists $\bar{n}$ in U with $x \in \Pi(\bar{n})$. For every $j \geq 0$, one can find $x(j)$ in $\Pi(\varphi_j(\bar{n}))$ with $|x - x(j)| \leq 2^{-j}$. Then, by the definition of m,

$$m\big(B(x, 2^{-j+1})\big) \geq m\big(B(x(j), 2^{-j})\big)$$
$$\geq \big(2^{j+1}(j+1)! N\big(\Pi(\varphi_j(\bar{n})), 2^{-j}B_2\big)\big)^{-1}.$$

Note that, for every $\varepsilon > 0$,

$$N\big(\Pi(\varphi_j(\bar{n})), \varepsilon B_2\big) \leq N\big(\Pi(\bar{n}), \varepsilon B_2\big).$$

Hence, by (15.23) (and $\mathcal{E}' \subset B_2$), the announced result (15.22) is clearly established. This concludes the explicit construction of a majorizing measure on ellipsoids that we wanted to present.

15.7 Cotype of the Canonical Injection $\ell_\infty^N \to L_{2,1}$

In this section, we develop a consequence of Sudakov's minoration for Rademacher averages (Theorem 4.15) to the cotype properties of the injection $\ell_\infty^N \to L_{2,1}$. The result is due to S. J. Montgomery-Smith [MS1].

Let S be a compact metric space and let $C(S)$ be the Banach space of all continuous functions on S equipped with the sup-norm $\|\cdot\|_\infty$. Consider an operator T from $C(S)$ to a Banach space B. We denote by $C_2(T)$ the Rademacher cotype 2 constant of T, i.e. $C_2(T)$ is the smallest number C such that

$$\Big(\sum_i \|T(x_i)\|^2\Big)^{1/2} \leq C\mathbb{E}\Big\|\sum_i \varepsilon_i x_i\Big\|_\infty$$

for all finite sequences (x_i) in $C(S)$. In particular, we have

(15.29)
$$\Big(\sum_i \|T(x_i)\|^2\Big)^{1/2} \leq C \sup_{s \in S} \sum_i |x_i(s)|.$$

This is expressed by saying that T is $(2,1)$-summing (cf. [Pie]). Hence, if $T : C(S) \to B$ is of cotype 2, it is $(2,1)$-summing (i.e. satisfies (15.29)) with a $(2,1)$-summing constant less than or equal to $C_2(T)$.

We recall that for a measurable function x on a probability space (S, Σ, μ), we can define a quasi-norm by

$$\|x\|_{2,1} = \int_0^\infty \big(\mu(|x| > t)\big)^{1/2} dt.$$

This quasi-norm defines the Lorentz space $L_{2,1}(\mu)$. It is known (and easy to see, cf. [S-W]) that $\|\cdot\|_{2,1}$ is equivalent to a norm and, for simplicity, we will use below that $\|\cdot\|_{2,1}$ as defined above behaves as a norm.

The canonical injection $C(S) \to L_{2,1}(\mu)$ (or $L_\infty(\mu) \to L_{2,1}(\mu)$), for any μ, is $(2,1)$-summing. Indeed, by Cauchy-Schwarz,

$$\|x\|_{2,1}^2 = \Big(\int_0^{\|x\|_\infty} \big(\mu(|x| > t)\big)^{1/2} dt\Big)^2$$
$$\leq \|x\|_\infty \int_0^\infty \mu(|x| > t) dt = \|x\|_\infty \int |x| d\mu.$$

Hence, if (x_i) is a finite sequence in $C(S)$,

$$\sum_i \|x_i\|_{2,1}^2 \le \sum_i \|x_i\|_\infty \int |x_i| d\mu \le \sup_s \left(\sum_i |x_i(s)| \right)^2 .$$

[G. Pisier [Pi17] has shown conversely that an operator $T : C(S) \to B$ that is $(2,1)$-summing (i.e. satisfying (15.29)) factors through $L_{2,1}(\mu)$, i.e. there exists a probability measure μ on S such that for all x in $C(S)$,

$$\|T(x)\| \le KC\|x\|_{2,1} ,$$

where K is a numerical constant.]

It is a natural question whether the canonical injection $C(S) \to L_{2,1}(\mu)$ is also of cotype 2. The answer to this question turns out to be no [Ta10]. Nonetheless, we have the following rather remarkable result.

Theorem 15.20. *Consider a probability space (S, Σ, μ) where Σ is assumed to be finite and with N atoms. Then, the cotype 2 constant $C_2(j)$ of the canonical injection $j : L_\infty(\mu) \to L_{2,1}(\mu)$ satisfies*

$$C_2(j) \le K \log(1 + \log N)$$

where $K > 0$ is a numerical constant.

When μ is uniform measure on N points, it can be shown that $C_2(j) \ge K^{-1}(\log(1 + \log N))^{1/2}$. We conjecture that, in Theorem 15.20, the term $\log(1 + \log N)$ can be replaced by $(\log(1 + \log N))^{1/2}$. From the subsequent proof, this would be the case if the conjecture on Rademacher processes presented at the end of Chapter 12 were true.

Proof. Our proof of Theorem 15.20 relies on Sudakov's minoration for Rademacher processes (Theorem 4.15) that can be reformulated as follows.

Lemma 15.21. *Consider a finite sequence (x_i) in $L_\infty(S, \Sigma, \mu)$, where Σ is finite, such that $\mathbb{E}\| \sum_i \varepsilon_i x_i \|_\infty \le 1$. Then, for all k, one can find a subalgebra Σ' of Σ that has at most 2^{2^k} atoms and Σ'-measurable functions x_i' on S such that, for all i, $x_i = x_i' + y_i + z_i$ where*

$$\mathbb{E}\left\| \sum_i \varepsilon_i x_i' \right\|_\infty \le 1 ,$$

$$\left\| \sum_i |y_i| \right\|_\infty = \sup_s \sum_i |y_i(s)| \le K_1$$

and

$$\left\| \sum_i z_i^2 \right\|_\infty = \sup_s \sum_i z_i(s)^2 \le K_1 2^{-k/2}$$

for some numerical constant $K_1 > 0$.

Proof. There is no loss of generality to assume that S is finite and that Σ is the algebra of all its subsets. Assume that $(x_i) = (x_i)_{i \leq M}$. By Theorem 4.15, the subset $\{(x_i(s))_{i \leq M} \; ; \; s \in S\}$ of $\mathbb{R}^M$ can be covered by 2^{2^k} translates of the ball $K_1(B_1 + 2^{-k/2}B_2)$ where $B_1 = B_1^M$ and $B_2 = B_2^M$ respectively are the ℓ_1 and ℓ_2 unit balls of $\mathbb{R}^M$. Thus, we can find a partition of S in at most 2^{2^k} sets A such that, given A, there exists $s_A \in S$ with

$$\forall s \in A, \quad \left(x_i(s) - x_i(s_A)\right)_{i \leq M} \in K_1(B_1 + 2^{-k/2}B_2).$$

For s in A, set $x_i'(s) = x_i(s_A)$. The conclusion is then obvious. $\square$

As a consequence of this lemma, note that by the triangle inequality

$$(15.30) \qquad
\begin{aligned}
\left(\sum_i \|x_i\|_{2,1}^2\right)^{1/2} & \\
\leq \left(\sum_i \|x_i'\|_{2,1}^2\right)^{1/2} &+ \left(\sum_i \|y_i\|_{2,1}^2\right)^{1/2} + \left(\sum_i \|z_i\|_{2,1}^2\right)^{1/2}.
\end{aligned}$$

Since $\|\sum_i |y_i|\|_\infty \leq K_1$, the fact that j is $(2,1)$-summing already indicates that

$$\left(\sum_i \|y_i\|_{2,1}^2\right)^{1/2} \leq K_1.$$

The main objective will be to establish the following fact.

Lemma 15.22. *If Σ has N atoms, then*

$$\left(\sum_i \|x_i\|_{2,1}^2\right)^{1/2} \leq K_2 \left(\log(N+1)\right)^{1/2} \left\|\sum_i x_i^2\right\|_\infty^{1/2}$$

for all finite sequences (x_i) of functions on (S, Σ, μ) where $K_2 > 0$ is a numerical constant.

Once this has been established, it follows, together with Lemma 15.21 and (15.30), that if $\mathbb{E}\|\sum_i \varepsilon_i x_i\|_\infty \leq 1$, for all k,

$$(15.31) \qquad
\begin{aligned}
\left(\sum_i \|x_i\|_{2,1}^2\right)^{1/2} & \\
\leq \left(\sum_i \|x_i'\|_{2,1}^2\right)^{1/2} &+ K_1 + K_1 K_2 2^{-k/2} \left(\log(N+1)\right)^{1/2}
\end{aligned}$$

where (x_i') are functions which are measurable with respect to a subalgebra Σ' of Σ with at most 2^{2^k} atoms and satisfy $\mathbb{E}\|\sum_i \varepsilon_i x_i'\|_\infty \leq 1$. If Σ has less than $2^{2^{k+1}}$ atoms, then (15.31) reads as

$$\left(\sum_i \|x_i\|_{2,1}^2\right)^{1/2} \leq \left(\sum_i \|x_i'\|_{2,1}^2\right)^{1/2} + K_3$$

for some numerical constant K_3. An iterative use of this property shows that if (x_i) is a finite sequence in $L_\infty(S, \Sigma, \mu)$, where Σ has less than 2^{2^k} atoms, with $\mathbb{E}\|\sum_i \varepsilon_i x_i\| \leq 1$, then, for some numerical constant K,

$$\left(\sum_i \|x_i\|_{2,1}^2\right)^{1/2} \leq K(k+1).$$

This shows Theorem 15.20.

Thus, we are left with the proof of Lemma 15.22.

Proof of Lemma 15.22. It relies on two observations. First, if x is in $L_{2,1}(\mu)$, and each atom of μ has mass $\geq a > 0$, then

(15.32) $$\|x\|_{2,1} \leq 2\left(1 + \log(a^{-1/2})\right)^{1/2}\|x\|_2 .$$

Indeed, assuming that $\|x\|_2 = 1$, we have $\|x\|_\infty \leq a^{1/2}$. Hence,

$$\|x\|_{2,1} \leq 1 + \int_1^{a^{-1/2}} \left(t\mu(|x| > t)\right)^{1/2} \frac{dt}{\sqrt{t}}$$
$$\leq 1 + \left(\int_0^\infty t\mu(|x| > t)\,dt\right)^{1/2} \left(\int_1^{a^{-1/2}} \frac{dt}{t}\right)^{1/2}$$

and (15.32) follows.

The second observation is as follows. Let ν be the probability measure on (S, Σ) that assigns mass $1/N$ to each atom of Σ. Let $\theta = \frac{1}{2}(\mu + \nu)$. Since $\mu(|x| > t) \leq 2\theta(|x| > t)$, we have $\|x\|_{L_{2,1}(\mu)} \leq \sqrt{2}\|x\|_{L_{2,1}(\theta)}$. θ assigns mass $\geq 1/2N$ to each atom of Σ. Thus, by (15.32),

$$\|x\|_{L_{2,1}(\theta)} \leq 2\left(1 + \log(\sqrt{2N})\right)^{1/2}\|x\|_{L_2(\theta)} .$$

When $\sum_i x_i(s)^2 \leq 1$ for all s, $\sum_i \|x_i\|_{L_2(\theta)}^2 \leq 1$ from which the conclusion then clearly follows. The lemma is established.

15.8 Miscellaneous Problems

In this last section, we present various problems about (or in relation to) the topics developed in this book. Some of them have already been explained in their context so that we only briefly recall them here.

Problem 1. Does inequality (1.1) (Chapter 1) of A. Ehrhard [Eh1]

$$\Phi^{-1}\left(\gamma_N(\lambda A + (1-\lambda)B)\right) \geq \lambda\Phi^{-1}\left(\gamma_N(A)\right) + (1-\lambda)\Phi^{-1}\left(\gamma_N(B)\right)$$

hold for all *Borel* sets A, B in $\mathbb{R}^N$ (and not only convex sets)?

Problem 2. Consider a Gaussian measure μ on a separable Banach space of unit ball B_1. Consider the function on $\mathbb{R}_+$, $F(t) = \mu(tB_1)$. It follows from (1.2) that $\log F$ is concave; thus it has right and left derivatives at every point. It is shown in [By] that these derivatives are equal so that F is differentiable (the proof of [Ta2] is erroneous). One may wonder how regular F is. Actually, in all the examples that we know the function F has an analytic extension in a sector $|\arg z| < \theta$ (think for example of Wiener measure on $C[0,1]$). This is a fascinating fact, since a priori one would think that the regularity of F should be related to the speed at which $F(t)$ goes to zero when $t \to 0$.

Problem 3. Consider a locally convex Hausdorff topological vector space E and μ a Radon measure on E equipped with its Borel σ-algebra that is Gaussian in the sense that the law of every continuous linear functional is Gaussian under μ. It is known that there is a compact metrizable set K with $\mu(K) > 0$. But does there exist a compact *convex* set K for which $\mu(K) > 0$? An equivalent formulation is the following. Consider the canonical Gaussian measure γ_N on $\mathbb{R}^N$ and a compact set $B \subset \mathbb{R}^N$ such that $\gamma_N(B) > 0$. Let E be the linear space generated by B, i.e. $E = \bigcup_n B_n$ where

$$B_n = \underbrace{[-n,n]B + [-n,n]B + \cdots + [-n,n]B}_{n \text{ times}}$$

(and $[-n,n]B = \{\lambda x \,;\, |\lambda| \leq n,\ x \in B\}$, $D + D = \{x + y \,;\, x,y \in D\}$). Is it true that for some compact *convex* set A with $\gamma_N(A) > 0$, we have $A \subset E$? It is not difficult to see that this problem is equivalent to the following. Does there exist n with the following property: whenever $B \subset \mathbb{R}^N$ is such that $\gamma_N(B) \geq 1/2$, then B_n contains a convex set A such that $\gamma_N(A) \geq 1/2$. It does not seem to be known whether $n = 3$ works.

Problem 4. In relation with Corollary 8.8 (Chapter 8), are type 2 spaces the only spaces B in which the conditions $\mathbb{E}(\|X\|^2 / LL\|X\|) < \infty$ and $\mathbb{E}f(X) = 0$, $\mathbb{E}f^2(X) < \infty$ for all f in B' are also sufficient for X to satisfy the bounded (say) LIL? It can be seen from Proposition 9.19 that a Banach space with this property is necessarily of type $2 - \varepsilon$ for all $\varepsilon > 0$ (see [Pi3]).

Problem 5. Theorem 8.10 completely describes the cluster set $C(S_n/a_n)$ of the LIL sequence in Banach spaces. In particular, using this result, Theorem 8.11 provides a complete picture of the limits in the LIL when $S_n/a_n \to 0$ in probability. One might wonder what these limits become when (S_n/a_n) is only bounded in probability, that is when X only satisfies the bounded LIL. What is in particular $\Lambda(X) = \limsup_{n\to\infty} \|S_n\|/a_n$? Example 8.12 suggests that this investigation might be difficult.

Problem 6. Try to characterize Banach spaces in which every random variable satisfying the CLT also satisfies the LIL. We have seen in Chapter 10, after Theorem 10.12, that cotype 2 spaces have this property, while conversely if a Banach space B satisfies this implication, B is necessarily of cotype $2 + \varepsilon$ for every $\varepsilon > 0$ [Pi3]. Are cotype 2 spaces the only ones?

Problem 7. Theorem 10.10 indicates that in a Banach space B satisfying the inequality $\mathrm{Ros}(p)$ for some $p > 2$, a random variable X satisfies the CLT if and only if it is pregaussian and $\lim_{t \to \infty} t^2 \mathbb{P}\{\|X\| > t\} = 0$. Is it true that, if, in a Banach space B, these best possible necessary conditions for the CLT are also sufficient, then B is of $\mathrm{Ros}(p)$ for some $p > 2$? This could be analogous to the law of large numbers and the type of a Banach space (Corollary 9.18).

Problem 8. More generally on Chapter 10 (and Chapter 14), try to understand when an infinite dimensional random variable satisfies the CLT. Of course, this is in relation with one of the main questions of Probability in Banach spaces: how to achieve efficiently tightness of a sum of independent random variables in terms for example of the individual summands?

Problem 9. Try to characterize almost sure boundedness and continuity of Gaussian chaos processes of order $d \geq 2$. See Section 11.3 and Section 15.2 in this chapter for more details and some (very) partial results. As conjectured in Section 15.2, an analog of Sudakov's minoration would be a first step in this investigation. Recently, some positive results in this direction have been obtained by the second author [Ta16].

Problem 10. Recall the problem described in Section 15.6 of this chapter of the explicit construction of a majorizing measure on $\mathrm{Conv}\, T$ when there is a majorizing measure on T.

Problem 11. Almost sure boundedness and continuity of Gaussian processes are now understood via the tool of majorizing measures (Section 12.1). Try now to understand boundedness and continuity of p-stable processes when $1 \leq p < 2$. In particular, since the necessary majorizing measure conditions of Section 12.2 are no longer sufficient when $p < 2$, what are the additional conditions to investigate? From the series representation of stable processes, this question is closely related to Problem 8. The paper [Ta12] describes some of the difficulties in such an investigation.

Problem 12. Is it possible to characterize boundedness (and continuity) of Rademacher processes as conjectured in Section 12.3?

Problem 13. Is there a minimal Banach algebra $\overline{B}$ with $A(G) \not\subset \overline{B} \not\subset C(G)$ on which all Lipschitz functions of order 1 operate? What is the contribution to this question of the algebra $\widetilde{B}$ discussed at the end of Section 13.2? Concerning this algebra $\widetilde{B}$, try to describe it from an Harmonic Analysis point of view as was done for $C_{\mathrm{a.s.}}$ by G. Pisier [Pi6].

Problem 14. In the random Fourier series notations of Chapter 13, is it true that

$$\mathbb{E}\sup_\gamma \left\|\sum_\gamma \xi_\gamma x_\gamma \gamma(t)\right\| \le K\left(\mathbb{E}\left\|\sum_\gamma \xi_\gamma x_\gamma\right\| + \sup_{\|f\|\le 1} \mathbb{E}\sup_{t\in V}\left|\sum_\gamma f(x_\gamma)\xi_\gamma\gamma(t)\right|\right)$$

for every (finite) sequence (x_γ) in a Banach space B when (ξ_γ) is either a Rademacher sequence (ε_γ) or a standard p-stable sequence (θ_γ), $1 < p < 2$ (and also $p = 1$ but for moments less than 1)? The constant K may depend on V as in the Gaussian case (Corollary 13.15).

Notes and References

Theorem 15.2 originates in the work of V. D. Milman on almost Euclidean subspaces of a quotient (cf. [Mi2], [Mi-S], [Pi18]). V. D. Milman used indeed a version of this result to establish the remarkable fact that if B is a Banach space of dimension N, there is a subspace of a quotient of B of dimension $[c(\varepsilon)N]$ which is $(1 + \varepsilon)$-isomorphic to a Hilbert space. A. Pajor and N. Tomczak-Jaegermann [P-TJ] improved Milman's estimate and established Theorem 15.2 using the isoperimetric inequality on the sphere and Sudakov's minoration. The first proof presented here is the Gaussian version of their argument and is taken from [Pi18]. The second proof is due to Y. Gordon [Gor4] with quantative improvements kindly communicated to us by the author.

Proposition 15.5 was shown to us by G. Pisier. As indicated in the text, some positive answers to the conjectures presented in Section 15.2 were recently obtained in [Ta16].

Section 15.3 presents a different proof of the sharp inequality that J. Bourgain [Bour] uses in his deep investigation on $\Lambda(p)$-sets.

Theorem 15.8 is taken from the work by J. Bourgain and L. Tzafriri [B-T1] (see also [B-T2] for more recent informations). The simplification in the proof of Proposition 15.10 was noticed by J. Bourgain at the light of some arguments used in [Ta15].

Embedding subspaces of L_p into ℓ_p^N was considered in special cases in [F-L-M], [J-S1], [Pi12], [Sch3], [Sch4] (among others). In particular, a breakthrough was made in [Sch4] by G. Schechtman who used empirical distributions to obtain various early general results. Schechtman's method was refined and combined with deep facts from the Banach space theory by J. Bourgain, J. Lindenstrauss and V. D. Milman [B-L-M]. In [Ta15], a simple random choice argument is introduced that simplifies the probabilistic part of the proofs of [B-L-M]. The crucial Lemma 15.15 is taken from the work by D. Lewis [Lew]. Theorem 15.12 was obtained in this way in [Ta15]. It is not known if the K-convexity constant $K(E)$ is necessary. The entropy computations of the proof of Theorem 15.13 are taken from [B-L-M].

The proof of the existence of a majorizing measure on ellipsoids was the first step of the second author on the way of his general solution of the majorizing measure question (Chapter 12). A refinement of this result (with a simplified proof) can be found in [Ta21], where it is shown to imply several very sharp discrepancy results.

The results of Section 15.7 are due to S. J. Montgomery-Smith [MS1] (that contains further developments). The proofs are somewhat simpler than those of [MS1] and are taken from [MS-T].

References

[Ac1]   A. de Acosta: Existence and convergence of probability measures in Banach spaces. Trans. Amer. Math. Soc. *152*, 273–298 (1970)

[Ac2]   A. de Acosta: Stable measures and seminorms. Ann. Probab. *3*, 865–875 (1975)

[Ac3]   A. de Acosta: Asymptotic behavior of stable measures. Ann. Probab. *5*, 494–499 (1977)

[Ac4]   A. de Acosta: Exponential moments of vector valued random series and triangular arrays. Ann. Probab. *8*, 381–389 (1980)

[Ac5]   A. de Acosta: Strong exponential integrability of sums of independent B-valued random vectors. Probab. Math. Statist. *1*, 133–150 (1980)

[Ac6]   A. de Acosta: Inequalities for B-valued random variables with applications to the strong law of large numbers. Ann. Probab. *9*, 157–161 (1981)

[Ac7]   A. de Acosta: A new proof of the Hartman-Wintner law of the iterated logarithm. Ann. Probab. *11*, 270–276 (1983)

[A-G]   A. de Acosta, E. Giné: Convergence of moments and related functionals in the central limit theorem in Banach spaces. Z. Wahrscheinlichkeitstheor. Verw. Geb. *48*, 213–231 (1979)

[A-K]   A. de Acosta, J. Kuelbs: Some results on the cluster set $C(\{S_n/a_n\})$ and the LIL. Ann. Probab. *11*, 102–122 (1983)

[A-S]   A. de Acosta, J. Samur: Infinitely divisible probability measures and the converse Kolmogorov inequality in Banach spaces. Studia Math. *66*, 143–160 (1979)

[A-A-G] A. de Acosta, A. Araujo, E. Giné: On Poisson measures, Gaussian measures and the central limit theorem in Banach spaces. Advances in Probability, vol. 4. Dekker, New York 1978, pp. 1–68

[A-K-L] A. de Acosta, J. Kuelbs, M. Ledoux: An inequality for the law of the iterated logarithm. Probability in Banach Spaces IV, Oberwolfach 1982. Lecture Notes in Mathematics, vol. 990. Springer, Berlin Heidelberg 1983, pp. 1–29

[Ad]   R. J. Adler: An introduction to continuity, extrema, and related topics for general Gaussian processes. (1989), notes to appear

[Ald]   D. Aldous: A characterization of Hilbert space using the central limit theorem. J. London Math. Soc. *14*, 376–380 (1976)

[Ale1] K. Alexander: Probability inequalities for empirical processes and a law of the iterated logarithm. Ann. Probab. *12*, 1041–1067 (1984)

[Ale2] K. Alexander: The central limit theorem for empirical processes on Vapnik-Červonenkis classes. Ann. Probab. *15*, 178–203 (1987)

[Ale3] K. Alexander: Characterization of the cluster set of the LIL sequence in Banach spaces. Ann. Probab. *17*, 737–759 (1989)

[Ale4] K. Alexander: Unusual cluster sets for the LIL sequence in Banach spaces. Ann. Probab. *17*, 1170–1185 (1989)

462   References

[A-T]   K. Alexander, M. Talagrand: The law of the iterated logarithm for empirical processes on Vapnik-Červonenkis classes. J. Multivariate Anal. *30*, 155–166 (1989)

[Al1]   J.-C. Alt: La loi des grands nombres de Prokhorov dans les espaces de type *p*. Ann. Inst. H. Poincaré *23*, 561–574 (1987)

[Al2]   J.-C. Alt: Une forme générale de la loi forte des grands nombres pour des variables aléatoires vectorielles. Probability Theory on Vector Spaces, Lancut (Poland) 1987. Lecture Notes in Mathematics, vol. 1391. Springer, Berlin Heidelberg 1989, pp. 1–15

[Al3]   J.-C. Alt: Sur la loi des grands nombres de Nagaev en dimension infinie. Probability in Banach spaces VII, Oberwolfach 1988. Progress in Probability, vol. 21. Birkhäuser, Basel 1990, pp. 13–30

[A-G-Z]   N. T. Andersen, E. Giné, J. Zinn: The central limit theorem under local conditions: the case of Radon infinitely divisible limits without Gaussian components. Trans. Amer. Math. Soc. *309*, 1–34 (1988)

[A-G-O-Z] N. T. Andersen, E. Giné, M. Ossiander, J. Zinn: The central limit theorem and the law of the iterated logarithm for empirical processes under local conditions. Probab. Theor. Rel. Fields *77*, 271–305 (1988)

[An]   T. W. Anderson: The integral of a symmetric unimodal function over a symmetric convex set and some probability inequalities. Proc. Amer. Math. Soc. *6*, 170–176 (1955)

[Ar-G1]   A. Araujo, E. Giné: On tails and domains of attraction of stable measures on Banach spaces. Trans. Amer. Math. Soc. *248*, 105–119 (1979)

[Ar-G2]   A. Araujo, E. Giné: The central limit theorem for real and Banach space valued random variables. Wiley, New York 1980

[A-G-M-Z] A. Araujo, E. Giné, V. Mandrekar, J. Zinn: On the accompanying laws in Banach spaces. Ann. Probab. *9*, 202–210 (1981)

[Az]   R. Azencott: Grandes déviations et applications. Ecole d'Eté de Probabilités de St-Flour 1978. Lecture Notes in Mathematics, vol. 774. Springer, Berlin Heidelberg 1979, pp. 1–176

[A-V]   T. A. Azlarov, N. A. Volodin: The law of large numbers for identically distributed Banach space valued random variables. Theor. Probab. Appl. *26*, 584–590 (1981)

[Azu]   K. Azuma: Weighted sums of certain dependent random variables. Tohoku Math. J. *19*, 357–367 (1967)

[Ba]   A. Badrikian: Séminaire sur les fonctions aléatoires linéaires et les mesures cylindriques. Lecture Notes in Mathematics, vol. 139. Springer, Berlin Heidelberg 1970

[B-C]   A. Badrikian, S. Chevet: Mesures cylindriques, espaces de Wiener et fonctions aléatoires gaussiennes. Lecture Notes in Mathematics, vol. 379. Springer, Berlin Heidelberg 1974

[Ba-T]   A. Baerenstein II, B. A. Taylor: Spherical rearrangements, subharmonic functions and *-functions in *n*-space. Duke Math. J. *43*, 245–268 (1976)

[Bar1]   M. T. Barlow: Continuity of local times for Lévy processes. Z. Wahrscheinlichkeitstheor. Verw. Geb. *69*, 23–35 (1985)

[Bar2]   M. T. Barlow: Necessary and sufficient conditions for the continuity of local times of Lévy processes. Ann. Probab. *16*, 1389–1427 (1988)

[B-H]   M. T. Barlow, J. Hawkes: Application de l'entropie métrique à la continuité des temps locaux des processus de Lévy. C. R. Acad. Sci. Paris *301*, 237–239 (1985)

[B-Y]   M. Barlow, M. Yor: Semi-martingale inequalities via the Garsia-Rodemich-Rumsey lemma, and applications to local times. J. Funct. Anal. *49*, 198–229 (1982)

[Bea]   B. Beauzamy: Introduction to Banach spaces and their geometry. North-Holland, Amsterdam 1985

[Be]   A. Beck: A convexity condition in Banach spaces and the strong law of large numbers. Proc. Amer. Math. Soc. *13*, 329–334 (1962)

[Bec]   W. Beckner: Inequalities in Fourier analysis. Ann. Math. *102*, 159–182 (1975)

[Bel]   Y. K. Belaev: Continuity and Hölder's conditions for sample functions of stationary Gaussian processes. Proc. of the Fourth Berkeley Symposium on Math. Statist. and Probab., vol. 2, 1961, pp. 22–33

[Ben]   G. Bennett: Probability inequalities for sums of independent random variables. J. Amer. Statist. Assoc. *57*, 33–45 (1962)

[Beny]   Y. Benyamini: Two point symmetrization, the isoperimetric inequality on the sphere and some applications. Texas Functional Analysis Seminar 1983-84. The University of Texas, 1984

[Ber]   E. Berger: Majorization, exponential inequalities, and almost sure behavior of vector valued random variables. (1989), to appear in Ann. Probab.

[B-P]   C. Bessaga, A. Pełczyński: On bases and unconditional convergence of series in Banach spaces. Studia Math. *17*, 151–164 (1958)

[Bi]   P. Billingsley: Convergence of probability measures. Wiley, New York 1968

[Bin]   N. H. Bingham: Variants on the law of the iterated logarithm. Bull. London Math. Soc. *18*, 433–467 (1986)

[Bon]   A. Bonami: Etude des coefficients de Fourier des fonctions de $L^p(G)$. Ann. Inst. Fourier *20*, 335–402 (1970)

[Bo1]   C. Borell: Convex measures on locally convex spaces. Ark. Math. *12*, 239–252 (1974)

[Bo2]   C. Borell: The Brunn-Minskowski inequality in Gauss space. Invent. math. *30*, 207–216 (1975)

[Bo3]   C. Borell: Gaussian Radon measures on locally convex spaces. Math. Scand. *38*, 265–284 (1976)

[Bo4]   C. Borell: Tail probabilities in Gauss space. Vector Space Measures and Applications, Dublin 1977. Lecture Notes in Mathematics, vol. 644. Springer, Berlin Heidelberg 1978, pp. 71–82

[Bo5]   C. Borell: On the integrability of Banach space valued Walsh polynomials. Séminaire de Probabilités XIII. Lecture Notes in Mathematics, vol. 721. Springer, Berlin Heidelberg 1979, pp. 1–3

[Bo6]   C. Borell: On polynomials chaos and integrability. Probab. Math. Statist. *3*, 191–203 (1984)

[Bo7]   C. Borell: On the Taylor series of a Wiener polynomial. Seminar Notes on multiple stochastic integration, polynomial chaos and their integration. Case Western Reserve Univ., Cleveland, 1984

[Bou]   P. Boulicaut: Fonctions de Young et continuité des trajectoires d'une fonction aléatoire. Ann. Inst. Fourier *24*, 27–47 (1974)

[Bour]   J. Bourgain: Bounded orthogonal systems and the $\Lambda(p)$-set problem. Acta Math. *162*, 227–246 (1989)

[B-M]   J. Bourgain, V. D. Milman: Dichotomie du cotype pour les espaces invariants. C. R. Acad. Sci. Paris *300*, 263–266 (1985)

[B-T1]   J. Bourgain, L. Tzafriri: Invertibility of "large" submatrices with applications to the geometry of Banach spaces and harmonic analysis. Israel J. Math. *57*, 137–224 (1987)

[B-T2]   J. Bourgain, L. Tzafriri: Restricted invertibility of matrices and applications. Analysis at Urbana II. Lecture Notes Series, vol. 138. London Math. Soc. 1989, pp. 61–107

[B-L-M]   J. Bourgain, J. Lindenstrauss, V. D. Milman: Approximation of zonoids by zonotopes. Acta Math. *162*, 73–141 (1989)

[B-DC-K] J. Bretagnolle, D. Dacunha-Castelle, J.-L. Krivine: Lois stables et espaces L^p. Ann. Inst. H. Poincaré *2*, 231–259 (1966)

[Br]   H. D. Brunk: The strong law of large numbers. Duke Math. J. *15*, 181–195
          (1948)

[B-Z]   Y. D. Burago, V. A. Zalgaller: Geometric inequalities. Springer, Berlin
          Heidelberg 1988. [Russian edn.: Nauka, Moscow, 1980]

[Bu]   D. L. Burkholder: Martingales and Fourier analysis in Banach spaces.
          Probability and Analysis, Varenna (Italy) 1985. Lecture Notes in Mathe-
          matics, vol. 1206. Springer, Berlin Heidelberg 1986, pp. 61–108

[By]   T. Byczkowski: On the density of 0-concave seminorms on vector spaces.
          (1989)

[C-R-W]   S. Cambanis, J. Rosiński, W. A. Woyczyńsky: Convergence of quadratic
          forms in p-stable variables and θ_p-radonifying operators. Ann. Probab. *13*,
          885–897 (1985)

[C-M]   R. H. Cameron, W. T. Martin: The orthogonal development of non linear
          functionals in series of Fourier-Hermite polynomials. Ann. Math. *48*, 385–
          392 (1947)

[Ca]   B. Carl: Entropy numbers, s-numbers and eigenvalue problems. J. Funct.
          Anal. *41*, 290–306 (1981)

[C-P]   B. Carl, A. Pajor: Gelfand numbers of operators with values in a Hilbert
          space. Invent. Math. *94*, 479–504 (1988)

[Car]   R. Carmona: Tensor product of Gaussian measures. Vector Space Measures
          and Applications, Dublin 1977. Lecture Notes in Mathematics, vol. 644.
          Springer, Berlin Heidelberg 1978, pp. 96–124

[Che]   H. Chernoff: A measure of asymptotic efficiency for tests hypothesis based
          on the sum of observations. Ann. Math. Statist. *23*, 493–507 (1952)

[Ch1]   S. Chevet: Un résultat sur les mesures gaussiennes. C. R. Acad. Sci. Paris
          284, 441–444 (1977)

[Ch2]   S. Chevet: Séries de variables aléatoires gaussiennes à valeurs dans $E \hat{\otimes}_\varepsilon F$.
          Applications aux produits d'espaces de Wiener abstraits. Séminaire sur
          la Géométrie des Espaces de Banach 1977-78. Ecole Polytechnique, Paris
          1978

[C-T1]   S. A. Chobanyan, V. I. Tarieladze: A counterexample concerning the CLT
          in Banach spaces. Probability Theory on Vector Spaces, Trzebieszowice
          (Poland) 1977. Lecture Notes in Mathematics, vol. 656. Springer, Berlin
          Heidelberg 1978. pp. 25–30

[C-T2]   S. A. Chobanyan, V. I. Tarieladze: Gaussian characterization of certain
          Banach spaces. J. Multivariate Anal. *7*, 183–203 (1977)

[Da]   M. M. Day: Normed linear spaces, 3rd edn. Springer, Berlin Heidelberg
          1973

[De]   J. Delporte: Fonctions aléatoires presque sûrement continues sur un inter-
          valle fermé. Ann. Inst. H. Poincaré *1*, 111–215 (1964)

[De-St]   J.-D. Deuschel, D. Stroock: Large deviations. Academic Press, 1989

[D-F]   P. Diaconis, D. Freedman: A dozen de Finetti-style results in search of a
          theory. Ann. Inst. H. Poincaré *23*, 397–423 (1987)

[D-U]   J. Diestel, J. J. Uhl: Vector measures. Providence, 1977

[Do]   M. Donsker: An invariance principle for certain probability limit theorems.
          Mem. Amer. Math. Soc., vol. 6. Providence, 1951

[Doo]   J. L. Doob: Stochastic processes. Wiley, New York 1953

[Du1]   R. Dudley: The sizes of compact subsets of Hilbert space and continuity
          of Gaussian processes. J. Funct. Anal. *1*, 290–330 (1967)

[Du2]   R. Dudley: Sample functions of the Gaussian process. Ann. Probab. *1*,
          66–103 (1973)

[Du3]   R. Dudley: Probabilities and metrics. Lecture Notes Series, vol. 45. Mate-
          matisk Inst. Aarhus, 1976

[Du4]   R. Dudley: Central limit theorems for empirical measures. Ann. Probab.
          6, 899–929 (1978)

[Du5]   R. Dudley: A course on empirical processes. Ecole d'Eté de Probabilités de St-Flour 1982. Lecture Notes in Mathematics, vol. 1097. Springer, Berlin Heidelberg 1984, pp. 2–142

[D-P]   R. Dudley, W. Philipp: Invariance principle for sums of Banach space valued random elements and empirical processes. Z. Wahrscheinlichkeitstheor. Verw. Geb. 62, 509–552 (1983)

[D-S]   R. Dudley, V. Strassen: The central limit theorem and ε-entropy. Probability and Information Theory. Lecture Notes in Mathematics, vol. 89. Springer, Berlin Heidelberg 1969, pp. 224–231

[D-St]   R. Dudley, D. Stroock: Slepian's inequality and commuting semigroups. Séminaire de Probabilités XXI. Lecture Notes in Mathematics, vol. 1247. Springer, Berlin Heidelberg 1987, pp. 574–578

[D-HJ-S] R. Dudley, J. Hoffmann-Jorgensen, L. A. Shepp: On the lower tail of Gaussian seminorms. Ann. Probab. 7, 319–342 (1979)

[Dun-S]   N. Dunford, J. T. Schwartz: Linear operators, vol. I. Interscience, 1958

[D-D]   M. Durst, R. Dudley: Empirical processes, Vapnik-Červonenkis classes and Poisson processes. Probab. Math. Statist. 1, 109–115 (1981)

[Dv]   A. Dvoretzky: Some results on convex bodies and Banach spaces. Proc. Symp. on Linear Spaces, Jerusalem, 1961, pp. 123–160

[D-R]   A. Dvoretzky, C. A. Rogers: Absolute and unconditional convergence in normed linear spaces. Proc. Nat. Acad. Sci. U.S. A. 36, 192–197 (1950)

[Eh1]   A. Ehrhard: Symétrisation dans l'espace de Gauss. Math. Scand. 53, 281–301 (1983)

[Eh2]   A. Ehrhard: Inégalités isopérimétriques et intégrales de Dirichlet gaussiennes. Ann. Scient. Ec. Norm. Sup. 17, 317–332 (1984)

[Eh3]   A. Ehrhard: Eléments extrémaux pour les inégalités de Brunn-Minskowski gaussiennes. Ann. Inst. H. Poincaré 22, 149–168 (1986)

[Eh4]   A. Ehrhard: Convexité des mesures gaussiennes. Thèse de l'Université de Strasbourg, 1985

[E-F]   A. Ehrhard, X. Fernique: Fonctions aléatoires stables irrégulières. C. R. Acad. Sci. Paris 292, 999–1001 (1981)

[Fe1]   W. Feller: An introduction to probability theory and its applications, vol. I and II (1967 and 1971). Wiley, New York

[Fe2]   W. Feller: An extension of the law of the iterated logarithm to variables without variances. J. Math. Mechanics 18, 343–355 (1968)

[Fer1]   X. Fernique: Continuité des processus gaussiens. C. R. Acad. Sci. Paris 258, 6058–6060 (1964)

[Fer2]   X. Fernique: Intégrabilité des vecteurs gaussiens. C. R. Acad. Sci. Paris 270, 1698–1699 (1970)

[Fer3]   X. Fernique: Régularité des processus gaussiens. Invent. math. 12, 304–320 (1971)

[Fer4]   X. Fernique: Régularité des trajectoires des fonctions aléatoires gaussiennes. Ecole d'Eté de Probabilités de St-Flour 1974. Lecture Notes in Mathematics, vol. 480. Springer, Berlin Heidelberg 1975, pp. 1–96

[Fer5]   X. Fernique: Evaluation de processus gaussiens composés. Probability in Banach Spaces, Oberwolfach 1975. Lecture Notes in Mathematics, vol. 526. Springer, Berlin Heidelberg 1976, pp. 67–83

[Fer6]   X. Fernique: Continuité et théorème central limite pour les transformées de Fourier aléatoires du second ordre. Z. Wahrscheinlichkeitstheor. Verw. Geb. 42, 57–66 (1978)

[Fer7]   X. Fernique: Régularité de fonctions aléatoires non gaussiennes. Ecole d'Eté de Probabilités de St-Flour 1981. Lecture Notes in Mathematics, vol. 976. Springer, Berlin Heidelberg 1983, pp. 1–74

[Fer8]   X. Fernique: Sur la convergence étroite des mesures gaussiennes. Z. Wahrscheinlichkeitstheor. Verw. Geb. 68, 331–336 (1985)

[Fer9]   X. Fernique: Gaussian random vectors and their reproducing kernel Hilbert spaces. Technical report, University of Ottawa, 1985

[Fer10]   X. Fernique: Régularité de fonctions aléatoires gaussiennes stationnaires à valeurs vectorielles. Probability Theory on Vector Spaces, Lancut (Poland) 1987. Lecture Notes in Mathematics, vol. 1391. Springer, Berlin Heidelberg 1989, pp. 66–73

[Fer11]   X. Fernique: Sur la régularité de certaines classes de fonctions aléatoires. Probability in Banach spaces 7, Oberwolfach 1988. Progress in Probability, vol. 21. Birkhäuser, Basel 1990, pp. 83–92

[Fer12]   X. Fernique: Fonctions aléatoires dans les espaces lusiniens. (1989), to appear in Expositiones Math.

[Fer 13] X. Fernique: Régularité de fonctions aléatoires gaussiennes à valeurs vectorielles. Ann. Probab. 18, 1739–1745 (1990)

[Fer14]   X. Fernique: Sur la régularité de certaines fonctions aléatoires d'Ornstein-Uhlenbeck. Ann. Inst. H. Poincaré 26, 399–417 (1990)

[F-L-M]   T. Figiel, J. Lindenstrauss, V. D. Milman: The dimensions of almost spherical sections of convex bodies. Acta Math. 139, 52–94 (1977)

[F-M1]   R. Fortet, E. Mourier: Les fonctions aléatoires comme éléments aléatoires dans les espaces de Banach. Studia Math. 15, 62–79 (1955)

[F-M2]   R. Fortet, E. Mourier: Résultats complémentaires sur les éléments aléatoires prenant leurs valeurs dans un espace de Banach. Bull. Sci. Math. 78, 14–30 (1965)

[Fr]   P. Frankl: On the trace of finite sets. J. Comb. Theory 34, 41–45 (1983)

[F-N]   D. K. Fuk, S. V. Nagaev: Probability inequalities for sums of independent random variables. Theor. Probab. Appl. 16, 643–660 (1971)

[Ga]   P. Gaenssler: Empirical processes. Inst. Math. Statist. Lecture Notes Monograph Series, vol. 3, 1983

[Gar]   A. M. Garsia: Continuity properties of Gaussian processes with multidimensional time parameter. Proc. of the Sixth Berkeley Symposium on Math. Statist. and Probab., vol. 2, 1971, pp. 369–374

[G-R]   A. M. Garsia, E. Rodemich: Monotonicity of certain functionals under rearrangements. Ann. Inst. Fourier 24, 67–117 (1974)

[G-R-R]   A. M. Garsia, E. Rodemich, H. Rumsey Jr.: A real variable lemma and the continuity of paths of some Gaussian processes. Indiana Math. J. 20, 565–578 (1978)

[Gi]   D. P. Giesy: On a convexity condition in normed linear spaces. Trans. Amer. Math. Soc. 125, 114–146 (1966)

[Gin]   E. Giné: On the central limit theorem for sample continuous processes. Ann. Probab. 2, 629–641 (1974)

[G-Z1]   E. Giné, J. Zinn: Central limit theorems and weak laws of large numbers in certain Banach spaces. Z. Wahrscheinlichkeitstheor. Verw. Geb. 62, 323–354 (1983)

[G-Z2]   E. Giné, J. Zinn: Some limit theorems for empirical processes. Ann. Probab. 12, 929–989 (1984)

[G-Z3]   E. Giné, J. Zinn: Lectures on the central limit theorem for empirical processes. Probability and Banach Spaces, Zaragoza (Spain) 1985. Lecture Notes in Mathematics, vol. 1221. Springer, Berlin Heidelberg 1986, pp. 50–113

[G-Z4]   E. Giné, J. Zinn: Bootstrapping general empirical measures. Ann. Probab. 18, 851–869 (1990)

[G-Z5]   E. Giné, J. Zinn: Gaussian characterization of uniform Donsker classes of functions. (1989), to appear in Ann. Probab.

[G-M-Z]   E. Giné, V. Mandrekar, J. Zinn: On sums of independent random variables
          with values in L_p, $2 \le p < \infty$. Probability in Banach spaces II, Ober-
          wolfach 1978. Lecture Notes in Mathematics, vol. 709. Springer, Berlin
          Heidelberg 1979, pp 111–124

[G-Ma-Z] E. Giné, M. B. Marcus, J. Zinn: A version of Chevet's theorem for stable
          processes. J. Funct. Anal. 63, 47–73 (1985)

[Gn-K]   B. V. Gnedenko, A. N. Kolmogorov: Limit distributions for sums of inde-
          pendent random variables. Addison-Wesley, 1954

[Go]   V. Goodman: Characteristics of normal samples. Ann. Probab. 16, 1281–
          1290 (1988)

[G-K1]   V. Goodman, J. Kuelbs: Rates of convergence for increments of Brownian
          motion. J. Theor. Probab. 1, 27–63 (1988)

[G-K2]   V. Goodman, J. Kuelbs: Rates of convergence for the functional LIL. Ann.
          Probab. 17, 301–316 (1989)

[G-K3]   V. Goodman, J. Kuelbs: Rates of clustering of some self similar Gaussian
          processes. (1990), to appear in Probab. Theor. Rel. Fields

[G-K-Z]   V. Goodman, J. Kuelbs, J. Zinn: Some results on the LIL in Banach space
          with applications to weighted empirical processes. Ann. Probab. 9, 713–
          752 (1981)

[Gor1]   Y. Gordon: Some inequalities for Gaussian processes and applications.
          Israel J. Math. 50, 265–289 (1985)

[Gor2]   Y. Gordon: Gaussian processes and almost spherical sections of convex
          bodies. Ann. Probab. 16, 180–188 (1988)

[Gor3]   Y. Gordon: Elliptically contoured distributions. Probab. Theor. Rel. Fields
          76, 429–438 (1987)

[Gor4]   Y. Gordon: On Milman's inequality and random subspaces which escape
          through a mesh in R^n. Geometric aspects of Functional Analysis, Is-
          rael Seminar 1986/87. Lecture Notes in Mathematics, vol. 1317. Springer,
          Berlin Heidelberg 1988, pp. 84–106

[Gr1]   M. Gromov: Paul Lévy's isoperimetric inequality. Preprint I.H.E.S. (1980)

[Gr2]   M. Gromov: Monotonicity of the volume of intersection of balls. Geometric
          Aspects of Functional Analysis, Israel Seminar 1985/86. Lecture Notes in
          Mathematics, vol. 1267. Springer, Berlin Heidelberg 1987, pp. 1–4

[G-M]   M. Gromov, V. D. Milman: A topological application of the isoperimetric
          inequality. Amer. J. Math. 105, 843–854 (1983)

[Gro]   L. Gross: Logarithmic Sobolev inequalities. Amer. J. Math. 97, 1061–1083
          (1975)

[Haa]   U. Haagerup: The best constants in the Khintchine inequality. Studia
          Math. 70, 231–283 (1982)

[Ha]   M. Hahn: Conditions for sample continuity and the central limit theorem.
          Ann. Probab. 5, 351–360 (1977)

[H-K]   M. Hahn, M. Klass: Sample continuity of square integrable processes. Ann.
          Probab. 5, 361–370 (1977)

[Har]   L. H. Harper: Optimal numbering and isoperimetric problems on graphs.
          J. Comb. Theor. 1, 385–393 (1966)

[H-W]   P. Hartman, A. Wintner: On the law of the iterated logarithm. Amer. J.
          Math. 63, 169–176 (1941)

[He1]   B. Heinkel: Mesures majorantes et le théorème de la limite centrale dans
          $C(S)$. Z. Wahrscheinlichkeitstheor. Verw. Geb. 38, 339–351 (1977)

[He2]   B. Heinkel: Relation entre théorème central-limite et loi du logarithme
          itéré dans les espaces de Banach. Z. Wahrscheinlichkeitstheor. Verw. Geb.
          41, 211–220 (1979)

[He3]   B. Heinkel: Mesures majorantes et régularité de fonctions aléatoires. As-
          pects Statistiques et Aspects Physiques des Processus Gaussiens, St-Flour
          1980. Colloque C.N.R.S., vol. 307. Paris 1981, pp. 407–434

[He4]   B. Heinkel: Majorizing measures and limit theorems for c_0-valued random variables. Probability in Banach Spaces IV, Oberwolfach 1982. Lecture Notes in Mathematics, vol. 990. Springer, Berlin Heidelberg 1983, pp. 136–149

[He5]   B. Heinkel: On the law of large numbers in 2-uniformly smooth Banach spaces. Ann. Probab. *12*, 851–857 (1984)

[He6]   B. Heinkel: Une extension de la loi des grands nombres de Prohorov. Z. Wahrscheinlichkeitstheor. Verw. Geb. *67*, 349–362 (1984)

[He7]   B. Heinkel: Some exponential inequalities with applications to the central limit theorem in $C[0,1]$. Probability in Banach Spaces 6, Sandbjerg (Denmark) 1986. Progress in Probability, vol. 20. Birkhäuser, Basel 1990, pp. 162–184

[He8]   B. Heinkel: Rearrangements of sequences of random variables and exponential inequalities. (1988), to appear in Probab. Math. Statist.

[Ho]   W. Hoeffding: Probability inequalities for sums of bounded random variables. J. Amer. Statist. Assoc. *58*, 13–30 (1963)

[HJ1]   J. Hoffmann-Jørgensen: Sums of independent Banach space valued random variables. Aarhus Univ. Preprint Series 1972/73, no. 15 (1973)

[HJ2]   J. Hoffmann-Jørgensen: Sums of independent Banach space valued random variables. Studia Math. *52*, 159–186 (1974)

[HJ3]   J. Hoffmann-Jørgensen: Probability in Banach spaces. Ecole d'Eté de Probabilités de St-Flour 1976. Lecture Notes in Mathematics, vol. 598. Springer, Berlin Heidelberg 1976, pp. 1–186

[HJ4]   J. Hoffmann-Jørgensen: Probability and Geometry of Banach spaces. Functional Analysis, Dubrovnik (Yugoslavia) 1981. Lecture Notes in Mathematics, vol. 948. Springer, Berlin Heidelberg 1982, pp. 164–229

[HJ5]   J. Hoffmann-Jørgensen: The law of large numbers for non-measurable and non-separable random elements. Colloque en l'honneur de L. Schwartz. Astérisque, vol. *131*. Herman, Paris 1985, pp. 299–356

[HJ-P]   J. Hoffmann-Jørgensen, G. Pisier: The law of large numbers and the central limit theorem in Banach spaces. Ann. Probab. *4*, 587–599 (1976)

[Ib]   I. Ibragimov: On smoothness conditions for trajectories of random functions. Theor. Probab. Appl. *28*, 240–262 (1983)

[I-M-M-T-Z] I. Iscoe, M. B. Marcus, D. McDonald, M. Talagrand, J. Zinn: Continuity of l^2-valued Ornstein-Uhlenbeck processes. Ann. Probab. *18*, 68–84 (1990)

[I-N]   K. Itô, M. Nisio: On the convergence of sums of independent Banach space valued random variables. Osaka Math. J. *5*, 35–48 (1968)

[Ja1]   N. C. Jain: Central limit theorem in a Banach space. Probability in Banach Spaces, Oberwolfach 1975. Lecture Notes in Mathematics, vol. 526. Springer, Berlin Heidelberg 1976, pp. 113–130

[Ja2]   N. C. Jain: Central limit theorem and related questions in Banach spaces. Proc. Symp. in Pure Mathematics, vol. XXXI. Amer. Math. Soc., Providence 1977, pp. 55–65

[Ja3]   N. C. Jain: An introduction to large deviations. Probability in Banach Spaces V, Medford (U.S. A.) 1984. Lecture Notes in Mathematics, vol. 1153. Springer, Berlin Heidelberg 1985, pp. 273–296

[J-M1]   N. C. Jain, M. B. Marcus: Central limit theorem for $C(S)$-valued random variables. J. Funct. Anal. *19*, 216–231 (1975)

[J-M2]   N. C. Jain, M. B. Marcus: Integrability of infinite sums of vector-valued random variables. Trans. Amer. Math. Soc. *212*, 1–36 (1975)

[J-M3]   N. C. Jain, M. B. Marcus: Continuity of sub-gaussian processes. Advances in Probability, vol. 4. Dekker, New York 1978, pp. 81–196

[J-O]   N. C. Jain, S. Orey: Vague convergence of sums of independent random variables. Israel J. Math. *33*, 317–348 (1979)

[Jaj]   R. Jajte: On stable distributions in Hilbert spaces. Studia Math. *30*, 63–71 (1968)

[J-S1]   W. B. Johnson, G. Schechtman: Embedding ℓ_p^m into ℓ_1^n. Acta Math. *149*, 71–85 (1982)

[J-S2]   W. B. Johnson, G. Schechtman: Remarks on Talagrand's deviation inequality for Rademacher functions. Texas Functional Analysis Seminar 1988-89. The University of Texas, 1989

[J-S-Z]   W. B. Johnson, G. Schechtman, J. Zinn: Best constants in moments inequalities for linear combinations of independent and exchangeable random variables. Ann. Probab. *13*, 234–253 (1985)

[Ju]   D. Juknevichiené: Central limit theorems in the space $C(S)$ and majorizing measures. Lithuanian Math. J. *26*, 186–193 (1986)

[Ka1]   J.-P. Kahane: Some random series of functions. Heath Math. Monographs, 1968. Cambridge Univ. Press, 1985, 2nd edn.

[Ka2]   J.-P. Kahane: Une inégalité du type de Slepian et Gordon sur les processus gaussiens. Israel J. Math. *55*, 109–110 (1986)

[Kan]   M. Kanter: Probability inequalities for convex sets. J. Multivariate Anal. *6*, 222–236 (1976)

[Ko]   A. N. Kolmogorov: Über das Gesetz des iterieten Logarithmus. Math. Ann. *101*, 126–135 (1929)

[Kô]   N. Kôno: Sample path properties of stochastic processes. J. Math. Kyoto Univ. *20*, 295–313 (1980)

[K-S]   W. Krakowiak, J. Szulga: Random linear forms. Ann. Probab. *14*, 955–973 (1986)

[K-R]   M. A. Krasnoselsky, Y. B. Rutitsky: Convex functions and Orlicz spaces. Noordhof, 1961

[Kr]   J.-L. Krivine: Sous-espaces de dimension finie des espaces de Banach réticulés. Ann. Math. *104*, 1–29 (1976)

[Kue1]   J. Kuelbs: A representation theorem for symmetric stable processes and stable measure on H. Z. Wahrscheinlichkeitstheor. Verw. Geb. *26*, 259–271 (1973)

[Kue2]   J. Kuelbs: The law of the iterated logarithm and related strong convergence theorems for Banach space valued random variables. Ecole d'Eté de Probabilités de St-Flour 1975. Lecture Notes in Mathematics, vol. 539. Springer, Berlin Heidelberg 1976, pp. 225–314

[Kue3]   J. Kuelbs: A strong convergence theorem for Banach space valued random variables. Ann. Probab. *4*, 744–771 (1976)

[Kue4]   J. Kuelbs: Kolmogorov's law of the iterated logarithm for Banach space valued random variables. Illinois J. Math. *21*, 784–800 (1977)

[Kue5]   J. Kuelbs: Some exponential moments of sums of independent random variables. Trans. Amer. Math. Soc. *240*, 145–162 (1978)

[Kue6]   J. Kuelbs: When is the cluster set of S_n/a_n empty ? Ann. Probab. *9*, 377–394 (1981)

[K-D]   J. Kuelbs, R. Dudley: Log log laws for empirical measures. Ann. Probab. *8*, 405–418 (1980)

[K-Z]   J. Kuelbs, J. Zinn: Some stability results for vector valued random variables. Ann. Probab. *7*, 75–84 (1979)

[Ku]   H. H. Kuo: Gaussian measures in Banach spaces. Lecture Notes in Mathematics, vol. 436. Springer, Berlin Heidelberg 1975

[Kw1]   S. Kwapień: Isomorphic characterization of inner product spaces by orthogonal series with vector valued coefficients. Studia Math. *44*, 583–595 (1972)

[Kw2]   S. Kwapień: On Banach spaces containing c_0. Studia Math. *52*, 187–190 (1974)

[Kw3]   S. Kwapień: A theorem on the Rademacher series with vector valued co-
        efficients. Probability in Banach Spaces, Oberwolfach 1975. Lecture Notes
        in Mathematics, vol. 526. Springer, Berlin Heidelberg 1976, pp. 157–158

[Kw4]   S. Kwapień: Decoupling inequalities for polynomial chaos. Ann. Probab.
        15, 1062–1071 (1987)

[K-S]   S. Kwapień, J. Szulga: Hypercontraction methods for comparison of mo-
        ments of random series in normed spaces. (1988), to appear in Ann.
        Probab.

[La]   M. T. Lacey: Laws of the iterated logarithm for the empirical characteristic
        function. Ann. Probab. 17, 292–300 (1989)

[L-S]   H. J. Landau, L. A. Shepp: On the supremum of a Gaussian process.
        Sankhyà A32, 369–378 (1970)

[LC]   L. Le Cam: Remarques sur le théorème limite central dans les espaces
        localement convexes. Les Probabilités sur les Structures Algébriques,
        Clermont-Ferrand 1969. Colloque C.N.R.S. Paris 1970, pp. 233–249

[Led1] M. Ledoux: Loi du logarithme itéré dans $C(S)$ et fonction caractéristique
        empirique. Z. Wahrscheinlichkeitstheor. Verw. Geb. 60, 425–435 (1982)

[Led2] M. Ledoux: Sur les théorèmes limites dans certains espaces de Banach
        lisses. Probability in Banach Spaces IV, Oberwolfach 1982. Lecture Notes
        in Mathematics, vol. 990. Springer, Berlin Heidelberg 1983. pp. 150–169

[Led3] M. Ledoux: A remark on the central limit theorem in Banach spaces. Prob-
        ability Theory on Vector Spaces III, Lublin (Poland) 1983. Lecture Notes
        in Mathematics, vol. 1080. Springer, Berlin Heidelberg 1984, pp. 144–151

[Led4] M. Ledoux: Sur une inégalité de H.P. Rosenthal et le théorème limite
        central dans les espaces de Banach. Israel J. Math. 50, 290–318 (1985)

[Led5] M. Ledoux: The law of the iterated logarithm in uniformly convex Banach
        spaces. Trans. Amer. Math. Soc. 294, 351–365 (1986)

[Led6] M. Ledoux: Gaussian randomization and the law of the iterated logarithm
        in type 2 Banach spaces. Unpublished manuscript (1985)

[Led7] M. Ledoux: Inégalités isopérimétriques et calcul stochastique. Séminaire
        de Probabilités XXII. Lecture Notes in Mathematics, vol. 1321. Springer,
        Berlin Heidelberg 1988, pp. 249–259

[Led8] M. Ledoux: A remark on hypercontractivity and the concentration of mea-
        sure phenomenon in a compact Riemannian manifold. Israel J. Math. 69,
        361–370 (1990)

[L-M]   M. Ledoux, M. B. Marcus: Some remarks on the uniform convergence of
        Gaussian and Rademacher Fourier quadratic forms. Geometrical and Sta-
        tistical Aspects of Probability in Banach Spaces, Strasbourg 1985. Lecture
        Notes in Mathematics, vol. 1193. Springer, Berlin Heidelberg 1986, pp. 53–
        72

[L-T1] M. Ledoux, M. Talagrand: Conditions d'intégrabilité pour les multiplica-
        teurs dans le TLC banachique. Ann. Probab. 14, 916–921 (1986)

[L-T2] M. Ledoux, M. Talagrand: Characterization of the law of the iterated
        logarithm in Banach spaces. Ann. Probab. 16, 1242–1264 (1988)

[L-T3] M. Ledoux, M. Talagrand: Un critère sur les petites boules dans le
        théorème limite central. Probab. Theor. Rel. Fields 77, 29–47 (1988)

[L-T4] M. Ledoux, M. Talagrand: Comparison theorems, random geometry and
        some limit theorems for empirical processes. Ann. Probab. 17, 596–631
        (1989)

[L-T5] M. Ledoux, M. Talagrand: Some applications of isoperimetric methods
        to strong limit theorems for sums of independent random variables. Ann.
        Probab. 18, 754–789 (1990)

[LP1]   R. LePage: Log log laws for Gaussian processes. Z. Wahrscheinlichkeits-
        theor. Verw. Geb. 25, 103–108 (1973)

[LP2]   R. LePage: Multidimensional infinitely divisible variables and processes. Part I: stable case. Technical Report 292, Stanford Univ. (1981). Probability Theory on Vector Spaces IV, Lancut (Poland) 1987. Lecture Notes in Mathematics, vol. 1391. Springer, Berlin Heidelberg 1989, pp. 153–163

[LP-W-Z] R. LePage, M. Woodroofe, J. Zinn: Convergence to a stable distribution via order statistics. Ann. Probab. *9*, 624–632 (1981)

[Lé1]   P. Lévy: Théorie de l'addition des variables aléatoires. Gauthier-Villars, Paris 1937

[Lé2]   P. Lévy: Problèmes concrets d'analyse fonctionnelle. Gauthier-Villars, Paris 1951

[Lew]   D. Lewis: Finite dimensional subspaces of L_p. Studia Math. *63*, 207–211 (1978)

[Li]   W. Linde: Probability in Banach spaces – Stable and infinitely divisible distributions. Wiley, New York 1986

[L-P]   W. Linde, A. Pietsch: Mappings of Gaussian cylindrical measures in Banach spaces. Theor. Probab. Appl. *19*, 445–460 (1974)

[Li-T1]   J. Lindenstrauss, L. Tzafriri: Classical Banach spaces I. Springer, Berlin Heidelberg 1977

[Li-T2]   J. Lindenstrauss, L. Tzafriri: Classical Banach spaces II. Springer, Berlin Heidelberg 1979

[MC-T1]   T. R. McConnell, M. S. Taqqu: Decoupling inequalities for multilinear forms in independent symmetric random variables. Ann. Probab. *14*, 943–954 (1986)

[MC-T2]   T. R. McConnell, M. S. Taqqu: Decoupling of Banach-valued multilinear forms in independent Banach-valued random variables. Probab. Theor. Rel. Fields *75*, 499–507 (1987)

[MK]   H. P. McKean: Geometry of differential space. Ann. Probab. *1*, 197–206 (1973)

[M-Z]   V. Mandrekar, J. Zinn: Central limit problem for symmetric case: convergence to non-Gaussian laws. Studia Math. *67*, 279–296 (1980)

[Ma1]   M. B. Marcus: Continuity and the central limit theorem for random trigonometric series. Z. Wahrscheinlichkeitstheor. Verw. Geb. *42*, 35–56 (1978)

[Ma2]   M. B. Marcus: Weak convergence of the empirical characteristic function. Ann. Probab. *9*, 194–201 (1981)

[Ma3]   M. B. Marcus: Extreme values for sequences of stable random variables. Proceedings of NATO Conference on Statistical Extremes, Vimeiro (Portugal) 1983. Reidel, Dordrecht 1984

[Ma4]   M. B. Marcus: ξ-radial processes and random Fourier series. Mem. Amer. Math. Soc., vol. 368. Providence, 1987

[M-P1]   M. B. Marcus, G. Pisier: Random Fourier series with applications to harmonic analysis. Ann. Math. Studies, vol. 101. Princeton Univ. Press, 1981

[M-P2]   M. B. Marcus, G. Pisier: Characterizations of almost surely continuous p-stable random Fourier series and strongly stationary processes. Acta Math. *152*, 245–301 (1984)

[M-P3]   M. B. Marcus, G. Pisier: Some results on the continuity of stable processes and the domain of attraction of continuous stable processes. Ann. Inst. H. Poincaré *20*, 177–199 (1984)

[M-R]   M. B. Marcus, J. Rosen: Sample paths properties of the local times of strongly symmetric Markov processes via Gaussian processes. (1990)

[M-S1]   M. B. Marcus, L. A. Shepp: Continuity of Gaussian processes. Trans. Amer. Math. Soc. *151*, 377–391 (1970)

[M-S2]   M. B. Marcus, L. A. Shepp: Sample behavior of Gaussian processes. Proc. of the Sixth Berkeley Symposium on Math. Statist. and Probab., vol. 2, 1972, pp. 423–441

[M-T]   M. B. Marcus, M. Talagrand: Chevet's theorem for stable processes II. J. Theor. Probab. *1*, 65–92 (1988)

[M-W]   M. B. Marcus, W. A. Woyczyński: Stable measures and central limit theorems in spaces of stable type. Trans. Amer. Math. Soc. *251*, 71–102 (1979)

[M-Zi]   M. B. Marcus, J. Zinn: The bounded law of the iterated logarithm for the weighted empirical distribution process in the non i.i.d. case. Ann. Probab. *12*, 335–360 (1984)

[Ma-Pl]   J. K. Matsak, A. N. Plitchko: Central limit theorem in Banach space. Ukrainian Math. J. *40*, 234–239 (1988)

[Mau1]   B. Maurey: Espaces de cotype p, $0 < p \leq 2$. Séminaire Maurey-Schwartz 1972-73. Ecole Polytechnique, Paris 1973

[Mau2]   B. Maurey: Théorèmes de factorisation pour les opérateurs linéaires à valeurs dans les espaces L^p. Astérisque, vol. 11. Herman, Paris 1974

[Mau3]   B. Maurey: Constructions de suites symétriques. C. R. Acad. Sci. Paris *288*, 679–681 (1979)

[Mau4]   B. Maurey: Some deviation inequalities. (1990)

[Mau-Pi]   B. Maurey, G. Pisier: Séries de variables aléatoires vectorielles indépendantes et géométrie des espaces de Banach. Studia Math. *58*, 45–90 (1976)

[Me]   P. A. Meyer: Probabilités et Potentiel. Herman, Paris 1966

[Mi1]   V. D. Milman: New proof of the theorem of Dvoretzky on sections of convex bodies. Funct. Anal. Appl. *5*, 28–37 (1971)

[Mi2]   V. D. Milman: Almost Euclidean quotient spaces of subspaces of finite dimensional normed spaces. Proc. Amer. Math. Soc. *94*, 445–449 (1986)

[Mi3]   V. D. Milman: The heritage of P. Lévy in geometric functional analysis. Colloque Paul Lévy sur les processus stochastiques. Astérisque, vol. 157–158. Herman, Paris 1988, pp. 273–302

[Mi-P]   V. D. Milman, G. Pisier: Gaussian processes and mixed volumes. Ann. Probab. *15*, 292–304 (1987)

[Mi-S]   V. D. Milman, G. Schechtman: Asymptotic theory of finite dimensional normed spaces. Lecture Notes in Mathematics, vol. 1200. Springer, Berlin Heidelberg 1986

[Mi-Sh]   V. D. Milman, M. Sharir: A new proof of the Maurey-Pisier theorem. Israel J. Math. *33*, 73–87 (1979)

[MS1]   S. J. Montgomery-Smith: The cotype of operators from $C(K)$. Ph. D. Thesis, Cambridge 1988

[MS2]   S. J. Montgomery-Smith: The distribution of Rademacher sums. Proc. Amer. Math. Soc. *109*, 517–522 (1990)

[MS-T]   S. J. Montgomery-Smith, M. Talagrand: The Rademacher cotype of operators from ℓ_∞^N. (1989)

[Mo]   E. Mourier: Eléments aléatoires dans un espace de Banach. Ann. Inst. H. Poincaré *13*, 161–244 (1953)

[Na1]   S. V. Nagaev: On necessary and sufficient conditions for the strong law of large numbers. Theor. Probab. Appl. *17*, 573–582 (1972)

[Na2]   S. V. Nagaev: Large deviations of sums of independent random variables. Ann. Probab. *7*, 745–789 (1979)

[N-N]   C. Nanopoulos, P. Nobelis: Etude de la régularité des fonctions aléatoires et de leurs propriétés limites. Séminaire de Probabilités XII. Lecture Notes in Mathematics, vol. 649. Springer, Berlin Heidelberg 1978, pp. 567–690

[Nel]   E. Nelson: The free Markov field. J. Funct. Anal. *12*, 211–227 (1973)

[Ne1]   J. Neveu: Bases mathématiques du calcul des probabilités. Masson, Paris 1964

[Ne2]   J. Neveu: Processus aléatoires gaussiens. Presses de l'Université de Montréal, 1968

[Ne3]   J. Neveu: Martingales à temps discret. Masson, Paris 1972

[Os]   R. Osserman: The isoperimetric inequality. Bull. Amer. Math. Soc. *84*, 1182–1238 (1978)

[Oss]   M. Ossiander: A central limit theorem under metric entropy with L_2-bracketing. Ann. Probab. *15*, 897–919 (1987)

[Pa]   A. Pajor: Sous-espaces ℓ_1^n des espaces de Banach. Travaux en cours. Herman, Paris 1985

[P-TJ]   A. Pajor, N. Tomczak-Jaegermann: Subspaces of small codimension of finite dimensional Banach spaces. Proc. Amer. Math. Soc. *97*, 637–642 (1986)

[Par]   K. R. Parthasarathy: Probability measures on metric spaces. Academic Press, New York 1967

[Pau]   V. Paulauskas: On the central limit theorem in the Banach space c_0. Probab. Math. Statist. *3*, 127–141 (1984)

[P-R1]   V. Paulauskas, A. Rachkauskas: Operators of stable type. Lithuanian Math. J. *24*, 160–171 (1984)

[P-R2]   V. Paulauskas, A. Rachkauskas: Approximation in the central limit theorem – Exact results in Banach spaces. Reidel, Dordrecht 1989

[Pe]   V. V. Petrov: Sums of independent random variables. Springer, Berlin Heidelberg 1975

[Pie]   A. Pietsch: Operator ideals. North-Holland, Amsterdam 1980

[Pi1]   G. Pisier: Sur les espaces qui ne contiennent pas de ℓ_n^1 uniformément. Séminaire Maurey-Schartz 1973-74. Ecole Polytechnique, Paris 1974

[Pi2]   G. Pisier: Sur la loi du logarithme itéré dans les espaces de Banach. Probability in Banach Spaces, Oberwolfach 1975. Lecture Notes in Mathematics, vol. 526. Springer, Berlin Heidelberg 1976, pp. 203–210

[Pi3]   G. Pisier: Le théorème de la limite centrale et la loi du logarithme itéré dans les espaces de Banach. Séminaire Maurey-Schwartz 1975-76. Ecole Polytechnique, Paris 1976

[Pi4]   G. Pisier: Les inégalités de Khintchine-Kahane d'après C. Borell. Séminaire sur la Géométrie des Espaces de Banach 1977-78. Ecole Polytechnique, Paris 1978

[Pi5]   G. Pisier: Ensembles de Sidon et espaces de cotype 2. Séminaire sur la Géométrie des Espaces de Banach 1977-78. Ecole Polytechnique, Paris 1978

[Pi6]   G. Pisier: Sur l'espace de Banach des séries de Fourier aléatoires presque sûrement continues. Séminaire sur la Géométrie des Espaces de Banach 1977-78. Ecole Polytechnique, Paris 1978

[Pi7]   G. Pisier: A remarkable homogeneous Banach algebra. Israel J. Math. *34*, 38–44 (1979)

[Pi8]   G. Pisier: Un théorème de factorisation pour les opérateurs linéaires entre espaces de Banach. Ann. Scien. Ec. Norm. Sup. *13*, 23–43 (1980)

[Pi9]   G. Pisier: Conditions d'entropie assurant la continuité de certains processus et applications à l'analyse harmonique. Séminaire d'Analyse Fonctionnelle 1979-1980. Ecole Polytechnique, Paris 1980

[Pi10]   G. Pisier: De nouvelles caractérisations des ensembles de Sidon. Mathematical Analysis and Applications. Adv. Math. Suppl. Studies *7B*, 686–725 (1981)

[Pi11]   G. Pisier: Holomorphic semi-groups and the geometry of Banach spaces. Ann. Math. *115*, 375–392 (1982)

[Pi12]   G. Pisier: On the dimension of the ℓ_p^n-subspaces of Banach spaces, for $1 \leq p < 2$. Trans. Amer. Math. Soc. *276*, 201–211 (1983)

[Pi13]   G. Pisier: Some applications of the metric entropy condition to harmonic analysis. Banach spaces, Harmonic Analysis and Probability, Univ. of Connecticut 1980-81. Lecture Notes in Mathematics, vol. 995. Springer, Berlin Heidelberg 1983, pp. 123–154

[Pi14] G. Pisier: Remarques sur les classes de Vapnik-Červonenkis. Ann. Inst. H. Poincaré *20*, 287–298 (1984)

[Pi15] G. Pisier: Factorization of linear operators and geometry of Banach spaces. Regional Conf. Series in Math., AMS *60*, 1986

[Pi16] G. Pisier: Probabilistic methods in the geometry of Banach spaces. Probability and Analysis, Varenna (Italy) 1985. Lecture Notes in Mathematics, vol. 1206. Springer, Berlin Heidelberg 1986, pp. 167–241

[Pi17] G. Pisier: Factorization of operators through $L_{p,\infty}$ or $L_{p,1}$ and noncommutative generalizations. Math. Ann. *276*, 105–136 (1986)

[Pi18] G. Pisier: The volume of convex bodies and Banach space geometry. Cambridge Univ. Press, 1989

[P-Z]   G. Pisier, J. Zinn: On the limit theorems for random variables with values in the spaces L_p $(2 \leq p < \infty)$. Z. Wahrscheinlichkeitstheor. Verw. Geb. *41*, 289–304 (1978)

[Po]   H. Poincaré: Calcul des probabilités. Gauthier-Villars, Paris 1912

[Pol]   D. Pollard: Convergence of stochastic processes. Springer, Berlin Heidelberg 1984

[Pr1]   C. Preston: Banach spaces arising from some integral inequalities. Indiana Math. J. *20*, 997–1015 (1971)

[Pr2]   C. Preston: Continuity properties of some Gaussian processes. Ann. Math. Statist. *43*, 285–292 (1972)

[Pro1] Y. V. Prokhorov: Convergence of random processes and limit theorems in probability theory. Theor. Probab. Appl. *1*, 157–214 (1956)

[Pro2] Y. V. Prokhorov: Some remarks on the strong law of large numbers. Theor. Probab. Appl. *4*, 204–208 (1959)

[Re]   P. Révész: The laws of large numbers. Academic Press, New York 1968

[R-T1] W. Rhee, M. Talagrand: Martingale inequalities and NP-complete problems. Math. Operation Res. *12*, 177–181 (1987)

[R-T2] W. Rhee, M. Talagrand: Martingale inequalities, interpolation and NP-complete problems. Math. Operation Res. *14*, 189–202 (1989)

[R-T3] W. Rhee, M. Talagrand: A sharp deviation inequality for the stochastic Traveling Salesman Problem. Ann. Probab. *17*, 1–8 (1989)

[R-S]   V. A. Rodin, E. M. Semyonov: Rademacher series in symmetric spaces. Anal. Math. *1*, 207–222 (1975)

[Ros]   H. P. Rosenthal: On the subspaces of L_p $(p > 2)$ spanned by sequences of independent random variables. Israel J. Math. *8*, 273–303 (1970)

[Ro1]   J. Rosiński: Remarks on Banach spaces of stable type. Probab. Math. Statist. *1*, 67–71 (1980)

[Ro2]   J. Rosiński: On stochastic integral representation of stable processes with sample paths in Banach spaces. J. Multivariate Anal. *20*, 277–302 (1986)

[Ro3]   J. Rosiński: On series representations of infinitely divisible random vectors. Ann. Probab. *18*, 405–430 (1990)

[Sa-T] G. Samorodnitsky, M. S. Taqqu: Stable random processes. Book to appear

[Sa]   N. Sauer: On the density of families of sets. J. Comb. Theor. *13*, 145–147 (1972)

[Sch1] G. Schechtman: Random embeddings of Euclidean spaces in sequence spaces. Israel J. Math. *40*, 187–192 (1981)

[Sch2] G. Schechtman: Lévy type inequality for a class of finite metric spaces. Martingale theory in Harmonic Analysis and Banach spaces, Cleveland 1981. Lecture Notes in Mathematics, vol. 939. Springer, Berlin Heidelberg 1982, pp. 211–215

[Sch3] G. Schechtman: Fine embedding of finite dimensional subspaces of L_p, $1 \leq p < 2$, into ℓ_1^m. Proc. Amer. Math. Soc. *94*, 617–623 (1985)

[Sch4] G. Schechtman: More on embedding subspaces of L_p in ℓ_r^n. Compositio Math. *61*, 159–170 (1987)

[Sch5]   G. Schechtman: A remark concerning the dependence in ε in Dvoret-zky's theorem. Geometric Aspects of Functional Analysis, Israel Seminar 1987/88. Lecture Notes in Mathematics, vol. 1376. Springer, Berlin Heidelberg 1989, pp. 274–277

[Schm]   E. Schmidt: Die Brunn-Minkowskische Ungleichung und ihr Spiegelbild sowie die isoperimetrische Eigenschaft der Kugel in der euklidischen und nichteuklidischen Geometrie. Math. Nach. *1*, 81–157 (1948)

[Schr]   M. Schreiber: Fermeture en probabilité des chaos de Wiener. C. R. Acad. Sci. Paris *265*, 859–862 (1967)

[Schü]   C. Schütt: Entropy numbers of diagonal operators between symmetric Banach spaces. J. Approx. Theory *40*, 121–128 (1984)

[Schw1]   L. Schwartz: Les applications 0-radonifiantes dans les espaces de suites. Séminaire Schwartz 1969-70. Ecole Polytechnique, Paris 1970

[Schw2]   L. Schwartz: Radon measures. Oxford Univ. Press, 1973

[Schw3]   L. Schwartz: Geometry and probability in Banach spaces. Lecture Notes in Mathematics, vol. 852. Springer, Berlin Heidelberg 1981

[Sh]   S. Shelah: A combinatorial problem : stability and order for models and theories in infinitary langages. Pacific J. Math. *41*, 247–261 (1972)

[Si1]   Z. Šidák: Rectangular confidences regions for the means of multivariate normal distributions. J. Amer. Statist. Assoc. *62*, 626–633 (1967)

[Si2]   Z. Šidák: On multivariate normal probabilities of rectangles : their dependence on correlations. Ann. Math. Statist. *39*, 1425–1434 (1968)

[Sk1]   A. V. Skorokod: Limit theorems for stochastic processes. Theor. Probab. Appl. *1*, 261–290 (1956)

[Sk2]   A. V. Skorokod: A note on Gaussian measures in a Banach space. Theor. Probab. Appl. *15*, 519–520 (1970)

[Sl]   D. Slepian: The one-sided barrier problem for Gaussian noise. Bell. System Tech. J. *41*, 463–501 (1962)

[Slu]   E. Slutsky: Alcuno propozitioni sulla teoria delle funzioni aleatorie. Giorn. Inst. Italiano degli Attuari *8*, 193–199 (1937)

[S-W]   E. Stein, G. Weiss: Introduction to Fourier analysis on Euclidean spaces. Princeton Univ. Press, 1971

[Sto]   W. Stout: Almost sure convergence. Academic Press, New York 1974

[St1]   V. Strassen: An invariance principle for the law of the iterated logarithm. Z. Wahrscheinlichkeitstheor. Verw. Geb. *3*, 211–226 (1964)

[St2]   V. Strassen: A converse to the law of the iterated logarithm. Z. Wahrscheinlichkeitstheor. Verw. Geb. *4*, 265–268 (1966)

[S-V]   D. Stroock, S. R. S. Varadhan: Multidimensional diffusion processes. Springer, Berlin Heidelberg 1979

[Su1]   V. N. Sudakov: Gaussian measures, Cauchy measures and ε-entropy. Soviet Math. Dokl. *10*, 310–313 (1969)

[Su2]   V. N. Sudakov: Gaussian random processes and measures of solid angles in Hilbert spaces. Soviet Math. Dokl. *12*, 412–415 (1971)

[Su3]   V. N. Sudakov: A remark on the criterion of continuity of Gaussian sample functions. Proceedings of the Second Japan-USSR Symposium on Probability Theory. Lecture Notes in Mathematics, vol. 330. Springer, Berlin Heidelberg 1973, pp. 444–454

[Su4]   V. N. Sudakov: Geometric problems of the theory of infinite-dimensional probability distributions. Trudy Mat. Inst. Steklov, vol. 141, 1976

[S-T]   V. N. Sudakov, B. S. Tsirel'son: Extremal properties of half-spaces for spherically invariant measures. J. Soviet. Math. *9*, 9–18 (1978) [translated from Zap. Nauch. Sem. L.O.M.I. *41*, 14–24 (1974)]

[Sz]   S. Szarek: On the best constant in the Khintchine inequality. Studia Math. *58*, 197–208 (1976)

[Ta1]   M. Talagrand: Mesures gaussiennes sur un espace localement convexe. Z. Wahrscheinlichkeitstheor. Verw. Geb. *64*, 181–209 (1983)

[Ta2]   M. Talagrand: Sur l'intégrabilité des vecteurs gaussiens. Z. Wahrscheinlichkeitstheor. Verw. Geb. *68*, 1–8 (1984)

[Ta3]   M. Talagrand: The Glivenko-Cantelli problem. Ann. Probab. *15*, 837–870 (1987)

[Ta4]   M. Talagrand: Donsker classes and random geometry. Ann. Probab. *15*, 1327–1338 (1987)

[Ta5]   M. Talagrand: Regularity of Gaussian processes. Acta Math. *159*, 99–149 (1987)

[Ta6]   M. Talagrand: Donsker classes of sets. Probab. Theor. Rel. Fields *78*, 169–191 (1988)

[Ta7]   M. Talagrand: The structure of sign invariant GB-sets and of certain Gaussian measures. Ann. Probab. *16*, 172–179 (1988)

[Ta8]   M. Talagrand: Necessary conditions for sample boundedness of p-stable processes. Ann. Probab. *16*, 1584–1595 (1988)

[Ta9]   M. Talagrand: An isoperimetric theorem on the cube and the Khintchine-Kahane inequalities. Proc. Amer. Math. Soc. *104*, 905–909 (1988)

[Ta10]   M. Talagrand: The canonical injection from $C([0,1])$ into $L_{2,1}$ is not of cotype 2. Contemporary Math. (Banach Space Theory) (1989) *85*, 513–521 (Banach Space Theory)

[Ta11]   M. Talagrand: Isoperimetry and integrability of the sum of independent Banach space valued random variables. Ann. Probab. *17*, 1546–1570 (1989)

[Ta12]   M. Talagrand: On subsets of L^p and p-stable processes. Ann. Inst. H. Poincaré *25*, 153–166 (1989)

[Ta13]   M. Talagrand: Sample boundedness of stochastic processes under increments conditions. Ann. Probab. *18*, 1–49 (1990)

[Ta14]   M. Talagrand: Characterization of almost surely continuous 1-stable random dom Fourier series and strongly stationary processes. Ann. Probab. *18*, 85–91 (1990)

[Ta15]   M. Talagrand: Embedding subspaces of L_1 into ℓ_1^N. Proc. Amer. Math. Soc. *108*, 363–369 (1990)

[Ta16]   M. Talagrand: Sudakov-type minoration for Gaussian chaos. (1989)

[Ta17]   M. Talagrand: Necessary and sufficient conditions for sample continuity of random Fourier series and of harmonic infinitely divisible processes. (1989), to appear in Ann. Probab.

[Ta18]   M. Talagrand: Regularity of infinitely divisible processes. (1990)

[Ta19]   M. Talagrand: A new isoperimetric inequality. (1990), to appear in Geometric Aspects of Functional Analysis, Israel Seminar 1989/90. Lecture Notes in Mathematics. Springer, Berlin Heidelberg

[Ta20]   M. Talagrand: Supremum of some canonical processes. (1990)

[Ta21]   M. Talagrand: Discrepancy and matching theorems via majorizing measures. (1990)

[Ta22]   M. Talagrand: (1990), in preparation

[TJ1]   N. Tomczak-Jaegermann: Dualité des nombres d'entropie pour des opérateurs à valeurs dans un espace de Hilbert. C. R. Acad. Sci. Paris *305*, 299–301 (1987)

[TJ2]   N. Tomczak-Jaegermann: Banach-Mazur distances and finite dimensional operator ideals. Pitman, 1988

[To]   Y. L. Tong: Probability inequalities in multivariate distributions. Academic Press, New York 1980

[Ts]   V. S. Tsirel'son: The density of the maximum of a Gaussian process. Theor. Probab. Appl. *20*, 847–856 (1975)

[Va1]   N. N. Vakhania: Sur une propriété des répartitions normales de probabilités dans les espaces ℓ_p $(1 \leq p < \infty)$ et H. C. R. Acad. Sci. Paris *260*, 1334–1336 (1965)

[Va2]   N. N. Vakhania: Probability distributions on linear spaces. North-Holland, Amsterdam 1981 [Russian edn.: Tbilissi, 1971]

[V-T-C]   N. N. Vakhania, V. I. Tarieladze, S. A. Chobanyan: Probability distributions on Banach spaces. Reidel, Dordrecht 1987

[V-C1]   V. N. Vapnik, A. Y. Chervonenkis: On the uniform convergence of relative frequencies of events to their probabilities. Theor. Probab. Appl. *16*, 264–280 (1971)

[V-C2]   V. N. Vapnik, A. Y. Chervonenkis: Necessary and sufficient conditions for the uniform convergence of means to their expectations. Theor. Probab. Appl. *26*, 532–553 (1981)

[Var]   S. R. S. Varadhan: Limit theorems for sums of independent random variables with values in a Hilbert space. Sankhyà *A24*, 213–238 (1962)

[Varb]   D. Varberg: Convergence of quadratic forms in independent random variables. Ann. Math. Statist. *37*, 567–576 (1966)

[We]   M. Weber: Analyse infinitésimale de fonctions aléatoires. Ecole d'été de Probabilités de St-Flour 1981. Lecture Notes in Mathematics, vol. 976. Springer, Berlin Heidelberg 1982, pp. 381–465

[Wer]   A. Weron: Stable processes and measures; a survey. Probability Theory on Vector Spaces III, Lublin (Poland) 1983. Lecture Notes in Mathematics, vol. 1080. Springer, Berlin Heidelberg 1984, pp. 306–364

[Wie]   N. Wiener: The homogeneous chaos. Amer. Math. J. *60*, 897–936 (1930)

[Wi]   R. Wittmann: A general law of the iterated logarithm. Z. Wahrscheinlichkeitstheor. Verw. Geb. *68*, 521–543 (1985)

[Wo1]   W. A. Woyczyński: Random series and laws of large numbers in some Banach spaces. Theor. Probab. Appl. *18*, 350–355 (1973)

[Wo2]   W. A. Woyczyński: Geometry and martingales in Banach spaces. Part II: independent increments. Advances in Probability, vol. 4. Dekker, New York 1978, pp. 267–517

[Wo3]   W. A. Woyczyński: On Marcinkiewicz-Zygmund laws of large numbers in Banach spaces and related rates of convergence. Probab. Math. Statist. *1*, 117–131 (1980)

[Yo]   M. Yor: Introduction au calcul stochastique. Séminaire Bourbaki. Astérisque, vol. 92–93. Herman, Paris 1982, pp. 275–292

[Yu1]   V. V. Yurinskii: Exponential bounds for large deviations. Theor. Probab. Appl. *19*, 154–155 (1974)

[Yu2]   V. V. Yurinskii: Exponential inequalities for sums of random vectors. J. Multivariate Anal. *6*, 473–499 (1976)

[Zi1]   J. Zinn: A note on the central limit theorem in Banach spaces. Ann. Probab. *5*, 283–286 (1977)

[Zi2]   J. Zinn: Inequalities in Banach spaces with applications to probabilistic limit theorems: a survey. Probability in Banach Spaces III, Medford (U.S.A.) 1980. Lecture Notes in Mathematics, vol. 860. Springer, Berlin Heidelberg 1981, pp. 324–329

[Zi3]   J. Zinn: Comparison of martingale difference sequences. Probability in Banach spaces V, Medford (U.S. A.) 1984. Lecture Notes in Mathematics, vol. 1153. Springer, Berlin Heidelberg 1985, pp. 451–457

[Zi4]   J. Zinn: Universal Donsker classes and type 2. Probability in Banach Spaces 6, Sandbjerg (Denmark) 1986. Progress in Probability, vol. 20. Birkhäuser, Basel 1990, pp. 283–288

Subject Index

Ergebnisse der Mathematik und ihrer Grenzgebiete, 3. Folge

A Series of Modern Surveys in Mathematics

Springer-Verlag
Berlin
Heidelberg
New York
London
Paris
Tokyo
Hong Kong
Barcelona

Volume 14: **M. R. Goresky, R. D. MacPherson**

Stratified Morse Theory

1988. XIV, 272 pp. 84 figs.
ISBN 3-540-17300-5

Volume 15: **T. Oda**

Convex Bodies and Algebraic Geometry

An Introduction to the Theory of Toric Varieties
1987. VIII, 212 pp. 42 figs. ISBN 3-540-17600-4

Volume 16: **G. van der Geer**

Hilbert Modular Surfaces

1988. IX, 291 pp. 39 figs. ISBN 3-540-17601-2

Volume 17: **G. A. Margulis**

Discrete Subgroups of Semisimple Lie Groups

1990. IX, 388 pp. ISBN 3-540-12179-X

Volume 18: **A. E. Brouwer, A. M. Cohen, A. Neumaier**

Distance-Regular Graphs

1989. XVII, 495 pp. ISBN 3-540-50619-5

Volume 19: **I. Ekeland**

Convexity Methods in Hamiltonian Mechanics

1990. X, 247 pp. 4 figs. ISBN 3-540-50613-6

Contents: Introduction. - Linear Hamiltonian Systems. - Convex Hamiltonian Systems. - Fixed-Period Problems: The Sublinear Case. - Fixed-Period Problems: The Superlinear Case. - Fixed-Energy Problems. - Open Problems. - Bibliography. - Index.

Volume 20: **A. I. Kostrikin**

Around Burnside

1990. XII, 255 pp. ISBN 3-540-50602-0

Contents: Preface. - Introduction. - The Descent to Sandwiches. - Local Analysis to thin Sandwiches. - Proof of the Main Theorem. - Evolution of the Method of Sandwiches. - The Problem of Global Nilpotency. - Finite p-Groups and Lie Algebras. - Appendix I. - Appendix II. - Epilogue. - References. - Author Index. - Subject Index. - Notation.

Volume 21: **S. Bosch, W. Lütkebohmert, M. Raynaud**

Néron Models

1990. X, 325 pp. 4 figs. ISBN 3-540-50587-3

Contents: Introduction. - What Is a Néron Model? - Some Background Material from Algebraic Geometry. - The Smoothening Process. - Construction of Birational Group Laws. - From Birational Group Laws to Group Schemes. - Descent. - Properties of Néron Models. - The Picard Functor. - Jacobians of Relative Curves. - Néron Models of Not Necessarily Proper Algebraic Groups. - Bibliography. - Subject Index.

Volume 22: **G. Faltings, C.-L. Chai**

Degeneration of Abelian Varieties

1990. XII, 316 pp.
ISBN 3-540-52015-5

Springer-Verlag
Berlin
Heidelberg
New York
London
Paris
Tokyo
Hong Kong
Barcelona

M. Aigner Combinatorial Theory ISBN 978-3-540-61787-7
A. L. Besse Einstein Manifolds ISBN 978-3-540-74120-6
N. P. Bhatia, G. P. Szegő Stability Theory of Dynamical Systems ISBN 978-3-540-42748-3
J. W. S. Cassels An Introduction to the Geometry of Numbers ISBN 978-3-540-61788-4
R. Courant, F. John Introduction to Calculus and Analysis I ISBN 978-3-540-65058-4
R. Courant, F. John Introduction to Calculus and Analysis II/1 ISBN 978-3-540-66569-4
R. Courant, F. John Introduction to Calculus and Analysis II/2 ISBN 978-3-540-66570-0
P. Dembowski Finite Geometries ISBN 978-3-540-61786-0
A. Dold Lectures on Algebraic Topology ISBN 978-3-540-58660-9
J. L. Doob Classical Potential Theory and Its Probabilistic Counterpart ISBN 978-3-540-41206-9
R. S. Ellis Entropy, Large Deviations, and Statistical Mechanics ISBN 978-3-540-29059-9
H. Federer Geometric Measure Theory ISBN 978-3-540-60656-7
S. Flügge Practical Quantum Mechanics ISBN 978-3-540-65035-5
L. D. Faddeev, L. A. Takhtajan Hamiltonian Methods in the Theory of Solitons
        ISBN 978-3-540-69843-2
I. I. Gikhman, A. V. Skorokhod The Theory of Stochastic Processes I ISBN 978-3-540-20284-4
I. I. Gikhman, A. V. Skorokhod The Theory of Stochastic Processes II ISBN 978-3-540-20285-1
I. I. Gikhman, A. V. Skorokhod The Theory of Stochastic Processes III ISBN 978-3-540-49940-4
D. Gilbarg, N. S. Trudinger Elliptic Partial Differential Equations of Second Order
        ISBN 978-3-540-41160-4
H. Grauert, R. Remmert Theory of Stein Spaces ISBN 978-3-540-00373-1
H. Hasse Number Theory ISBN 978-3-540-42749-0
F. Hirzebruch Topological Methods in Algebraic Geometry ISBN 978-3-540-58663-0
L. Hörmander The Analysis of Linear Partial Differential Operators I – Distribution Theory
        and Fourier Analysis ISBN 978-3-540-00662-6
L. Hörmander The Analysis of Linear Partial Differential Operators II – Differential
        Operators with Constant Coefficients ISBN 978-3-540-22516-4
L. Hörmander The Analysis of Linear Partial Differential Operators III – Pseudo-
        Differential Operators ISBN 978-3-540-49937-4
L. Hörmander The Analysis of Linear Partial Differential Operators IV – Fourier
        Integral Operators ISBN 978-3-642-00117-8
K. Itô, H. P. McKean, Jr. Diffusion Processes and Their Sample Paths ISBN 978-3-540-60629-1
M. Karoubi K-Theory ISBN 978-3-540-79889-7
T. Kato Perturbation Theory for Linear Operators ISBN 978-3-540-58661-6
S. Kobayashi Transformation Groups in Differential Geometry ISBN 978-3-540-58659-3
K. Kodaira Complex Manifolds and Deformation of Complex Structures ISBN 978-3-540-22614-7
M. Ledoux, M. Talagrand Probability in Banach Spaces - Isoperimetry and Processes ISBN 978-3-642-20211-7
Th. M. Liggett Interacting Particle Systems ISBN 978-3-540-22617-8
J. Lindenstrauss, L. Tzafriri Classical Banach Spaces I and II ISBN 978-3-540-60628-4
R. C. Lyndon, P. E Schupp Combinatorial Group Theory ISBN 978-3-540-41158-1
S. Mac Lane Homology ISBN 978-3-540-58662-3
C. B. Morrey Jr. Multiple Integrals in the Calculus of Variations ISBN 978-3-540-69915-6
D. Mumford Algebraic Geometry I – Complex Projective Varieties ISBN 978-3-540-58657-9
O. T. O'Meara Introduction to Quadratic Forms ISBN 978-3-540-66564-9
G. Pólya, G. Szegő Problems and Theorems in Analysis I – Series. Integral Calculus.
        Theory of Functions ISBN 978-3-540-63640-3
G. Pólya, G. Szegő Problems and Theorems in Analysis II – Theory of Functions. Zeros.
        Polynomials. Determinants. Number Theory. Geometry
        ISBN 978-3-540-63686-1
W. Rudin Function Theory in the Unit Ball of C^n ISBN 978-3-540-68272-1
S. Sakai C*-Algebras and W*-Algebras ISBN 978-3-540-63633-5
C. L. Siegel, J. K. Moser Lectures on Celestial Mechanics ISBN 978-3-540-58656-2
T. A. Springer Jordan Algebras and Algebraic Groups ISBN 978-3-540-63632-8
D. W. Stroock, S. R. S. Varadhan Multidimensional Diffusion Processes ISBN 978-3-540-28998-2
R. R. Switzer Algebraic Topology: Homology and Homotopy ISBN 978-3-540-42750-6
A. Weil Basic Number Theory ISBN 978-3-540-58655-5
A. Weil Elliptic Functions According to Eisenstein and Kronecker ISBN 978-3-540-65036-2
K. Yosida Functional Analysis ISBN 978-3-540-58654-8
O. Zariski Algebraic Surfaces ISBN 978-3-540-58658-6